JETS AT ALL SCALES

IAU SYMPOSIUM No. 275

INTERNATIONAL ASTRONOMICAL UNION

UNION ASTRONOMIQUE INTERNATIONALE

JETS AT ALL SCALES

PROCEEDINGS OF THE 275th SYMPOSIUM OF THE INTERNATIONAL ASTRONOMICAL UNION HELD IN BUENOS AIRES, ARGENTINA SEPTEMBER 13–17, 2010

Edited by

Gustavo E. Romero

Instituto Argentino de Radioastronomía (CCT La Plata-CONICET), Villa Elisa, Argentina

Rashid A. Sunyaev

MPI for Astrophysik, Germany, and Space Research Institute, Russia

and

Tomaso M. Belloni

INAF, Osservatorio Astronomico di Brera, Italy

CAMBRIDGE
UNIVERSITY PRESS

CAMBRIDGE UNIVERSITY PRESS
The Edinburgh Building, Cambridge CB2 2RU, UnitedKingdom
40 West 20th Street, New York, NY 10011–4211, USA
10 Stamford Road, Oakleigh, Melbourne 3166, Australia

First published 2011

Printed in the United Kingdom at the University Press, Cambridge

Typeset in System LaTeX 2_ε

A catalogue record for this book is available from the British Library

Library of Congress Cataloguing in Publication data

This journal issue has been printed on FSC-certified paper and cover board. FSC is an
independent, non-governmental, not-for-profit organization established to promote the
responsible management of the worlds forests. Please see www.fsc.org for information.

ISBN 9780521766074 hardback
ISSN 1743-9213

Table of Contents

OPENING TALK

Part 1. Jets basic issues and physical proccesses

Poster papers of Part 1

Part 2. Active galactic nuclei

Poster papers of Part 2

Part 3. Microquasars

Poster papers of Part 3

Part 4. Gamma-ray bursts

Part 5. Young stellar objects

Preface

The IAU Symposium No. 275 on *Jets at all Scales* was held in Buenos Aires city, Argentina, in September 2010. Out of 187 registered participants, more than 150 from 31 countries met at the Novotel in the traditional Calle Corrientes of Buenos Aires to discuss the latest results on astrophysical jets and outflows.

The first ideas for this Symposium appeared at discussions among participants of the 7th Microquasar Workshop, entitled *Microquasars and Beyond,* held in Foca, Izmir, Turkey, on September, 2008. The series of Microquasar Workshops had by then extended over more than a decade and started to attract participants from far beyond the relatively small community of researchers on galactic binary systems. Comparisons between the jets of microquasars and those presented by other astrophysical objects like gamma-ray bursts, active galactic nuclei, and young stellar objects were becoming more and more common in these meetings. The time seemed to have arrived for a much larger meeting that could gather outstanding researchers from all these different fields to discuss in length the similarities and differences among all types of jets, as well as the underlying physics.

The opportunity came in 2009 with the endorsement and sponsoring by IAU Commissions and Divisions and subsequent approval by the IAU Executive Committee. The final proposal was written during rainy and cold days in Paris, with the input and help of several members of the by-then proposed SOC. The task of assembling the programme was a challenge that required many consultations to many members of the different research fields involved. The final result reflects, I think, a good balance of different topics related to the production, collimation, propagation, interaction, and radiative properties of jets on all scales. Both new theoretical and observational results of high impact were presented at the Symposium. The discussions, in part reflected in this book, were highly motivating and constructive. Many of them occurred during extensive coffee breaks, posters sessions, and in the nearby cafés of Buenos Aires. Their effect, I am sure, will appear in many forthcoming publications.

The meeting was also an occasion to celebrate Félix Mirabel's 65th birthday and pay tribute to his outstanding contributions to our current knowledge of jets. Félix has been a source of inspiration for all of us that have worked with him and had the privilege to share long scientific discussions on the most varied topics. His permanent action fostering high-energy astrophysics in South America deserves a particular mention.

It has past a while since the last IAU Symposium was held in Argentina. This new occasion helped to promote a research field that is growing very fast in several South American countries. It was also a good opportunity to show how strong is the female astrophysical community in Argentina: 17 out of 25 Argentinians that attended the Symposium were women. I doubt that such a rate can be matched in any other country!

It is a great pleasure to acknowledge the financial support of our sponsors listed on page *xiv* of these Proceedings, and the active support of the members of the LOC in realizing the numerous details always associated with such a symposium, in particular Ileana Andruchow (FCAyG, UNLP) and Florencia L. Vieyro (IAR, CONICET). Both were far beyond their duty to take care of every detail of the meeting. Very special thanks go to Matías M. Reynoso (IFIMAR, CONICET) for his essential help to prepare the manuscript of this book and the careful transcription of the discussion sheets. I, personally, remain grateful to Carlos Hatcherian and Marina Piranian, from Booking Travel SRL, for their help and kind assistance. I want also to thank all the members of my group for their participation and support. The strongest support, nonetheless, came from my family: thanks Paula and Blumi. Probably my daughter will never forget that she spent her 12th birthday in the Welcome Cocktail of a Symposium on astrophysics.

Let us hope that this book will help to motivate further discussions on astrophysical jets and inspire new symposia devoted to the topic, initiating a new series.

Gustavo E. Romero, chair, SOC and LOC
Heidelberg, December 8, 2010

THE ORGANIZING COMMITTEE

Scientific

T. Belloni (INAF, Italy)
S. Corbel (CEA Saclay, France)
M. Gilfanov (MPA, Germany)
J. Greiner (MPE, Germany)
A. Levinson (Tel Aviv Univ., Israel)

S. Mineshige (Kyoto Univ., Japan)
G.E. Romero -Chair- (IAR, Argentina)
R.A. Sunyaev (MPA, Germany)

A.J. Castro-Tirado (IAA-CSIC, Spain)
E. Gallo (MIT, USA)
E. M. de Gouveia Dal Pino (IAG-USP, Brazil)
E. Kalemci (Sabanci Univ., Turkey)
S. Markoff (Univ. of Amsterdam, The Netherlands)
J.M. Paredes (Univ. of Barcelona, Spain)
R.M. Sambruna (NASA/GSFC, USA)
J. Wilms (Univ. of Erlangen-Nuremberg, Germany)

Local

I. Andruchow
S.A. Cora
L.J. Pellizza
M.M. Reynoso
F.L. Vieyro

S.A. Cellone
J.A. Combi
M. Orellana
G.E. Romero -Chair-

Acknowledgements

The symposium is sponsored and supported by the IAU Divisions X (Radio Astronomy), XI (Space & High Energy Astrophysics).

The Local Organizing Committee operated under the auspices of the
Instituto Argentino de Radioastronomía.

Funding by the
International Astronomical Union,
CONICET (Argentina),
ANCyT (Argentina),
CLAF,
and
Booking Travel S.R.L.,
is gratefully acknowledged.

CONFERENCE PHOTOGRAPH

Participants

Ileana **Andruchow**, IALP/FCAGLP - CONICET/UNLP, Argentina — andru@fcaglp.unlp.edu.ar
Anabella T. **Araudo**, IAR - CONICET, Argentina — aaraudo@fcaglp.unlp.edu.ar
Keiichi **Asada** ASIAA, Taiwan — asada@asiaa.sinica.edu.tw
Maarten **Baes**, Universiteit Gent, Belgium — maarten.baes@ugent.be
Tiara **Battich**, FCAGLP-UNLP, Argentina — battich@carina.fcaglp.unlp.edu.ar
Wlodek **Bednarek**, University of Lodz, Department of Astrophysics, Poland — bednar@astro.phys.uni.lodz.pl
Paula **Benaglia**, Instituto Argentino de Radioastronomía, Villa Elisa, Argentina — paula@irma.iar.unlp.edu.ar
Valenti **Bosch-Ramon**, Universitat de Barcelona, Spain — vbrcat@yahoo.com
Omer **Bromberg**, The Hebrew University Jerusalem, Israel — omer@wise.tau.ac.il
Leila Magdalena **Calcaferro**, FCAGLP-UNLP, Argentina — leila_1385@hotmail.com
Anderson **Caproni**, Universidade Cruzeiro do Sul, Brazil — anderson.caproni@cruzeirodosul.edu.br
Alberto J. **Castro-Tirado**, IAA-CSIC, Spain — ajct@iaa.es
Sergio **Cellone**, FCAGLP, UNLP & IALP, CONICET-UNLP Argentina — scellone@fcaglp.unlp.edu.ar
Anna Lisa **Celotti**, SISSA, Italy — celotti@sissa.it
Tao **Chen**, CEA Service d'Astrophysique, France — tao.chen@cea.fr
Yoon Young **Chun**, Sabanci University, Turkey
Giuseppe **Cimo**, Joint Institute for VLBI in Europe, Netherlands — cimo@jive.nl
Jorge Ariel **Combi**, IAR, Argentina — jcombi@fcaglp.unlp.edu.ar
Sofia Alejandra **Cora**, Instituto de Astrofisica de La Plata, Argentina — sacora@fcaglp.unlp.edu.ar
Stephane **Corbel**, University Paris Diderot & CEA Saclay, France — stephane.corbel@cea.fr
Mickael **Coriat**, CEA-Saclay / Univ. Paris Diderot, France — mickael.coriat@cea.fr
Camila Anahi **Correa**, FCAGLP-UNLP, Argentina — camilacorrea@carina.fcaglp.unlp.edu.ar
David **Cseh**, CEA Service d'Astrophysique, France — david.cseh@cea.fr
Thomas **Dauser**, Remeis-Observatory & ECAP, Bamberg, Germany — Thomas.Dauser@sternwarte.uni-erlangen.de
Fabio **De Colle**, Astron. & Astrophys. Dep., University of California, USA — fabio@ucolick.org
Elisabete **de Gouveia Dal Pino**, University of Sao Paulo - IAG-USP, Brazil — dalpino@astro.iag.usp.br
Maria V. **del Valle**, IAR - CONICET, Argentina — maria@iar-conicet.gov.ar
Charles **Dermer**, Naval Research Laboratory, USA — charles.dermer@nrl.navy.mil
Salome **Dibi**, Astronomical Institute "Anton Pannekoek", Netherlands — s.rousselle@uva.nl
Samia **Drappeau**, Astronomical Institute "Anton Pannekoek", The Netherlands — s.drappeau@uva.nl
Richard **Dubois**, SLAC National Accelerator Laboratory, USA — richard@slac.stanford.edu
Stephen **Eikenberry**, University of Florida Research Foundation — eikenberry@astro.ufl.edu
Dimitrios **Emmanoulopoulos**, University of Southampton, UK — D.Emmanoulopoulos@soton.ac.uk
Sergei **Fabrika**, Special Astrophysical Observatory, Russia — fabrika@sao.ru
Diego **Falceta-Goncalves**, Universidade Cruzeiro do Sul, Brazil — diego.goncalves@cruzeirodosul.edu.br
Heino **Falcke**, Radboud University Nijmegen/ASTRON, The Netherlands — h.falcke@astro.ru.nl
Junhui **Fan**, Center for Astrophysics, Guangzhou University, China — jhfan-cn@yahoo.com.cn
Christian **Fendt**, Max Planck Institute for Astronomy, Germany — fendt@mpia.de
Jonathan **Ferreira**, Laboratoire d'Astrophysique de Grenoble, France — Jonathan.Ferreira@obs.ujf-grenoble.fr
Luigi **Foschini**, INAF - Osservatorio Astronomico di Brera, Italy — luigi.foschini@brera.inaf.it
Giovanni **Fossati**, Rice University, USA — gfossati@rice.edu
Christian **Fromm**, Max Planck Institute for Radio Astronomy, Bonn, Germany — cfromm@mpifr.de
Elena **Gallo**, University of Michigan, USA — egallo@mit.edu
Federico **García**, FCAGLP-UNLP, Argentina — fgarcia@carina.fcaglp.unlp.edu.ar
Walter **Gear**, CF, UK — Walter.Gear@astro.cf.ac.uk
Giancarlo **Ghirlanda**, INAF-Osservatorio Astronomico di Brera, Italy — giancarlo.ghirlanda@brera.inaf.it
Gabriele **Ghisellini**, INAF-Osservatorio Astronomico di Brera, Italy — gabriele.ghisellini@brera.inaf.it
Gabriele **Giovannini**, Instituto di Radioastronomia, Italy — ggiovann@ira.inaf.it
Nectaria **Gizani**, Hellenic Open University, Greece — ngizani@eap.gr
Yolanda **Gomez**, CRyA-UNAM, Mexico — y.gomez@crya.unam.mx
Pierre **Guillard**, Spitzer Science Center, Caltech, USA — guillard@ipac.caltech.edu
Philip **Hardee**, University of Alabama, USA — phardee@bama.ua.edu
John **Hawley**, University of Virginia, USA — jh8h@virginia.edu
Masaaki **Hayashida**, KIPAC/SLAC, USA — mahaya@slac.stanford.edu
Martin **Huarte Espinosa**, University or Rochester & Cambridge Cavendish Astrophysics Group, USA — martinhe@pas.rochester.edu
Naoki **Isobe**, Kyoto University, Japan — n-isobe@kusastro.kyoto-u.ac.jp
Agnieszka **Janiuk**, Copernicus Astronomical Center, Polish Academy of Sciences, Poland — agnes@camk.edu.pl
Sarka **Jiraskova**, Radboud University Nijmegen, the Netherlands — sarka@astro.ru.nl
Emrah **Kalemci**, Sabanci University, Tuzla, Turkey — ekalemci@sabanciuniv.edu
John **Kirk**, Max-Planck-Institut für Kernphysik, Heidelberg, Germany — John.Kirk@mpi-hd.mpg.de
Shinji **Koide**, Kumamoto University, Japan — koidesin@sci.kumamoto-u.ac.jp
Karri **Koljonen**, Alto University Metsähovi Radio Observatory, Finland — karri.koljonen@gmail.com
Ruben **Krasnopolsky**, Academia Sinica, Taiwan (R.O.C.) — ruben@asiaa.sinica.edu.tw
Wolfgang **Kundt**, Argelander Institute of Bonn University, Germany — wkundt@astro.uni-bonn.de
Magdalena **Kunert-Bajraszeska**, Torun Centre for Astronomy, Poland — magda@astro.uni.torun.pl
Nick **Kylafis**, University of Crete and FORTH, Greece — kylafis@physics.uoc.gr
Alvaro **Labiano**, European Space Agency, Spain — alabiano@sciops.esa.int
Tatiana **Larchenkova**, Astro Space Center P.N.Lebedev Physical Institute, Russia — tanya@lukash.asc.rssi.ru
Laurits **Leedjärv**, Tartu Observatory, Estonia — leed@aai.ee
Amir **Levinson**, Dep. of Phys. Faculty of Exact Sc., Tel Aviv University, Israel — Levinson@wise.tau.ac.il
Diego **Lopez-Camara Ramirez**, Instituto de Ciencias Nucleares, UNAM, Mexico — diego.lopez@nucleares.unam.mx
Alexander **Lutovinov**, Space Research Institute, Russia — aal@iki.rssi.ru
Yuri **Lyubarsky**, Physics Department, Ben-Gurion University, Israel — lyub@bgu.ac.il
Dipankar **Maitra**, University of Michigan, USA — dmaitra@umich.edu
Julien **Malzac**, CESR (CNRS/Universit Toulouse), France — malzac@cesr.fr
Sera **Markoff**, Astronomical Institute "Anton Pannekoek" University of Amsterdam — s.b.markoff@uva.nl
Francesco **Massaro**, Harvard - Smithsonian Astrophysicsl Observatory, USA — fmassaro@cfa.harvard.edu
David **Meier**, Jet Propulsion Laboratory, California Institute of Technology, USA — dlm@jpl.nasa.gov
Attila **Meszaros**, Charles University, Prague, Czech Republic — meszaros@cesnet.cz
Claire **Michaut**, LUTH - Observatoire de Paris, France — Claire.Michaut@obspm.fr
Simone **Migliari**, European Space Agency, Spain — smigliari@sciops.esa.int
James **Miller-Jones**, University of Valencia, Spain — jmiller@nrao.edu
Petar **Mimica**, University of Valencia, Spain — Petar.Mimica@uv.es
Felix **Mirabel**, CONICET & CEA, Argentina & France — mirabel@iafe.uba.ar
Danilo **Morales Texeira**, IAG-USP, Brazil — danilo@astro.iag.usp.br
Ludmila **Nazarova** EuroAsian Astron. Socc, Univ. pr.13, Moscow, Russia — lsnazarova@rambler.ru
Joseph **Nielsen**, Harvard, USA — jneilsen@head.cfa.harvard.edu

Ken-Ichi **Nishikawa**, UAH/CSPAR, USA — ken-ichi.nishikawa-1@nasa.gov
Mariana **Orellana**, UV, Chile / FCAG-UNLP, Argentina, Chile — morellana@fcaglp.unlp.edu.ar
Zsolt **Paragi**, JIVE, Netherlands — zparagi@jive.nl
Joseph M. **Paredes**, Universitat de Barcelona, Spain — jmparedes@ub.edu
Asaf **Pe'er**, Space Telescope Science Institute, USA — apeer@stsci.edu
Leonardo J. **Pellizza**, Institute for Astronomy and Space Physics, CONICET/UBA, Argentina — pellizza@iafe.uba.ar
Carolina **Pepe**, IAFE-UBA-CONICET, Argentina — carolina.pepe@gmail.com
Pérez, Daniela, FCAGLP, University of La Plata, Argentina — danielaperez@iar.unlp.edu.ar
Cintia **Peri**, FCAGLP - Universidad Nacional de La Plata, Argentina — cintia@carina.fcaglp.unlp.edu.ar
Manel **Perucho-Pla**, Universitat de Valncia, Spain — manel.perucho@uv.es
Pierre-Olivier **Petrucci**, LAOG, France — pierre-olivier.petrucci@obs.ujf-grenoble.fr
Tsvi **Piran**, The Hebrew University of Jerusalem, Israel — tsvi@phys.huji.ac.il
Richard **Plotkin** University of Amsterdam, Netherlands — r.m.plotkin@uva.nl
Peter **Polko** Astronomical Institute "Anton Pannekoek", the Netherlands — P.Polko@uva.nl
Almudena **Prieto**, IAC (Instituto Astrofisica Canarias), Spain — aprieto@iac.es
Daniel **Proga**, Department of Physics & Astronomy, University of Nevada, Las Vegas — dproga@physics.unlv.edu
Andreas **Quirrenbach**, Landessternwarte Heidelberg, Germany — A.Quirrenbach@lsw.uni-heidelberg.de
Farid **Rahoui**, Harvard-Smithonian Center for Astrophysics, France — frahoui@cfa.harvard.edu
Luis **Reyes**, KICP - University of Chicago, United States — luis.c.reyes@gmail.com
Matías M. **Reynoso**, University of Mar del Plata & IFIMAR- CONICET, Argentina — matias_reynoso@yahoo.com
Rogemar **Riffel** Departamento de Física - UFSM, Brasil — rogemar@smail.ufsm.br
Luis **Rodriguez**, CRyA, UNAM, Mexico — l.rodriguez@crya.unam.mx
Gustavo E. **Romero**, Instituto Argentino de Radioastronomía (IAR - CONICET), Argentina — romero@iar-conicet.gov.ar
David **Russell**, University of Amsterdam, Netherlands — d.m.russell@uva.nl
Rita **Sambruna**, NASA/GSFC, USA — Rita.M.Sambruna@nasa.gov
Celia **Sanchez-Fernandez**, INTEGRAL Science Operations Center, ESA, Spain — celia.sanchez@sciops.esa.int
Frank **Schinzel**, Max-Planck-Institut fuer Radioastronomie, Germany — schinzel@mpifr-bonn.mpg.de
Hiromi **Seta**, Saitama University, Japan — seta@heal.phy.saitama-u.ac.jp
Tariq **Shahbaz**, Instituto de Astrofisica de Canarias, Spain
Marek **Sikora**, Copernicus Astronomical Center, Poland — sikora@camk.edu.pl
Paolo **Soleri**, Kapteyn Astronomical Institute, University of Groningen, The Netherlands — soleri@astro.rug.nl
Marina Soledad **Sosa**, FCAGLP-UNLP, Argentina — marina@carina.fcaglp.unlp.edu.ar
Alejandra E. **Suárez**, FCAGLP, UNLP, Argentina — suarezal@carina.fcaglp.unlp.edu.ar
Ovidiu **Tesileanu**, University of Bucharest, Romania — ovidiu.tesileanu@gmail.com
Francesco **Tombesi**, NASA/GSFC, USA — tombesi@iasfbo.inaf.it
Gagik **Tovmassian**, Instituto de Astronomia, UNAM, Mexico — gag@astrosen.unam.mx
Valeriu **Tudose**, ASTRON, Netherlands — tudose@astron.nl
Yoshihiro **Ueda**, Kyoto University, Japan — ueda@kusastro.kyoto-u.ac.jp
Mauri **Valtonen**, University of Turku, Finland — mvaltonen2001@yahoo.com
Pieter **van Oers**, University of Southampton, United Kingdom — pvo1g09@soton.ac.uk
Silvia **Vicente**, ESA/ESTEC/RSSD, Holland — svicente@rssd.esa.int
Florencia **Vieyro**, IAR - CONICET, Argentina — florenciavieyro@gmail.com
Gabriela S. **Vila**, IAR - CONICET, Argentina — gvila@iar-conicet.gov.ar
Emma **Whelan**, Laboratoire dAstrophysique de lObservatoire de Grenoble, France — whelane@obs.ujf-grenoble.fr
Feng **Yuan**, Shanghai Astronomical Observatory, China — fyuan@shao.ac.cn
Lorena **Zibecchi**, FCAGLP, Argentina — lorenazibecchi@hotmail.com
Janusz **Ziolkowski**, Copernicus Astronomical Center, Poland — jz@camk.edu.pl
Juan Antonio **Zurita Heras**, AIM Paris Saclay, France — juan-antonio.zurita-heras@cea.fr

OPENING TALK

Jets at all Scales
Proceedings IAU Symposium No. 275, 2011
G. E. Romero, R. A. Sunyaev & T. Belloni, eds.

© International Astronomical Union 2011
doi:10.1017/S1743921310015589

Stellar black holes: Cosmic history and feedback at the dawn of the universe

I. Felix Mirabel[1,2]

[1] CEA-Saclay, IRFU/DSM/Service d'Astrophysique. 91191 Gif sur Yvette. France
[2] IAFE-UBA-CONICET. cc 67, suc. 28. (C1428) Buenos Aires. Argentina
email: `felix.mirabel@cea.fr`

Abstract. Significant historic cosmic evolution for the formation rate of stellar black holes is inferred from current theoretical models of the evolution of massive stars, the multiple observations of compact stellar remnants in the near and distant universe, and the cosmic chemical evolution. The mean mass of stellar black holes, the fraction of black holes/neutron stars, and the fraction of black hole high mass X-ray binaries (BH-HMXBs)/solitary black holes increase with redshift. The energetic feedback from large populations of BH-HMXBs form in the first generations of star burst galaxies has been overlooked in most cosmological models of the reionization epoch of the universe. The powerful radiation, jets, and winds from BH-HMXBs heat the intergalactic medium over large volumes of space and keep it ionized until AGN take over. It is concluded that stellar black holes constrained the properties of the faintest galaxies at high redshifts. I present here the theoretical and observational grounds for the historic cosmic evolution of stellar black holes. Detailed calculations on their cosmic impact are presented elsewhere (Mirabel, Dijkstra, Laurent, Loeb, & Pritchard 2011).

Keywords. Black holes, microquasars, cosmology

1. The dark ages

Motivated by a talk of Rashid Sunyaev at the 7th Microquasar Workshop, I became interested in exploring the possible role of black holes of stellar mass in the early cosmic evolution, and in particular, on how feedback from accretion of the remnants of massive stars could have affected the intergalactic medium during the dark ages. The so called "dark ages" of the universe began about 400,000 years after the Big Bang as matter cooled down and space became filled with neutral hydrogen for hundreds of millions years. How most of the matter in the universe became again ionized (reionized) in less than a billon year is a question of topical interest in cosmology. The recent detection of the explosion of a massive star (Salvaterra R. *et al.* 2009) at $z{\sim}8.2$ and the observation of galaxies (Bouwens *et al.* 2010) up to $z{\sim}8$ support the idea that the ultraviolet radiation from massive stars in the first galaxies played an important role in the reionization. However, from the luminosity density of the most distant galaxies it has been claimed that the UV flux available from massive stars may not have been enough to keep fully reionized the universe (Bouwens *et al.* 2010). Because X-rays have a longer mean free path than the ultraviolet photons, it is proposed that accreting stellar black holes provided heating and secondary ionizations over large volumes of space (Mirabel *et al.* 2011).

2. Formation of black holes by implosion: cosmic historic evolution of BH-HMXBs

Theoretical models show that the evolution and final fate of massive stars strongly depend on the initial metallicity and rotation. Stars with low metal content and initial masses of a few tens of solar masses collapse directly as black holes, with no energetic supernova natal kicks (Heger *et al.* 2003; Meynet & Maeder 2005). On the other hand, recent hydrodynamic simulations of the formation of stars with low metal content (Krumholz *et al.* 2009; Turk *et al.* 2009, Stacy *et al.* 2009) show that a substantial fraction of these stars form as small multiple systems dominated by binaries with typical masses of tens of solar masses. Therefore, from current theoretical models it is inferred that the majority of high mass stellar binaries of low metallicity should remain gravitationally bound after the implosion, ending as BH-HMXBs, which are known to be powerful sources of X-rays, massive winds, and relativistic jets (microquasars, see Mirabel & Rodríguez 1999). In the context of these models and the cosmic evolution of metallicity and star formation evolution it is then expected that: *1) the mass of stellar black holes, 2) the fraction of black holes/neutron stars, and 3) the fraction of black hole binaries/solitary black holes should increase with redshift. Therefore, the formation rate of BH-HMXBs must have been significantly larger in the early universe than in later epochs.*

Formation rate of stellar black holes as function metallicity

How massive stars evolve and die depends on their initial mass, metal content, angular momentum, and whether they are born in isolation or in multiple systems (Meynet & Maeder 2005). Despite these complexities the expected cosmic evolution of BH-HMXBs mentioned above is consistent with the following observations of stellar black holes and neutron stars in the near and distant universe.

a) In agreement with the theoretical expectations (Heger *et al.* 2003, Meynet & Maeder 2005), the masses of black holes in high mass x-ray binaries determined dynamically seem to be a decreasing function of the metallicity of the host galaxy (Crowther *et al.* 2010). The black holes in the high mass binaries M 33 X-7, NGC 300 X-1 and IC10 X-1 which are in small galaxies of low metallicity, have masses in the range of 16 to 30 solar masses, which are larger than the mass of any known stellar black hole in the Milky Way and Andromeda galaxies.

b) It has been proposed that most ultraluminous X-ray sources (ULXs) are BH-HMXBs that contain black holes of several tens of solar masses in low metallicity environments, accreting in a slightly critical regime (Zamperi & Roberts 2009). It is found that the occurrence rate of ULXs per unit galaxy mass in nearby galaxies is a decreasing function of the mass of the host galaxy (Swartz *et al.* 2009), namely, of its metal content. One extreme case in the local universe is that of the metal-poor ($Z < 0.05Z_\odot$) ring Cartwheel galaxy, where it is estimated (Mapelli *et al.* 2009) that more than $\sim$100 stellar black holes of $30 - 80$ solar masses might have been generated via direct collapse (implosion) during the last 10^7 yr. The X-ray luminosity of this small galaxy of low metallicity is 10^{42} erg/s, which rivals that of low luminosity AGN.

c) The kinematics in three dimensions of Galactic black hole binaries relative to their birth place provides evidence for black hole formation by implosion. So far, the space kinematics has been determined for five Galactic black hole binaries. The kinematics of the three black holes that have masses equal or larger than 10 solar masses suggest that they have been form directly, without very energetic supernovae, whereas the two black holes with less than 10 solar masses were form with energetic natal explosions. In fact, the microquasars GRO J1655-40 and XTE 1118+48 which host respectively black holes

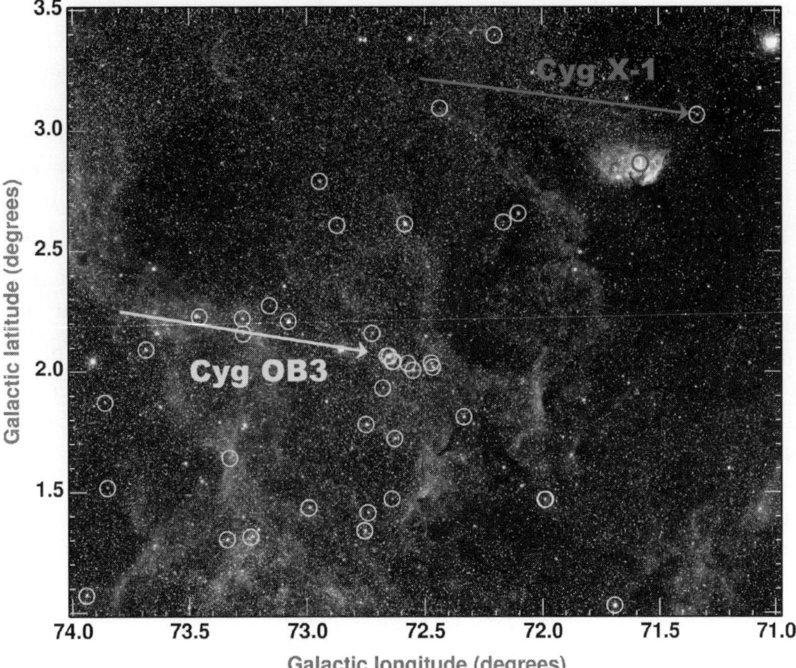

Figure 1. The kinematics represented in this optical image of the sky around the black hole x-ray binary Cygnus X-1 and the association of massive stars Cygnus OB3 shows that Cygnus X-1 remained anchored at its birth place in the association Cygnus OB3. The red arrow represents the motion in the sky of the radio counterpart of Cygnus X-1 for the past 0.5 million years. The yellow arrow the average Hipparcos motion of the massive stars of Cygnus OB3 (circled in yellow) for the past 0.5 million years. Despite the different observational techniques used to determine the proper motions, Cygnus X-1 moves in the sky as Cygnus OB3. At a distance of 2 kpc, the space velocity of Cygnus X-1 relative to that of Cygnus OB3 is < 9 km/s, from which it is inferred that not more than one solar mass could have been suddenly ejected during the formation of the black hole in a natal supernova. From Mirabel, I.F. & Rodrigues (2003).

of 5 and 7 solar masses have runaway space velocities relative to their environment of 112 ± 18 km/s and 177 ± 33 km/s, respectively (Mirabel *et al.* 2002, Mirabel *et al.* 2001). On the contrary, Cyg X-1 which contains a black hole of $\sim$10 solar masses remained anchored at its birth place in the association Cyg OB3, and not more than one solar mass could have been suddenly ejected during a possible natal supernova (Mirabel & Rodrigues, 2003).

The low mass X-ray binaries GRS 1915+105 and V404 Cyg that contain black holes of 14 and 12 solar masses respectively have both peculiar velocities relative to their environment of several tens of km/s that lie on the plane of the Galaxy and are mostly in radial direction towards the Galactic Centre. However, their velocity components perpendicular to the Galactic plane are very small, 10 ± 4 km/s for GRS 1915+105 (Dhawan *et al.* 2007) and 0.2 ± 3 km/s for V404 Cyg (Miller-Jones *et al.* 2009). The kinematics of pulsars show that natal kicks have no preferential direction, and in this context the large peculiar velocities on the plane of the Galaxy of these two low mass X-ray binaries with black holes of more than 10 solar masses are most probably due to Galactic diffusion, rather than to energetic natal kicks.

d) The theoretical expectation that very massive stars with high metal content may end as neutron stars instead of black holes (Meynet & Maeder 2005) is consistent with observations of some newly formed neutron stars. In fact, young neutron stars observed as Soft Gamma Ray Repeaters (SGRs) and Anomalous X-ray Pulsars (AXPs) are found in young clusters of massive stars. SGR 1806-20 (Mirabel *et al.* 2000, Figer *et al.* 2005) and AXP CXOU J1647-45 are associated with parent clusters of massive stars, the latter being inside the young cluster Westerlund 1 (Muno *et al.* 2006). These clusters are in the inner Galaxy and have metal contents larger than solar. Assuming coeval formation of the most massive stars in the clusters and the progenitors of the young neutron stars, lower limits of 40 and 50 solar masses for the masses of the progenitors of these neutron stars are inferred (Figer *et al.* 2005, Muno *et al.* 2006).

e) Gamma Ray Bursts of long duration (LGRBs) mark the formation of black holes by the collapse of massive stars. Although a fraction of dark GRBs may require local extinction columns of $Av > 1$mag, the majority of the hosts of LGRBs are faint, irregular galaxies of limited chemical evolution (Le Floc'h *et al.* 2003, Fruchter *et al.* 2006). GRB 060505 and GRB 060614 were observed (Della Valle *et al.* 2006, Fynbo *et al.* 2006, Gal-Yam *et al.* 2006) with no luminous SNe, and one possible explanation -among others (Gehrels *et al.* 2006)- is black hole formation by direct collapse.

f) A number of supernovae, classified as core collapse Type II show extremely low expansion velocity and an extraordinarily small amount of ^{56}Ni in the ejecta (Zampieri *et al.* 2003). These SNe are under-energetic with respect to a typical Type II supernova and may originate from the explosion of a massive progenitor in which the rate of early infall of stellar material on the collapsed core is large. Events of this type could form a black hole remnant, giving rise to significant fallback with late-time accretion and relatively small kicks. Recently it has been observed a low-energy core-collapse supernova without hydrogen envelope (Valenti *et al.* 2009), suggesting a link of faint supernovae with long-duration gamma-ray bursts.

g) There is increasing evidence for enhanced LGRB rate at $z > 3$, as expected from the increase of specific star formation rate with decreasing metallicity (Daigne *et al.* 2006, Kistler *et al.* 2008, Qin *et al.* 2010).

3. Feedback from accreting stellar black holes

As in AGN, feedback from accreting stellar black holes is observed in the form of energetic radiation, powerful jets, and massive winds.

Energetic radiation: Super-Eddington radiation that last relatively short times is observed in outbursts of black hole novae, which have low-mass stellar donors (e.g. Nova Muscae, V404 Cyg). BH-HMXBs can exhibit a somewhat steady form of super-Eddington radiation of up to 10^{41}erg/s. This is observed in the extragalactic ULXs, which often have spectra that resemble the very high (super-soft) state observed in some Galactic black hole binaries (e.g. GRS 1915+105). The majority of ULXs exhibit a complex curvature spectrum which can be modeled by a cool disk component together with a power law which breaks above 3 keV, probably due to a cool, optically thick corona produced by super-Eddington accretion flows (Gladstone *et al.* 2009).

On the other hand, the satellites Fermi and Agile detected gamma-ray flares from two high mass X-ray binaries: from the microquasar Cygnus X-3 and from Cygnus X-1. TeV flares may have been observed with the Cherenkov telescope MAGIC. This high energy emission from microquasars has been modeled in the context of jet leptonic and hadronic models (Vila & Romero 2010, Romero 2008).

Powerful jets: Steady large-scale jets have been observed in several Galactic black hole binaries (e.g. 1E 1740.7-2942; GRS 1758-258; SS 433), which are viewed as small scale analogues of the extragalactic FR II radio galaxies. The existence of powerful super-Eddington jets of low radiation efficiency has been revealed by the multiwavelength observations of the HMXB SS 433, where the jets entrain atomic nuclei with velocities of 0.26 c, have mechanical energies $> 10^{39}$ erg/s, and are capable of blowing laterally the nebula W50 up to distances of tens of parsecs. Large scale bow-shocked nebula produced by powerful dark jets have been observed in the BH-HMXB Cygnus X-1 (Gallo *et al.* 2005), and more recently in the HMXB S26 in the galaxy NGC 7793 (Pakull *et al.* 2010). The mechanical energy injected by S26 is $> 10^{40}$ erg/s, showing that the overall energy injected by these HMXB microquasars during their whole lifetime can be several orders of magnitude that of the photonic and baryonic energy from a typical core collapse supernova. Further evidence for powerful jets of low radiation efficiency has been obtained by the observations in the X-rays with Chandra of moving jets in the black hole binaries XTE J1550-564 (Corbel *et al.* 2002) & H 1743-32 (Corbel *et al.* 2005). These observations show in real time the formation of double-lobe X-ray and radio lobes. The X-rays are produced by synchrotron mechanism, implying that electrons are accelerated up to TeV energies by shocks in the moving lobes at parsec distances from the BHXRBs.

Massive winds: The connection between winds, jets and x-ray emission has been observed with unprecedented detail in GRS 1915-105 (Neilsen *et al.* 2011). The X-ray observations reveal ejections from the inner disk in the form of jets followed by winds from the outer disk that are massive enough to quench the jet and produces transitions in the X-ray overall output. Non-relativistic massive outflows in the form of winds of ionized and neutral gas have now been observed in accreting black holes of all mass scales.

4. Discussion and results

Ionization and thermal history of the Intergalactic Medium

Early star-forming galaxies as the ones recently discovered with the Hubble Space Telescope up to $z{\sim}8$ (Bouwens *et al.* 2010), when the universe was only ${\sim}800$ millon years old, must have caused the reionization of the intergalactic medium (IGM). However, it is currently believed that the ultraviolet radiation from the massive stars in those galaxies was enough to produce and keep over large volumes of space most of the IGM reionized and heated to temperatures of ${\sim}10^4$ K (Roberston *et al.* 2010). This belief resides in the assumption that the escape fraction into the IGM of UV photons $f_{\rm esc} = 0.1 - 0.2$ as in galaxies at $z{\sim}3$. Furthermore, the recombination rate depends -besides on the hydrogen density- on the temperature of the IGM. Since the mean free path of the X-rays in a neutral medium is much larger than that of the UVs and X-rays are capable of producing multiple ionizations, Mirabel *et al.* (2011) propose that the X-ray radiation from the large population of BH-HMXBs in the firsts star burst galaxies ionized and heated to temperatures of ${\sim}10^4$ K the IGM at large distances in the low density regions, bringing down the recombination rate, and therefore keeping the whole IGM ionized.

The idea that black holes may have played a role in the reionization era of the universe was developed previously by Madau *et al.* (2004). Motivated by the early reports from the Wilkinson Microwave Probe (WMAP) of a large optical depth to Thomson scattering that would have implied a very early reionization epoch, it was proposed a scenario where the universe was first reionized by intermediate black holes at $z > 20$. However, a more accurate determination from WMAP3 later lead to a revision of the optical depth downward (Page *et al.* 2007), to a value consistent with reionization significantly later, at the epoch when the first stars were form. On the other hand, it has been shown (Alvarez *et al.* 2009, Milosavljevic *et al.* 2009) that feedback from accretion to solitary black holes significantly affects further inflow and the consequent injection of radiation and high energy particles to the surrounding medium. Therefore, if X-rays played a role in the heating and reionization of the Intergalactic Medium (IGM), in the context of the current simulations of the first generations of massive stars (Krumholz *et al.* 2009, Turk *et al.* 2009, Stacy *et al.* 2009), and the observations of stellar black holes in the near and distance universe, the most realistic alternative to quasi-radial, Bondi-like accretion would be accretion to black hole stellar remnants in the first generations of high mass binary stars, namely, BH-HMXBs.

On the other hand, current observational results indicate that galaxies -and therefore stellar black holes- were formed before supermassive black holes. In this context, heating of the IGM was first caused and maintained by accreting stellar black holes, before this role was taken over by AGN.

Stellar black holes constrain the properties of dwarf galaxies

The apparent disparity between the number of dwarf galaxies predicted by the Cold Dark Matter model of the universe, with the number of low mass galaxies observed so far in the halo of the Galaxy is a subject of topical interest in cosmology. Power *et al.* (2009) pointed out the possible implications of X-ray binaries in primordial globular clusters for the reionization, and therefore, galaxy formation at high redshifts.

As shown by Mirabel *et al.* (2011), once the IGM is heated to a temperature of 10^4 K, dark matter halos with masses below $10^9 M_\odot$ no longer can accrete IGM material because the temperature of the infalling gas increases by an extra order of magnitude as its density increases on its way into these galaxies. In that regime, only gaseous halos with virial temperatures above 10^5 K could have accreted fresh IGM gas and converted it to stars. The census of dwarf galaxy satellites of the Milky Way requires a related suppression in the abundance of low-mass galaxies relative to low-mass dark matter halos (Alvarez *et al.* 2009). *The thermal history of the IGM therefore has a direct impact on the properties of the faintest galaxies at high redshifts as well as the smallest dwarf galaxies in the local universe.*

It is interesting to note that black holes of different mass scales have a role in galaxy formation. Feedback from supermassive black holes halt star formation, quenching the unlimited mass growth of massive galaxies (Fabian 2009). Feedback from stellar black holes in HMXBs during the reionization epoch suppress the number of dwarf galaxies with masses $< 10^9 M_\odot$. Therefore, BH-HMXBs in the early universe are an important ingredient to reconcile the apparent disparity between the observed number of dwarf galaxies in the Galactic halo with the number of low mass galaxies predicted by the Cold Dark Matter model of the universe.

References

Alvarez, M. A. Wise, J. H., & Abel, T. 2009, *ApJ* 701, L133

Bouwens, R. J. *et al.* 2010, *ApJ* 709, L16

Corbel, S. *et al.* 2002, *Science* 298, 196

Corbel, S. *et al.* 2005, *ApJ*, 632, 504

Crowther, P. A. *et al.* 2010, *MNRAS*, 403, L41

Daigne, F. & Rossi, E. M., Mochkovitch, R. 2006, *MNRAS* 372, 1034

Della Valle, M. *et al.* 2006, *Nature*, 444, 1050

Dhawan, V., *et al.* 2007, *ApJ* 668, 430

Fabian, A. C. 2009, *Proc. of IAU Symposium 267, B.M. Peterson, R.S. Somerville, & T. Storchi-Bergmann, eds.*

Figer, D. F. 2005, *ApJ* 622, L49

Fruchter, A. S. *et al.* 2006, *Nature* 441, 463

Fynbo, J. P. U. *et al.* 2006, *Nature* 444, 1047

Gal-Yam, A. *et al.* 2006, *Nature* 444, 1053

Gallo, E. *et al.* 2005, *Nature* 436, 819

Gehrels, N. *et al.* 2006, *Nature* 444, 1044

Gladstone, J. C., Roberts, T. P., & Done, C. 2009, *MNRAS* 397, 1836

Heger, A. *et al.* 2003, *ApJ* 591, 288

Kistler, M. D. *et al.* 2008, *ApJ* 673, L119

Krumholz, M. R. *et al.* 2009, *Science* 323, 754

Le Floc'h, E. *et al.* 2003, *A&A* 400, 499

Madau, P. *et al.* 2004, *ApJ* 604, 484

Mapelli, M., Colpi, M., & Zampieri, L. 2009, *MNRAS* 395, L71

Meynet, G. & Maeder, A. 2005, *A&A* 429, 581

Miller-Jones, J. C. A. *et al.* 2009, *ApJ* 706, L230

Milosavljevic, M. *et al.* 2009, *ApJ* 698, 766

Mirabel, I. F. & Rodríguez, L. F. 1999, *Ann. Rev. Astron. Astrophys.* 37, 409

Mirabel, I. F., Fuchs, Y., & Chaty, S. 2000, *AIP Conf. Proc.* 526, 814

Mirabel, I. F. *et al.* 2001, *Nature* 413, 139

Mirabel, I. F. *et al.* 2002, *A&A* 395, 595

Mirabel, I. F. & Rodrigues, I. 2003, *Science* 300, 1119

Mirabel, I. F., Dijkstra, M., Laurent, Ph., Loeb, A., & Pritchard, J. R. 2011, submitted to *A&A*

Muno, M. P. *et al.* 2006, *ApJ* 636, L41

Neilsen, J., Lee, J. C., & Remillard, R. 2011, *Proc. of IAU Symposium 275*, G. E. Romero, R. A. Sunyaev & T. Belloni, eds.

Page, L. *et al.* 2007, *ApJ* Supp. 170, 335

Pakull, M. W., Soria, R., & Motch, C. 2010, *Nature* 466, 209

Power, C. *et al.* 2009, *MNRAS* 364, 1146

Qin, S. F. 2010, *MNRAS* 406, 558

Roberston, B. E. *et al.* 2010 *Nature* 468, 49

Romero, G. E. 2008, *Rev. Mex. Astron. Astrof.* 33, 82

Salvaterra, R. *et al.* 2009., *Nature* 461, 1258

Stacy, A., Greif, T. H., & Bromm, V. 2010, *MNRAS* 403, 45

Swartz, D. A., Tennat, A. F., Soria, R. 2009, *ApJ* 703, 159

Turk, M. J., Abel, T., & O'Shea, B. 2009, *Science* 325, 601

Valenti, S. *et al.* 2009, *Nature* 459, 674

Vila, G. S., & Romero, G. E. 2010, *MNRAS* 403, 1457

Zampieri, L. *et al.* 2003, *MNRAS* 338, 711

Zamperi, L. & Roberts, T. P. 2009, *MNRAS* 400, 677

Discussion

PE'ER: Where would you draw the line between jet (continuous) and "blobs" ejection as seen in GRS 1915; is there evidence for jet there at all?

MIRABEL: In GRS 1915+105 we have imaged with the VLBA a compact "continuous" jet long of tens of astronomical units and velocities $< 0.2\,c$ associated to the low-hard state. The sudden transitions from the low-hard state to the high soft state mark the onset of discrete and bright condensations or shocks that move away in the form of collimated jets at apparent superluminal motions.

KYLAFIS: In the slide that you showed about GRS 1915, the optical should appear also. Was it detected?

MIRABEL: No, because GRS 1915 is on the Galactic plane at > 6 kpc and there are about 30 magnitudes of optical absorption along the line of sight.

SAMBRUNA: In AGN (and GRS 1915) disk winds are observed (see F. Tombesi contribution) that most likely contribute to feedbak on large scales.

MIRABEL: I fully agree. In fact, besides the observations of ionized gas with Chandra reported in this meeting by Neilsen *et al.*, recent observations of Mrk 231 revealed massive outflows of molecular gas. The ions moving at $0.26\,c$ in SS433 are from winds that have been entrained and accelerated by the jets.

KAWAI: Half of GRBs are so-called "dark" GRBs with no optical counterpart. They may be formed in metal-rich environments. Thus, low-metalicity is not an essential condition for GRB formation.

MIRABEL: In a recent study, Perley *et al.* conclude that the source of obscuring dust is local to the vicinity of the GRB progenitor and may be highly unevenly distributed within the host galaxy. In fact, my former PhD student Emeric Le Floc'h did not detect with Spitzer any of the host galaxies of dark GRBs, which implies that these are not globally speaking, highly dusty galaxies. The production of dust may occur rapidly in associations of massive stars and the GRB could take place after other massive stars have already enriched locally the interstellar medium.

FENDT: You have been introducing the "universal model for jet-disk coupling" for BH. Why should it not work for neutron stars? Central masses are similar, also the disks should be similar (& simirlaly relativistic).

MIRABEL: It may work for neutron stars, white dwarfs, and young stars. But I prefer to leave a more detailed answer to this question to the speakers that will discuss this issue later in the week.

RODRÍGUEZ: To correct for these two failures of the Cold Dark Matter model, do you need stellar-mass or supermassive black holes, or both?

MIRABEL: Both. Supermassive black holes may account for the absence of masive cusps in the central regions of galaxies. Stellar black holes in the early universe would have heated the IGM to temperatures of about 10^4 K which would have impeded the formation of large numbers of low-mass galaxies by baryonic accretion onto small haloes of dark matter.

NISHIKAWA: What causes the time lag between infrared and radio emission in GRS 1915+105?

MIRABEL: The time delay is due to the fact that in an adiabaticaly expanding plasma cloud, the radiation becomes transparent to longer wavelenths at later times.

Part 1. Jets basic issues and physical proccesses

Jets at all Scales
Proceedings IAU Symposium No. 275, 2011
G. E. Romero, R. A. Sunyaev & T. Belloni, eds.

© International Astronomical Union 2011
doi:10.1017/S1743921310015590

The formation of relativistic cosmic jets

David L. Meier[1]

[1]MS 169-506, Jet Propulsion Laboratory, California Institute of Technology,
4800 Oak Grove Drive, Pasadena, CA 91109, USA
email: David.L.Meier@jpl.nasa.gov

Abstract. I review current ideas on the launching, acceleration, collimation and propagation of relativistic jets and the influence of strong magnetic fields in the process. Recently, several important elements of the entire jet "engine" structure have been shown to play key roles in the production of an astrophysical jet. Depending on the type of system, these include the spin of the central black hole, the thermal and/or magnetic state of the accretion flow, the presence of a re-collimation point in the jet outflow far away from the central object, and the behavior of MHD shocks and kink instabilities in the final jet. While these physical processes probably are at work in all types of relativistic jets (and many even in more benign stellar outflows), I shall concentrate on ones produced by lower luminosity black hole sources, both in active galactic nuclei and in X-ray binaries. I also will discuss the connection between the theoretical concepts and the large body of observational data now available on these systems.

Keywords. black hole physics, accretion, MHD, galaxies: jets, X-ray binaries: jets

---

1. Introduction

Despite the tremendous theoretical and numerical progress made in the past >40 years since the first models of pulsar "winds" as outflows driven by rotating magnetic fields, many astronomers believe that we still have no idea how cosmic jets are produced. Actually, we really *do* know a lot about the subject, and we have very good ideas about most of the missing pieces of the puzzle. The purpose of this paper is to outline what we do know, what we strongly suspect, and what aspects of jet production remain a mystery.

While this work will concentrate on the production of *relativistic* jets, much of it (especially Sections 3, 4, and 5) are relevant to jets on all scales. But, before beginning the discussion, it is very important to define for the reader what we mean by a "cosmic jet" (even though this definition is clear to most attendees of this conference). A cosmic jet is an extraordinarily *large scale*, coordinated, virtually universally *bipolar* and highly supersonic outflow of plasma from the region near a central gravitating star. Jets frequently, but not universally, involve accretion of outside material onto that star. The acceleration and collimation zone (ACZ) alone is typically $100 - 1000$ time the "injection" radius where most of the outflow originates, and the latter can be larger than the star itself. For example, while the sun no longer produces such a large-scale jet, it almost certainly did so when it was a protostar 4.6 Gyr ago. The ACZ extended perhaps $1 - 10$ AU above the protoplanetary disk ecliptic plane, and the length of the full jet was a fair fraction of a parsec. Such jets dwarf, by $6 - 7$ orders of magnitude in size and power, phenomena such as solar flare jets and even coronal mass ejections.

Below we cover five issues in the launching, acceleration, and collimation of jets. We begin by discussing the role of the central rotating object itself and end with the behavior of the freely-propagating jet flow, far from the central object and causally disconnected from the jet engine itself.

2. The Role of the Central Rotating Star

Our discussion of the role of the central rotating, gravitating object in launching the jet will be oriented specifically toward the relativistic jet case. In fact, we shall deal only with black holes and not with neutron stars in this review. The main question here is, how important is the rotating black hole in driving the jet (*i.e.*, the Blandford-Znajek process [Blandford & Znajek 1977, hereinafter BZ]) compared to the rotating accretion disk (*i.e.*, the Blandford-Payne process [Blandford & Payne 1982, hereinafter BP]).

Black holes alone cannot support magnetic fields. While a spinning and charged (*i.e.*, Kerr-Newman) black hole can have a magnetic moment, the amount of charge expected on astrophysical black holes is extremely small (BZ). Therefore, any black hole magnetosphere must be anchored in the accretion inflow itself, where strong currents can be maintained (BZ; Punsly & Coroniti 1990).

Closed vs. open magnetospheres. In a steady state we expect two different types of magnetosphere: closed (see, *e.g.*, Uzdensky 2005), with no open field lines penetrating the black hole horizon, and open (see, *e.g.*, Garofalo 2009), with significant amounts of magnetic flux deposited on the horizon. While both types of magnetosphere can extract angular momentum from the black hole, the ability of each in driving a jet is quite different. The closed magnetosphere deposits black hole angular momentum into the accretion disk, where it can be transferred radially outward through via viscous processes. The open magnetosphere, on the other hand, extracts angular momentum to vertical infinity in a large-scale torsional Alfvén wave (a Poynting flux jet). To first order, then, open black hole magnetospheres should produce jets, while closed ones should not.

Strong vs. weak shear and the magnetic tower effect. Whether or not a closed magnetosphere will remain so (and not produce a jet) will, in turn, depend on the amount of shear between the black hole and the accretion disk — specifically between the horizon and the point on the accretion disk where the other footpoint of the closed field loop is anchored. If the shear is high, the magnetic tower process (Lynden-Bell 1996; Meier *et al.* 1997; Romanova *et al.* 1998; Uzdensky & MacFadyen 2006) will cause the closed field loops to inflate vertically and break, converting a closed magnetosphere into an open, jet-producing one. Since most of the field loop outer footprints lie near the disk innermost stable circular orbit (ISCO), we will approximate the shear as simply the absolute value of the angular velocity difference between the horizon and the ISCO, and normalize the value to that of a non-rotating Schwarzschild black hole

$$\Sigma \;=\; 6^{3/2} \left| (j/2r_{\rm h}) - 1/(r_{\rm isco}^{3/2} + j) \right| \tag{2.1}$$

where $r_{\rm h}$ and $r_{\rm isco}$ are the standard horizon and ISCO radii in units of the gravitational radius $r_g = GM/c^2$ and are functions of the normalized black hole spin $j = Jc/GM^2$, which ranges between -1 and $+1$.

The strongest shear occurs for black holes spinning in the *opposite* sense to the accretion disk ($-1 < j < 0$) and for a small range of fairly rapid prograde holes ($0.75 < j < 0.99$). The weakest shear occurs for $0 < j < 0.75$ and for $j > 0.99$ (Meier & Garofalo 2010), with zero shear occurring at $j \approx 0.36$ and $j = 1.0$. Therefore, *black-hole-driven BZ jets are most likely to form in retrograde black hole systems, while prograde holes* (with an order of magnitude or less shear) *are least likely to produce BZ jets.*

Jet power vs. black hole spin. In addition to determining the likelihood of producing a jet, black hole spin also determines the *power* of that jet, and this process operates in the same sense: retrograde black holes produce the strongest jets by $1.5 - 2$ orders of magnitude (Garofalo 2009; Garofalo *et al.* 2010). This effect is due to magnetic flux in the plunging region ($r < r_{\rm isco}$) rapidly accreting onto the black hole. That is, flux that would have been located in the "gap" region between the horizon and ISCO instead ends up on the horizon. Retrograde black holes have large gaps, up to $8\,GM/c^2$ *vs.* as low as $\sim 0\,GM/c^2$ for maximum prograde Kerr holes. Since the jet power is proportional to the square of the horizon flux (BZ), this leads to much stronger BZ jets from retrograde systems than prograde ones.

The combined effect of shear and horizon magnetic flux creates a gross spin asymmetry in the production of jets by spinning black holes: *retrograde holes should produce strong, radio loud sources, while prograde holes should produce radio quiet sources, or even radio silent ones in the nearly zero shear cases.*

A modified spin paradigm. These results suggest a "spin paradigm" that differs significantly from that suggested previously (Wilson & Colbert 1995; Meier 1999). While rapid black hole spin is necessary for producing a radio loud source, it is not sufficient. That rapid spin also must be *retrograde*. Rapid prograde spin produces radio quiet sources. In terms of the Sikora *et al.* (2007) picture of radio loud and quiet sources, one would conclude that Fanaroff & Riley class I sources (FR Is), broad line radio radio galaxies (BLRGs), and radio-loud quasars (FR IIs & RLQs) have primarily retrograde black holes. On the other hand, low-ionization nuclear emission-line regions (LINERs), Seyfert galaxies, and Palomar-Green (PG) quasars have primarily prograde black holes.

Solution to the spin paradox. Under the old spin paradigm (rapid spin produces powerful radio jets) a paradox arose. Because few powerful FR II radio sources exist at the present epoch, the old paradigm implied that black holes now spin slowly. However, optical statistical studies have concluded that black holes now spin *rapidly* (Yu & Tremaine 2002; Elvis *et al.* 2002). Furthermore, the few sources that have measured rapid prograde spin are radio silent (Iwasawa *et al.* 1996; Brenneman & Reynolds 2006; Fabian *et al.* 2009), while powerful radio sources with measured ISCOs show a very large gap/plunging region (Kataoka *et al.* 2007; Sambruna *et al.* 2009). This all can be explained by the new paradigm: black holes do indeed rotate rapidly now, but in a prograde sense, rendering most radio quiet. And it is quite possible that radio loud objects reveal a large gap because they have retrograde spin.

In short, therefore, black hole systems (at least supermassive ones) show strong evidence of black-hole-driven BZ jets when the spin state is of the correct sign (retrograde).

3. The Role of the Rotating Accretion Disk

Because jet outflows also can be driven by rotating, magnetized accretion disks (BP; Li *et al.* 1992; Vlahakis *et al.* 2000; Vlahakis & Konigl 2003), it is appropriate to ask if many of the jets emanating from black hole systems are disk, rather than black hole, driven. Indeed, after an extensive study of microquasars in X-ray binary systems, Fender *et al.* (2010) concluded that in the X-ray hard state there is no evidence for any correlation between the jets produced and the reported spin measurements of the black hole in those binary systems. While this is indeed an interesting result, it is far too early to conclude that black hole spin plays no role in X-ray binary systems for the following reasons:

- Absence of evidence is not necessarily evidence of absence.
- Black hole spin measurements of the same source vary greatly; *e.g.*, GRS 1915+105 measurements range from 0 to 1.
- Just because a jet property clearly changes with the state of the accretion flow does not mean that the jet is driven by the rotating disk. Recall (see above) that the BZ process depends on the state of the magnetosphere, which is anchored in, and created by, the accretion flow.

- A black-hole-spin-driven jet may be very relativistic ($\gamma > 10$) and, therefore, highly beamed. Unless an observer views such a jet within a few degrees of its axis, it may be largely invisible to observers but nevertheless important in the overall source energetics.

Nevertheless, it is important to take stock of accretion disk jet launching theory and attempt to compare with observations, particularly those in hard accretion states.

Jets from advection dominated accretion flows (ADAFs). The ADAF is the leading model for hard state accretion flows, so it is important to assess its ability to produce the jets observed in hard state objects. McClintock & Remillard (2006) published both photon and power density spectra (PDS) of several microquasars in the hard state (their Figs. 4.11 & 4.12). While there are several accretion models for producing a hard power-law photon spectrum up to 100 keV, the PDS affords a glimpse into the turbulent state of the accretion flow that cannot be seen in the photon spectrum alone. If we assume that the observed power spectrum is composed of many individual power spectra at different disk radii R, then the power contributed at a frequency $f = V_{turb}/H$ (where the velocity of the largest eddy is proportional to the sound speed $V_{turb} \propto c_s \propto R^{-1/2}$ and and the disk half-thickness $H \propto R$ in an ADAF model) is given by $d\mathcal{P} = 2\pi R\,2H\,dR\,\rho\,V_{turb}^2$. The total power per unit frequency then will vary as

$$d\mathcal{P}/df \;\propto\; f^{-4/3} \tag{3.1}$$

Most of the McClintock & Remillard power density spectra have slopes between -1.1 and -1.4 and extend up to several hundred Hz — very similar to what is expected from an ADAF whose turbulence extends down to the ISCO of a $10\,M_\odot$ black hole.

A very nice demonstration of how ADAF-like accretion flows can drive jets was performed in a *tour de force* simulation by McKinney (2006). Like many prior magneto-rotational instability (MRI) simulations (*e.g.*, McKinney & Gammie 2004; Hawley & Krolik 2006), McKinney's generated a collimated outflow from a hot, geometrically thick, radiatively inefficient, magnetized accretion flow. However, in this case he allowed the simulation to continue for $\sim 10^4$ gravitational times ($\tau_g = GM/c^3$) and the jet to propagate to about $10^4\,r_g$. Both observationally and theoretically, therefore, ADAFs or similar flows appear to be a good explanation for steady jets observed in the hard state — especially for those sources that show PDS up to hundreds of Hz.

Jets from magnetically dominated accretion flows (MDAFs). For some of McClintock & Remillard's sources, most notably GRO J1655-40 in its hard state and GRS 1915+105 in its plateau state, the PDS has a sharp cutoff in turbulent noise above a few Hz plus a strong low-frequency quasi-periodic oscillation (LF QPO) just below the noise cutoff.† It has been suggested (Meier 2005; Fragile & Meier 2009) that such a PDS can be best explained by the formation of a strongly-magnetized region inside a classical ADAF, with a size $r_{\mathrm{mdaf}} \sim 10\,r_{\mathrm{isco}}$. A semi-rigid, rotating inner magnetosphere would account for both the noise bandwidth cutoff *and* the strong QPO at only a few Hz.

Jet mini-suppression in the hard state. MDAFs also predict a moderate suppression of the jet power of order $r_{\mathrm{mdaf}}/r_{\mathrm{isco}} \approx 10$, since the ADAF jet in this case would be launched from a shallower region of the black hole potential well (10 *vs.* $1\,r_{\mathrm{isco}}$). When we factor in the observed jet radio power as a function of jet total power ($L_R \propto L_J^{1.3-1.4}$, Migliari & Fender 2006), this predicts *radio* power suppression between ADAFs and MDAFs of order $\sim 20-25$.

---

† Interestingly, GRS 1915+105 and XTE J1550-564 show similar bandwidth-limited noise and LF QPO in the very high/soft power-law state [McClintock & Remillard 2006, Fig. 4.15], which may be the super-Eddington, radiation-pressure-dominated equivalent of the ADAF state.

Fender *et al.* (2010) have discovered a new class of microquasars, which they call "outliers", that lie below the classical fundamental plane, with radio powers suppressed by a factor of ~ 20 compared with other hard-state microquasars with the same X-ray luminosity.‡ (See also Gallo's and Soleri & Fender's articles in this volume.) Curiously, and perhaps not coincidentally, GRO J1655-40 and XTE 1550-564 are outliers, and GRS 1915+105 lies at the intersection of the outlier trend line and the fundamental plane. *It is suggested here, therefore, that outliers are related to sources with bandwidth-limited PDS noise and may be related to accretion flows with MDAFs in the centers of their ADAFs.* More work is needed to see if most sources in an "outlier" state also have bandwidth-limited noise when in that state. Why some sources show these features and why some do not is a complete mystery at present.

Jet suppression in the soft state. Suppression of the jet radio power by a factor of 50 or more occurs when sources transition from a hard state to a soft one (Fender 2001). This author (Meier 2001) suggested that this suppression occurs because the magnetic field vertical component (needed to drive a BP jet from the disk) is significantly smaller in cool, geometrically thin accretion disks than in hot, geometrically thick ADAFs. From standard disk/ADAF models, one predicts a suppression factor of $\sim 100 - 200$, consistent with observations. *This suggests that all strong accretion disk driven (BP) black hole jets are produced only when the source is in a geometrically thick accretion state.*

At first, this conjecture seems to be in contradiction with the generation of strong, "ballistic" and explosive jets when sources transition from the hard state to the soft state (Fender *et al.* 2004). The two can be reconciled, however, if we recall that at high accretion rates disks may be thermally and secularly unstable (Shakura & Sunyaev 1976; Pringle 1976) and may exhibit limit-cycle behavior in which the disk temporarily becomes hot and geometrically thick, drains quickly, and then refills on an accretion time scale (Szuszkiewicz & Miller 2001). In this case, a jet still would be generated in the geometrically thick hot state as it drains (the explosive jet) followed by a long period of recovery in a cooler, geometrically thin state. For GRS 1915+105 the predicted cycle time is ~ 1000 seconds, similar to the observed cycle time for the explosive jets when this source transitions from a hard to a soft state.

4. Jet Acceleration and Collimation

Once launched from the rotating compact object or from the rotating accretion flow, the outgoing magnetized plasma will be accelerated and collimated by the strong, twisting global magnetic field. Compared to the launching processes, the physics of jet acceleration and collimation is much better understood, due in no small part to the pioneering work of Blandford & Payne (1982). They outlined the basic principles of MHD jet production assuming a steady state, axisymmetric, and spherically self-similar flow to turn a 3-dimensional, time-dependent problem into a 1-dimensional one.

Many things have been said about the self-similar assumption made by BP, both favorable and unfavorable. This assumption is not the last word in jet theory by any means. More realistic and more complicated (but related) models and simulations (such as McKinney 2006) will follow in the coming years. Self-similarity also is not the physically implausible assumption that some have capriciously claimed. It simply is a convenient way to express external confinement of the greater jet engine by a poorly understood external medium (*e.g.*, by the interstellar medium). With surprisingly little external

‡ These also sometimes are called "radio quiet black holes", which is an inappropriate name, as it evokes memory of the radio quiet quasars, whose radio powers are $4 - 6$ *orders of magnitude* weaker than radio loud ones!

gas pressure confinement, a large-scale rotating magnetosphere can attain a pseudo-self-similar structure interior to that external medium and external to the rotational axis.

An important parameter in these models is the run of current (and therefore magnetic field) with spherical radius r, which can be specified along the equator (*i.e.*, near the accretion flow) in cylindrical coordinates

$$F \;\equiv\; \partial \ln B/\partial R \,+\, 2 \tag{4.1}$$

When $F > 1$ (often called the *forward current regime*), the current carried by the jet *increases* with disk radius R; when $F < 1$ (the *return current regime*), the current decreases with R; and when $F = 1$, there is *no* current carried by the jet normal to the disk, only that carried in the radial r direction. For all types of rotating, turbulent accretion flows (ADAF, slim [super-Eddington, radiation-pressure-dominated] disk, Shakura and Sunyaev outer and middle regions) $F \approx 0.75 - 0.81$ (the return current regime). So, for most accretion flow jets, return current is carried primarily by the portion of the jet that emanates from near the disk inner edge (shortly outside the ISCO), but inside that portion there must be a non-self-similar region that transitions to a forward current solution to keep the current on the rotation axis finite.

MHD wind/jet models are very similar in mathematical structure to the Parker hydrodynamic (HD) wind, just more complex because the equations are solved in the spherical θ direction, rather than in r, and because there are three characteristic waves and phase speeds instead of one. When the magnetic field is much stronger than the plasma dynamical forces (*i.e.*, plasma $\beta_{\rm p} = p_{\rm gas}/p_{\rm mag} < 1$), the three modes are

• Transverse Alfvén waves, which propagate along field lines with $V_{\rm phase} = V_A \cos\chi$, $V_A = (B/4\pi\rho)^{1/2}$, and χ being the angle between the magnetic field and the propagation direction and ρ being the plasma mass density.

• Longitudinal, *fast* magnetosonic (pressure) waves, which propagate fastest normal to the field at $V_F \approx (V_A^2 + c_s^2)^{1/2}$.

• Longitudinal, *slow* magnetosonic waves, which propagate at $V_{\rm phase} = \pm(c_s\,V_A)^{1/2}$ $\cos\chi$ — best along the field ($\chi = 0$) and not at all normal to it.

In a very nice paper Bogovalov (1994) outlined the physics of MHD winds in the general (non-self-similar) case. There are a total of *six* important surfaces in the flow:

• Three "critical surfaces" (the fast magnetosonic, slow magnetosonic, and cusp surfaces), across which information propagation in a given wave mode changes from being omni-directional to occurring along specific directions or *caustics*. We will not discuss these surfaces much in this paper.

• Three "separatrix surfaces", across which information propagation in a particular wave mode *further* changes from having components upstream and downstream in the plasma flow to having only one of these components. These are the Alfvén surface (AS), the fast magnetosonic separatrix surface (FMSS), and the slow magnetosonic separatrix surface (SMSS). Of these three, the FMSS is the most important and plays the same role as the sonic surface in the Parker wind. *The FMSS is the magnetosonic horizon, beyond which no information can propagate back to the jet engine that accelerates the flow.*†

In a time-independent mathematical model, the separatrix surfaces become *singular* surfaces, where both numerator and denominator of the wind acceleration equation vanish simultaneously. And, if that theory also is 1-dimensional, the singular surfaces become simple singular points along a representative streamline. In the strong-field, MHD case, these points have the names "modified slow point" (MSP), "Alfvén point" (AP), and

---

† In the Parker HD wind model there is only one critical and one separatrix surface, and these coincide to form a single sonic surface where $V = c_s$.

"modified fast point" (MFP). Setting the wind equation denominator to zero at these points generally determines the location of each surface; setting the numerator to zero determines the flow internal boundary (or *regularity*) conditions on those surfaces; and applying l'Hospital's rule to the ratio of numerator and denominator determines how the flow conditions cross these surfaces. Any viable steady wind/jet solution of the MHD flow must pass through *all three* separatrix/singular surfaces in a smooth manner.

Implications for jet acceleration and collimation. In an accelerating jet flow the physics is different in at different stages of the acceleration. From the launch point near the accretion disk or black hole up to the SMSS the flow will be accelerated primarily by gas pressure forces. Beyond that surface the plasma flow is too fast for hydrodynamic forces to act, but not for the rotating magnetic field (torsional Alfvén wave) to continue the acceleration up to the AS. By this time the magnetic field has developed a strong toroidal component, and magnetic pressure $(-d(B_\phi^2/8\pi)/dZ)$ assists in the acceleration. But soon the flow exceeds even the fast magnetosonic speed.

However, observed jet speeds far exceed their magnetosonic speeds. What produces the remaining acceleration? The self-similar theory reveals the answer: pinch forces $(-B_\phi^2 \cot\theta/4\pi r)$ accelerate the flow forward and toward the axis until the θ component of the velocity exceeds the fast magnetosonic speed (*i.e.*, $-V_\theta > V_F$) when it crosses the FMSS. There are several important points about this process:

• If the FMSS lies near the axis ($\theta_{\rm FMSS} << 1$), then the jet speed $V_Z \approx V_r >> |V_\theta| \sim V_F$; *i.e.*, the jet speed far exceeds the fast magnetosound speed.

• In fact, if the parameter $\sigma \approx (V_A^2/V_0 c)(R_0^2/R_L^2) >> 1$ (where V_0 and R_0 are the velocity and radius at the base of a jet streamline, and R_L is the light cylinder radius), very relativistic jet speeds can be attained (Michel 1969; Li *et al.* 1992; Polko *et al.* 2010).

• Since the flow toward the axis is supermagnetosonic, it will bounce or shock on the axis in less than a magnetosound crossing time.

• However, because the FMSS is the magnetosonic horizon, any shock or other feature in the jet flow there will not affect the ACZ. *Beyond the FMSS the jet is causally disconnected from the engine that created it.*

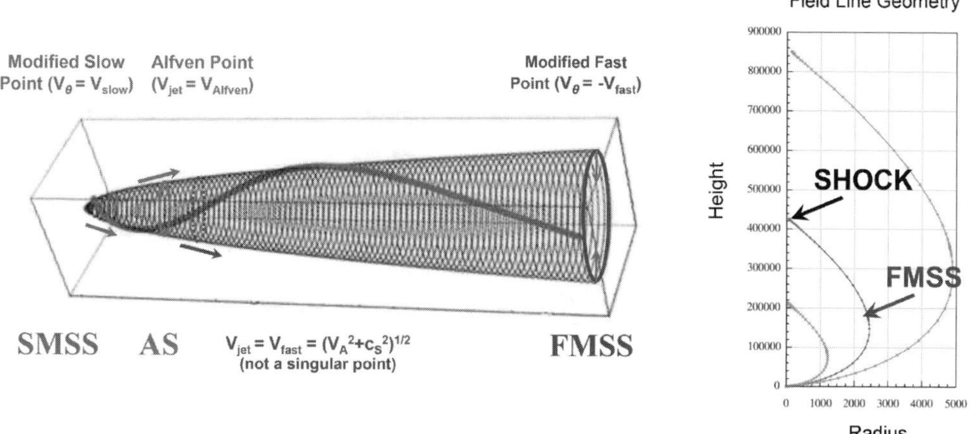

Figure 1. Two views of an accelerating jet model. Left: diagram of the jet magnetic field (tight coil), example plasma trajectory (heavy grey line), and schematic positions of the singular points/surfaces (SMSS, AS, FMSS). Right: three poloidal field/stream lines from a relativistic self-similar model that cross all three singular points, showing the position of the FMSS and possible shock. Also note that the cylindrical radial axis is stretched by a factor of 100 over the vertical axis.

The first work to produce jet models that cross all three separatrix surfaces is Polko *et al.* (2010) (see also Polko's paper in this volume). The recollimation of Polko's field/ stream lines toward the axis is quite evident. Recollimation is also evident in the non-relativistic models of Vlahakis *et al.* (2000) and in the relativistic simulation of McKinney (2006). These cases are all examples of a well-known device used in terrestrial jet engines to power supersonic aircraft — the convergent nozzle. The pinching effect of the convergent nozzle accelerates jet exhaust to supersonic speeds, producing one or more recollimation shocks in the outflow (which is also causally disconnected from the jet engine itself). In an F-16 supersonic fighter, for example, the nozzle is constructed with a tapered steel iris, instead of strong magnetic field, but the principle is essentially the same.

5. The Case for a Collimation Shock Feature in Most Cosmic Jets

Observational evidence. It is quite possible that many, if not most, cosmic jets that we observe posses a collimation feature, perhaps even a shock, shortly after their flow becomes causally disconnected from the central engine. There are, in fact, many observational clues and some strong evidence that this is indeed the case.

The first clue comes from images of FR II radio lobes and from early HD and MHD simulations of jet propagation, which show that radio lobes are consistent with the jets that feed them being weakly magnetized (Norman *et al.* 1982) — well below equipartition — and inconsistent with the jet structures expected from strongly magnetized flows.† Yet, as we have discussed above, jets are believed to be accelerated and collimated by *strong* magnetic fields. If so, then the initial FR II jet created in the galactic nucleus must have lost its strong field somehow — probably closer to the black hole than we usually probe with VLBI (*i.e.*, $<< 10^4 \, r_g \approx 1$ pc).

In FR I type objects the collimation feature may be farther out and may already have been imaged in some sources. Cheung *et al.* (2007) report a speed of $< 0.25 \, c$ for the d component of HST-1 in M87, which lies $\sim 10^6 \, r_g$ (120 pc) from the central black hole. Yet other components before and after this one show apparent speeds of $4 - 6 \, c$ (see Asada *et al.*, this volume). This argues for the HST-1d region possessing a collimation shock, where the character of the relativistic jet could undergo significant changes. Similarly, BL Lac shows a strong shock at $\sim 10^5 \, r_g$, in which a γ-ray flare was observed (Marscher *et al.* 2008).

Stellar mass black hole jets also show possible evidence of collimation features. γ-ray burst jets indicate possible "violent dissipation at $\sim 1000 \, r_g$" (see Lyubarsky, this volume). And most convincingly, broad band analyses of X-ray binary jets indicate the presence of shocks that accelerate particles to relativistic energies at a distance of $100 - 1000 \, r_g$ from the black hole (Markoff & Nowak 2004; and Markoff, this volume).

Predicted observational features. What behavior might we expect at an MHD recollimation shock? Surprisingly, MHD collimation shocks were studied with non-relativistic simulations in the 1980s (Clarke *et al.* 1986; Lind *et al.* 1989, hereinafter LPMB). LPMB Figs. 6 & 9 show a strong, toroidal magnetic field dominated, super-magnetosonic jet injected from the left and a strong pinch shock at $Z \approx 200$. For $Z > 200$ the jet flow considerably broadens, by almost an order of magnitude, but remains forward-flowing, although slowed to trans-magnetosonic speeds in a "nose-cone" structure. In the region $Z \sim 150 - 200$ a magnetic chamber-like structure forms, ejecting multiple jet pulses at speeds greater than the slow magnetosonic speed, creating pairs of forward- and reverse-slow shocks that propagate within the nose cone. This general structure was confirmed

---

† Note that this does not preclude the hot spots themselves [the jet Mach disks] from enhancing the magnetic field locally up to equipartition via MHD turbulence or some other process.

for the relativistic case by Komissarov (1999). This collimation feature will have consequences for the downstream flow in the jet on kiloparsec scales and greater. If it is strong (*e.g.*, a shock that randomizes and dissipates the strong magnetic field), a weakly-magnetized, hydrodynamic-like flow could result (as in FR IIs). However, if the feature is relatively weak (*e.g.*, a bounce that largely preserves the strong toroidal field), then the downstream flow could continue to be current-carrying and subject to current-driven instabilities, such as helical kinks, *etc.* (Nakamura & Meier 2004). Nakamura *et al.* (2010) argue that 1) the magnetic field in this feature remains strong ($\beta_p \sim 1$ in the HST-1c component and ~ 0.1 downstream), and 2) the wiggles in the M87 jet well downstream of HST-1 are due to current-driven (and not Kelvin-Helmholtz) instabilities. The implication is that some FR I jets may remain current-carrying on the kiloparsec scale, even though FR II jets appear to be weakly-magnetized flows throughout most of their length.

6. Summary and Conclusions

There are some observational indications that black hole spin may play a significant role in producing jets seen in supermassive systems (active galactic nuclei [AGN]). Both observationally and theoretically, it appears that powerful, BZ-type jets may be produced by rapidly-rotating, *retrograde* black holes (spinning in the opposite direction to their accretion flow), while prograde ones produce radio quiet or silent sources. Not only is this "modified spin paradigm" a possible explanation for the radio loud/quiet dichotomy, it also is consistent with our current knowledge of the spin state of AGN.

In stellar-mass (X-ray binary) systems, however, it is much less clear if black hole spin plays any role. While this could indicate a fundamental difference between stellar and supermassive systems, it also could be due to gross errors in spin measurements or to observational bias against detecting the highly beamed relativistic jet expected from the BZ process.

It is clear that the rotating accretion flow does play an important role in the production of relativistic jets. And growing evidence, both observational and theoretical, indicates that jets from accretion disks are launched from a hot, geometrically thick ADAF-like state. A dramatic factor of ~ 100 decrease in the power of this jet (when the accretion flow goes into a cool, geometrically thin state) continues to be a viable model for the suppression of X-ray binary jets in the soft state. However, we also have identified a (factor of ~ 20) mini-suppression that can occur in the *hard* state when some ADAFs develop an MDAF near the black hole. This may be a viable model for hard state "outliers", if a strong correlation can be established between outliers and sources possessing bandwidth-limited noise and a strong LF QPO.

Jet acceleration and collimation is very likely achieved via rotating magnetic fields anchored either in the accretion flow or in the black hole horizon itself. While this certainly is a 3-dimensional process, self-similar jet theory can elucidate much of the physics. Specifically, different forces conspire at different stages to accomplish the acceleration: gas pressure below the slow point, magneto-centrifugal effects up to the Alfvén point, magnetic pressure up to the fast point, and pinch forces up to the modified fast point. Beyond that the flow becomes causally disconnected from the engine, and at the same time is converging toward the rotation axis at a speed greater than magnetosonic. It is, therefore, likely that the jet flow will bounce, or even shock, after it becomes a free, causally disconnected jet.

There also are numerous observational indications that many, perhaps most, cosmic jets under go a shock or other collimation feature. Depending on the strength of this feature the jet could remain current-carrying (and unstable to helical kink modes) or dissipate its magnetic field and become a weakly-magnetized, hydrodynamic-like flow. These processes may play a role in the morphological structure of FR I and II sources.

Acknowledgement. This research was carried out at the Jet Propulsion Laboratory, California Institute of Technology, under contract with the National Aeronautics and Space Administration.

References

Blandford, R. D. & Payne, D. G. 1982, *MNRAS*, 199, 883 (BP)

Blandford, R. D. & Znajek, R. L. 1977, *MNRAS*, 179, 433 (BZ)

Bogovalov S. V. 1994, *MNRAS*, 270, 721

Brenneman, R. D. & Reynolds, R. L. 2006, *ApJ*, 652, 1028

Cheung, C. C. *et al.* 2007, *ApJ*, 663, L65

Clarke, D. A. *et al.* 1986, *ApJ*, 311, L63

Elvis, M. *et al.* 2002, *ApJ*, 565, L75

Fabian, A. C. *et al.* 2009, *Nature*, 459, 540

Fender, R. P. 2001, in *Black Holes in Binaries and Galactic Nuclei* (Heidelberg: Springer-Verlag), ed. L. Kaper, E. P. J. van den Huevel, & P. A. Woudt, p. 193

Fender, R. P. *et al.* 2004, *MNRAS*, 355, 1105

Fender, R. P. *et al.* 2010, *MNRAS*, 406 1425

Fragile, P. C. & Meier, D. L. 2009, *ApJ*, 693, 771

Garofalo, D. 2009, *ApJ*, 699, 400

Garofalo, D. *et al.* 2010, *MNRAS*, 406, 975

Hawley, J. F. & Krolik, J. H. 2006, *ApJ*, 61, 103

Iwasawa, D. *et al.* 1996, *MNRAS*, 282, 1038

Kataoka, J. *et al.* 2007, *PASJ*, 59, 279

Komissarov, S. S. 1999, *MNRAS*, 308, 1069

Li, Z.-Y. *et al.* 1992, *ApJ*, 394, 459

Lind, K. R. *et al.* 1989, *ApJ*, 344, 89

Lynden-Bell, D. 1996, *MNRAS*, 279, 389

Markoff, J. C. & Nowak, C. F. 2004, *ApJ*, 609, 976

McClintock, J. E. & Remillard, R. A. 2006, in *Compact stellar X-ray sources* (Cambridge: Cambridge Univ. Press), ed. W. Lewin & M. van der Klis, p. 157

McKinney, J. C. 2006, *MNRAS*, 368, 1561

McKinney, J. C. & Gammie, C. F. 2004, *ApJ*, 611, 977

Meier, D. L. *et al.* 1997, *Nature*, 388, 350

Meier, D. L. 1999, *ApJ*, 522, 753

Meier, D. L. 2005, *Ap & SS*, 300, 55

Meier, D. L. & Garofalo, D. 2010, in preparation.

Michel, F. C. 1969, *ApJ*, 158, 727

Migliari, S. & Fender, R. P. 2006, *MNRAS*, 366, 79

Nakamura, M. & Meier, D. L. 2004, *ApJ*, 617, 123

Nakamura, M. *et al.* 2010, *ApJ*, 721, 1783

Polko, P. *et al.* 2010, *ApJ*, 723, 1343

Pringle, J. E. 1976, *MNRAS*, 177, 65

Punsly, B. & Coroniti, F. V. 1990, *ApJ*, 354, 583

Romanova, M. M. *et al.* 1998, *ApJ*, 500, 703

Sambruna, R. M. *et al.* 2009, *ApJ*, 700, 1473

Shakura, N. I. & Sunyaev, R. A. 1976, *MNRAS*, 175, 613

Sikora, M. *et al.* 2007, *ApJ*, 658, 815

Szuszkiewicz, E. & Miller, J. C. 2001, *MNRAS*, 328, 36

Uzdensky, D. A. 2005, *ApJ*, 620, 889

Uzdensky, D. A. & MacFadyen, A. I. 2006, *ApJ*, 647, 1192

Vlahakis, N. *et al.* 2000, *MNRAS*, 318, 417

Vlahakis, N. & Konigl, A. 2003, *ApJ*, 596, 1080
Wilson, A. S. & Colbet, E. J. M. 1995, *ApJ*, 438, 62
Yu, Y. & Tremaine, S. 2002, *MNRAS*, 335, 965

Discussion

PE'ER: You spoke about Blandford-Znajek mechanism for jet formation. However, we know that jets exist in neutron stars, and possibly white dwarfs systems. So, what is the jet launching mechanism in these systems?

MEIER: First, as Dr. Gallo reported, there is yet no evidence that BH spin is important in microquasars. So, for microquasars it is possible that the jets are disk driven (Blandford-Payne process). I would argue, however, that there is some evidence thath the BZ process is important in radio loud AGNs. In the case of mocroquasars with neutron stars, a disk-driven jet (BP process) is a reasonable model. However, these systems also are quenched at $\dot{M} \sim 0.01$, but not by a factor of 30 or more. Instead, the jet partially recovers and continues to increase in strength with $\dot{M}$, albeit at a level perhaps 10 times lower than for $\dot{M} < 0.01$. For this range ($\dot{M} < 0.01$), I would suggest that this weaker jet is powered by the rotating neutron star and the "propeller" mechanism.

GHISELLINI: You supposed that jets were likely to be formed at low $\dot{m}$, but recent results on blazars suggest that they exist at all $\dot{m}$, and that the jet power is proportional to the accretion rate

MEIER: My suggestion was that jets are produced when the accretion flow is in a hard, geometrically thick state. This does not mean that the source is not accreting at a high state, only that for a short period o time the source enters a temporarily hot accretion phase to eject a jet component, but then returns to a softer disk state (which temporarily turns off the jet again). This all would be possible if the accretion flow were unsteady and the disk wew undergoing disk instabilities - just like those instabilities expected as the accretion rate approaches Eddington.

SIKORA: What can be the mechanism/scenario of loading of Blandford-Znajek jets by protons, presence of which in extragalactic jets is indicated by several independent observations?

MEIER: Mass loading, of course, is a big problem in the theory of jets of all kinds. Some possibilities are loading from the accretion flow into the launched jet, loading by collision of the jet with the broad line clouds or other material in the ISM, a entrainment of material at a collimation shock, or entrainment at the interface between the jet and the ambient medium. There also may be a wind surounding the jet that could carry baryons which would be entrained downstream. I am not currently familiar with the observational constraints to judge which of these are plausible and which are not.

Jets at all Scales
Proceedings IAU Symposium No. 275, 2011
G. E. Romero, R. A. Sunyaev & T. Belloni, eds.

© International Astronomical Union 2011
doi:10.1017/S1743921310015607

Relativistic jets at high energies

Amir Levinson[1]

[1] School of Physics and Astronomy, Tel Aviv University
Tel Aviv 69978, Israel
email: `levinson@wise.tau.ac.il`

Abstract. A recent progress in the study of γ-ray jets is reviewed, with a focus on some theoretical interpretations of the VHE emission from M87, and possibly other misaligned blazars; the connection between the GeV breaks exhibited by bright LAT blazars and opacity sources in the broad line region; the consequences of the detection of GeV emission from GRBs to models of magnetic outflows; and the implications of the thermal emission observed is some GRBs to dissipation of the outflow bulk energy.

Keywords. Jets, blazars, gamma ray bursts.

1. Introduction

Observations by Fermi and the various TeV experiments, and advances in numerical techniques have led to a progress in our understanding of relativistic jets: i) In M87, combined VLBA and TeV data (Acciari *et al.* 2009) seem to indicate that the TeV emission is produced on horizon scales by either some magnetospheric process, or at the base of the VLBA jet. The observational constraints raise interesting questions about the structure of the BH magnetosphere and the jet formation mechanism, that appear to be relevant also to other TeV AGNs. ii) The LAT spectrum of blazars provide now a better than before probe of opacity sources on sub-parsec scales. The data reveal relationship between source power and spectral features that can be interpreted in terms of BLR properties. iii) Combined Fermi and TeV observations of VHE blazars may be used to probe intergalactic magnetic fields (e.g., Neronov *et al.* 2010; Tavecchio *et al.* 2010). Attempts to derive constrains on the IGMF have been published, and although inconclusive yet, the results demonstrate the potential in pursuing further these efforts. iv) Fermi LAT detections of several GRBs indicate very high Lorentz factors in outflows having opening angles larger than the causality scale. Those detections triggered recent theoretical studies and numerical simulations of magnetic outflows. The conclusion emerging from these studies is that processes beyond ideal MHD are crucial. Hydrodynamical outflows, perhaps driven by neutrino annihilation, may provide an alternative. The nature of the GeV emission in GRBs is yet an open issue. v) In several bursts a prominent, quasi-thermal spectral component has been detected, challenging the "standard model". It opens up the issue of dissipation and emission during the prompt phase. In particular, part of the emission is likely produced behind relativistic radiation mediated shocks.

In what follows, these issues are discussed in some greater detail.

2. GeV-TeV emission from AGNs

Over 700 AGNs, most of which are blazars, are listed in the first Fermi LAT catalogue (Abdo, *et al.* 2010c). Around 3 dozens AGNs have been detected by various TeV experiments at energies above 100 GeV, with a spectrum extending up to 10 TeV in some cases. At least 50 percents of the TeV sources have GeV counterparts. The LAT spectrum of many bright Fermi blazars exhibits curvature or breaks at GeV energies (Abdo *et al.* 2010b), conceivably associated with attenuation by radiation emitted from highly ionized clouds in the broad line region (BLR).

Eleven non-blazar AGNs, - 7 FRI radio galaxies and 4 FRII radio sources - exhibit VHE ($>$ 100 MeV) emission (Abdo, *et al.* 2010a), and some of them, e.g., M87 (Wagner *et al.* 2009), Cen A (Aharonian, *et al.* 2009), are also TeV sources. According to the unified model, those radio galaxies are misaligned blazars, and the question then arises if the VHE photons seen represent side emission from the jet, or have a different origin. In the former case, the high luminosities measured in some of the objects may imply surprisingly small Doppler factors, or extremely high jet power.

It is conceivable that the misaligned blazars exhibit spectral components that are difficult to detect in blazars because they are overwhelmed by the beamed emission from jet. Emission from a starved magnetosphere is one possibility (Levinson 2000). Recent VLBA and TeV observations of M87 (Acciari *et al.* 2009) strongly motivates reconsideration of this scenario. Below we propose that the variable TeV emission observed in M87 could be a manifestation of the jet formation process.

2.1. *Magnetospheric TeV emission and jet formation in M87*

Combined VLBA and TeV observations of M87 reveal a rapidly varying TeV emission that appears to be associated with the m.a.s VLBA jet. The rapid flaring activity of the TeV source, with timescales $t = 1t_{day}$ day as low as 1-2 days, implies a source size of $d \sim 4r_s t_{day}$ for a black hole mass $M_{BH} = 4 \times 10^9$ solar masses. This, and the fact that the TeV emission appears to be correlated with the VLBA jet but not with emission from larger scales (and in particular HST-1), motivates the consideration that the observed TeV rays originate from the black hole magnetosphere. A plausible magnetospheric process discussed in the literature is curvature and/or IC emission by particles, either hadrons or leptons, accelerating in a vacuum gap of a starved magnetosphere. An alternative explanation is emission from small regions located at larger radii, $r \sim 100r_g$, as, e.g., in the misaligned minijets model of Gainnios *et al.* (2010).

The presence of the VLBA jet implies that a force-free (or ideal MHD) flow is established on scales $< 100r_g$, so that the magnetosphere is anticipated to be screened in the sense that the invariant $\mathbf{E} \cdot \mathbf{B}$ nearly vanishes everywhere. However, as in pulsar theory, there must be a plasma source that compensates for the loss of particles which escape the system (both, to infinity and across the horizon) along the open magnetic field lines in the polar region. The nature of this plasma source is poorly understood at present.

The injection of charges into the magnetosphere may be associated with the accretion process. Direct feeding seems unlikely, as charged particles would have to cross magnetic field lines on timescale shorter than the accretion time in order to reach the polar outflow. Free neutrons, if exist, decay over length scale much smaller than the gravitational radius of the black hole. Annihilation of MeV photons produced in the radiative inefficient flow, may provide the required charge density, depending on accretion rate and other details.

A possible plasma source is cascade formation in starved magnetospheric regions. The size of the gap then depends on the conditions in the magnetosphere and the pair production opacity. As shown elsewhere (Levinson 2000), vacuum breakdown by back-reaction is unlikely, as it requires magnetic field strength in excess of a few times 10^5 G, higher than the equipartition value for Eddington accretion. Pair production via absorption of TeV photons by the ambient radiation field is more likely. However, the spectrum of the VHE photons observed by HESS extends up to 10 TeV, and the assumption that these photons originate from the magnetosphere (or even the VLBA jet) implies that the

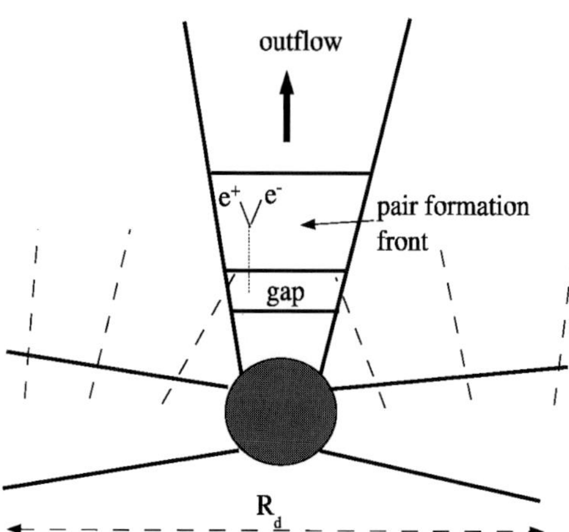

Figure 1. Schematic representation of the magnetosphere structure: A vacuum gap of height $h < r_g$ accelerates particles (electrons or positrons) to high Lorentz factors. The gap is exposed to soft radiation emitted from a source of size $R_d >> r_g$. Curvature emission and inverse Compton scattering of ambient radiation produce VHE photons with a spectrum extending up to 10^4 TeV. Photons having energies below a few TeV can escape freely to infinity. Interactions of IC photons having energies well above 10 TeV with the ambient radiation initiate pair cascades just above the gap, leading to a large multiplicity. A force free outflow is established just above the pair formation front, and appears as the VLBA jet. Intermittencies of the cascade process give rise to the observed variability of the TeV emission, and the resultant force-free outflow, as indicated by the morphological changes of the VLBA jet.

pair production opacity at these energies must not exceed unity. This raises the question whether pair cascades in the magnetosphere can at all account for the multiplicity required to establish a force-free flow.

In an upcoming paper (Levinson & Rieger, 2010, in preparation) it is shown that inverse Compton scattering (IC) of ambient photons by electrons accelerated in the gap can lead to a large multiplicity, in excess of 10^3, while still allowing photons at energies of up to a few TeV to freely escape the system. The seed electrons are generated by the MeV emission from the RIAF. This requires that the ambient radiation source contributing the opacity will have a characteristic size $R_d \sim 10^2 r_g$. Preliminary results indicate that the electromagnetic cascade is initiated by IC photons having much higher energies, $\sim 10^3$ TeV, for which the $\gamma\gamma$ optical depth is much larger. It is found that the gap width is not much smaller than $0.1 r_g$. The luminosity of the VHE photons produced in the gap, roughly a fraction $(h/r_g)^2$ of the maximum BZ power, can easily account for the TeV luminosity observed by HESS and Fermi. Any intermittencies of the cascade formation process would naturally lead to variability of both, the magnetospheric TeV emission and the resultant force-free flow, as observed. A schematic illustration of the model is presented in figure 1.

This model may be applicable also to some of the other misaligned VHE blazars.

2.2. GeV breaks and opacity sources in blazars

The nature of opacity sources in blazars is an issue of considerable interest, particularly in regards to the location of the VHE emission zones. Much efforts have been devoted over the years to identify different opacity sources, and detailed calculations of $\gamma\gamma$ attenuation by these sources are presented in the literature. Recent Fermi observations have led to reconsideration of the effect of the broad line region on the GeV spectrum of blazars.

Strong departure from a single power law appears to be a common feature in FSRQs (Abdo *et al.* 2010b). The break energies, typically lying in the energy range 2-10 GeV, are too low to be due to absorption by Lyα radiation. Recently, it has been pointed out (Putanen & Stern 2010) that the LAT spectrum of FSRQs can be well reproduced by a double absorber model. Detailed numerical calculations of the BLR spectrum for different ionization parameters indicates that the observed GeV breaks can be accounted for by γ-ray absorption through pair production on He II and H I recombination continua. The exact location of the break energy depends on the ionization parameter, as described in Putanen & Stern (2010). The fact that GeV breaks are seen mainly in the brightest blazars is naturally explained by the scaling of the BLR size with luminosity, $R_{BLR} \propto L^{1/2}$, obtained from reverberation mapping techniques (Kaspi *et al.* 2007), that implies $\tau_{\gamma\gamma} \propto L^{1/2}$.

3. VHE emission and dissipation in GRBs

Over a dozen GRBs, both long and short, have been detected thus far by LAT onboard Fermi. The origin of the GeV emission is yet an open issue. The flux level of the LAT emission and its long lasting light curve, suggest that this component is produced during the afterglow phase, although other explanations have been offered. For sources that show sub-second variations of the GeV flux, e.g., 080916C, opacity arguments yield high Lorentz factors, $\Gamma \sim 10^3$. While various factors render those estimates uncertain (e.g., Yuan-Chuan *et al.* 2010), it seems difficult to avoid the conclusion that the outflow reaches high Lorentz factors. Such high Lorentz factors posses a great challenge to jet models, as discussed in the following.

Another interesting result is that in some GeV bursts, e.g., GRB 090902B (Pe'er *et al.* 2010), a prominent quasi-thermal spectral component has been detected, whereas in others, e.g., 080916C, a fit by a Band spectrum is claimed to be satisfactory, and there appears to be no indication for thermal emission. This, presumably, reflects the conditions in the dissipation region. Below it is argued that in bursts that exhibit a thermal component a substantial fraction of the outflow bulk energy dissipates below the photosphere, in relativistic radiation mediated shocks, whereas in other bursts most of the energy dissipates above the photosphere, behind collisionless shocks.

3.1. *Formation and acceleration of magnetic jets*

Extraction of rotational energy from a rapidly spinning black hole by magnetic fields is a widely discussed jet production mechanism (e.g., Levinson & Eichler 1993; Lyutikov & Blandford 2003; Giannios & Spruit 2005). Up to 30 percents of the black hole mass can be extracted from a maximally rotating hole, which is more than sufficient to account for the energetics of essentially all GRBs. The rate at which the energy is extracted depends on the strength of the magnetic field in the vicinity of the horizon. Recent numerical simulations confirm that a Blandford-Znajek process does operate in Kerr spacetime, and that for reasonable assumptions about the progenitor star, sufficiently large power can be extracted in the form of an outflow. Moreover, a very low baryonic content is anticipated along horizon threading field lines, allowing, in principle, high asymptotic Lorentz factors of the polar outflow. Pressure support by the external medium may give rise to collimation of the central outflow (Bromberg & Levinson 2007; Lyubarsky 2009). Magnetic outflows have been considered also in the context of the unexpected paucity of optical flashes seen in GRBs (e.g., Zhang & Kobayashi 2005; Mimica *et al.* 2009), and it has been shown that even moderate magnetization ($\sigma \sim 0.3$) would strongly suppress the reverse shock. The question remains as to how the internal shocks that produce the prompt emission form under these conditions. Alternative explanations for the paucity of optical flashes, e.g., an early onset of a R-T instability (Levinson, 2009, 2010a), have been offered recently.

An key issue in the theory of magnetically dominated outflows is magnetic energy dissipation. The process by which magnetic energy is converted into kinetic energy has not been identified yet. Stationary magnetic outflows allow, in general, only partial conversion of magnetic-to-kinetic energy. For a split monopole field acceleration ceases at an asymptotic Lorentz factor $\Gamma_\infty \sim \sigma_0^{1/3}$, and magnetization $\sigma_\infty \sim \sigma_0^{2/3}$, where σ_0 is the initial magnetization of the expanding shell. A better conversion can be achieved if the outflow is collimated into a small opening angle, $\theta \simeq \Gamma^{-1}$ (Komissarov *et al.* 2009). Such small opening angles are problematic, particularly for the GeV bursts, for which $\Gamma_\infty \sim 500 - 1000$ is inferred. Corking from a star that, within the framework of the collapsar model, may be relevant to long bursts can alleviate the latter condition (Tchekhovskoy *et al.*, 2010; Komissarov *et al.* 2010). However, even then σ_∞ cannot be much smaller than unity (Lyubarski 2010). Furthermore, there is evidence for extreme Lorentz factors also in short bursts, where strong deconfinement is not anticipated.

Recently it has been shown (Granot *et al.*, 2010; hereafter GKS10; Lyutikov 2010a,b) that time-dependent effects may play a crucial role in the acceleration of a magnetized flow. Unlike a stationary flow, for which acceleration ceases at $\Gamma_\infty \sim \sigma_0^{1/3}$, an impulsive spherical shell expelled by a central source continues accelerating, even after loosing causal contact with the engine, until reaching nearly complete conversion of magnetic energy into bulk kinetic energy. The terminal Lorentz factor of a shell expanding in vacuum is $\Gamma_\infty \simeq \sigma_0$. During the acceleration phase the major fraction of the shell energy is contained in a layer of width $2r_0$, where r_0 is the initial shell's width, bounded between the front of a rarefaction wave reflected from the central source and the head of the shell. The average Lorentz factor of the shell, roughly equals the Lorentz factor of the fluid at the rarefaction front, evolves as $< \Gamma > \propto t^{1/3}$. Once the shell enters the coasting phase its width starts growing and its magnetization continues to drop, until nearly full conversion is accomplished.

However, for sufficiently high values of the initial magnetization σ_0 the evolution of the system is significantly altered by the ambient medium well before the shell reaches its coasting phase (Levinson 2010b). The maximum Lorentz factor of the shell is limited to values well below σ_0. To be concrete, for a shell of initial energy $E = 10^{52}E_{52}$ erg and size $r_0 = 10^{12}T_{30}$ cm expelled into a medium having a uniform density n_i the Lorentz factor is limited to $\Gamma_{max} \simeq 180(E_{52}/T_{30}^3 n_i)^{1/8}$ in the high sigma limit. The reverse shock and any internal shocks that might form if the source is fluctuating are shown to be very weak. The restriction on the Lorentz factor is more severe for shells propagating in a stellar wind. Intermittent ejection of multiple, thin shells does not seem to help, as even in vacuum unlikely small duty cycle is required in order for shells to collide after reaching the coasting phase. Such episodes are expected to produce a smooth, relatively fast rising slowly decaying (power law) light curve, even in a multi-shell scenario. Events like GRB080916C and GRB090510 (Abdo *et al.* 2009a,b) are not easily accounted for by the impulsive high-sigma shell model.

The main conclusion from those recent studies of magnetic outflows is that processes beyond ideal MHD, as might occur in e.g., a striped wind model, are required.

3.2. *Thermal emission and relativistic radiation mediated shocks*

Shocks that form by overtaking collisions can dissipate energy at radii $r_d > \Gamma^2 c\delta t$, where Γ is the Lorentz factor of the slow shell and $\delta t \geqslant r_s/c$ is time interval between ejections of the two consecutive shells. In blazars and microquasars with $\Gamma \sim 1 - 50$ dissipation by

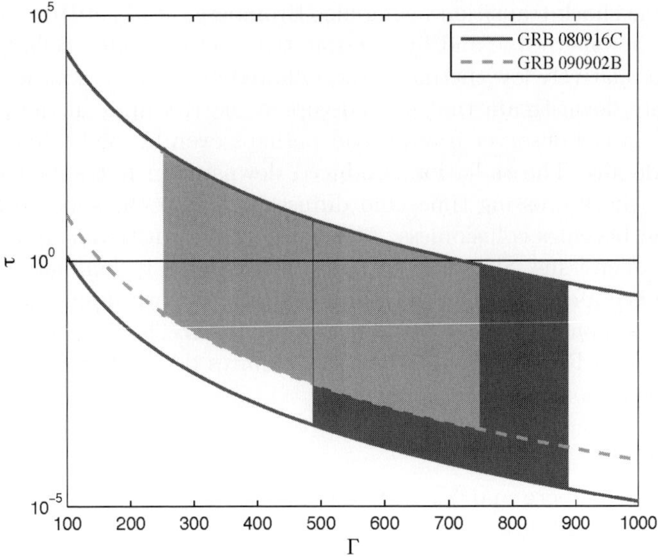

Figure 2. Optical depth at the shells' collsion radius as a function of Lorentz factor, for different values of δt, the time interval between ejections of the two consecutive shells. The upper curves correspond to the dynamical time of a 3 solar mass black hole, $\delta t \simeq 10^{-4}$ s. The lower curves are limits from observations, with $\delta t = 0.5$ s for GRB080916C, and $\delta t = 0.05$ s for GRB090902B. The range of Γ in each source represents various estimates taken from the literature (Bromberg *et al.* 2010).

internal shocks is expected close to the BH, consistent with (but not necessarily implied by) the short durations of strong flares observed in these objects, particularly in TeV blazars.

In GRBs $r_d > 10^5 r_s$ or so for the Lorentz factors envisaged. A rough condition for internal shocks to form above the photosphere is $\Gamma > 200 L_{52}^{1/5} \delta t_{-3}^{-1/5}$, where L_{52} is the burst luminosity in units of 10^{52} erg/s, and $\delta t_{-3} = \delta t/(1\text{ms})$. Results of more exact calculations (Bromberg *et al.* 2010) are exhibited in figure 2, where the Thompson optical depth at the radius of shock formation, τ, is plotted agains the flow Lorentz factor Γ. The two colored areas correspond to a range of δt and Γ for GRB 080916C (blue) and GRB 090902B (green), as explained in the figure caption.

As shown, in case of GRB 080916C internal shocks are expected to form above the photosphere for shells accelerated to $\Gamma > 700$. Such shocks are probably collisionless, and can Fermi accelerate particles to nonthermal energies. Slower shells ejected over time intervals of the order of the dynamical time will collide below the photosphere, at a moderate optical depth $\tau < 100$. The parameter space explored in Figure 2 indicates that in this burst the majority of the explosion energy dissipates above the photosphere, behind collisionless shocks, so that a thermal component may be very weak or absent.

In GRB 090902B, on the other hand, the major fraction of the energy is likely to dissipate below the photosphere, albeit in a region of moderate optical depth, $\tau < 300$, as seen in Figure 2. Only sufficiently slow shells that have large separations, such $r_s/\delta t$ is much larger than the anticipated duty cycle of the engine, will collide above the photosphere to form collisionless shocks. The shocks that form below the photosphere, where the Thomson depth exceeds unity, are mediated by Compton scattering (Bromberg & Levinson 2008; Katz *et al.* 2010). Under conditions anticipated in GRBs, such relativistic radiation mediated shocks (RRMS) convect enough radiation upstream to render photon production in the shock transition negligible (Bromberg, *et al.*, 2010), unlike the case of shock breakout in supernovae and hypernovae (Katz *et al.* 2010). Bulk Comptonization then produces a relatively low thermal peak, followed by a broad, nonthermal component in the immediate downstream that extends up to the KN limit in the shock frame (or up to $\sim \Gamma m_e c^2$ in the observer frame), and perhaps even beyond (Budnik *et al.* 2010), depending on details. The radiation produced downstream is trapped for times much longer than the shock crossing time, and diffuses out after the shock breaks out of the photosphere and becomes collisionless. At what depth downstream equilibrium is established is yet an open issue. Since the enthalpy downstream is dominated by radiation, a full equilibrium is not expected for the moderate optical depths found above. We contend that a thermal component plus a hard tail, as seen in GRB 090902B, can be emitted from the downstream of a RRMS. A full treatment requires detailed transfer calculations, or Monte-Carlo simulations.

Acknowledgement

I thank Omer Bromberg and Yuri Lyubarsky for enlightening discussions. This work was supported by an ISF grant for the Israeli Center for High Energy Astrophysics.

References

Abdo, A. A., *et al.* 2009a, *Science*, 323, 1688

Abdo, A. A., *et al.* 2009b, *arXiv0908.1832*

Abdo A. A. *et al.*, 2010a *ApJ*, 720, 912

Abdo A. A. *et al.*, 2010b *ApJ*, 710, 1271

Abdo A. A. *et al.*, 2010c *ApJS*, 188, 405

Acciari, V. A. *et al.* 2009, *Science*, 325, 444

Aharonian, F., *et al.* (H.E.S.S. Collaboration) 2009, *ApJ*, 695, L40

Bromberg, O. & Levinson, A. 2007, *ApJ*, 671, 678

Bromberg, O., Mikolitzky, Z., & Levinson, A. 2010, (to be submitted)

Budnik, R. *et al.* 2010, *arXiv:1005.0141*

Giannios D. & Spruit H., 2005, *A&A*, 430, 1

Giannios, D., Uzdensky, D. A., & Begelman, M. C., 2010, *MNRAS*, 402, 1649

Granot, J., Komissarov S., & Spitkovsy, A., 2010, *arXiv:1004.0959*

Kaspi, S. *et al.* 2007, *ApJ*, 659, 997

Katz, B. Budnik, R. & Waxman, E. 2010, *ApJ*, 716, 781

Levinson, A. 2000, *Phys. Rev. Lett*, 85, 912

Levinson, A. 2009, *ApJ*, 705, 213

Levinson, A. 2010, *GApFD*, 104, 84

Levinson, A. 2010, *ApJ*, 720, 1490

Levinson A. & Eichler D., 1993, *ApJ*, 418, 386

Levinson, A. & Bromberg, O. *Phys. Rev. Lett.*, **100** (2008) 131101

Lyubarsky, y. 2009, *ApJ*, 698, 1570

Lyubarsky, y. 2010, *MNRAS*, 402, 353

Lyutikov M. & Blandford R. D. 2003, *arXiv:astro-ph/0312347*

Lyutikov M. 2010a, *arXiv:1004.2428*

Lyutikov M. 2010b, *arXiv:1004.2429*

Mimica, P., Giannios, D., & Aloy, M. A., 2009, A&A, 494, 879

Neronov, A. *et al.* 2010, *ApJ*, 719, L130

Peer, A. *et al.* 2010, arXiv:1007.2228

Poutanen, J. & Stern, B. 2010, *ApJ*, 717, L118

Tavecchio, F. *et al.* 2010, *MNRAS*, 406, L70

Tchekhovskoy, A., Narayan, R., & McKinney, J. C. 2010, *New Astron.*

Wagner, R. M. *et al.* 2009, *Proc. ICRC, arXiv:0907.1465*
Yuan-Chuan, Z., Yi-Zhong, F., & Piran, T. 2010 *arXiv:1008.2253*
Zhang, B. & Kobayashi, S. 2005, *ApJ*, 628, 315

Discussion

ROMERO: Can you please elaborate a bit about the possible implications of "dark" blazars for UHECRs?

LEVINSON: Confinement of an UHECR requires certain magnetic field. This gives a lower limit on the Poynting flux, and, hence, on the jet power. The implied jet power is very high, and if the radiative efficiency of the jet is not small, the source must be bright. There is no evidence for such bright sources within the GZK sphere, so the only way around it is to assume "dark" sources.

KIRK: You stated that the acceleration of a magnetic shell is inhibited by an ambient medium. Is it possible that in a GRB the first shell evacuates a channel and subsequent shells are accelerated to high Lorentz factors in a very low density environment?

LEVINSON: It is in principle possible if the duty cycle is extremely small, which I find unlikely. The point is that even in vacuum, for reasonable separation between shells they will collide while still highly magnetized, because the fron moves faster than the inner regions.

PE'ER: How well is the Lorentz factor in M87 constrained, and can Lorentz aberration help resolve the "small scale" problem of the TeV emission?

LEVINSON: Not well, SL motions imply $\Gamma > 6$ on certain scales. However, the viewing angle is not small, $\sim 20° - 30°$, so the Doppler factor is always small.

Jets at all Scales
Proceedings IAU Symposium No. 275, 2011
G. E. Romero, R. A. Sunyaev & T. Belloni, eds.

© International Astronomical Union 2011
doi:10.1017/S1743921310015619

General relativistic plasmas around rotating black holes

Shinji Koide[1]

[1]Faculty of Science, Kumamoto University, 2-39-1 Kurokami, Kumamoto, 860-8555, Japan
email: koidesin@sci.kumamoto-u.ac.jp

Abstract. We show one of possible directions of general relativistic MHD (GRMHD) calculations of astrophysical relativistic jet formation. This was motivated by observations of radio knot ejections associated with high energy flares of an AGN in M87 and micro-quasars. We introduce a modified version of a solar flare model for the phenomenon. In the model, we have to consider a process beyond the ideal GRMHD. Especially, we focus on the special features of general relativistic plasmas around rotating black holes.

Keywords. black hole physics, gravitation, MHD, relativity, plasmas

1. Introduction

Numerical simulations of general relativistic MHD (GRMHD) have played important roles in investigation of plasma dynamics around rapidly rotating black holes (e.g., Koide, Shibata, and Kudoh 2006; McKinney 2006). For example, Koide *et al.* (2006) revealed that magnetic flux tubes bridging between an ergosphere and an accretion disk around a black hole, called "magnetic bridges", are unstable; they are twisted by the plasma in the ergosphere rapidly due to the frame-dragging effect, and are elongated by the magnetic pressure above the disk. During this process, the plasma is lifted outwardly with the expanding magnetic bridges to become a jet. McKinney (2006) performed a longer-term GRMHD simulation with an assumption of initial weak poloidal magnetic field in a disk around a rapidly rotating black hole. The numerical result showed the magnetic field is amplified due to the magnetorotational instability (MRI) and the Ω effect of magnetized plasma shear flow, and the amplified strong magnetic field is formed near the black hole to launch a jet. The jet is accelerated by the magnetic field over a long distance to become relativistic far from the black hole. It is believed that this simulation revealed the formation mechanism of a relativistic jet, while no corresponding analytic model is not clarified yet. We suppose that in the numerical simulation, the mechanism of magnetic bridges between the ergosphere and the disk played a main role for the launch of the jet. It is noted that zero resistivity (ideal MHD condition) was assumed in all of these GRMHD simulations. Within the framework of ideal GRMHD, the energy from the plasmas in the ergosphere is stored in the anti-parallel elongated magnetic flux tubes, and the energy is not released. In fact, the ideal GRMHD simulations show no explosive phenomena due to the energy release; they show gradual phenomena only in the longer term simulation (McKinney 2006). On the other hand, if we go beyond the ideal GRMHD, the magnetic energy of anti-parallel magnetic fields can be released suddenly by resistive phenomena.

Recent observations of M87 with X-ray/γ-ray and radio showed the correlation of a high energy flare of nucleus of M87 and the radio knot ejection from the nucleus (Acciari *et al.* 2009). A similar correlation was also found in micro-QSOs more frequently (Mirabel and Rodrigus 2005). These explosive phenomena associated with knot ejection can not be explained within the ideal GRMHD. Similar correlation between a high energy flare and a knot ejection is found in a solar flare. The solar flare is explained by the magnetic reconnection model of elongated magnetic arcades (Shibata *et al.* 1995). The elongated magnetic flux tubes are formed by shear convection flow of the photosphere at the foot points of the magnetic flux tubes. In the flux tubes, a large amount of energy is stored because of very high magnetic Reynolds number and is released suddenly due to the magnetic reconnection with anomalous resistivity after the thin current sheet is formed. A similar model was considered to explain a flare and knot ejection of a proto-star. A non-relativistic MHD numerical simulation of interaction between plasma of a disk and corona around a proto-star and dipole magnetic field of the proto-star was performed by Hayashi, Shibata, and Matsumoto (1996). The simulation showed that magnetic bridges between the proto-star and the disk are twisted by the disk plasma rotating around the proto-star and the magnetic pressure increases gradually. The magnetic pressure gradient elongates the magnetic flux tube radially. Eventually, the thin current sheet is formed in the anti-parallel magnetic field just above the disk. In this thin current sheet, the magnetic reconnection is caused by the anomalous resistivity and the magnetic energy is released rapidly. This model of a proto-star seems to be directly applicable to the plasma dynamics in the black hole magnetosphere. In the magnetosphere around the rotating black hole, the magnetic bridges between the ergosphere and the disk are twisted by the plasma in the ergosphere mainly because of the frame-dragging effect. This model may explain the correlation between the flare and the knot ejection from M87 and micro-QSOs.

In section 2, we summarize the results of ideal GRMHD simulations of jet formation in black hole magnetospheres. In section 3, we mention the observation of correlation between the flare and the knot ejection with respect to M87 and introduce a model of magnetic reconnection for the observation. We discuss the equations required by the model beyond the ideal GRMHD. In section 4, we introduce a set of generalized GRMHD equations. In section 5, we show some peculiar phenomena suggested by the generalized GRMHD equations briefly. In section 6, we present summary and some discussions.

2. Ideal GRMHD simulations of jet formation

In models of black hole magnetospheres, we have to consider the interaction between the plasmas and magnetic field around a black hole. It is well known that the interaction between the plasmas and magnetic field often causes complex phenomena. In such case, numerical simulations provide strong tools to find the solution of the phenomena. The simplest treatment of such interaction is given by GRMHD equations. Recently, numerical simulations of GRMHD had been performed by some groups. Most of them assumed the electric resistivity is zero (ideal GRMHD), while a few simulations of relativistic MHD with resistivity (resistive RMHD) were performed within special relativistic framework (Watanabe & Yokoyama 2006). Here, we review the results of ideal GRMHD simulations of the jet formation in the black hole magnetospheres.

In the black hole magnetosphere, there may be several types of magnetic field lines. Some of open magnetic field lines cross the disk around the black hole, while some of them cross the black hole. Some of the closed magnetic field lines are anchored to the disk. Here, we focus on the closed magnetic field lines, called "magnetic bridges", between the ergosphere and black hole accretion disk. Plasmas in the ergosphere rotate faster than the local light speed because of the frame-dragging effect around the black hole. Usually, angular velocity of the plasma in the ergosphere is much faster than that of the accretion disk around the black hole. The rotation of the plasma in the ergosphere twists the magnetic bridges between the ergosphere and the disk rapidly when the black hole rotates rapidly. The magnetic pressure increases in the magnetic bridges quickly. This rapid increase in the magnetic pressure will brow off the plasma around the ergosphere and will form the relativistic jet (Fig. 1, panels (a) and (b)). It is expected that thin current sheet is formed in the elongated magnetic bridges and a large amount of magnetic energy is stored (Fig. 1 (c)). When the current sheet becomes thin and the current density exceeds a critical value, the magnetic reconnection is caused and a large amount of magnetic energy is released (Fig. 3 (d)). The upward reconnection jet pushes and accelerates a plasmoid and the downward jet may heat up the inner edge of the disk. The ejection of the plasmoid and the heating of the disk edge may correspond to the knot ejection and high energy flare, respectively.

A numerical simulation of this type of jet formation was performed by Koide, Shibata, & Kudoh (2006). The simulation showed that the magnetic bridges are twisted by the plasma in the ergosphere and the disk. The twist by the plasma in the ergosphere is dominant and the magnetic pressure increases rapidly near the ergosphere. The magnetic

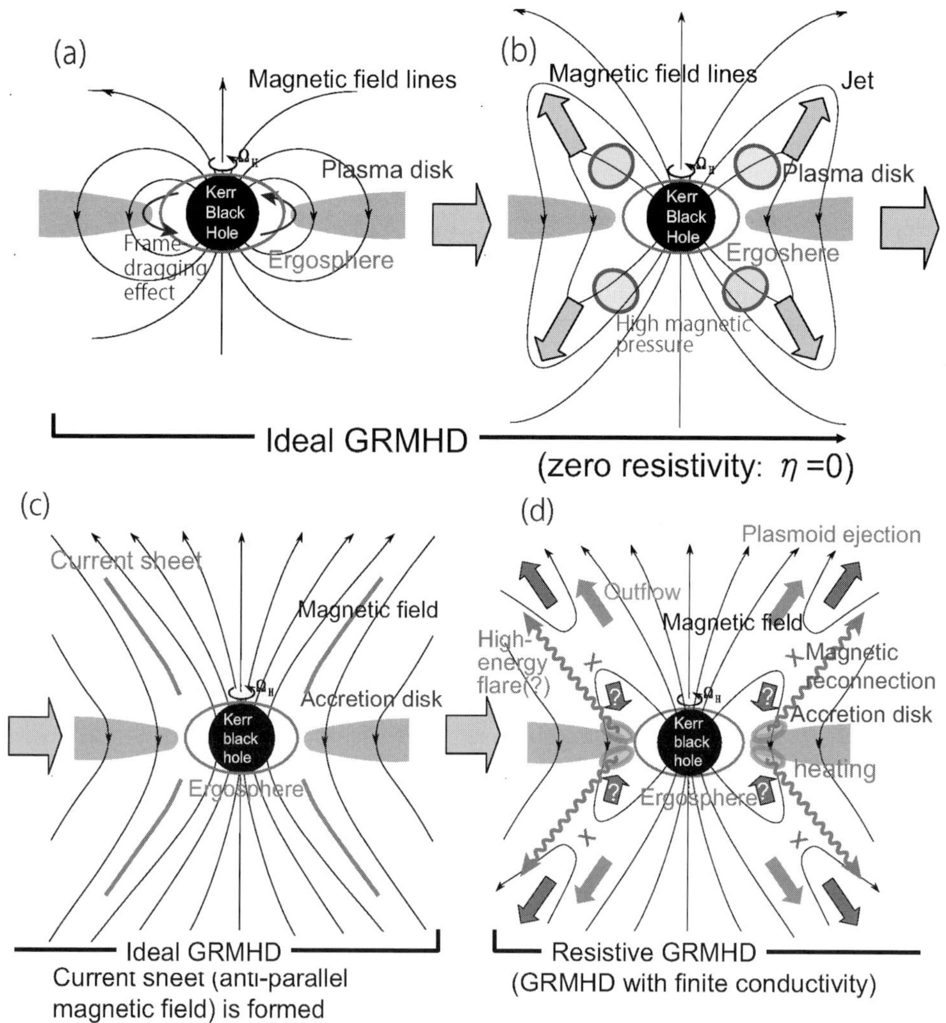

Figure 1. Dynamics of magnetic bridges between the ergosphere and accretion disk around the rapidly rotating black hole.

pressure gradient brows off the plasma around the ergosphere radially. The radial outflow is collimated by the magnetic tension to become a jet. At $t = 110\ \tau_S$, the maximum velocity of the jet is $\sim 0.6\,c$, sub-relativistic, where c is the light speed in vacuum, $\tau_S \equiv r_S/c$ is the time unit, and r_S is the Schwarzschild radius. In this calculation, we set a strong magnetic field as the initial condition of magnetic field, where the 4-Alfven velocity is $0.6c$. In such a strong magnetic field case, the small-scale MRI is suppressed by the magnetic tension and then no small-scale turbulence is formed. In this situation, the analysis of the simulation is relatively easy because of the simple structure without turbulence of MRI. This enable us to find the thin current sheet formation clearly at the trough of the magnetic pressure, $B_\phi^2 = 0$, above the disk (Figs. 2 and 3). The trough of magnetic field exists along the marginal surface of the magnetic flux surface (Fig. 3 (a)). The magnetic pressure gradient pushes the plasma toward the trough of the magnetic pressure and forms thin current sheet naturally (Fig. 3 (b)). Then, a thin current sheet is formed at the trough of the magnetic pressure as a natural consequence.

Similar thin current sheet formation was also found in the non-relativistic MHD simulation of magnetic bridges between a proto-star and a disk as mentioned in section 1 (Hayashi *et al.* 1996). They, furthermore, showed the formation of fast ejection of the plasmoid due to the fast magnetic reconnection at the current sheet with anomalous resistivity. This mechanism of magnetic reconnection is applicable around the current sheet near the black hole suggested by the results of the ideal GRMHD simulations. When this magnetic reconnection mechanism occurs around the rapidly rotating black hole, the plasmoid ejection will be observed as a knot-ejection and the heating of inner edge of the disk due to the collision with the downward reconnection flow will be observed as a high energy flare (Fig. 1 (d)). It is noted that if the outflow velocity around the current sheet is larger than the Alfven velocity at the region, the downward reconnection flow can not reach to the inner edge of the disk (this is expressed by question marks in Fig. 1 (d)). In such case, the collision between reconnection outflow and disk plasma can not happen and the heating of the inner edge of the disk is not realized. In this case, the plasmoid is ejected without the heating of the inner disk. In the numerical result of Hayashi*et al.*(1996), they did not find the heating of the inner edge of the disk. This may be explained by the above process. We also mention that, in such case, the direction of the reconnection outflow is almost the same as that of the bulk plasma outflow driven

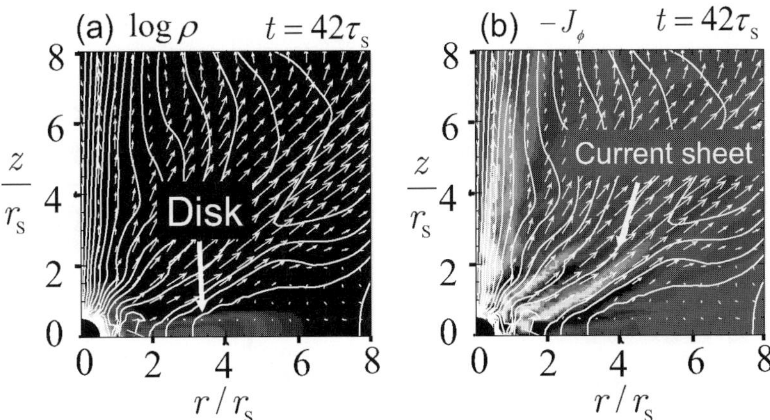

Figure 2. Density ρ and azimuthal component of current density J_ϕ around the rotating black hole at $t = 42\tau_S$ of the GRMHD simulation by Koide *et al.* (2006).

by the magnetic pressure. It is expected that the magnetic reconnection accelerates the magnetically driven outflow faster more than a factor of 2 with 4-velocity.

On the other hand, a much longer-term, much wider-region simulation of ideal GRMHD was performed by McKinney (2006). The time duration reached to $t = 7000\ \tau_S$ and the spatial range spreads to the area of $r \leqslant 10^4 r_S$. The numerical result showed that the jet is accelerated gradually along the jet and its velocity reaches to the Lorentz factor $\Gamma \approx 7$ at $r = 1000\ r_S$. His numerical simulation also showed that the density ρ and magnetic field $B_\perp$ decrease monotonically along the jet as the distance from the black hole. Because intensity of synchrotron radiation is roughly estimated by $P \propto \rho B_\perp^2$, this suggests the radio image of the jet is monotonically continuous and has no knot within the ideal GRMHD models.

3. A model beyond ideal GRMHD

Multi-wavelength observations of M87 with radio, X-ray, and γ-ray showed the correlation between the high energy (γ-ray) flare of the nucleus (accretion disk around the central black hole) and the knot ejection from the nucleus (Acciari *et al.* 2009). The knot ejection was observed by the subtraction of the average image over 11 period between January and August 2007 from the image of one period on 5 May 2008 of VLBA. No knot was observed in the subtraction from the image on 25 February 2008 just before the high energy flare. The former subtracted image also shows that the radio emission from the radio core is dominant during the radio flare. The core is identified to the M87 nucleus because of no spatial shift within $6r_S$ during the radio flare. Similar correlation between a high energy flare and a radio knot ejection was observed in micro-QSOs more frequently (Mirabel & Rodriguez 1998). These drastically time-varying high energy phenomena correlating the flare and the knot ejection may be explained by the magnetic reconnection model mentioned in Sect. 2.

To investigate the magnetic reconnection model in the black hole magnetosphere, the simplest approximation of the plasma is given by GRMHD equations with finite resistivity, called "standard resistive GRMHD". However, the standard resistive GRMHD has a problem with respect to causality. The problem comes from the "standard" relativistic Ohm's law, $U^\nu F^\mu{}_\nu = \eta[J^\mu - (U^\nu J_\nu)U^\mu]$, where U^μ is the 4-velocity, J^μ is the 4-current density, $F_{\mu\nu}$ is the electromagnetic field tensor, and η is the resistivity. Here, the electric

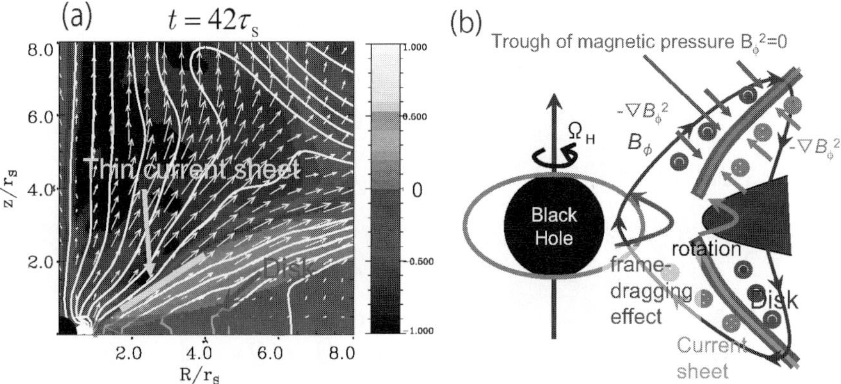

Figure 3. Azimuthal component of magnetic field B_ϕ and the location of current sheet (thick solid line) at $t = 42\ \tau_S$ of the GRMHD simulation by Koide *et al.* (2006) (a) and the formation mechanism of the thin current sheet (b).

field is given by $E_i = F_{i0}$ and the magnetic field is $F_{ij} = \sum_k \epsilon_{ijk} B_k$ (ϵ_{ijk} is the Levi-Civita symbol), where the alphabetic index (i, j, k) runs from 1 to 3. Hereafter, we use the unit system where light speed is unity and the energy densities of electric field $\boldsymbol{E}$ and magnetic field $\boldsymbol{B}$ are given by $E^2/2$ and $B^2/2$ in the Minkowski space-time, respectively.

In the ideal GRMHD case, the electromagnetic wave never propagates. On the other hand, in the resistive GRMHD case, the electromagnetic wave can propagate. The dispersion relation of the wave in uniform, rest plasma is $\omega^2 + \frac{i}{\eta}\omega - k^2 = 0$, where ω and k are the complex frequency and the wave number, respectively. The group velocity of the wave, $v_{\mathrm{g}} = \frac{\partial \omega}{\partial k} = \frac{k}{\sqrt{k^2 - (2\eta)^{-2}}} > 1$, is always greater than the light speed. This means the breakdown of causality. To solve this problem, we reconsider the GRMHD equations based on the general relativistic two-fluid equations (Koide 2010a).

4. Generalized GRMHD equations

We derive generalized GRMHD equations based on general relativistic two-fluid equations. For simplicity, we assumed that the plasma is composed of two fluids, where one fluid consists of positively charged particles with mass m_+ and electric charge e, and the other fluid consists of negatively charged particles with mass m_- and charge $-e$. The relativistic equations of the two fluids and the Maxwell equations are

$$\nabla_\nu (n_\pm U_\pm^\nu) = 0, \tag{4.1}$$

$$\nabla_\nu (h_\pm U_\pm^\mu U_\pm^\nu) = -\nabla^\mu p_\pm \pm e n_\pm g^{\mu\sigma} U_\pm^\nu F_{\sigma\nu} \pm R^\mu, \tag{4.2}$$

$$\nabla_\nu {}^*F^{\mu\nu} = 0, \qquad \nabla_\nu F^{\mu\nu} = J^\mu, \tag{4.3}$$

where variables with subscripts, plus $(+)$ and minus $(-)$, are those of the fluid of positively charged particles and of the fluid of negative particles, respectively. Here, $n_\pm$ is the proper particle number density, $p_\pm$ is the proper pressure, $h_\pm$ is the relativistic enthalpy density, ∇_μ is the covariant derivative, ${}^*F^{\mu\nu}$ is the dual tensor of $F_{\mu\nu}$, and R^μ is the frictional 4-force density between the two fluids.

To derive one-fluid equations of the plasma, we define the average and difference variables as follows:

$$\rho = m_+ n_+ \gamma_+' + m_- n_- \gamma_-', \qquad n = \frac{\rho}{m}, \tag{4.4}$$

$$p = p_+ + p_-, \qquad \Delta p = p_+ - p_-, \tag{4.5}$$

$$U^\mu = \frac{1}{\rho}(m_+ n_+ U_+^\mu + m_- n_- U_-^\mu), \qquad J^\mu = e(n_+ U_+^\mu - n_- U_-^\mu), \tag{4.6}$$

$$h = n^2 \left(\frac{h_+}{n_+^2} + \frac{h_-}{n_-^2} \right), \qquad \Delta h = \frac{mn^2}{2} \left(\frac{h_+}{m_+ n_+^2} - \frac{h_-}{m_- n_-^2} \right), \tag{4.7}$$

$$h^\ddagger = h - \Delta\mu\Delta h, \qquad \Delta h^\ddagger = \Delta\mu h - \frac{1 - 3\mu}{2\mu}\Delta h, \tag{4.8}$$

where $\gamma_\pm'$ is the Lorentz factor of the two fluids observed by the local center-of-mass frame of the plasma and $m = m_+ + m_-$ is regarded as the typical mass of the plasma particle. Hereafter, a prime is used to denote the variables of the center-of-mass frame of the plasma. Using the above variables, the normalized reduced mass, $\mu = m_+ m_-/m^2$, and the normalized mass difference of the positively and negatively charged particles, $\Delta\mu = (m_+ - m_-)/m$, we obtain

$$\nabla_\nu (\rho U^\nu) = 0, \tag{4.9}$$

$$\nabla_\nu T^{\mu\nu} = 0, \tag{4.10}$$

$$\frac{1}{ne}\nabla_\nu\left(\frac{\mu h}{ne}Q^{\mu\nu}\right) = \frac{1}{2ne}\nabla^\mu(\Delta\mu p - \Delta p) + \left(U^\nu - \frac{\Delta\mu}{ne}J^\nu\right)F^\mu{}_\nu - \eta\left[J^\mu - \rho'_e(1+\Theta)U^\mu\right],$$

$$(4.11)$$

where $\rho'_e = -J^\nu J_\nu$ is the charge density observed by the plasma local rest frame, and the energy-momentum tensor $T^{\mu\nu}$ and "charge-current density tensor" $Q^{\mu\nu}$ are given by

$$T^{\mu\nu} \equiv pg^{\mu\nu} + hU^\mu U^\nu + \frac{\mu h^\ddagger}{(ne)^2}J^\mu J^\nu + \frac{2\mu\Delta h}{ne}(U^\mu J^\nu + J^\mu U^\nu) + F^\mu{}_\sigma F^{\mu\sigma} - \frac{1}{4}g^{\mu\nu}F^{\kappa\lambda}F_{\kappa\lambda},$$

$$(4.12)$$

$$Q^{\mu\nu} \equiv \frac{en}{h}\left[\frac{h^\ddagger}{ne}(U^\mu J^\nu + J^\mu U^\nu) + 2\Delta h U^\mu U^\nu - \frac{\Delta h^\sharp}{(ne)^2}J^\mu J^\nu\right].$$

$$(4.13)$$

Here, we used the relation, $R^\mu = -\eta ne[J^\mu - \rho'_e(1+\Theta)U^\mu]$, where Θ is the rate of equipartition with respect to the thermalized energy due to friction (see Appendix A of Koide (2009)).

These equations show the inertia effect and momentum of the electric charge and current (3rd and 4th terms of the right-hand side of Equation (4.12) and left-hand side of Equation (4.11)), and thermo-electromotive force (1st term of right-hand side of Equation (4.11)), and Hall effect (2nd term in bracket of factor of 2nd term of right-hand side of Equation (4.11)). When we neglect these terms, there equations reduce to the standard resistive GRMHD equations. Then, we call these equations "generalized GRMHD equations". It is noted that if

$$\eta < (2.5 \sim 3)\frac{m}{e\rho}\sqrt{\mu h^\dagger},$$

$$(4.14)$$

we can prove that causality is kept (Koide 2009). The condition (4.14) is identical to the definition of plasma, that is, the plasma parameter is much larger than unity, where we used the Spitzer resistivity (Bellan 2006).

Now, we found the basic equations for the resistive GRMHD calculation within causality (Koide 2010a). These equations are much complicated compared to the standard resistive GRMHD, thus the calculations with these equations are more difficult.

5. Peculiar phenomena suggested by generalized GRMHD equations

We mention two peculiar phenomena suggested by the generalized GRMHD equations. When we write the generalized Ohm's law among the generalized GRMHD equations using 3+1 formalism, we find new types of electromotive forces due to gravitation, shear of frame-dragging, and centrifugal force of the current. With respect to the gravitational electromotive force $\boldsymbol{E}_\mathrm{grv}$, we obtain a formula, $\boldsymbol{E}_\mathrm{grv} = \frac{2\mu h^\ddagger}{(ne)^2\gamma}\nabla(\ln\alpha)\rho^\dagger_e$, where α is the lapse function and $\rho^\dagger_e = Q^{00}$ is a kind of electric density (Koide 2010a). Then, if charge separation occurs at the current sheet between the azimuthal anti-parallel magnetic field near the black hole, the gravitational electromotive force may cause the magnetic reconnection without the resistivity, while frame dragging and centrifugal electromotive forces never change the topology of the magnetic field configuration.

We also mention a result of linear analysis of charge separation in the disk plasma rotating circularly around the rotating black hole. It shows that the charge separation is stable outside of the inner most stable orbit (ISCO) and then plasma oscillation is found. On the other hand, inside of the ISCO, the charge separation becomes unstable when the plasma frequency is small enough (Koide 2010b). The charge separation induced by this instability may cause a magnetic reconnection due to the gravitational electromotive force without resistivity. The inflow of the charged disk due to the instability into the black hole may charge up the black hole.

6. Summary and dreams in future

After the brief review of the results of ideal GRMHD simulations, we showed a model beyond the ideal GRMHD. Especially, we focus on the resistive GRMHD to simulate the magnetic reconnection in the black hole magnetosphere. This magnetic reconnection model comes from the modified model of a solar flare, and it can explain the correlation between the high energy flares and the radio knot ejections of M87 and micro-QSOs (Acciari *et al.* 2009; Mirabel & Rodrguez 1998). With respect to the basic equations beyond the ideal GRMHD within causality, we introduced the generalized GRMHD equations. We showed two peculiar phenomena suggested by these equations. To investigate the plasma dynamics in the black hole magnetosphere using the generalized GRMHD equations, we have to use new numerical calculation technique of these equations. The equations of continuity and energy and momentum conservation are similar to those of the ideal GRMHD equations. However, the generalized Ohm's law is drastically different from the Ohm's law of the standard resistive GRMHD equations. In the generalized Ohm's law, we have to consider the time derivative of current density, which is neglected in the standard GRMHD. This term is important to keep causality. Then, we have to develop entirely new numerical method for it. Unfortunately, we have no idea of the numerical calculation yet. This is our important, urgent task.

Once we develop the numerical code of the generalized GRMHD with a high magnetic Reynolds number, we could test the magnetic reconnection model of the flare-knot ejection correlation phenomenon around the black holes. In the calculation, we have to set the magnetic Reynolds number so large that a large amount of magnetic energy in the elongated magnetic bridges is stored, while the resistivity must exceed the numerical resistivity. It is noted that the numerical resistivity is not negligible in the recent ideal GRMHD simulations so that thin current sheet was diffused to reduce the magnetic energy stored in the magnetic bridges artificially. In such case, explosive phenomena expected in the very high magnetic Reynolds number case are suppressed. To calculate an explosive phenomena, we have to perform numerical simulations with very small numerical resistivity before we perform the resistive GRMHD calculation with large magnetic Reynolds number. Numerical simulations of peculiar phenomena suggested by the generalized GRMHD equations are also interesting.

Acknowledgements

We thank Mika Koide for her helpful comments on this manuscript. This work was supported in part by the Science Research Fund of the Japanese Ministry of Education, Culture, Sports, Science, and Technology.

References

Acciari, V. A. *et al.* 2009, Science, 325, 444.
Bellan, P. M. 2006, *Fundamentals of Plasma Physics* (Cambridge Univ., Cambridge).
Hayashi, M. R., Shibata, K., & Matsumoto, R. 1996, *ApJ*, 468, L37.
Koide, S. 2009, *ApJ*, 696, 2220.
Koide, S. 2010a, *ApJ*, 708, 1459.
Koide, S. 2010b, arXiv:1009.6041.
Koide, S., Shibata, K., & Kudoh, T. 2006, *Phys. Rev. D*, 74, 044005.

McKinney, J. C. 2006, *Mon. Not. R. Acad. Soc.*, 368, 1561.
Mirabel, I. F. & Rodriguez, L. F. 1998, Nature, 392, 673.
Shibata, K., Masuda, S., Shimojo, M., Hara, H., Yokoyama, T., Tsuneta, S., Kosugi, T., Ogawara, Y. 1995, *ApJ*, 451, L83.
Watanabe, N. & Yokoyama, T. 2006, *ApJ*, 647, L123.

Discussion

KIRK: I agree that two-fluid generalization of GRMHD is very important. In particular, the charge-separation instability you described is potentially crucial. Do you expect that this or other instabilities will be strongly affected by non-axisymmetric perturbations?

KOIDE: That is an important question. However, I did not perform a linear analysis with the non-axisymmetric perturbation. I would like to perform the linear analysis in the future.

YUAN: Can you comment on the existence of the "magnetic bridge" initial configuration?

KOIDE: During accretion, the inner accretion flow loses angular momentum, plunging into the BH, and resulting in the formation of this configuration.

NISHIKAWA: The idea of charge-separation instability may be a mechanism of jet formation seen in GRPIC simulations in Watson & Wikshikawa (2010)?

KOIDE: I think the charge-separation instability causes a current parallel to the equatorial plane only. Then, it does not drive a jet nor an outflow.

DE GOUVEIA DAL PINO: In order to have efficient reconnection events if you inject weak turbulence in the reconnection site, this eventually speeds up reconnection making it very fast even with small levels of resistivity as suggested by Vishniac & Lazarian (1999) and recently tested by 3D numerical MHD simulations by Kowal *et al.* (2009).

KOIDE: That is interesting. That mechanism would realize the anomalous resistivity in the current sheets formed in black hole magnetospheres.

Jets at all Scales
Proceedings IAU Symposium No. 275, 2011
G. E. Romero, R. A. Sunyaev & T. Belloni, eds.

© International Astronomical Union 2011
doi:10.1017/S1743921310015620

The stability of astrophysical jets

Philip E. Hardee[1]

[1]Department of Physics & Astronomy, Gallalee Hall,
The University of Alabama, Tuscaloosa, AL 35487 USA
email: phardee@bama.ua.edu

Abstract. Jets are produced by young stellar objects (**YSOs**), by black hole binary star system "microquasars" (μ**QSOs**), by active galactic nuclei (**AGN**), are associated with neutron stars and pulsar wind nebulae (**PWNe**), and are thought responsible for the gamma-ray bursts (**GRBs**). An understanding of these outflows must include how they are launched and collimated into jets, and how they propagate to large distances. Jets be they Poynting flux and/or kinetic flux dominated are current driven (**CD**) and/or Kelvin-Helmholtz (**KH**) velocity shear driven unstable. Here I present some of the work that is leading to a better understanding of the properties required for the observed relative stability of astrophysical jets.

Keywords. Galaxies: jets, ISM: jets and outflows, (magnetohydrodynamics:) MHD, instabilities, relativity

1. Introduction

Jetted outflows are associated with YSOs, e.g., Reipurth & Bally (2001), μQSOs, e.g., Mirabel & Rodríguez (1999), AGN, e.g., Urry & Padovani (1995); Ferrari (1998); Meier, Koide & Uchida (2001), neutron stars and PWNe, e.g., Crab nebula jet (Weisskopf *et al.* 2000) and are thought responsible for the GRBs, e.g., Zhang & Mészáros (2004); Piran (2005); Mészáros (2006). Jets are subject to current driven instability (**CDI**) where jets are magnetically dominated, and when kinetically dominated subject to the Kelvin-Helmholtz instability (**KHI**) driven by velocity shear between the jet and the surrounding medium. Instability is predicted theoretically from the linearized magnetofluid equations or, when the electromagnetic fields exceed the rest mass energy, the force-free "Poynting-flux" approximation. Thus, in astrophysical systems one might expect stability conditions to provide tools for selection between launching and propagation configurations satisfying the MHD equations governing jet flow. Jet structures attributed to instability are observed in laboratory experiments, e.g., Hsu & Bellan (2002), in numerical simulations, e.g., Hardee *et al.* (1997), and in astrophysical jets.

In YSO systems the development of KHI driven structures strongly depends on the effects of radiative cooling. In general, radiative cooling appropriate to the ISM leads to enhanced stability relative to a non-radiatively cooled adiabatic jet (Massaglia *et al.* 1992; Micono *et al.* 1998). Note that strong magnetic fields are predicted to reduce the effects of radiative cooling, e.g., Hardee & Stone (1997); Hardee, Stone & Rosen (1997). For a recent review of KHI in YSO jets see the article by Trussoni (2009). While radiative cooling is not thought to be important on the relativistic jets there are still some interesting similarities in morphology between YSO jets and AGN jets, e.g., compare the knots, twists, asymmetries and bow shock structures in HH 212 (Zinnecker *et al.* 1998) and HH 111 (Reipurth *et al.* 1997; Reipurth *et al.* 1999) with the knots and twists in 3C 273 (Jester *et al.* 2005; Uchiyama *et al.* 2006) and the twists and bow shock structures in Hercules A (Dreher & Feigelson 1984). Thus, the following discussion restricted to non-radiative KHI and CDI work will also be relevant to the YSO jets.

Twisted structures are observed in many AGN jets on sub-parsec, parsec and kiloparsec scales, e.g., 3C 120 (Gómez *et al.* 2001), 3C 273 (Lobanov & Zensus 2001) and M 87 (Lobanov, Hardee, & Eilek 2003). The propagation, radial structure and growth or damping of instability produced jet structures depends on whether the structure is magnetically (CDI) or kinetically (KHI) induced. Non-relativistic and relativistic simulations of magnetized jet formation and/or propagation have shown helical structures attributed to CDI, e.g., (Lery *et al.* (2000); Nakamura *et al.* (2001); Ouyed *et al.* (2003); Nakamura & Meier (2004); Nakamura *et al.* (2007); Moll *et al.* (2008); Moll(2009); McKinney & Blandford (2009); Carey & Sovinec (2009). Non-relativistic and relativistic simulations of kinetically dominated jets have shown that helical structures may be attributed to KHI, e.g., Hardee *et al.* (1997); Hardee *et al.* (2001); Hardee & Hughes (2003). Large scale helical twisting may also be triggered by precession of the jet ejection axis (Begelman *et al.* 1980). It is still not clear whether current driven, velocity shear driven or jet precession is responsible for the observed structures, or whether these different processes are responsible for the observed twisted structures at different spatial scales.

2. Kelvin-Helmholtz instability

The KHI properties of jets in the kinetically dominated regime have been reviewed in various astrophysical contexts, e.g., Birkinshaw (1991a); Ferrari (1998); Hardee (2004); Hardee (2006); Trussoni (2008); Trussoni (2009); Keppens *et al.* (2009). Extensive analyses have been performed for fluid jets, e.g., Ferrari *et al.* (1978); Hardee (1979); Birkinshaw (1984); Hardee (1987); Hardee (2000), considered the effects of a velocity shear layer (Birkinshaw 1991b), and rotation (Bodo *et al.* 1996). Additional investigations added magnetic fields parallel to the flow velocity, e.g., Ferrari *et al.* (1980); Ferrari *et al.* (1981); Ferrari & Trussoni (1983); Ray (1981); Londrillo (1985); Trussoni *et al.* (1988); Hardee (2007), and considered the addition of toroidal magnetic fields, e.g., Cohn (1983); Feidler & Jones (1984); Appl & Camenzind (1992). Relativistic jet simulations have found growth rates, e.g., Perucho *et al.* (2004a); Perucho *et al.* (2004b); Mizuno *et al.* (2007), and jet structure, e.g., Hardee *et al.* (1998); Hardee *et al.* (2001); Agudo *et al.* (2001); Hardee & Hughes (2003), predicted theoretically.

Perturbations to a cylindrical jet can be considered to consist of Fourier components of the form $f_1(r, \phi, z, t) = f_1(r) \exp[i(kz \pm n\phi - \omega t)]$ where flow is along the z-axis and n is an integer azimuthal wavenumber. The normal modes involve pinch ($n = 0$), helical ($n = 1$), elliptical ($n = 2$), triangular ($n = 3$), etc. (Hardee 1979) jet distortions and $+n$ and $-n$ specify the sense of spatial twist. Pinch structures can be triggered by flow variation associated with the central engine or jet overpressure. Twisted structures and the sense of twist can be induced by precession or orbital motion of the jet base, by rotation of the jet fluid, or by the helicity of the magnetic field.

For a super-magnetosonic jet with uniform density, pressure and velocity (top hat profile), and separated by a sharp boundary from an equal pressure uniform external environment, each normal mode formally consists of a fundamental and an infinite number of body mode solutions with different radial structure (see Fig. 5 in Hardee (2006); Fig. 13 in Hardee *et al.* (1997); Fig. 7 in Hardee *et al.* (2001)). The helical and higher order fundamental modes produce high pressure helically twisted filaments near to the jet's surface. The accompanying body modes produce helically twisted high pressure filaments in the jet's interior. The maximum fundamental mode growth rate when compared to the accompanying maximum body mode growth rates is less for the pinch, comparable for helical, and greater for elliptical and higher order normal modes. Spatial growth lengths scale with the Lorentz factor, magnetosonic Mach number and jet radius. In the subsonic but super-Alfvénic limit the body modes are stable but the jet surface remains unstable.

On astrophysical jets we expect a velocity shear layer to develop between the jet core and the external medium. This shear layer significantly modifies the growth of wavelengths comparable to the shear layer thickness, e.g., Ferrari *et al.* (1982); Ray (1982); Birkinshaw 1991b; Urpin (2002). There can be enhanced instability at short shear layer resonant wavelengths but, in general, growth rates are reduced by the presence of velocity shear, e.g., Perucho *et al.* (2005); Perucho *et al.* (2007). Still the growth rates are significant and the development of distortions to the jet's surface promotes mixing of the external medium into the jet fluid with the potential for disrupting the jet. Interestingly the analytically predicted shear layer resonances, which result in the formation of shocks and heating of the shear layer in numerical simulations, serves to stabilize the layer during non-linear evolution. The importance of the shear-layer resonant modes relies not only on their dominance among solutions of the linearized problem, but also on the fact that those jets for which the resonant modes are most important, do not disrupt and are very stable during the nonlinear evolution. It is noted that the steadiness of jets developing the shear layer resonant modes makes them firm candidates to remain collimated through long distances. Hence results would point to high Lorentz factor, highly supersonic jets as forming the FR II type jets, whereas the FR I type jets would be found in the opposite limit of slower and smaller Mach number jets.

The slow growth of large scale helical and elliptical twisting is not suppressed by the development of a velocity shear layer (Birkinshaw 1991b) and ultimately large scale helical twisting and to a lesser extent elliptical distortion can lead to catastrophic mixing and disruption of collimated jet flow, see Rosen *et al.* (1999). However, jets can remain highly collimated and avoid disruption by mixing by a multitude of effects associated with jet expansion. In the absence of magnetic fields, spatial growth of KHI is slowed significantly by increase in the Lorentz factor, increase in the Mach number and increase in the jet radius as spatial growth lengths scale with the Lorentz factor, the Mach number and the jet radius. Jet expansion also serves to push destructive helical twisting to slower growing longer wavelengths (relative to the jet radius). Along expanding jets waves can be advected down the jet to where their wavelength is shorter than the local fastest growing "resonant" wavelength. In addition to growth rate reduction in this shorter wavelength regime, the helical twist saturates at an amplitude that does not lead to significant surface distortion and accompanying mixing (Xu *et al.* 2000).

In the presence of magnetic fields with magnetic and thermal pressures not too far from equipartition, linear KHI is modified partly by replacing sound speeds and accompanying Mach numbers with magnetosonic speeds and accompanying Mach numbers but significant magnetic helicity can enhance KHI or CDI depending on field helicity and strength Baty (2005). Additionally, the non-linear development of KHI can be significantly modified. Equipartition or weaker magnetic fields aligned with the velocity in a shear layer have been shown to be capable of stabilizing a shear profile through non-linear saturation, e.g., Frank *et al.* (1996); Jones *et al.* (1997); Keppens & Tóth (2000); Ryu *et al.* (2000). The development of mixing can also be slowed by a helically twisted equipartition magnetic field configuration (Rosen *et al.* 1999). The toroidal magnetic field component reduces development of the elliptical and higher order normal modes, and simulations performed by Baty & Keppens (2002) indicate a mode-mode interaction between KHI and CDI that prevents the KH vorticies from fully developing. Mixing is considerably reduced and only minimal reduction in collimation occurs. While mixing can never be completely suppressed on the KH unstable jet, effects of mixing on jet propagation also will be minimized if the jet core is denser than the immediately surrounding medium and entrains only lighter material, e.g., a hot tenuous shear layer or the tenuous radio lobes associated with FR II type radio sources like Cygnus A.

Numerical simulations have been used to study the development of KH pinch structures driven by velocity fluctuation in parsec-scale jets (Agudo *et al.* 2001), and twisted structures on AGN jets have been used to estimate the physical conditions in jets and in their surroundings Perucho *et al.* (2009). While this technique requires knowing jet and pattern speeds, the approach has been applied to the superluminal motions and accelerations of components along curved trajectories in the 3C 345 jet (Hardee 1987, Steffen *et al.* 1995), the milliarcsec twisted structure of the 3C 120 jet (Hardee *et al.* 2005), the twisted emission threads in the 3C 273 jet (Perucho *et al.* 2006) and the M 87 jet (Hardee & Eilek 2010), and the transverse jet structure in 0836+710 (Perucho & Lobanov 2007). Comparison between the 3C 120 results and M 87 results is particularly interesting as the derived conditions are very different. The 3C 120 results (Hardee *et al.* 2005) indicate modest jet acceleration and cooling expansion along the inner kpc jet, effects that would serve to explain the relative stability and ~ 100 kpc jet length. On the other hand, the M 87 results (Hardee & Eilek 2010) indicate considerable jet deceleration and heating along the kpc jet, effects that would serve to explain the observed destabilization beyond a few kpc.

3. Current driven instability

It is thought that the relativistic jets produced from rotating bodies (neutron stars, black holes and accretion disks) are powered and collimated hydromagnetically, e.g., Blandford (2000). A toroidal magnetic field is wound up in such outflows and in the far zone becomes dominant because the poloidal field falls off faster with expansion and distance. In configurations with strong toroidal magnetic field, the CD kink mode is unstable. This instability excites large-scale helical motions that can strongly distort or even disrupt the system. For static cylindrical force-free equilibria, the Kruskal-Shafranov criterion for instability, $|B_\phi/B_p| > 2\pi R/z$, indicates that the instability develops if the length of the plasma column is long enough for the field lines to go around the cylinder at least once, e.g., Bateman (1978). Non-relativistic theoretical work (Appl *et al.* 2000; Baty 2005) and numerical simulations (Lery *et al.* 2000; Baty & Keppens 2002) show that CDI growth rates exceed KHI growth rates for sufficient magnetic field strength and helicity. CDI modes become dominant at increased helicity as the instability results from an imbalance between destabilizing toroidal and stabilizing poloidal magnetic field.

Poynting flux dominated outflows arise on magnetic field lines threading the horizon of a rotating black hole, e.g., Blandford & Znajek (1977); Blandford & Payne (1982). Jet acceleration and collimation is generally attributed to the toroidal magnetic field acting in concert with an external confining medium. For relativistic jet flow $\gamma_j >> 1$ implies $E_{kinetic} >> W_j c^2 \equiv \{\rho_j + [\Gamma/(\Gamma - 1)]P_j/c^2\}c^2$. Here γ_j is the jet Lorentz factor, Γ is the adiabatic index, and W is the enthalpy. For a magnetically produced jet we expect the jet speed $v_j \sim v_{Alfvén}$ and $\gamma_j \sim \gamma_A \sim [V_A/c] \equiv [B^2/(4\pi W_j c^2)]^{1/2} >> 1$ so there will be relatively low mass loading of the magnetic field lines. Here B is the magnetic field measured in the jet rest frame and the jet is Poynting-flux dominated. In this regime the sound speed is less than the Alfvén wave speed, i.e., $a_s^2 \equiv \Gamma P_j/W_j < v_A^2 \equiv V_A^2/(1+V_A^2/c^2)$. A magnetically dominated jet with a uniform poloidal field component is KH stable when $\gamma_j < \gamma_A[1 + (W_j/W_e)(V_{A,p}^2/c^2)]^{1/2}$ where $V_{A,p} \equiv [B_p^2/(4\pi W_j)]^{1/2}$ and W_e is the enthalpy of an unmagnetized external medium. Additional stabilization is provided by the presence of a significantly magnetized sheath or wind even if the jet core is super-Alfvénic (Hardee

2007; Mizuno *et al.* 2007). The addition of a toroidal field component is not likely to alter this stability condition so we expect the Poynting-flux jet to be KH stable.

For magnetically dominated flows the magnetic field is likely to approximate a force-free configuration. For relativistic force-free static configurations, analytical analyses of CDI criteria typically have been performed in the force-free Poynting-flux limit. The CDI kink mode, analogous to the KHI helical mode, is the most dangerous instability in configurations with toroidal magnetic field. This instability excites large scale helical motions that may disrupt the system. Cylindrical force-free jets are kink stable if the poloidal field is independent of the radius (Istomin & Pariev 1994, Istomin & Pariev 1996), but cylindrical force-free jets are kink unstable if the poloidal field decreases with the radius (Begelman 1998; Lyubarskii 1999). In the case of non-relativistic rotation, force-free jets are kink unstable if $|B_\phi/B_p| > |\Omega|R/c$ (Tomimatsu *et al.* 2001). All of these analyses are performed in a static reference frame, either infinite in extent or with a jet confined by rigid walls (Narayan *et al.* 2009). Provided the instability grows well inside the velocity shear surface defining a jet one might expect a CDI kink to move with $\gamma_{kink}v_{kink} \approx \gamma_j v_j$ and with "temporal" growth rate $\omega_i \propto (\gamma_{kink}R)^{-1}$ (Lyubarskii 1999). As a result we expect growth lengths $\ell_e \gtrsim \gamma_j R$ and for a typical jet with opening angle $\theta \sim few/\gamma_j$ the growth of the kink mode is relatively slow (Narayan *et al.* 2009). In a more realistic case, fluid inertia, relativistic rotation and velocity shear could significantly affect the instability criteria and the spatial growth of CDI. Unfortunately, no general stability analysis has been performed for magnetically dominated jets. Nevertheless, near to the central engine it is anticipated that CDI is the dominant instability.

The effects of magnetic helicity, jet density and jet flow profiles can be explored via numerical simulations. For static relativistic configurations (or more generally rigidly moving flows considered in the proper reference frame) with magnetic energy density comparable to or greater than the plasma energy density, Mizuno *et al.* (2009) found that the kink develops as predicted by linear theory but does not disrupt the initial force-free helical magnetic field. Growth rates are reduced by reduction in the Alfvén speed. Growth and morphological differences in the non-linear limit include a continually growing slender helically twisted column wrapped by magnetic field for magnetic helicity decreasing with radius, but growth nearly ceases for magnetic helicity increasing with radius. In all cases the characteristic time for the instability to affect strongly the initial structure for constant magnetic helicity is roughly $\tau \sim 100(a/v_A)$, where a/v_A is the Alfvén crossing time across the characteristic radius, a, of the force-free magnetic field.

Mizuno *et al.* (2010) find that the introduction of a sub-Alfvénic weakly relativistic velocity shear surface has profound consequences for kink growth, propagation and the associated flow field. For the velocity shear surface well inside the characteristic radius, transverse growth is similar to the static case and the plasma flows through a growing non-moving kink. For the velocity shear surface well outside the characteristic radius, initial growth is similar to that of a static case and the kink is advected with the flow until the growing kink approaches the velocity shear surface. For velocity shear radii on the order of the characteristic radius there is a more intimate interaction between the growing kink and the flow field. In general, growth is slowed, the kink propagates more slowly than the flow and slows further as growth continues. Thus, there is some flow through the moving kink and more so as the kink grows and slows.

For constant magnetic pitch and declining density profile the characteristic time is roughly $\tau \sim 10\tau_e$, with values for τ_e being dependent on the structure of the undisturbed state. In a jet context our perturbations remain static or can move with the flow frame depending on the location, R_j, of the velocity shear surface. In order to check whether the instability would affect a jet flow, one has to compare τ with a propagation time and present results suggest a scaling like $\tau \sim 10\tau_e \equiv 10\gamma_k^\alpha \tau_e^{st}$ with $3 \geqslant \alpha \geqslant 1$ for a moving kink with $R_j \gtrsim a$. In this case the condition for the instability to affect the jet structure might be written as $z > A\gamma_k^\alpha(v_k/c)a$, where we set $10\tau_e^{st} \equiv A(a/c)$ and $0 < v_k \leqslant v_j$ is a function of R_j/a and is sensitively dependent on the location of the velocity shear surface provided $R_j/a << 10$. This result suggests that the characteristic spatial scale for kink development could be very much shorter or longer than for a kink simply advected with $v_k = v_j \sim c$ for which $z \gtrsim 100\gamma_j a$. Non-relativistic analytic and numerical studies (Carey & Sovinec 2009) indicate that jet rotation can provide a stabilizing influence.

The 3D relativistic jet generation simulation performed by McKinney & Blandford (2009) indicates relatively rapid, less than 25 gravitational radii, but non-disruptive kink development over 500 gravitational radii. Our simulations for static and moving kinks suggest that the non-disruptive kink development in the jet generation simulation could be a result of the velocity shear surface located less or on order of the characteristic magnetic radius and a transverse density increase. This combination results in a slowly moving kink and rapid initial spatial development. Non-linear growth would be slowed by the transverse density increase, accompanying transverse Alfvén speed decline and increased Alfvén crossing time, associated with the confining denser slowly moving sheath found in jet generation simulations. Non-linear growth would also be significantly slowed by the jet expansion seen in jet generation simulations. Of course, a proper investigation of spatial growth requires stability simulations designed to study spatial kink development using realistic profiles.

It has been proposed that CDI converts Poynting flux to plasma energy flux and leads to a transition to kinetic flux jets. VLBA observations of the M 87 jet (Walker *et al.* 2008, Walker *et al.* 2009) with a resolution of 0.21×0.43 mas provide an observational opportunity to constrain the Poynting flux region. The luminosity distance of M 87 is $D = 16.7$ Mpc (Mei *et al.* 2007) and the mass of the central black hole is $(6.0 \pm 0.5) \times 10^9\ M_\odot$ (Gebhardi & Thomas 2009), at this distance. The Schwarzschild radius $R_s \sim 1.8 \times 10^{15}$ cm and 0.5 mas corresponds to $\sim 70R_s$ in the sky plane. VLBA resolution is $\approx 230R_s$ along the jet at a viewing angle of 15° consistent with the $\sim 6c$ (Biretta *et al.* 1999) superluminal motions at a few hundred parsecs (HST-1). The VLBA structure is consistent with radio and optical kinetically dominated twisted structure at 10 parsec to kiloparsec scales (Hardee & Eilek 2010). This suggests Poynting flux to plasma energy flux conversion within a few hundred gravitational radii of the black hole and places severe requirements on the jet structure that would allow CDI driven magnetic reconnection and reconfiguration on these spatial scales.

4. Conclusion

On relativistic jets CDI and KHI can produce twisted global structures mixed with structure induced by non-uniformity of the flow. In AGN Poynting flux to plasma energy flux conversion is likely restricted to the acceleration and collimation region at scales less than a few hundred gravitational radii from the black hole. The role of CDI in Poynting flux to plasma energy flux conversion is not yet determined but will depend on: (1) the global and/or sub-structure of the magnetic field in the acceleration region, (2) the gradients associated with jet acceleration and expansion, (3) the Lorentz factor of the flow, and (4) the global spine/sheath structure.

For jets to propagate to large distances and avoid disruptive mixing instigated by KHI they employ a variety of strategies to reduce entrainment. Stabilizing influences that reduce entrainment include: (1) strong or suitably ordered weaker magnetic fields, (2) suitable transverse gradients, (3) gradients such as those associated with jet acceleration or expansion, (4) high jet density (relative to the cocooning medium), and (5) high Lorentz factor and/or high Mach number.

Acknowledgements

P. Hardee acknowledges support from NASA (NNX08AG83G) and NSF (AST-0908010).

References

Agudo, I. *et al.* 2001, *ApJL*, 549, L183
Appl, S. & Camenzind, M. 1992, *A&A*, 256, 354
Appl, S., Lery, T., & Baty, H. 2000, *A&A*, 355, 818
Bateman, G. 1978, *MHD Instabilites*, (Cambridge, Mass., MIT Press), p.270
Baty, H. 2005, *A&A*, 430, 9
Baty, H. & Keppens, R. 2002, *ApJ*, 580, 800
Begelman, M. C. 1998, *ApJ*, 493, 291
Begelman, M. C., Blandford, R. D. & Rees, M. J. 1980, *Nature*, 287, 307
Biretta, J. A., Sparks, W. B., & Macchetto, F. 1999, *ApJ*, 520, 621
Birkinshaw, M. 1984, *MNRAS*, 208, 887
Birkinshaw, M. 1991, *Beams and Jets in Astrophysics*, ed. P. Hughes, p.278
Birkinshaw, M. 1991, *MNRAS*, 252, 73
Blandford, R. D. 2000, *Phil.Trans.Roy.Soc. A*, 358, 1
Blandford, R. D. & Payne, D. G. 1982, *MNRAS*, 199, 883
Blandford, R D. & Znajek, R. L. 1977, *MNRAS*, 179, 433
Bodo, G., Rosner, R., Ferrari, A., & Knoblock, E. 1996, *ApJ*, 470, 797.
Carey, C. S., & Sovinec, C. R. 2009, *ApJ*, 699, 362
Cohn, H. 1983, *ApJ*, 269, 500
Dreher, J., & Feigelson, E. 1984, *Nature*, 308, 43
Feidler, R. & Jones, T. W. 1984, *ApJ*, 283, 532
Ferrari, A. 1998, *ARA&A*, 36, 539
Ferrari, A., Massaglia, S., & Trussoni, E. 1982, *MNRAS*, 198, 106
Ferrari, A. & Trussoni, E. 1983, *MNRAS*, 125, 179
Ferrari, A., Trussoni, E., & Zaninetti, L. 1978, *A&A*, 64, 43
Ferrari, A., Trussoni, E., & Zaninetti, L. 1980, *MNRAS*, 193, 469
Ferrari, A., Trussoni, E., & Zaninetti, L. 1981, *MNRAS*, 196, 105
Frank, A., Jones, T. W., Ryu, D., & Gaalaas, J. B. 1996, *ApJ*, 460, 777
Gebhardt, K. & Thomas, J. 2009, *ApJ*, 700, 1690
Gómez, J. L. *et al.* 2001, *ApJ*, 561, L161
Hardee, P. E. 1979, *ApJ*, 234, 47
Hardee, P. E. 1987, *ApJ*, 318, 78
Hardee, P. E. 2000, *ApJ*, 533, 176
Hardee, P. E. 2004, *ApSS*, 293, 117
Hardee, P. E. 2006, *AIP Conf. Series 856*, eds. Hughes & Bregman, p.57
Hardee, P. E. 2007, *ApJ*, 664, 26
Hardee, P. E., Clarke, D. A., & Rosen, A. 1997, *ApJ*, 485, 533
Hardee, P. & Eilek, J. 2010, in preparation
Hardee, P. E. & Hughes, P. A. 2003, *ApJ*, 583, 116.
Hardee, P. E., Hughes, P. A., Rosen, A., & Gomez, E. A. 2001, *ApJ*, 555, 744.
Hardee, P. E., Rosen, A., Hughes, P. A., & Duncan, G. C. 1998, *ApJ*, 500, 599
Hardee, P. E., Walker, R. C., & Gómez, J. L. 2005, *ApJ*, 620, 646
Hardee, P. E. & Stone, J. M., 1997, *ApJ*, 483, pp.121-135 (1997)
Hardee, P. E., Stone, J. M., & Rosen, A., 1997, *IAU Symp. 182*, eds. Malbet & Castets, p.132
Hsu, S. C. & Bellan, P. M. 2002, *MNRAS*, 334, 257
Istomin, Y. N. & Pariev, V. I. 1994, *MNRAS*, 267, 629

Istomin, Y. N. & Pariev, V. I. 1996, *MNRAS*, 281, 1
Jester, S., Röser, H.-J., Meisenheimer, K., & Perley, R. 2005, *A&A*, 431, 447
Jones, T. W., Gaalaas, J. B., Ryu, D., & Frank, A. 1997, *ApJ*, 482, 230
Keppens, R. & Tóth, G. 2000, *Phys.Plasmas*, 6, 1461
Keppens, R., Meliani, Z., Baty, H., & van der Holst, B. 2009, *Lect. Notes Phys.*, 791, 179
Lery, T., Baty, H., & Appl, S. 2000, *A&A*, 355, 1201
Lobanov, A., Hardee, P., & Eilek, J. 2003, *New Science Reviews*, 47, 629
Lobanov, A. P. & Zensus, J. A. 2001, Science, 294, 128
Londrillo, P. 1985, *A&A*, 145, 353
Lyubarskii, Y. E. 1999, *MNRAS*, 308, 1006
Massaglia, S., Trussoni, E., Bodo, G., Rossi, P., & Ferrari, A. 1992, *A&A*, 260, 243
McKinney, J. C. & Blandford, R. D., 2009, *MNRAS*, 394, L126
Mei *et al.* 2007, *ApJ*, 655, 144
Mészáros, P. 2006, *Rep.Prog.Phys.*, 69, 2259
Micono, M., Bodo, G., Massaglia, S., Rossi, P., & Ferrari, A. 1992, *A&A*, 260, 243
Meier, D. L., Koide, S., & Uchida, Y. 2001, Science, 291, 84
Mirabel, I F. & Rodríguez, L. F. 1999, *ARA&A*, 37, 409
Mizuno, Y., Hardee, P., & Nishikawa, K-I. 2007, *ApJ*, 662, 835
Mizuno, Y., Lyubarsky, Y., Nishikawa, K.-I., & Hardee, P. E. 2009, *ApJ*, 700, 684
Mizuno, Y., Hardee, P., & Nishikawa, K.-I. 2010, in preparation
Moll, R. 2009, *A&A*, 507, 1203
Moll, R., Spruit, H. C., & Obergaulinger, M. 2008, *A&A*, 492, 621
Nakamura, M., Uchida, Y., & Hirose, S. 2001, *New Astronomy*, vol.6, issue 2, 61
Nakamura, M., & Meier, D. L. 2004, *ApJ*, 617, 123
Nakamura, M., Li, H., & Li, S. 2007, *ApJ*, 656, 721
Narayan, R., Li, J., & Tchekhovskoy, A. 2009, *ApJ*, 697, 1681
Ouyed, R., Clarke, D. A., & Pudritz, R. E. 2003, *ApJ*, 582, 292
Perucho, M., Hanasz, M., Martí, J. M., & Sol, H. 2004a, *A&A*, 427, 415
Perucho, M., Hanasz, M., Martí, J. M., & Miralles, J. A. 2007, *Phys.Rev.E*, 75, 631
Perucho, M., Lobanov, A. P., & Kovalev, Y. Y. 2009, *ASPCS*, 402, 349
Perucho, M. & Lobanov, A. P. 2007, *A&A*, 469, 23
Perucho, M., Lobanov, A. P., Martí, J.-M., & Hardee, P. E. 2006, *A&A*, 456 , 493
Perucho, M., Martí, J. M., & Hanasz, M. 2004b, *A&A*, 427, 431
Perucho, M., Martí, J. M., & Hanasz, M. 2005, *A&A*, 443, 863
Piran, T. 2005, *Reviews of Modern Physics*, 76, 1143
Ray, T. P. 1981, *MNRAS*, 196, 195
Ray, T. P. 1982, *MNRAS*, 198, 617
Reipurth, B. & Bally, J. 2001, *ARA&A*, 39, 403
Reipurth, B., Hartigan, P., Heathcote, S., Morse, J., & Bally, J. 1997, *AJ*, 114, 757
Reipurth, B., Yu, K., Rodrguez, L., Heathcote, S., & Bally, J. 1999, *A&A*, 352, 83
Rosen, A., Hardee, P. E., Clarke, D. A., & Johnson, A. 1999, *ApJ*, 510, 136.
Ryu, D., Jones, T. W., & Frank, A. 2000, *ApJ*, 545, 475.
Steffen, W., Zensus, J. A., Krichbaum, T. P., Witzel, A., & Qian, S. J. 1995, *A&A*, 302, 335
Tomimatsu, A., Matsuoka, T., & Takahashi, M. 2001, *Phys.Rev.D*, 64, 123003
Trussoni, E., Massaglia, S., Bodo, G., & Ferrari, A. 1988, *MNRAS*, 234, 539
Trussoni, E. 2008, *Jets from Young Stars III - Numerical MHD and Instabilities*, eds. Massaglia
    et al. LNP 754 (Springer: Berlin Heidelberg), p.105
Trussoni, E. 2009, *Protostellar Jets in Context*, eds. Tsinganos, Ray & Stute (Springer: Berlin
    Heidelberg), p.285
Uchiyama, Y. *et al.* 2006, *ApJ*, 648, 910
Urpin, V. 2002, *A&A*, 385, 14
Urry, C. M. & Padovani, P. 1995, *PASP*, 107, 903
Walker, R. *et al.* 2008, *JPhCS*, 131, 012053
Walker, R. *et al.* 2009, *ASPCS*, 402, eds. Hagiwara *et al.*, (San Francisco, ASP), p. 227

Weisskopf, C. *et al.* 2000, *ApJ*, 536, L81
Xu, J., Hardee, P., & Stone, J. 2000, *ApJ*, 543, 161
Zhang, B. & Mészáros, P. 2004, *Int.J.Mod.Phys.*, A19, 2385
Zinnecker, H., McCaughrean, M., & Rayner, J. 1998, *Nature*, 394, 862

Discussion

SIKORA: What determines the polarity of twisted modes triggered by KH instabilities?

HARDEE: The polarity of twisted modes is determined by the rotation and associated heliciy of the flow and magnetic fields.

Jets at all Scales
Proceedings IAU Symposium No. 275, 2011
G. E. Romero, R. A. Sunyaev & T. Belloni, eds.

© International Astronomical Union 2011
doi:10.1017/S1743921310015632

Relativistic jets: Physics and simulations

John F. Hawley

Department of Astronomy, University of Virginia Charlottesville, VA 22904 USA
email: jh8h@virginia.edu

Abstract. Jets are one of the more dramatic and visible manifestations of black hole accretion. It is becoming increasingly accepted that magnetic fields underlie the mechanism behind the launching of relativistic jets. At the same time it is now appreciated that the fundamental driving mechanism behind accretion is due to magnetohydrodynamical processes and the stresses they engender. Global disk simulations in full general relativity have begun to reveal the radial dependence of disk stresses and the intimate connections between accretion and jet-launching. This talk will review the issues involved, the recent progress that has been made, and the challenges that remain.

Keywords. accretion, accretion disks, black hole physics, MHD

1. Introduction

The topic of this symposium is "Jets at All Scales" and, indeed, jets are seen on many scales and in a wide variety of astrophysical objects. Protostellar systems, binary stars, active galactic nuclei, quasars, neutron stars and white dwarfs, all have demonstrated the ability to generate jets. The ubiquity of the jet phenomenon argues that jets do not require particularly exotic circumstances in which to form. Jets appear to be the consequence of accretion, rotation, and magnetic fields (Livio 2000). When the jet's energy is substantial and the velocities relativistic, the jet most likely originates from a black hole system. After all, the environment near a black hole is nothing if not relativistic!

The dominant paradigms for jet launching follow the ideas outlined in the influential papers of Blandford & Znajek(1977) and Blandford & Payne(1982). In the Blandford-Payne (BP) model the power for the jet comes from the accretion disk. A large-scale poloidal magnetic field is anchored in and rotates with the disk. If the fieldlines are angled outward sufficiently with respect to the disk, there can be a net outward force on the matter. As matter is accelerated along the rotating fieldlines, its angular momentum increases still further, increasing the acceleration and driving an outflow.

In the BP model the energy driving the jet derives from the rotational energy in the accretion disk. But the central star in an accreting system can also rotate and drive a jet. Even a black hole, which lacks a solid surface, can do so. In the Blandford-Znajek (BZ) model the jet is powered by the rotating space-time of a black hole. In this model, radial magnetic field lines pass through the ergosphere and the horizon of the hole. Frame dragging by the rotating hole leads to the creation of toroidal field and hence a Poynting flux. The key to the BZ process is that within the ergosphere it is possible to have an electromagnetic flux with negative energy at infinity. This negative energy flux enters the hole, thereby reducing both the hole's mass-energy and angular momentum. The outgoing Poynting flux can have substantial positive energy derived from the black hole.

2. Magnetic Fields and Accretion

The common thread in both the BP and the BZ models is the presence of a large-scale poloidal field that can tap into the rotational energy of the system, either that in the disk or the black hole. The notion that jets are fundamentally a magnetic phenomenon fits in nicely with our emerging overall understanding of accretion itself. Accretion depends on the removal of angular momentum from fluid elements by some internal stress. Although differential rotation is linearly *stable* to hydrodynamic perturbations by the Rayleigh criterion, it is *destabilized* by the presence of a magnetic field (Balbus & Hawley 1991, 1998; Balbus 2003). The magnetorotational instability (MRI) leads to turbulence and Maxwell and Reynolds stresses. From linear theory we know that the MRI is present as long as the field is relatively weak which, for an accretion disk, means subthermal, i.e., has a plasma $\beta = P/P_{mag}$ greater than 1. The maximum growth occurs when $(\mathbf{k} \cdot \mathbf{v}_A)^2$ is comparable to Ω^2. This corresponds to a wavelength that is about the distance an Alfvén wave travels in an orbit. All orientations of magnetic field can be unstable, including nonaxisymmetric modes involving the toroidal field.

The presence of the linear MRI provides the means to destabilize the otherwise laminar orbital motion of the gas. The properties of the resulting fully developed MHD turbulence are not accessible to analytic theory, however, and we have to rely upon simulations to gain insight.

3. Simulations

For the most part, the theory of jet formation and black hole accretion is based on somewhat idealized and parameterized models that have characteristics consistent with the observed phenomena. One would prefer to begin with a very general set of well-defined equations and from them develop a model from first-principles. We are still a long way from that ideal, but we do know one thing: the equations of MHD plus gravity provide a minimal set of conservation equations that contain sufficient physics to describe the basic dynamics of the accretion process and of jet formation.

3.1. *Local Shearing Boxes*

The most detailed simulations of the MRI-driven turbulence that drive accretion have been done in a local approximation known as the shearing box (Hawley *et al.* 1995). These shearing box simulations have revealed a number of important properties of magnetized accretion. First, the MRI leads to turbulence which produces angular momentum transport. The important component of the stress tensor is

$$\tau_{r\phi} = -B_r B_\phi / 4\pi + \rho v_r \delta v_\phi \qquad (3.1)$$

where the first term is the Maxwell stress and the second is the Reynolds stress. In MRI turbulence the Maxwell stress always exceeds the Reynolds stress by about a factor of 4, and the total $r\phi$ component of the stress is proportional to the magnetic pressure, $\tau_{r\phi} = \alpha_{\mathrm{mag}} P_{\mathrm{mag}}$, where α_{mag} is the constant of proportionality. Blackman, Penna & Varnière (2008) summarized the results from a large number of shearing box simulations in the literature and showed that quite generally α_{mag} is nearly constant across a wide range of simulations, with a value ~ 0.4–0.5. The problem for constructing accretion disk models is that we still don't know what sets P_{mag} within a disk, except that it is likely to be subthermal, i.e., $\beta = P_{\mathrm{gas}}/P_{\mathrm{mag}} > 1$. Note that if one uses the conventional Shakura & Sunyaev (1973) α value to characterize the stress, a specific value of α implies a magnetization of $\beta = \alpha_{\mathrm{mag}}/\alpha$. Typical β values in shearing box simulations are ~ 10–100, corresponding to α values ~ 0.1–0.005.

One of the defining characteristics of the α disk model is that the same stress that accounts for the angular momentum transport also determines the local heating (Balbus & Papaloizou 1999). This requires that the orbital energy released by the stress go directly into the energy of the MHD turbulence, and cascade to small scales where it is thermalized promptly. This need not necessarily be the case; if energy is transported by waves or by the motion of strong fields (for example, field buoyantly rising into a corona) rather than rapidly dissipating as heat, things become nonlocal. This may happen under some circumstances, but shearing box simulations (e.g., Simon, Hawley & Beckwith 2009) find that turbulent dissipation occurs rapidly, on a timescale of order Ω^{-1}, roughly a turnover time of the largest scale turbulent eddies and consistent with the requirements of the α model. In the radiative models of Hirose *et al.* (2009) dissipation is local and the bulk of the energy is carried out of the disk by radiation. Only a small fraction is carried by magnetic flux.

One great advantage of the local model compared to global simulations is that is is easier to add more complex physics and to employ greater resolution. The most realistic shearing box simulations done to date are those that include radiation transport (e.g., Turner 2004; Hirose, Krolik & Stone 2006). One set of such simulations carried out by Hirose, Krolik & Blaes (2009) who answered a long-standing question in disk modeling. If one adopts the α formulation for the stress, radiation-dominated disks become thermally unstable (Lightman & Eardley 1974; Shakura & Sunyaev 1976). Essentially the stress, and hence disk heating, increase faster than the radiative cooling rate. This thermal instability goes away if the Shakura-Sunyaev stress formula is modified so that the stress is proportional to the gas pressure alone rather than the total pressure. Then an increase in radiation pressure does not lead to increased heating. It has been long suspected that instabilities such as the thermal instability are simply artifacts of the α prescription (Pringle 1981). This seems to be the case, at least for the local thermal instability. Hirose *et al.* (2009) demonstrated that although the stress is, on average, proportional to the total pressure, there is no thermal instability in radiation pressure supported disks. The reason is that stress determines the pressure, not the other way around. An increase in stress will lead to an increase in pressure, but an increase in pressure does not necessarily lead to an increase in magnetic energy and hence stress. Although this result rules out the local thermal instability, various global instabilities remain possible. The shearing box's ability to investigate global properties is quite restricted.

3.2. *Global Simulations - Jet Production*

One area where local simulations are of particularly limited use is in modeling jet formation; global simulations are essential. Some of the questions that might be addressed by simulations include:

- Does a particular jet model in fact produce a stable, sufficiently energetic jet?
- What magnetic field configurations and strengths lead to jet formation?
- Under what circumstances does an accreting system develop the required magnetic fields?

For example, the viability of the BP process has been demonstrated by following the evolution of a disk embedded in an initial vertical field (see the review by Pudritz *et al.* 2007). The questions are whether the necessary field configurations are stable over the long term, and whether they can arise naturally as a result of the accretion process.

For black holes, global simulations have produced reasonably powerful and stable jets consistent with the BZ mechanism, beginning from simple initial conditions. The key to jet formation is to establish the right kind of magnetic field, here a large-scale poloidal field along the axis of the black hole. The first GR simulations (De Villiers *et al.* 2005;

Hawley *et al.* 2006; McKinney 2006) start with dipole field loops contained entirely within an initial orbiting gas torus. Early in the simulation, inflow drags the inner part of these loops down to the hole creating an extended radial field. Relative shear along this field creates toroidal field that grows in amplitude. This drives a vertical expansion of the field generating a magnetic tower (Lynden-Bell 2006; Kato *et al.* 2004) which eventually resolves itself as a large-scale dipole field embedded on the black hole. The rotation of the black hole produces a Poynting flux moving outward at relativistic velocities. The power of the Poynting flux jet is determined by the spin of the black hole and the strength of the magnetic field. In the simulations the resulting jet power matches well with the predictions of the BZ model (McKinney & Gammie 2004; McKinney 2005). The Lorentz factor Γ in the jet depends on the mass loading. The Poynting flux jet has very little mass within it; the angular momentum within the gas is too large to enter the axial funnel. For numerical stability the simulations must impose a density floor and this can control the mass density within the funnel. However, jets with $\Gamma > 10$ have been produced (e.g., McKinney 2006).

Hawley & Krolik (2006) examined a series of simulations that employed black holes with rotations ranging from retrograde with $a/M = -0.9$ to prograde with $a/M = 0.99$. They found that for prograde models with $a/M \sim 0.9$ and above the jet energy can exceed the nominal energy associated with accretion as determined by the binding energy of the last stable orbit associated with that black hole spin. Both the increase in spin and an increase in the field strength within the funnel account for the strong increase in jet energy. For the retrograde model the sign of the angular momentum in the jet is consistent with that of the hole rather than the disk, establishing that the source of rotation is, indeed, the hole. The jet energy is greater than the nominal energy of accretion for the retrograde disk, but considerably lower than the power in the prograde $a/M = 0.9$ jet. This result is not consistent with the model discussed in the talk by David Meier (these proceedings) wherein the retrograde holes have the strongest jets. In these simulations the process producing the magnetic tower that leads to the funnel field is largely insensitive to the black hole spin. The depth of the last stable orbit helps to set the funnel field's strength, and once established that field is relatively constant. Ultimately the issue depends on the processes by which Nature loads a black hole with poloidal field, and many uncertainties remain in this topic.

In the simulations, unbound jets with significant mass flux arise only outside the centrifugal boundary surrounding the axis. These jets are accelerated by fields that remain within the accretion flow; pressure forces push gas up against the centrifugal barrier and outward, resulting in an unbound mass flux and an outgoing velocity of about $\sim 0.5c$ (De Villiers *et al.* 2005; Hawley *et al.* 2006). This is the only jet seen in simulations with nonrotating Schwarzschild holes.

These simulations succeed in producing a jet, and suggest several questions for further investigation. First, when will the necessary field get established on the black hole? Jet formation requires a consistent sense of vertical field to be brought down to the event horizon. This readily occurs for simulations that begin with a large dipole loop. But simulations by Beckwith *et al.* (2008) found that much weaker funnel fields result from quadrupolar field initial conditions, and no funnel field at all develops for toroidal field initial conditions. A strong jet seems to require a net vertical field in the disk midplane whose sign remains consistent for at least an inner-disk inflow time. In simulations with an initial quadrupole field, a funnel field can be established, but the presence of a current sheet in the corona between the black hole's axis and the equator makes conditions favorable for reconnection and field loss. In the dipole case, the current sheet is located at the equator inside the disk and the funnel field is relatively insulated against reconnection.

For purely toroidal field initial conditions there is MRI-driven turbulence, but the poloidal field coherence length tends to be relatively short. There is insufficient field amplification to create a magnetic tower and thereby load the funnel with field.

Assuming that a poloidal funnel field is established, what determines the strength of that field? Within the disk the basic requirements of the MRI suggest that the field will be subthermal, possibly with $\beta \sim 10$–100. If the funnel field were of comparable strength it would be difficult to drive a powerful jet, particularly for cool, thin disks. However, the field strength in the funnel may not be so limited. In the simulations, the magnetic pressure within the funnel is more or less comparable to the *total* pressure in the inner disk, and the field is correspondingly stronger. While this result seems entirely reasonable, it remains to be seen if it holds under all circumstances.

3.3. *Radial Advection of Field:*

Jets require the development of a large-scale poloidal field passing through the disk, the black hole, or both. How does such a field arise? Since the disk is fundamentally magnetic, it is possible that the MRI can act as a dynamo to generate the required field. In fact, stratified shearing box simulations have found that a mean toroidal field can be generated at the disk midplane and buoyantly rise out, to be replaced by a mean toroidal field of the opposite sign. This happens on 10 orbit timescale and is consistent with a simple α–Ω dynamo (Brandenburg *et al.* 1995; Stone *et al.* 1996; Hirose et a. 2006; Guan 2009; Shi et a. 2010; Gressel2010; Davis *et al.* 2010; Simon *et al.* 2010). The resulting field does not appear, however, to have the properties required for jets.

An alternative way to obtain a large-scale net field in the near-hole region is by bringing net field in from large radius. A field that is very weak far from the hole can be strongly amplified as it is concentrated by accretion. The issue is whether the rate at which the field diffuses through the matter exceeds the rate at which it is accreted (van Ballegooijen 1989; Lubow *et al.* 1994; Heyvaerts *et al.* 1996). This question is usually framed in terms of a competition between an effective turbulent viscosity, ν_t, and an effective magnetic diffusivity, η_t. The conventional assumption is that the turbulent viscosity will be comparable to diffusivity, and field advection will be inhibited. Local simulations by Guan & Gammie (2009) seem to support this viewpoint. Guan & Gammie examined the rate at which a sinusoidal vertical flux distribution decayed within MRI turbulence and found a diffusion time on order several tens of Ω^{-1}, comparable to the accretion time.

Beckwith *et al.* (2009) carried out global simulations to directly address the question of whether net vertical flux can be advected inward by MRI-driven turbulence. In this simulation an initial isolated, uniform vertical field passes through a gas torus. At the beginning of the simulation the field in the low density region outside of the disk expands. Within the disk the field is unstable to the MRI. Turbulence develops on an orbital timescale and the disk begins to accrete. While the magnetic flux does diffuse somewhat within the disk, it is not carried in efficiently with the accretion flow. In fact, much of the initial vertical flux within the disk remains near its initial location. Nevertheless, the black hole rapidly acquires a net dipole field. Instead of being brought in by the disk, this field is carried in by infall of low density and low angular momentum material in the corona surrounding the disk. These results suggest two points. First, field is most effectively transported by an inflow driven not by turbulent stresses, but by large-scale magnetic torques operating in low density gas (e.g., Rothstein & Lovelace 2008). Second, net flux is a global, not a local concept. Here, the acquisition of a net dipole on the hole requires both field to be carried down to the black hole, and reconnection across the equator to change the global topology.

3.4. *The Spin of the Hole*

As several talks at this Symposium have indicated, there are important issues related to the spin of black holes, including whether or not we can measure black hole spin from observed disk or jet properties, and whether there is any observational evidence for the role of black hole spin in powering jets (see, e.g., the talk by Elena Gallo).

Current methods of black hole spin measurement, including both continuum fitting and Fe Kα line fitting, rely on knowledge of the underlying disk and its inclination angle. These procedures rely on the assumption that the disk's inner edge coincides precisely with the innermost stable circular orbit (ISCO) which is itself precisely determined by a/M. In fact, the situation is likely to be a bit more complex than that. What constitutes a disk's inner edge, for example, depends on the property being examined (Krolik & Hawley 2002). There will be a location where from which the last photons emerge (the radiation edge, relevant to continuum fitting), a location where the optical depth drops below the point where photons can be reflected (the reflection edge, relevant to Fe Kα line modeling), a location where turbulent flow is replaced by laminar inflow (the turbulence edge), and the place where the internal stress goes to zero (the stress edge). The standard model (Novikov & Thorne 1973) assumes that all these edges are located at the ISCO. This may be problematic; magnetic fields can exert force over extended distances. Gammie(1999) and Krolik(1999) showed that, in principle, significant magnetic stress can be present within the ISCO, even for cold, thin disks. Global simulations have found that significant magnetic stress can continue right through and inside the ISCO (Hawley & Krolik 2001; Reynolds & Armitage 2001; Hawley & Krolik 2002; Machida & Matsumoto 2003; Gammie *et al.* 2004; Krolik *et al.* 2005).

The disks in these first simulations were all modestly thick, with $H/R \sim 0.15$. Because the magnetic stress might be a function disk temperature and hence H at the ISCO, simulations were carried out that included direct regulation of disk temperature and thickness. Noble, Krolik & Hawley (2009) used a simple cooling function to compute a disk with $H/R \sim 0.1$ accreting into an $a/M = 0.9$ hole and found enhanced stress that led to a 6% increase in luminosity at infinity over the conventional value. Shafee *et al.* (2008) found a reduced (but still nonzero) stress inside the ISCO for thin disks, as did Penna *et al.* (2010), with the net specific angular momentum accreted into the hole reduced by only a few percent below the ISCO value. A systematic study by Noble *et al.* (2010) considered different H/R values accreting into a Schwarzschild black hole and found that functionally the stress continues to rise through the ISCO, dropping to zero only right outside of the black hole horizon. The Maxwell stress showed no dependence on H, but the Reynolds stress at the ISCO did decline with disk thickness. Beckwith *et al.* (2008) found that the initial field topology could influence the stress at the ISCO. In particular, a simulation that began with a purely toroidal initial field barely had any enhanced stress at the ISCO. Penna *et al.* (2010) pointed out that the differences in stress levels seen in various thin disk simulations could be due to different initial magnetic field topologies and strengths.

Where does this leave us currently? It is clear that magnetic fields *can* increase the efficiency of accretion through enhanced stress at the ISCO, but the precise amount depends on both physical and numerical factors. The simple "zero stress at the ISCO" assumption is not universally applicable.

Simulations also have implications for the evolution of the black hole's spin if it is being determined by accretion. Spin evolution was considered by Thorne (1974) who showed that although accretion would tend to spin up the hole to its physical limit, the capture of photons from the disk, preferentially with retrograde orbits, limits the spin-up to

$a/M = 0.998$. Simulations have found that the upper limit could be considerably smaller for *magnetized* accretion (Gammie *et al.* 2004; McKinney & Gammie 2004; Krolik *et al.* 2005; Beckwith *et al.* 2008). Magnetic fields threading the horizon can carry angular momentum to infinity as the hole captures Alfvén waves with retrograde angular momentum. In the simulations of McKinney & Gammie (2004) the angular momentum flux into holes with different spins was measured and the transition point between net spin-up and spin-down was at $a/M \sim 0.93$. That number is specific to this set of simulations, but if this general idea holds then the most rapidly spinning (and hence most energetic) holes might be rare or nonexistent in Nature.

4. Summary

Simulations are currently playing an important role in investigating the physics of black hole accretion, from the stresses within the disks to the formation of jets. Local shearing box simulations have demonstrated that the MRI leads to turbulence that accounts for angular momentum transport in disks. While the stress is related to the pressure, it is not strictly determined by it. Rather, the stress is directly proportional to the *magnetic* pressure. What determines the magnetic pressure under different conditions remains a topic of continuing investigation.

Simulations have successfully demonstrated the creation and maintenance of Poynting flux jets from spinning black holes. These jets require a dipole field within the axial funnel of the hole. The manner in which such a field might generally be created and how it would be sustained remain open questions. It may be that net flux can be brought in from large radius by some aspect of the accretion process. It is also possible that the process is stochastic, simply dependent upon a random build up from field brought in by accretion. If so, jets would be intermittent on accretion timescales.

Simulations are providing new details regarding the properties of the near-hole region of the accretion disk. Magnetic stresses can operate through the ISCO, potentially altering the accretion efficiency and luminosity. By capturing Alfvén waves with retrograde angular momentum, magnetic fields have the ability to limit a/M in black holes whose spin is determined by accretion.

We have long known that gravity is a potential source of energy, and that angular momentum leads to disks that can efficiently extract and radiate that energy. But it turns out that it is the magnetic field that really makes things lively!

Acknowledgements

This work was supported by NASA grant NNX09AD14G and NSF grant AST-0908869. I thank collaborators Steve Balbus, Kris Beckwith, Jean-Pierre De Villiers, Xiaoyue Guan, Andrew Hamilton, Julian Krolik, Scott Noble, Jeremy Schnittman, and Jacob Simon. Computational resources were supplied by the TeraGrid, supported by the National Science Foundation.

References

Balbus, S. A. 2003, *Ann. Rev. Astron. Astrophys.*, 41, 555
Balbus, S. A. & Hawley, J. F. 1991, *ApJ*, 376, 214
— 1998, *Rev. Mod. Phys.*, 70, 1
Balbus, S. A. & Papaloizou, J. C. B. 1999, *ApJ*, 521, 650
Beckwith, K., Hawley, J. F., & Krolik, J. H. 2008, *ApJ*, 678, 1180
— 2009, *ApJ*, 707, 428

Blackman, E. G., Penna, R. F., & Varnière, P. 2008, *New Astron.*, 13, 244

Blandford, R. D. & Payne, D. G. 1982, *MNRAS*, 199, 883

Blandford, R. D. & Znajek, R. L. 1977, *MNRAS*, 179, 433

Brandenburg, A., Nordlund, A., Stein, R. F., & Torkelsson, U. 1995, *ApJ*, 446, 741

Davis, S. W., Stone, J. M., & Pessah, M. E. 2010, *ApJ*, 713, 51

De Villiers, J., Hawley, J. F., Krolik, J. H., & Hirose, S. 2005, *ApJ*, 620, 878

Gammie, C. F. 1999, *ApJ*, 522, L57

Gammie, C. F., Shapiro, S. L., & McKinney, J. C. 2004, *ApJ*, 602, 312

Gressel, O. 2010, *MNRAS*, 405, 41

Guan, X. 2009, PhD Thesis, Univ. of Illinois

Guan, X. & Gammie, C. F. 2009, *ApJ*, 697, 1901

Hawley, J. F., Gammie, C. F., & Balbus, S. A. 1995, *ApJ*, 440, 742

Hawley, J. F. & Krolik, J. H. 2001, *ApJ*, 548, 348

— 2002, *ApJ*, 566, 164

— 2006, *ApJ*, 641, 103

Heyvaerts, J., Priest, E. R., & Bardou, A. 1996, *ApJ*, 473, 403

Hirose, S., Krolik, J. H., & Blaes, O. 2009, *ApJ*, 691, 16

Hirose, S., Krolik, J. H., & Stone, J. M. 2006, *ApJ*, 640, 901

Kato, Y., Mineshige, S., & Shibata, K. 2004, *ApJ*, 605, 307

Krolik, J. H. 1999, *ApJ*, 515, L73

Krolik, J. H. & Hawley, J. F. 2002, *ApJ*, 573, 754

Krolik, J. H., Hawley, J. F., & Hirose, S. 2005, *ApJ*, 622, 1008

Lightman, A. P. & Eardley, D. M. 1974, *ApJ*, 187, L1

Livio, M. 2000, in American Institute of Physics Conference Series, edited by S. S. Holt, & W. W. Zhang, vol. 522 of American Institute of Physics Conference Series, 275

Lubow, S. H., Papaloizou, J. C. B., & Pringle, J. E. 1994, *MNRAS*, 267, 235

Lynden-Bell, D. 2006, *MNRAS*, 369, 1167

Machida, M. & Matsumoto, R. 2003, *ApJ*, 585, 429

McKinney, J. C. 2005, *ApJ*, 630, L5

— 2006, *MNRAS*, 368, 1561

McKinney, J. C. & Gammie, C. F. 2004, *ApJ*, 611, 977

Noble, S. C., Krolik, J. H., & Hawley, J. F. 2009, *ApJ*, 692, 411

— 2010, *ApJ*, 711, 959

Novikov, I. D. & Thorne, K. S. 1973, in *Black Holes (Les Astres Occlus)*, edited by C. De Witt (Gordon and Breach, New York), 343

Penna, R. F., McKinney, J. C., Narayan, R., Tchekhovskoy, A., Shafee, R., & McClintock, J. E. 2010, *MNRAS*, 1216

Pringle, J. E. 1981, *Ann. Rev. Astr. Astrophs.*, 19, 137

Pudritz, R. E., Ouyed, R., Fendt, C., & Brandenburg, A. 2007, in *Protostars and Planets V*, edited by B. Reipurth, D. Jewitt, & K. Keil, 277

Reynolds, C. S. & Armitage, P. J. 2001, *ApJ*, 561, L81

Rothstein, D. M. & Lovelace, R. V. E. 2008, *ApJ*, 677, 1221

Shafee, R., McKinney, J. C., Narayan, R., Tchekhovskoy, A., Gammie, C. F., & McClintock, J. E. 2008, *ApJ*, 687, L25

Shakura, N. I. & Sunyaev, R. A. 1973, *A&A*, 24, 337

— 1976, *MNRAS*, 175, 613

Shi, J., Krolik, J. H., & Hirose, S. 2010, *ApJ*, 708, 1716

Simon, J. B., Hawley, J. F., & Beckwith, K. 2009, *ApJ*, 690, 974

— 2010, *ApJ*, submitted. Astro-ph 1010.0005

Stone, J. M., Hawley, J. F., Gammie, C. F., & Balbus, S. A. 1996, *ApJ*, 463, 656

Thorne, K. S. 1974, *ApJ*, 191, 507

Turner, N. J. 2004, *ApJ*, 605, L45

van Ballegooijen, A. A. 1989, in *Accretion Disks and Magnetic Fields in Astrophysics*, edited by G. Belvedere, vol. 156 of Astrophysics and Space Science Library, 99

Discussion

NEILSEN: If you are suggesting that spin measurements are only of modest reliability beacuse of stress inside the ISCO, could you comment on recent simulations that show that this stress is due to stretching of the magnetic field, and that because the flow is laminar there is no dissipation?

HAWLEY: Additional stress at the ISCO is only one uncertainty in spin measurement and probably not the largest. If stess is due to the "field stretching" work is still done and energy is transferred. Nonzero stress at the ISCO has a significant effect outside of the ISCO (compared to the zero stress assumption). Of course, what will be seen at infinity depends on many factors including dissipation, emission, photon capture by the black hole, etc.

YUAN: Why the magnetic field dissipation timescale is the orbital time if it's intrinsically magnetic reconnection?

HAWLEY: Dissipation is occurring through a turbulent cascade. The eddy turnover time is of the order of Ω^{-1} in the turbulence. In simulations we assume thermalization at the grid scale. Detailed studies of MHD turbulence are required to know.

KOIDE: How large is the magnetic Reynolds number in your GRMHD simulations? If you consider the numerical resistivity, it depends on the Mach number of the simulations?

HAWLEY: In these simulations, magnetic reconnection occurs at the grid scale. Numerical reconnection is not directly analogous to Ohmic resistivity, so it is difficult to assign an specific Reynolds number. Simon *et al.* (ApJ 2009) measured an effective magnetic Reynolds number for the Athena code and obtained $R_n \sim 8000$ for 128 zone resolution with the R_n value proportional to Δx^{-2}.

MEIER: The most reliable BH spin determination are > 0.98. Also, Fender, Gallo, & Belloni (2010) studied several μQSRs and concluded that the BZ process is not active in these systems. Please comment.

HAWLEY: I suspect that the spin measurements are not reliable to the requested level. Also the BZ process dependes not just on the black hole spin, but also on the magnetic field topology and strength.

PE'ER: Given that MRI is most effective for weak B-fields, what is the typical ratio of Poynting energy to matter (kinetic) energy in the jet?

HAWLEY: Although the MRI field has $B^2 < P$ in the disk, the funnel field driving the Poynting flux of order $B^2 = P(\text{disk})$, that is overall pressure equilibrium. Under those circumnsances, we have found that Poynting flux energy can be comparable to the disk luminosity for large black hole spins.

Jets at all Scales
Proceedings IAU Symposium No. 275, 2011
G. E. Romero, R. A. Sunyaev & T. Belloni, eds.

© International Astronomical Union 2011
doi:10.1017/S1743921310015644

Hadronic jet models today

Marek Sikora[1]

[1]Nicolaus Copernicus Astronomical Center,
Bartycka 18, 00-716 Warsaw, Poland
email: sikora@camk.edu.pl

Abstract. The matter content of relativistic jets in AGNs is dominated by a mixture of protons, electrons, and positrons. During dissipative events these particles tap a significant portion of the internal and/or kinetic energy of the jet and convert it into electromagnetic radiation. While leptons – even those with only mildly relativistic energies - can radiate efficiently, protons need to be accelerated up to energies exceeding 10^{16-19} eV to dissipate radiatively a significant amount of energy via either trigerring pair cascades or direct synchrotron emission. Here I review various constraints imposed on the role of hadronic non-adiabatic cooling processes in shaping the high energy spectra of blazars. It will be argued that protons, despite being efficiently accelerated and presumably playing a crucial role in jet dynamics and dissipation of the jet kinetic energy to the internal energy of electrons and positrons, are more likely to remain radiatively passive in AGN jets.

Keywords. galaxies: active, (galaxies:) BL Lacertae objects: general, galaxies: jets, (galaxies:) quasars: general, gamma rays: theory

1. Introduction

Production of relativistic jets in AGN is very likely mediated by rotation of large scale magnetic fields in magnetosphere of black hole and/or accretion disk (Blandford 1976; Lovelace 1976; Phinney 1983; Camenzind 1986; McKinney & Blandford 2009). This leads to a formation of Poynting flux dominated outflows. Those in turn can at some point be converted to the matter dominated jets (Komissarov *et al.* 2007; Tchekhovskoy *et al.* 2009; Lyubarsky 2010; Komissarov 2010), with a terminal Lorentz factor $\Gamma \sim P_j/\dot{M}c^2$ where P_j is the rate of energy extraction from rotating BH and/or accretion disk and $\dot{M}$ is the mass flux. Depending on whether a jet is launched in the BH magnetosphere or by the accretion disk, the mass flux is expected to be dominated by electron/positron pairs or by protons.

Presence of protons in AGN jets is indicated by the low-energy cutoffs in the radio spectra of hot spots in radio-lobes (Blundell *et al.* 2006; Stawarz *et al.* 2007; Godfrey *et al.* 2009) and by circular polarization and Faraday Rotation in radio cores (Vitrishchak *et al.* 2008; Park & Blackman 2010). If so, such protons might manifest their presence also via radiative contribution to high energy spectra in blazars, by synchrotron emission of pair cascades triggered by photo-meson process and by direct synchrotron emission of protons and mesons (Mannheim & Biermann 1992; Mannheim 1993; Rachen & Mészáros 1998; Aharonian 2000). Whether such a contribution can be significant is the main issue of this presentation. We start with a short review of the leading hadronic models (§2); discuss efficiencies of the proton acceleration and cooling processes (§3); present observational constraints on hadronic contribution to blazar spectra (§4). And, finally, we discuss the role of protons they play in the leptonic models regarding the aspect of preheating electrons up to thermal proton energies required to participate in further stochastic acceleration process (§5). Main results are summarized in §6.

2. Hadronic models - basic features

2.1. *Luminous blazars - 'photo-meson' models*

Luminous blazars are hosted by quasars which, when observed away from the axis of the jet, form population of FRII type double radio sources. They are powered by accretion onto BHs with masses typically of the order of 10^9 solar masses and their accretion luminosities are in the range $10^{46} - 10^{47}$erg s^{-1}. However, when such jets are oriented close to the line of sight, these quasars are seen as being dominated by nonthermal radiation of jets, with the apparent luminosities 10^{48-49}erg s^{-1} and spectra dominated by the broad high energy component peaking in the 1-100 MeV band. This radiation shows high amplitude variability on time scales from years down to days and even hours, implying the strong dissipative events taking place not far from the base of a jet. There the energy densities of the magnetic and radiation fields are very large, providing conditions for acceleration of protons up to energies of 10^{9-10} GeV, and for cooling them via inelastic collisions with soft photons (Sikora *et al.* 1987). For typical background radiation fields most of proton energy is converted to mesons and this initiates processes which according to proposers of hadronic models are responsible for γ-ray production in luminous blazars (Mannheim & Biermann 1992).

These processes are dominated by the following channels. In approximately 90% collisions, the produced mesons are pions. In 2/3 of them they are the neural pions (π^0) and in 1/3 of them – the positive pions π^+. Neutral pions almost immediately decay into photons which in turn trigger pair-cascades driven by photon-photon pair production and their synchrotron radiation. Escaping radiation is the product of 3rd and 4th generation of pairs (radiation of the first two are totally converted to e^+e^--pairs). The resulting electromagnetic spectrum is predicted to form the high energy component peaked in the γ-ray band, with a high energy break at $h\nu_{br} \sim 10 - 30$ GeV, where $\tau_{\gamma\gamma \to e^+e^-} \simeq 1$, and low energy tail in the X-ray band with a slope $\alpha_X > 0.5$.

Such spectra can be affected, but not significantly, by electromagnetic output resulting from production of charged pions via processes $p\gamma \to n + \pi^+$ and $n\gamma \to p + \pi^-$. The charged pions decay producing muons ($\pi^\pm \to \mu^\pm + \bar{\nu}_\mu/\nu_\mu$), and the muons decay producing positrons/electrons ($\mu^\pm \to e^\pm + \bar{\nu}_e/\nu_e + \nu_\mu/\bar{\nu}_\mu$). The resulting electrons/positrons join the pair cascade triggered by the decays of the neutral pions and together with the synchrotron emission of muons (Rachen & Mészáros 1998) contribute an additional $\sim$ 6-16 % to the electromagnetic output of the photo-meson process.

2.2. *Low luminosity TeV BL Lac objects - proton-synchrotron models*

The low luminosity BL Lac objects are hosted by radio-galaxies of type FR I. Just as FR II radio sources, they are associated with giant elliptical galaxies, with central BH masses of the order of $10^9 M_\odot$, but with the jet powers at least 3 orders of magnitude lower than in the radio-loud quasars and extremely low accretion luminosities. Doppler boosted nonthermal radiation from their jets reaches TeV energies, and the luminosity peak of the high energy spectral component is located at GeV energies. Resulting from the low radiative environment, energy losses of ultra-relativistic protons in theses objects are presumably dominated by direct synchrotron emission of protons. This mechanism was suggested by Aharonian (2000) to be the primary source of γ-rays in low luminosity BL Lac objects. Main spectral features of such models, as is in the case of synchrotron radiation of electrons, are directly related to the magnetic field intensities and the injection function of relativistic particles (here of protons). They produce photons with average energies $\nu_{p,syn} = (2e/(3\pi m_p c))\gamma_p^2 B'\mathcal{D}$ and energy spectra with the slopes $\alpha = (q_p - 1)/2$ in the slow cooling regime and $\alpha = q_p/2$ in the fast cooling regime, where B' is the intensity of magnetic field in the jet co-moving frame, $\mathcal{D}$ is the Doppler factor, q_p is the index of the proton power-law injection function, $Q_p \propto \gamma_p^{-q_p}$, and α is the index of the radiation energy flux, $F_\nu \propto \nu^{-\alpha}$. Application of those formulae indicates that protons accelerated up to Lorentz factors $\gamma_p \sim 10^{10}$ and cooled in magnetic fields $B' \sim 100$ Gauss may produce spectra reaching TeV energies and with the luminosity peak located near $\nu_{p,syn,max}$. An investigation whether these parameters are feasible – given the constraints imposed on $\gamma_{p,max}$ by jets with the limited power and magnetisation – is given in §4.2.

3. Radiative efficiencies

3.1. *Time scales*

Acceleration.

Time scale of the proton acceleration as measured in the co-moving frame of the flow is

$$t'_{acc} = fR_L/c = \frac{m_p c}{e}\,\frac{f\gamma_p}{B'}\,, \tag{3.1}$$

where R_L is the Larmor radius, B' is the magnetic field intensity, and f is the parameter which in the case of shock acceleration depends on the spectrum of magnetic turbulence and on the velocity of the upstream-flow (Rieger *et al.* 2007) and for mildly relativistic shocks is expected to be at least of the order of 10.

Adiabatic losses

Assuming that cross-sectional radius of the source is of the order of the cross-sectional radius of the jet, R, relativistic plasma moving down the jet with a Lorentz factor Γ undergoes energy losses due to adiabatic expansion. For conical jets with the half-opening angle $\theta_j \equiv R/r < 1/\Gamma$, where r is the distance of the source in a jet, the time scale of the adiabatic losses is (Moderski *et al.* 2003)

$$t'_{ad} = \frac{3}{2}\,\frac{R}{(\theta_j \Gamma)c}\,. \tag{3.2}$$

Photo-meson process

Time scale of the energy losses via the photo-meson process can be estimated using the approximate formula (Begelman *et al.* 1990)

$$t'_{p\gamma} \sim \frac{1}{\langle \sigma_{p\gamma} K_{p\gamma}\rangle c n'_{ph}(\nu' > \nu'_{th})}\,, \tag{3.3}$$

where $\langle \sigma_{p\gamma} K_{p\gamma}\rangle \sim 0.7 \times 10^{-28}\,\mathrm{cm}^2$ is the product of the photo-meson cross-section and inelasticity parameter averaged over the resonant energy range, $n'_{ph}(\nu' > \nu'_{th}) = \int_{\nu'_{th}} n'_\nu \, dn'_\nu$, $h\nu'_{th} \simeq m_\pi c^2/\gamma_p$ is the threshold photon energy and m_π is the rest mass of the pion. The target radiation field is provided by the internal and external sources. The internal seed soft radiation is dominated by synchrotron emission of primary (directly accelerated) electrons, the external one – by re-scattered/reprocessed disk radiation. Approximating the broad synchrotron spectral component by a power-law function with the energy-flux index $\alpha = 1$ and denoting its luminosity by L_s, we have

$$n'_{ph(int)}(\nu' > \nu'_{th}) \sim \frac{L_s}{4\pi m_\pi c^3 R^2 \mathcal{D}^4}\,\gamma_p\,. \tag{3.4}$$

Spectra of external radiation fields are in turn narrow and therefore can be approximated by mono-energetic functions. Hence,

$$n'_{ph(ext)} \sim \frac{\xi L_d \Gamma}{4\pi c r^2\, h\nu_{ext}} = \frac{\xi L_d (\theta_j \Gamma)^2}{4\pi c R^2 \Gamma\, h\nu_{ext}} \quad \text{for } \gamma_p > m_\pi c^2/h\nu_{ext}\,, \tag{3.5}$$

and $n'_{ph,ext} = 0$ for $\gamma_p < m_\pi c^2/(\Gamma h \nu_{ext})$, where L_d is the luminosity of the accretion disk and ξ is the fraction of this luminosity rescattered/reprocesssed on a spatial scale corresponding with a distance r of the source in a jet.

Synchrotron emission

Time scale of the proton cooling via the synchrotron process is

$$t'_{p,syn} = \frac{3}{4} \left(\frac{m_p}{m_e}\right)^3 \frac{m_e c}{\sigma_T u'_B} \frac{1}{\gamma_p},\tag{3.6}$$

where $u'_B = B'^2/(8\pi)$ is the magnetic energy density.

3.2. *'Mono-energetic' efficiencies*

In order to illustrate efficiencies of the proton acceleration and cooling processes we introduce their dimensionless rates, as scaled by the adiabatic losses rates, $\tau_i \equiv t'^{-1}_i/t'_{ad}$. They are:

$$\tau_{acc} \simeq 4.8 \times 10^{-7} \frac{B'R}{f(\theta_j\Gamma)\gamma_p} \simeq 0.8 \frac{(\sigma/0.1)^{1/2} L_{j,47}^{1/2}}{(f/10)\,(\Gamma/10)\,(\theta_j\Gamma)} \frac{1}{\gamma_{p,10}},\tag{3.7}$$

$$\tau_{p\gamma}^{(int)} \sim 1.2 \times 10^{-36} \frac{L_s\,\gamma_p}{(\theta_j\Gamma)\,\mathcal{D}^4\,R} \simeq 12 \frac{L_{s,47}}{(\theta_j\Gamma)\,(\Gamma/10)^4 R_{16}} \gamma_{p,10},\tag{3.8}$$

$$\tau_{p\gamma}^{(ext)} \sim 2.8 \times 10^{-40} \frac{\xi L_d\,(\theta_j\Gamma)}{h\nu_{ext}\,\Gamma R} \simeq 0.2 \frac{(\xi L_d)_{45}\,(\theta_j\Gamma)}{(h\nu_{ext}/10\text{eV})\,(\Gamma/10)\,R_{16}},\tag{3.9}$$

and

$$\tau_{p,syn} = 1.1 \times 10^{-29} \frac{RB'^2\,\gamma_p}{\theta_j\Gamma} \simeq 0.3 \frac{(\sigma/0.1)L_{j,47}}{(\theta_j\Gamma)\,(\Gamma/10)^2\,R_{16}} \gamma_{p,10},\tag{3.10}$$

where $L_B \simeq cu'_B\pi R^2\Gamma^2 = \sigma L_j/(1+\sigma)$ is the magnetic energy flux, $L_j = L_M + L_B$ is the total jet power, $\sigma \equiv L_B/L_M$, and L_M is the energy flux of the rest mass. Assuming $\sigma < 1$ we use approximation $L_B \sim \sigma L_j$.

The 'scaled' quantities, $L_{j,47}$, $\gamma_{p,10}$, R_{16}, $L_{s,47}$, and $(\xi L_d)_{46}$ are defined in the usual way, i.e. $X_n \equiv X/10^n$.

3.3. *Maximal proton energies*

Maximal proton energies – if limited only by adiabatic losses – can be found from $\tau_{acc} = 1$ to be

$$\gamma_p(\tau_{acc} = 1) \simeq 8 \times 10^9 \frac{(\sigma/0.1)^{1/2} L_{j,47}^{1/2}}{(f/10)\,(\Gamma/10)\,(\theta_j\Gamma)},\tag{3.11}$$

Stronger limits are imposed if dominant energy losses are non-adiabatic, i.e. for $\tau_{cool} = \tau_{p\gamma} + \tau_{p,syn} > 1$. For energy losses dominated by photo-meson process with the target radiation field provided by internal sources or by proton-synchrotron radiation

$$\gamma_{p,max} = \gamma_p(\tau_{acc} = 1)\,\text{Min}[1; \sqrt{R/R_c}].\tag{3.12}$$

where in the 1st case (photo-meson process)

$$R_c^{(p\gamma)} \simeq 9.5 \times 10^{16} \frac{L_{s,47}\,(\sigma/0.1)^{1/2} L_{j,47}^{1/2}}{(\theta_j\Gamma)\,(\Gamma/10)^5\,(f/10)} \,[\text{cm}].\tag{3.13}$$

and in the 2nd case (proton-synchrotron radiation)

$$R_c^{(syn)} \simeq 2.3 \times 10^{15} \frac{(\sigma/0.1)^{3/2} L_{j,47}^{3/2}}{(f/10)\,(\theta_j\Gamma)\,(\Gamma/10)^3} \,[\text{cm}].\tag{3.14}$$

For energy losses dominated by photo-meson process with the target radiation field provided by external sources, $\tau_{p\gamma}^{(ext)}$ is predicted to be lower than unity for any proton energy and therefore $\gamma_{p,max}$ is not expected to be affected.

3.4. *Total efficiencies*

Radiative efficiency of a given cooling process can be estimated using formula

$$\eta_i \simeq \frac{\int_1^{\gamma_{p,max}} \eta_i(\gamma_p) Q_{\gamma_p} \gamma_p \, d\gamma_p}{\int_1^{\gamma_{p,max}} Q_{\gamma_p} \gamma_p \, d\gamma_p}, \tag{3.15}$$

where $\eta_i(\gamma_p) \simeq \mathrm{Min}[\tau_i; 1]$ and Q_{γ_p} is the proton injection function.

We calculate such efficiencies below assuming power-law injection of protons $Q_{\gamma_p} \propto \gamma_p^{-q_p}$ with $q_p = 2$. This specific value of the index is chosen because efficiencies obtained for $q_p = 2$ provide upper limits of efficiencies available for $q_p > 2$ being predicted by theoretical models of a diffusive acceleration of particles in mildly relativistic shocks and supported by the slopes of synchrotron spectra produced in the IR-Optical bands by primary (directly) accelerated electrons/positrons.

For energy losses dominated by photo-meson process with internally produced seed photons and by proton-synchrotron emission, we obtain

$$\eta_i \simeq \frac{1 + \ln\left(\gamma_{p,max}/\gamma_{p1}\right)}{\ln \gamma_{p,max}} \text{ for } R < R_c, \tag{3.16}$$

and $\eta_i < 1/\ln \gamma_{p,max}$ for $R > R_c$, where in case of 'internal' photo-meson process, R_c is given by Eq.(13) and

$$\gamma_{p1}^{(p\gamma)} = \gamma_p(\tau_{p\gamma}^{(int)} = 1) \simeq 8.3 \times 10^8 \frac{(\theta_j \Gamma)) \, (\Gamma/10)^4 \, R_{16}}{L_{s,47}}, \tag{3.17}$$

while in case of proton-synchrotron emission, R_c is given by Eq.(14) and

$$\gamma_{p1}^{(syn)} = \gamma_p(\tau_{p,syn}) \simeq 3.4 \times 10^{10} \frac{(\theta_j \Gamma) \, (\Gamma/10)^2 \, R_{16}}{(\sigma/0.1) L_{j,47}}. \tag{3.18}$$

For energy losses dominated by photo-meson process with externally produced seed photons, at any distance larger than

$$r = \frac{\Gamma R(\tau_{p,\gamma}^{(ext)} = 1)}{(\theta_j \Gamma)} \simeq 1.8 \times 10^{16} \frac{(h\nu_{ext}/10\mathrm{eV}) \, (\Gamma/10)^2}{(\xi L_d)_{45} \, (\theta_j \Gamma)^2} \text{ [cm]} \tag{3.19}$$

$\tau_{p,\gamma}^{(ext)} \leqslant 1$ and the efficiency is

$$\eta_{p\gamma}^{(ext)} \simeq \tau_{p\gamma}^{(ext)} \frac{\ln\left(\gamma_{p,max}/\gamma_{p,th}\right)}{\ln \gamma_{p,max}}, \tag{3.20}$$

where $\gamma_{p,th} = m_\pi c^2/(\Gamma h\nu_{ext}) \simeq 1.4 \times 10^6/((\Gamma/10)(h\nu_{ext}/10\mathrm{eV}))$.

4. Observational constraints

4.1. *Luminous blazars*

In order to account for the γ-ray luminosities of powerful blazars, radiative efficiency has to be

$$\eta_i > 0.3 \frac{L_{\gamma,48}}{(\Gamma/10)^2 \, (\eta_{diss}/0.3) \, L_{j,47}}, \tag{4.1}$$

64   M. Sikora

where η_{diss} is the fraction of the jet energy flux dissipated in the 'blazar zone'. As it can be verified using approximate formulae presented in §3.4, such efficiency is difficult to reach even for protons injected with the energy distribution slopes $q_p = 2$. For our fiducial parameters it is about 10 % for the photo-meson process with intenally produced seed photons and much less for others. For slope $q_p \sim 2.4$ the efficiency drops to $\eta \sim 10^{-3}$.

Hadronic models may have also problems to explain very hard X-ray spectra of luminous blazars. Those blazars often have slopes $\alpha_x < 0.5$ (see Table 1 in Sikora *et al.* 2009) and in order to explain such spectra by synchrotron radiation of secondary $e^\pm$ – products of the cascades powered by hadrons – one needs to assume inefficient cooling of ultra-relativistic electrons/positrons up to energies

$$\gamma_e > 8.0 \times 10^5 \frac{(h\nu/100\mathrm{keV})^{1/2}}{B'\,(\Gamma/10)}. \tag{4.2}$$

Inefficient cooling of such energetic electrons/positrons implies very weak magnetic fields and, therefore, puts strong constraints on the efficiency of the proton acceleration and on efficiency of photo-meson energy losses via limitation of the maximal proton energy. Furthermore, as Sikora *et al.* (2009) demonstrated, in order to avoid overproduction of X-rays in these magnetically weak sources by SSC radiation of primary electrons, it is necessary to assume the source sizes of the order of parsecs, and for such sources the efficiency of the photo-meson process is further reduced.

4.2. *Low luminosity BL Lac objects*

In these objects, because of low radiation energy densities – both in jets themselves and in the surroundings of the jet – the non-adiabatic energy losses of protons are presumably dominated by the proton-synchrotron mechanism. However, noting that such objects are hosted by weak radio-galaxies, with the jet powers $L_j \leqslant 10^{44}\mathrm{erg\ s^{-1}}$, efficiency of synchrotron-proton models is also expected to be strongly reduced because weaker magnetic fields. In order to keep them at reasonable level more compact sources must be assumed. However, even in the case of most relativistic protons, the required size of the source to provide sufficiently strong magnetic fields for efficient cooling is unreasonably small (see Eq. (3.14),

$$R_c \sim 0.7 \times 10^{11} \frac{(\sigma/0.1)^{3/2} L_{j,44}^{3/2}}{(f/10)\,(\theta_j\Gamma)\,(\Gamma/10)^3}\ [\mathrm{cm}]. \tag{4.3}$$

This is 3 orders less than the gravitational radius of the BH with mass $M_{BH} \sim 10^9 M_\odot$. Considering the minimal cross-sectional radius of a jet to be $R \sim 10^{15}\mathrm{cm}$, which for $\theta_j \sim 1/\Gamma$ corresponds with a distance 100 gravitational radii of the $10^9 M_\odot$ BH – required to be at least of this order to accelerate the jet up to $\Gamma \sim 10$ (Komissarov *et al.* 2007) – we can find using Eqs. (3.10) and (3.11) that

$$\tau_{p,syn}(\gamma_{p,max}) \simeq 0.7 \times 10^{-4} \frac{(\sigma/0.1)L_{j,44})^{3/2}}{(f/1)\,(\theta_j\Gamma)\,(\Gamma/10)^3\,R_{15}}. \tag{4.4}$$

This indicates that even for such extreme parameters as $f \sim 1$ and $\sigma \sim 1$ the efficiency is too small to explain γ-ray luminosities $L_\gamma \sim 10^{44}\mathrm{erg\ s^{-1}}$, unless one assumes very hard ($q_p < 1$) proton injection spectra and adopts significantly larger total jet power.

The main purpose of proton-synchrotron models was to explain the relatively stable shape of the TeV spectra in variable low luminosity BL Lac objects (Aharonian 2000). Obviously, the critical issue of such models is whether protons can reach sufficiently large energies to produce synchrotron spectra extending up to TeV energies. Combining formulae for average synchrotron photon energies, magnetic energy flux, and maximal proton energies (Eq. 3.11) gives

$$h\nu_{p,syn,max} \simeq 3 \times 10^{-5} \frac{(\sigma/0.1)^{3/2} L_{j,44}^{3/2}}{(f/10)^2 \, (\Gamma/10)^2 \, (\theta_j \Gamma)^2 \, R_{15}} \; [\text{TeV}].$$ (4.5)

One can see that spectra may extend to TeV energies only if assuming $f \sim 1$, $\sigma \sim 1$, and jet powers $L_j > 10^{45}$erg s^{-1}.

5. Protons in leptonic models

Disproving interpretation of high energy spectra of blazars produced via hadronic models does not disprove the presence of protons in AGN jets. They simply are expected to be radiatively inefficient but are likely to dominate the jet energy flux and strongly affect the dynamics of dissipation processes via shocks in the regions of low magnetization parameter ($\sigma < 0.1$). In this scenario, a large fraction of dissipated energy must be converted to relativistic electrons, otherwise leptonic models will be inefficient despite high radiative efficiencies of energetic electrons. However in order to get electrons to participate together with protons in the diffusive shock acceleration process, electrons must be first preheated up to average energy (to be strict – up to average momentum) of protons heated by randomization and compression in the shocked plasma. Several scenarios have been investigated for such electron preheating, both analytically and in PIC-simulations (Amato & Arons 2006; Amano & Hoshino 2009; Sironi & Spitkovsky 2010). Some of them indicate the formation of a power-law energy distribution with the slope $1 < q_e < 2$. If it is true, then the number of electrons joining protons in the stochastic acceleration process can be lower than the number of protons by significant factor. Hence, the high efficiency of blazar radiation may indicate that $q_e < 1$, or that there is a significant pair content.

In the ERC (External-Radiation-Compton) models (Sikora *et al.* 1994) of γ-ray production in luminous blazars such a low-energy tail should be observed in the 30keV–1 MeV spectral energy range. Unfortunately, these bands are observationally poorly covered, particularly above 20 keV. A number of blazars have been detected up to $\sim$ 50keV, by INTEGRAL (Beckmann *et al.* 2009), by Swift/BAT (Ghisellini *et al.* 2010), and by others (see Table 1 in Sikora *et al.* 2007). Some of them have spectral slopes corresponding with $1 < q_e < 2$. This, together with the location of the break in the energy range $1 - 10$MeV as suggested by CGRO/COMPTEL and OSSE observations (Zhang *et al.* 2005; McNaron-Brown *et al.* 1995), may indicate a moderate ($n_e/n_p \sim 10$) pair content, implied also by studies of the bulk-Compton and Compton-rocket effects (Sikora & Madejski 2000; Ghisellini & Tavecchio 2010). At energies < 20 keV most blazars have softer X-ray spectra, but they can result from the contribution of the SSC process and/or from the superposition of X-rays produced at different locations in a jet, such as the orphan X-ray outburst detected in 3C279 (Abdo *et al.* 2010).

6. Conclusions

For realistic jet powers ($L_j \leqslant L_{Edd}$ in luminous blazars and $L_j \leqslant 10^{-3} L_{Edd}$ in low luminosity BL Lac objects), limited magnetization ($\sigma \leqslant 0.1$ is required to allow formation of strong shocks), and moderately steep proton injection function ($q_p \geqslant 2$ is suggested by IR-optical spectra), the hadronic models *fail to*:

• reproduce γ-ray luminosities of blazars;

- explain formation of very hard X-ray spectra in luminous blazars;
- provide the spectral extension up to TeV energies in low luminosity blazars.

Nevertheless, as indicated by several independent observations, protons *are* present in AGN jets and presumably play key role in dissipation processes in shocks. In particular, they transfer a fraction of the dissipated energy to electrons/positrons helping them to reach threshold energies for further acceleration by stochastic mechanisms and produce the observed γ-ray spectra via the ERC and SSC scenarios. However, significant pair content may be required to achieve reasonable effiecincy of that energy transfer.

Acknowledgements

I thank G. Madejski, K. Nalewajko and L. Stawarz for helpful comments. The work was supported by Polish grant MNiSW NN203 301635.

References

Abdo, A. A., *et al.* 2010, *Nature*, 463, 919
Aharonian, F. A. 2000, *New Astron.*, 5, 377
Amato, E. & Arons, J. 2006, *ApJ*, 653, 325
Amano, T. & Hoshino, M. 2009, *ApJ*, 690, 244
Beckmann, V., Ricci, C., & Soldi, S. 2010, *arXiv:* 0912.2254
Begelman, M. C., Rudak, B., & Sikora, M. 1990, *ApJ*, 362, 38
Blandford, R. D. 1976, *MNRAS*, 176, 465
Blundell, K. M., Fabian, A. C., Crawford, C. S., *et al.* 2006, *ApJ*, 644, L13
Camenzind, M. 1986, A&A, 156, 137
Ghisellini, G., Della Ceca, R., Volonteri, M., *et al.* 2010, *MNRAS*, 405, 387
Ghisellini, G. & Tavecchio, F. 2010, *arXiv:* 1008.1982
Godfrey, L. E. H., Bicknell, G. V., Lovell, J. E. J., *et al.* 2009, *ApJ*, 695, 707
Komissarov, S. S. 2010, *arXiv:* 1006.2242
Komissarov, S. S., Barkov, M. V., Vlahakis, N., & Königl, A. 2007, *MNRAS*, 380, 51
Lovelace, R. V. E. 1976, *Nature*, 262, 649
Lyubrasky, Y. E. 2010, MNRAS, 402, 353
Mannheim 1993 1993, A&A, 269, 67
Mannheim, K. & Biermann. P. L. 1992, *A&A*, 253, L21
McKinney, J. C. & Blandford, R. D. 2009, *MNRAS*, 394, L126
McNaron-Brown, K., Johnson, W. N., Jung, G. V., *et al.* 1995, *Apj*, 451, 575
Moderski, R., Sikora, M., & Błażejowski, M. 2003, *A&A*, 406, 855
Park, K. & Blackman, E. G. 2010, *MNRAS*, 403, 1993
Phinney, E. E. 1983, PhD Thesis, Cambridge University
Rachen, J. P. & Mészáros, P. 1998, *Phys.Rev.D*, 58, 123005
Rieger, F. M., Bosch-Ramon, V., & Duffy, P. 2007, *ApSS*, 309, 119
Sikora, M., Begelman, M. C., & Rees, M. J. 1994, *ApJ*, 421, 153
Sikora, M., Kirk, J., Begelman, M., & Schneider, P. 1987, *ApJ*, 320, L81
Sikora, M. & Madejski 2000, *ApJ*, 534, 109
Sikora, M., Stawarz, L., Moderski, M., Nalewajko, K., & Madejski, G. M. 2009, *ApJ*, 704, 38
Sironi, L. & Spitkovsky, A. 2010, *arXiv:* 1009.0024
Stawarz, L., Cheung, C. C., Harris, D. E., & Ostrowski, M. 2007, *ApJ*, 662, 213
Tchekhovskoy, A., McKinney, J. C., & Narayan, R. 2009, ApJ, 699, 1789
Vitrishchak, V. M., Gabuzda, D. C., Algaba, *et al.* 2008, *MNRAS*, 391, 124
Zhang, S., Collmar, W., & Schönfelder, V. 2005 *A&A*, 444, 767

Discussion

BEDNAREK: Can curvature energy losses of protons be important in AGNs?

SIKORA: They cannot be. This is because following particle acceleration processes in shocks or reconnection regions, relativistic protons are injected with a broad distribution of pitch angles. Such protons (as well as all other charged particles) spiral around magnetic field lines rather than slide on them.

YUAN: There are two kinds of B field in a jet. One is a large scale helical field, another being turbulent in a shock front. When we calculate radiation spectrum, we use turbulent one, but when we calculate polarization, we seem to use the large scale ordered field.

SIKORA: Properly calculated polarization must take into account both, the tangled/turbulent magnetic fields compressed and/or generated in a shock, as well, as the large scale magnetic fields transmitted through the shock.

DERMER: Acceleration to high energies is faster for ions than protons. How do your conclusions change if you consider Fe rather than p?

SIKORA: For approximately solar abundances of AGN plasmas the number of heavy nuclei is $\gg Z$ times smaller than of protons. Therefore, despite the fact that heavy nuclei are accelerated Z times faster, they contribute to radiative processes much less than protons and our main conclusion that hadronic models cannot reproduce γ-ray luminosities of blazars remains valid.

PIRAN: Auger indicates that UHECRs are nuclei. It is much easier to accelerate nuclei. Nuclei will diffuse in the intergalactic field and can propagate only up to ~ 10 Mpc. This suggestes that Cen A is the main source of UHECRs if those are nuclei and if the intergalactic magnetic field is $> 10^{-9}$ G (Piran *et al.* 2010).

SIKORA: Yes, I agree that Cen A may contribute significantly to the observed UHECRs.

Jets at all Scales
Proceedings IAU Symposium No. 275, 2011
G. E. Romero, R. A. Sunyaev & T. Belloni, eds.

© International Astronomical Union 2011
doi:10.1017/S1743921310015656

The jet in the galactic center: An ideal laboratory for magnetohydrodynamics and general relativity

Heino Falcke[1,2], **Sera Markoff**[3], **Geoffrey C. Bower**[4], **Charles F. Gammie**[5,6], **Monika Mościbrodzka**[5], **and Dipankar Maitra**[7]

[1] Department of Astrophysics, Institute for Mathematics, Astrophysics and Particle Physics (IMAPP), Radboud University, Nijmegen, The Netherlands

[2] ASTRON, Oude Hoogeveensedijk 4, 7991 PD Dwingeloo, The Netherlands

[3] Astronomical Institute "Anton Pannekoek", University of Amsterdam, The Netherlands

[4] Astronomy Department & Radio Astronomy Lab, UC Berkeley, USA

[5] Department of Physics, University of Illinois, Urbana, Illinois, USA

[6] Astronomy Department, University of Illinois, Urbana, Illinois, USA

[7] Department of Astronomy, University of Michigan, Ann Arbor, Michigan, USA

Abstract. Of all possible black hole sources, the event horizon of the Galactic Center black hole, Sgr A*, subtends the largest angular scale on the sky. It is therefore a prime candidate to study and image plasma processes in strong gravity and it even allows imaging of the shadow cast by the event horizon. Recent mm-wave VLBI and radio timing observations as well as numerical GRMHD simulations now have provided several breakthroughs that put Sgr A* back into the focus. Firstly, VLBI observations have now measured the intrinsic size of Sgr A* at multiple frequencies, where the highest frequency measurements have approached the scale of the black hole shadow. Moreover, measurements of the radio variability show a clear time lag between 22 GHz and 43 GHz. The combination of size and timing measurements, allows one to actually measure the flow speed and direction of magnetized plasma at some tens of Schwarzschild radii. This data strongly support a moderately relativistic outflow, consistent with an accelerating jet model. This is compared to recent GRMHD simulation that show the presence of a moderately relativistic outflow coupled to an accretion flow Sgr A*. Further VLBI and timing observations coupled to simulations have the potential to map out the velocity profile from 5-40 Schwarzschild radii and to provide a first glimpse at the appearance of a jet-disk system near the event horizon. Future submm-VLBI experiments would even be able to directly image those processes in strong gravity and directly confirm the presence of an event horizon.

Keywords. accretion, black hole physics, gravitation, MHD, relativity, instrumentation: interferometers, instrumentation: high angular resolution, Galaxy: center, Galaxy: nucleus, galaxies: jets, quasars: general

---

1. Introduction

Based on an analogy to other active galactic nuclei (AGN) Lynden-Bell & Rees (1971) proposed to look for a compact radio source in the center of our own Milky Way, which was then discovered by Balick & Brown (1974) only a little later with the NRAO interferometer at Green Bank – barely beating the team at Westerbork (Ekers *et al.* 1975). This "compact radio source in the Galactic Center" became later known as "Sagittarius A*" (Sgr A*) and by now is the best constrained super-massive black hole candidate we know of to date (see Melia & Falcke 2001; Genzel *et al.* 2010, for a review), with a mass of $4 \times 10^6 M_\odot$ measured from stellar orbits (Schödel *et al.* 2002; Ghez *et al.* 2008; Gillessen *et al.* 2009). Sgr A* serves as an excellent example to understand black holes (and compact radio cores) at very low accretion rates and is a potentially exciting laboratory for the study of General Relativity (GR) itself. In fact, Falcke *et al.* (2000) argued that, what they called the "shadow of the black hole event horizon" could reasonably be detected with mm-wave very long baseline interferometry (VLBI) in this source and many papers have since illuminated various general relativistic effects that could be tested in this way (Broderick & Loeb 2006; Fish *et al.* 2009; Harko *et al.* 2009; Yuan *et al.* 2009; Johannsen & Psaltis 2010; Dexter *et al.* 2010).

However, ever since its discovery, a discussion raged about the exact nature of its emission processes and the region where the observable radiation is emitted. Already Reynolds & McKee (1980) suggested an origin of the radio emission in a jet or wind from a stellar-sized object. Indeed, the properties of Sgr A* resemble those of flat-spectrum compact radio cores in quasars (later also found in X-ray binaries), which were explained by Blandford & Königl (1979) as the $\tau=1$ surfaces (optical depth becomes unity) of powerful relativistic jets. This issue was revisited by Falcke *et al.* (1993); Falcke & Markoff (2000) who explained spectrum and size of Sgr A* as a scaled-down quasar jet from an "AGN on a starvation diet", i.e. a supermassive black hole with very low accretion rate ($\dot{M} < 10^{-7} M_\odot/\mathrm{yr}$). This semi-analytic approach was based based upon a modified Blandford-Königl model (Falcke 1996, taking the longitudinal pressure gradient into account) and the jet-disk symbiosis ansatz (Falcke & Biermann 1995, 1999), which introduced a very simple, yet until to day very effective linear scaling between jet-power and accretion disk rate for black holes $Q_{\mathrm{jet}} = q_{\mathrm{j}}\dot{M}_{\mathrm{disk}}$, where $q_{\mathrm{j}} \sim 3 - 10\%$. Based on the jet-disk symbiosis model, a number of concrete predictions were made (Falcke 1999a) that all have stood the test of time.

Alternatively, it was proposed that the radio emission – and subsequently the emission at other wavelengths – was produced in the accretion flow itself. While standard optically thick, geometrically thin accretion models started to fail in explaining the low infrared flux of Sgr A* (Falcke & Melia 1997; Coker *et al.* 1999), the picture of an optically thin, geometrically thick accretion flow, so-called advection-dominated or radiatively-inefficient accretion flows (ADAFs, RIAFs) started to emerge (Melia 1992; Narayan *et al.* 1998; Quataert & Gruzinov 2000). These models initially had much higher accretion rates, but this was brought down when polarization measurements, using Faraday rotation arguments, (Bower *et al.* 1999; Aitken *et al.* 2000; Bower *et al.* 2005; Marrone *et al.* 2007) showed a low particle density towards Sgr A*. This implied an accretion rate below $10^{-7} M_\odot/\mathrm{yr}$. Hence, our Galactic Center is likely starving as well as radiating inefficiently and producing a jet (Yuan *et al.* 2002).

2. Review of observational progress

In recent years, important progress has come from the detection of Sgr A* in X-rays and near-infrared (Baganoff *et al.* 2001; Genzel *et al.* 2003) and a lot of emphasis has been placed on variability studies of Sgr A* (Herrnstein *et al.* 2004; Falcke 1999b; Porquet *et al.* 2008; Mauerhan *et al.* 2005; Yusef-Zadeh *et al.* 2006; Marrone *et al.* 2008; Eckart *et al.* 2008; Meyer *et al.* 2008; Yusef-Zadeh *et al.* 2009; Dodds-Eden *et al.* 2009, to name just a few). The various campaigns have shown that flares in the near-infrared (NIR) and X-rays are simultaneous within minutes and hence come from the same region. Dodds-Eden *et al.* (2009) favor optically thin synchrotron radiation over inverse Compton as the dominant emission process to explain the remarkable similarity of a particularly bright flare in both bands. On the other hand radio/submm-wave and X-ray/NIR flares do not

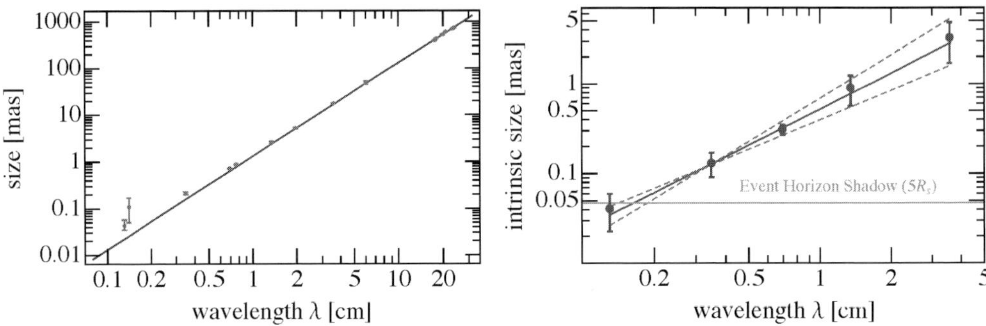

Figure 1. Left: measured major axis size of Sgr A* as function of wavelength measured by various VLBI experiments. Right: Derived intrinsic size of Sgr A* after subtraction of a λ^2 scattering law using all available VLBI data (see Falcke *et al.* (2009) for details). The dashed line indicates the systematic uncertainties due to different normalizations of the scattering law.

seem to be simultaneous and no generally accepted lag has been derived, despite various claims in the literature, which are ranging from several hours to minutes.

Very exciting has also been the progress based on radio observations, constraining any model – and in particular any jet model – much better. Bower *et al.* (2004) were the first to measure the *intrinsic* source size of Sgr A* at 43 GHz directly using Very Long Baseline Interferometry (VLBI), yielding a size of some 20 Schwarzschild radii (R_s) only. This was a true breakthrough, given that the structure of Sgr A* had been washed out due to interstellar scattering for 30 years since its discovery (yielding a measured size that decreases with λ^2 at low frequencies, Fig. 1, left). Further VLBI measurements then provided measurements up to 220 GHz (Shen *et al.* 2005; Doeleman *et al.* 2008) giving sizes down to $4R_s$! Hence, it is clear now that the intrinsic source size decreases with increasing frequency (Fig. 1, right) – an important prediction of the jet model. Direct comparison of the jet model with the actual VLBI data shows a consistency with the compact source size and structure, but does favor edge-on geometries (Markoff *et al.* 2007).

Additional information has come from the measurement of time lags in the radio (Yusef-Zadeh *et al.* 2006, 2008), where it was shown that 43 GHz radio flares precede 22 GHz flares by about 20 minutes. Given that the intrinsic size difference between these two frequencies is about 30 light minutes one arrives at the natural conclusion that the radio emitting plasma flows out with the speed of light (Falcke *et al.* 2009), which is now also suggested by the short-time variability of the source (Yusef-Zadeh *et al.* 2009).

Alternative models based on simple adiabatic expansion of "blobs" which expand sub-relativisitically (Yusef-Zadeh *et al.* 2006; Eckart *et al.* 2009) have only been fitted to the light curves and fail to take the VLBI sizes into account. They predict much smaller source sizes and would postulate two very different emission regions, while fits for adiabatic expansion in a jet (Maitra *et al.* 2009) do reproduce radio light curves and VLBI sizes very well.

In the following we will now focus on the the basic properties of emission from a jet and briefly discuss how future radio observations and modeling will provide us with deeper insight into the formation of jets close to the event horizon.

3. The basic jet model

Given the observational progress any model for Sgr A* nowadays has to explain the spectrum, size, and variability properties of the source. Here we will briefly explain the

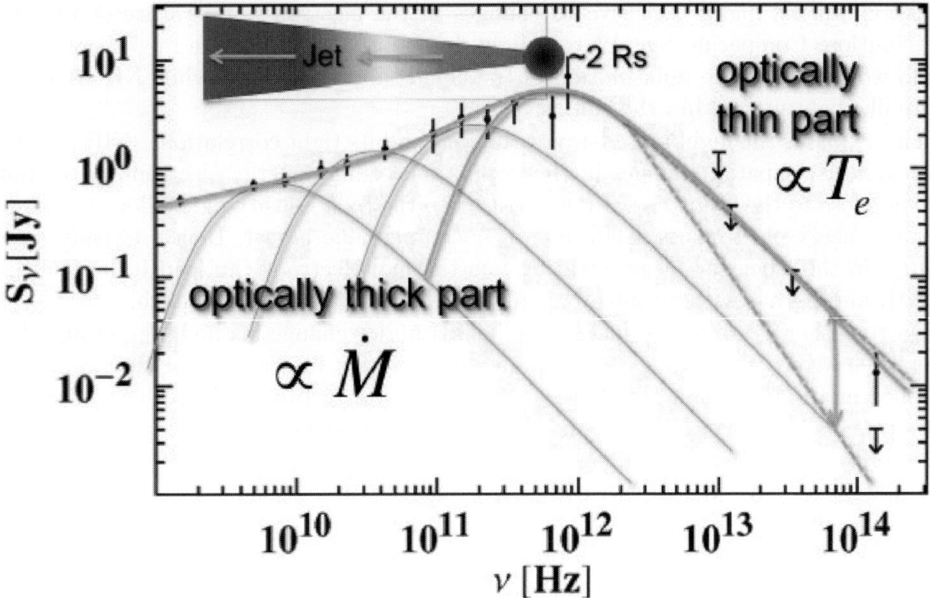

Figure 2. Spectrum of Sgr A* and schematic overview of various contributions from a jet.

physical background, how many of the observed properties arise quite naturally in a jet model. Figure 2 shows the radio spectrum of Sgr A* and a schematic view of the various jet model components. Assume a hot magnetized plasma is ejected close to the black hole from a compact nozzle and expands freely thereafter. The plasma in the nozzle will emit synchrotron radiation peaking between 100 GHz and a few THz due to optical depth effects (almost black body radiation), producing the "submm-bump". If the electron distribution is non-Maxwellian, i.e., has a high-energy power-law tail, optically thin synchrotron radiation will appear, that can easily extend into the near-infrared or even X-ray regime (which could also have a contribution from inverse Compton emission from the submm-bump).

When the plasma leaves the nozzle and expands, it will expand with sound speed and advance with the bulk flow speed, roughly filling a conical Mach cone. Like in a multi-color accretion disk, the synchrotron emission will then peak at lower and lower frequencies the further out the plasma gets. This will naturally lead to an optically thick, flat radio spectrum and a source size $\propto \nu^{-1}$ (Blandford & Königl 1979; Falcke & Biermann 1995). With the jet-disk scaling mentioned above one can then readily explain flux and size of Sgr A* to zeroth order (see Falcke 2003, for a text-book level description). However, a simple Blandford-Königl model is not really appropriate, since close to the black hole the pressure gradient – and likely other processes as well – will lead to an acceleration of the jet, yielding an inverted spectrum and a steeper than ν^{-1} frequency-size relation (Falcke 1996; Falcke & Markoff 2000).

To understand the variability behavior one has to remember that the radio/submm spectrum ($\lesssim$1 THz) is optically thick and thus originates in a superposition of emission components from different spatial regions, peaking at increasingly lower frequencies as their characteristic scale increases. The NIR/X-ray spectrum, on the other hand, is optically thin and comes from one, i.e., the smallest, spatial scale. Already Markoff *et al.* (2001) showed that the jet radio emission will mainly respond to changes in the density (and hence accretion rate, given the assumed jet-disk coupling), while the NIR/X-ray emission will be highly sensitive to changes in the electron temperature or power-law distribution. Comparing Sgr A* to a lightning storm on earth, one might say that the radio would reflect the bulk motion of the clouds and the winds, while NIR/X-rays are more like lightning within the clouds.

This explains the high NIR/X-ray variability and its tight correlation – NIR and X-ray both reflect co-spatial changes in the population of high-energy electrons in the nozzle region, close to the event horizon. Compared to the bulk density of the flow, the number of high-energy electrons is small and their cooling time is fast. Hence, already a small change in the efficiency of accelerating non-thermal electrons can lead to large changes in NIR and X-rays without affecting the overall energy budget too much.

On the other hand, to produce radio flares a major change in the bulk density of low-energy electrons is required, i.e., a change in accretion rate and total jet power. When induced at the foot point the density surge will have to propagate with the flow speed outwards (see Falcke *et al.* 2009; Maitra *et al.* 2009, for detailed modeling). This suggests lower relative flux changes compared to NIR, time lags in the sense that higher-frequencies lead lower ones, and not necessarily a tight connection between major radio/mm–flares on one hand and NIR/X-ray–flares on the other. Of course, it is conceivable that a sudden increase in accretion rate onto the black hole, which is seen as a radio flare, will also lead to more efficient dissipation and particle acceleration. However, in any case NIR/radio delays are likely not easy to interpret and should certainly not be used to estimate, e.g., time scales for adiabatic expansion. On the other hand, time lags within the radio/submm-wave regime could have high diagnostic powers, since they are directly related to the flow speed. Hence, VLBI size and radio monitoring data combined would provide unique information about the acceleration and formation of astrophysical jets and in particular the longitudinal velocity profile.

4. MHD modeling and event horizon imaging

Parallel to the semi-analytic modeling a new breed of MHD models has become available, some of which are now geared towards Sgr A*. These models basically start from first principles following the evolution of a hot magnetized plasma as it is accreted onto a black hole. Initially, jets did not form naturally in these simulations (Stone *et al.* 1996; Hawley 2000), especially when the initial configuration contained only a toroidal magnetic field, while simulations with random or poloidal fields seemed well to be able to produce jet-like outflows (e.g., Koide *et al.* 1999; Meier *et al.* 2001). Nowadays, most MHD simulations include poloidal magnetic field and hence show some evidence for jet-like outflows. Given the excellent observational constraints on scales close to the event horizon, Sgr A* has now become a useful testbed also for numerical MHD simulations (e.g., Ohsuga *et al.* 2005; Noble *et al.* 2007; Chan *et al.* 2009; Mościbrodzka *et al.* 2009; Dexter *et al.* 2010; Hilburn *et al.* 2010).

For example, Mościbrodzka *et al.* (2009) have performed simulations based on an axisymmetric version of the GRMHD code *harm*, applying relativistic radiative transfer (grmonty: Dolence *et al.* 2009) and GR ray tracing (ibothros: Noble *et al.* 2007) during post-processing. These 2D simulations do indeed also show a jet-like outflow, but it appears rather faint. Compared to the early pure GR ray tracing simulations (Falcke *et al.* 2000) the shadow of the black hole remains visible and a significant part of the parameter range of the models can be excluded. In their simulations, Mościbrodzka *et al.* (2009) find that the VLBI sizes and other properties favor edge-on orientations and a fast spinning black hole.

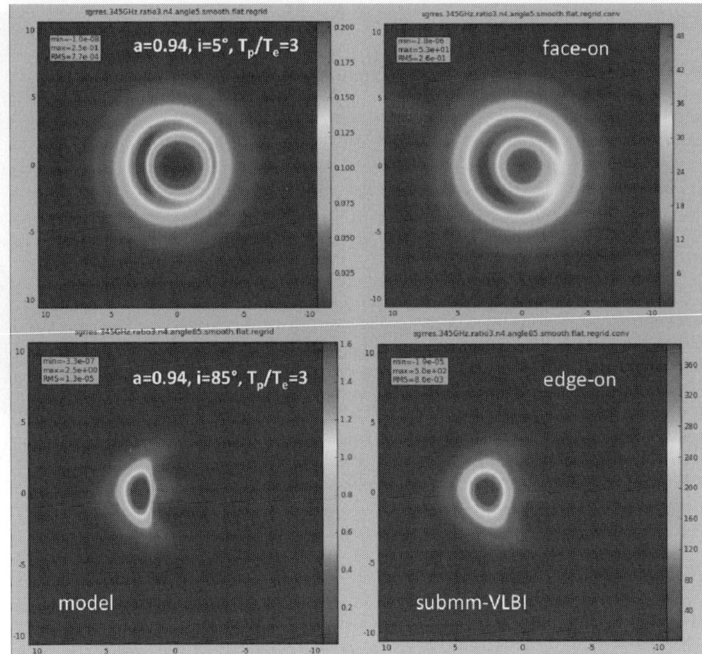

Figure 3. Left: Time-averaged appearance of 2D MHD accretion models at 345 GHz (bottom-left: best-fit model) from Mościbrodzka *et al.* (2009). Right: Reconstructed images with a putative global submm-VLBI array. Top: face-on geometry, Bottom: edge-on geometry. The black hole shadow is clearly visible for face-one orientation but much less so for edge-on.

To demonstrate the potential of this approach, we have performed some simple simulations of future VLBI observations (Figure 3). For this we have taken the predicted intensity distributions at 345 GHz (Fig. 3) from Mościbrodzka *et al.* (2009), smeared them with a Gaussian kernel according to the scattering law (taken from Falcke *et al.* 2009), and passed them through the interferometry simulation tool *simdata2* in *CASA* (McMullin *et al.* 2007). Frequencies and baselines were appropriately scaled to avoid numerical problems and noise contributions were neglected. The array layout (Fig. 4) consisted of a number of existing (sub)mm-wave facilities (CARMA, SMT, JCMT), telescopes which are planned, are under construction or are not yet equipped for VLBI (LLAMA, ALMA, LMT, SPT), and one putative site in Peru to fill in missing baselines.

One can see that indeed such an array would be well matched to the resolution needed to study event-horizon scale structures in Sgr A*. High spin and inclination will produce a rather compact emission region with a size of $\sim 2 - 4R_{\rm g}$ and a shadow that is difficult to see unless one achieves sufficient dynamic range ($\sim 100 : 1$).

It will be interesting to see how this picture changes as the simulations improve. Figure 4 (right) shows a snapshot of a single frame from a preliminary 3D GRMHD simulation with *harm*, again including ray tracing and radiation transport, for a black hole spin of a=0.9. While in the 2D simulation the jet is very dim, to the point that it is not even visible in the emission maps presented here, the jet becomes much more prominent in the 3D case. This is apparently due to an increased electron temperature in the jet, possibly coming from increased acoustic heating. Clearly, the electron temperature in disk and jet plays and important role for the appearance of Sgr A*, but it is also the least understood parameter. Such simulations provide a taste for the future direction the field can evolve to.

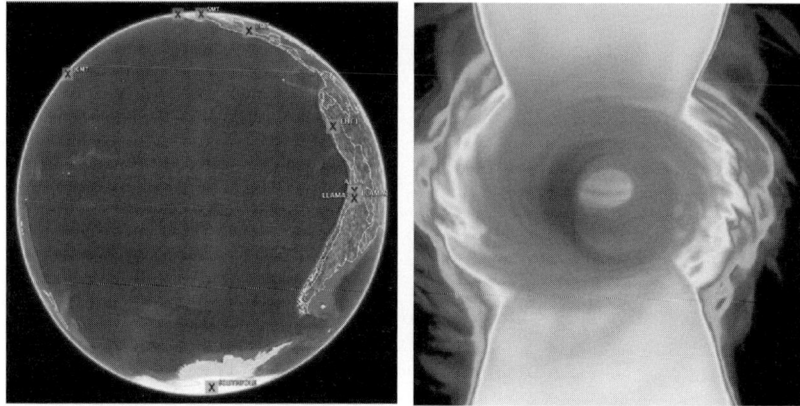

Figure 4. Left: Google-Earth view of the Earth with positions of the submm-VLBI stations used in the simulation as green boxes. Right: Snapshot of a 3D GRMHD simulation with similar parameters as in Fig. 3. Shown is the intensity at 345 GHz. The jet is much more visible, compared to the 2D case.

5. Conclusions

There is no other supermassive black hole with such a wealth of information. The properties are not different from many other radio cores in nearby low-power AGN when appropriately scaled to a very low accretion rate (Falcke & Biermann 1999; Nagar *et al.* 2005; Markoff *et al.* 2008). In fact, within the fundamental plane of black hole activity (Merloni *et al.* 2003; Falcke *et al.* 2004; Körding *et al.* 2006), that has been discussed at length in this meeting, it represents almost an extreme "off-state".

Spectrum, size, and radio time-lags and evolution in Sgr A* strongly suggest a jet-nature of the radio emission, with mildly relativistic outflow speeds. Since the the submm-wave emission comes from the direct proximity of the event horizon, Sgr A* is an ideal source to image the shadow of the event horizon *and* to find out how astrophysical jets are formed – for example by combining mm-VLBI observations and time-lag measurements.

References

Aitken, D. K., Greaves, J., Chrysostomou, A., *et al.* 2000, *ApJL*, 534, L173
Baganoff, F. K., Bautz, M. W., Brandt, W. N., *et al.* 2001, *Nature*, 413, 45
Balick, B. & Brown, R. L. 1974, *ApJ*, 194, 265
Blandford, R. D. & Königl, A. 1979, *ApJ*, 232, 34
Bower, G. C., Backer, D. C., Zhao, J. H., Goss, M., & Falcke, H. 1999, *ApJ*, 521, 582
Bower, G. C., Falcke, H., Herrnstein, R. M., *et al.* 2004, *Science*, 304, 704
Bower, G. C., Falcke, H., Wright, M. C., & Backer, D. C. 2005, *ApJL*, 618, L29
Broderick, A. E. & Loeb, A. 2006, MNRAS, 367, 905
Chan, C., Liu, S., Fryer, C. L., *et al.* 2009, *ApJ*, 701, 521
Coker, R., Melia, F., & Falcke, H. 1999, *ApJ*, 523, 642
Dexter, J., Agol, E., Fragile, P. C., & McKinney, J. C. 2010, *ApJ*, 717, 1092
Dodds-Eden, K., Porquet, D., Trap, G., *et al.* 2009, *ApJ*, 698, 676
Doeleman, S., Weintroub, J., Rogers, A. E. E., & *et al.* 2008, *Nature*, 455, 78
Dolence, J. C., Gammie, C. F., Mościbrodzka, M., & Leung, P. K. 2009, *ApJs*, 184, 387
Eckart, A., Baganoff, F. K., Morris, M. R., *et al.* 2009, A&A, 500, 935
Eckart, A., Baganoff, F. K., Zamaninasab, M., *et al.* 2008, A&A, 479, 625
Ekers, R. D., Goss, W. M., Schwarz, U. J., Downes, D., & Rogstad, D. H. 1975, *A&A*, 43, 159

Falcke, H. 1996, *ApJL*, 464, L67

Falcke, H. 1999a, in ASP Conf. Ser. 186: *The Central Parsecs of the Galaxy*, ed. H. Falcke, A. Cotera, W. Duschl, F. Melia, & M. J. Rieke (San Francisco: Astronomical Society of the Pacific), 148

Falcke, H. 1999b, in ASP Conf. Ser. 186: *The Central Parsecs of the Galaxy*, ed. H. Falcke, A. Cotera, W. Duschl, F. Melia, & M. J. Rieke (San Francisco: Astronomical Society of the Pacific), 113

Falcke, H. 2003, *Radio and X-ray emission from the Galactic black hole*, ed. Falcke, H. & Hehl, F. W. (IoP), 310–342

Falcke, H. & Biermann, P. L. 1995, *A&A*, 293, 665

Falcke, H. & Biermann, P. L. 1999, *A&A*, 342, 49

Falcke, H., Körding, E., & Markoff, S. 2004, *A&A*, 414, 895

Falcke, H., Mannheim, K., & Biermann, P. L. 1993, *A&A*, 278, L1

Falcke, H. & Markoff, S. 2000, *A&A*, 362, 113

Falcke, H., Markoff, S., & Bower, G. C. 2009, *A&A*, 496, 77

Falcke, H. & Melia, F. 1997, *ApJ*, 479, 740

Falcke, H., Melia, F., & Agol, E. 2000, *ApJL*, 528, L13

Fish, V. L., Doeleman, S. S., Broderick, A. E., Loeb, A., & Rogers, A. E. E. 2009, *ApJ*, 706, 1353

Genzel, R., Eisenhauer, F., & Gillessen, S. 2010, ArXiv e-prints

Genzel, R., Schödel, R., Ott, T., *et al.* 2003, *Nature*, 425, 934

Ghez, A. M., Salim, S., Weinberg, N. N., *et al.* 2008, *ApJ*, 689, 1044

Gillessen, S., Eisenhauer, F., Fritz, T. K., *et al.* 2009, *ApJL*, 707, L114

Harko, T., Kovács, Z., & Lobo, F. S. N. 2009, *Class. Quant. Grav.*, 26, 215006

Hawley, J. F. 2000, *ApJ*, 528, 462

Herrnstein, R. M., Zhao, J.-H., Bower, G. C., & Goss, W. M. 2004, AJ, 127, 3399

Hilburn, G., Liang, E., Liu, S., & Li, H. 2010, *MNRAS*, 401, 1620

Johannsen, T. & Psaltis, D. 2010, *ApJ*, 718, 446

Koide, S., Shibata, K., & Kudoh, T. 1999, *ApJ*, 522, 727

Körding, E., Falcke, H., & Corbel, S. 2006, *A&A*, 456, 439

Lynden-Bell, D. & Rees, M. J. 1971, *MNRAS*, 152, 461

Maitra, D., Markoff, S., & Falcke, H. 2009, *A&A*, 508, L13

Markoff, S., Bower, G. C., & Falcke, H. 2007, *MNRAS*, 379, 1519

Markoff, S., Falcke, H., Yuan, F., & Biermann, P. L. 2001, *A&A*, 379, L13

Markoff, S., Nowak, M., Young, A., *et al.* 2008, *ApJ*, 681, 905

Marrone, D. P., Baganoff, F. K., Morris, M. R., *et al.* 2008, *ApJ*, 682, 373

Marrone, D. P., Moran, J. M., Zhao, J.-H., & Rao, R. 2007, *ApJL*, 654, L57

Mauerhan, J. C., Morris, M., Walter, F., & Baganoff, F. K. 2005, *ApJL*, 623, L25

McMullin, J. P., Waters, B., Schiebel, D., Young, W., & Golap, K. 2007, in *Astronomical Society of the Pacific Conference Series*, Vol. 376, *Astronomical Data Analysis Software and Systems XVI*, ed. R. A. Shaw, F. Hill, & D. J. Bell, 127–

Meier, D. L., Koide, S., & Uchida, Y. 2001, Science, 291, 84

Melia, F. 1992, *ApJL*, 387, L25

Melia, F. & Falcke, H. 2001, *ARA&A*, 39, 309

Merloni, A., Heinz, S., & di Matteo, T. 2003, *MNRAS*, 345, 1057

Meyer, L., Do, T., Ghez, A., *et al.* 2008, *ApJL*, 688, L17

Mościbrodzka, M., Gammie, C. F., Dolence, J. C., Shiokawa, H., & Leung, P. K. 2009, *ApJ*, 706, 497

Nagar, N. M., Falcke, H., & Wilson, A. S. 2005, *A&A*, 435, 521

Narayan, R., Mahadevan, R., Grindlay, J. E., Popham, R. G., & Gammie, C. 1998, *ApJ*, 492, 554

Noble, S. C., Leung, P. K., Gammie, C. F., & Book, L. G. 2007, *Class. Quant. Grav.*, 24, 259

Ohsuga, K., Kato, Y., & Mineshige, S. 2005, *ApJ*, 627, 782

Porquet, D., Grosso, N., Predehl, P., *et al.* 2008, *A&A*, 488, 549

Quataert, E. & Gruzinov, A. 2000, *ApJ*, 539, 809

Reynolds, S. P. & McKee, C. F. 1980, *ApJ*, 239, 893
Schödel, R., Ott, T., Genzel, R., *et al.* 2002, *Nature*, 419, 694
Shen, Z.-Q., Lo, K. Y., Liang, M.-C., Ho, P. T. P., & Zhao, J.-H. 2005, *Nature*, 438, 62
Stone, J. M., Hawley, J. F., Gammie, C. F., & Balbus, S. A. 1996, *ApJ*, 463, 656
Yuan, F., Markoff, S., & Falcke, H. 2002, *A&A*, 383, 854
Yuan, Y., Cao, X., Huang, L., & Shen, Z. 2009, *ApJ*, 699, 722
Yusef-Zadeh, F., Bushouse, H., Wardle, M., *et al.* 2009, *ApJ*, 706, 348
Yusef-Zadeh, F., Roberts, D., Wardle, M., Heinke, C. O., & Bower, G. C. 2006, *ApJ*, 650, 189
Yusef-Zadeh, F., Wardle, M., Heinke, C., *et al.* 2008, *ApJ*, 682, 361

Discussion

FENDT: What is your heating model for the electrons? This is beyond MHD.

FALCKE: It is essentially local numerical dissipation.

RODRIGUEZ: Is there any chance of seeing the jet a few arcsec away from Sgr A* using an ultra-sensitive radio array like the EVLA?

FALCKE: Unfortunately does the Sgr A* size grow roughly with λ while the scattering size grows with λ^2, hence the core will always be hidden at lower frequencies. Perhaps one could see some interaction region further out, but teh Galactic Center is a very complicated region.

CORBEL: Do you have any idea why we live in a galaxy with the faintest LLAGN?

FALCKE: Estimates for the accretion rate based on environmental factors are all many orders of magnitude above the current value. Maybe we live in a special time. For example, the nearby supernova remnant Sgr A East has probably just swept over the black hole a few thousand years ago. Such local effects could easily disrupt a steady accretion flow.

DE GOUVEIA DAL PINO: Do you think that the time lag technique will be able to separate the complex motions that may occur at launching like, for instance, rotational velocity from the v_z component?

FALCKE: Good question – helical motion is indeed quite commonly observed in quasar jets with VLBI. Rotation will lead to differential beaming and hence will imprint a wobble as a function of frequency on top of the flare spectrum. Maybe with very sensitive broad-band monitoring, e.g. with ALMA and SKA in the future, might such subtleties be detectable.

Jets at all scales
Proceedings IAU Symposium No. 275, 2011
G. E. Romero, R. A. Sunyaev & T. Belloni, eds.

© International Astronomical Union 2011
doi:10.1017/S1743921310015668

Waves in Poynting-flux dominated jets

John G. Kirk and Iwona Mochol

Max-Planck-Institut für Kernphysik, Postfach 10 39 80, 69029 Heidelberg, Germany
email: john.kirk@mpi-hd.mpg.de, iwona.mochol@mpi-hd.mpg.de

Abstract. High-energy emission from blazars is thought to arise in a relativistic jet launched by a supermassive black hole. The rapid variability of the emission suggests that structure of length scale smaller than the gravitational radius of the central black hole is imprinted on the jet as it is launched, and modulates the radiation released after it has been accelerated to high Lorentz factor. We describe a mechanism which can account for the acceleration of the jet, and for the rapid variability of the radiation, based on the propagation characteristics of nonlinear waves in charge-starved, polar jets. These exhibit a delayed acceleration phase, that kicks-in when the inertia associated with the wave currents becomes important. The time structure imprinted on the jet at launch modulates the photons produced by the accelerating jet provided that the electromagnetic cascade in the black-hole magnetosphere is not prolific.

Keywords. MHD – plasmas – waves – BL Lacertae objects: general

---

1. Introduction

The blazar PKS 2155-304 has been observed to exhibit remarkably rapid variability at very high flux levels (Aharonian *et al.* 2007, HESS Collaboration 2010). In the most extreme flare, structure in the jet is implied that is roughly one hundred times smaller than the gravitational radius $r_\mathrm{g} = GM/c^2$, with M the black-hole mass (Begelman *et al.* 2008). This structure is presumably imprinted as the jet leaves the black-hole magnetosphere, suggesting that the axisymmetric, ideal MHD approximation might not provide an adequate description. In analogy with pulsar magnetospheres (Spitkovsky 2006) it seems reasonably to assume that a generalised form of the axisymmetric Blandford-Znajek mechanism (Blandford & Znajek 1977) causes the non-axisymmetric magnetosphere of a rotating black hole to drive a Poynting-flux dominated jet, part of which might propagate in the low-density funnel around the rotation axis. In the following, we develop this idea by examining the propagation characteristics of nonlinear electromagnetic waves above the polar regions of a rotating black hole.

2. Jet parameters

Matter accreting onto a rotating black hole fails to penetrate into a conical region or "funnel" around the rotation axis (De Villiers & Hawley 2003; McKinney & Gammie 2004), where the density is presumably dominated by an electron-positron plasma produced in an electromagnetic cascade. The physical conditions in this outflowing plasma can be described by three dimensionless parameters: (i) the nonlinearity or strength parameter $a = a_0 \, (c/\omega r)$, where $\omega/2\pi$ is the wave frequency and

$$a_0 = \left[4\pi e^2 L/\left(m^2 c^5 \Omega_\mathrm{s}\right)\right]^{1/2} = 3.4 \times 10^{14} L_{46}^{1/2} \qquad (2.1)$$

(L is the luminosity in solid angle Ω_s, L_{46} the "4π" luminosity in units of 10^{46} erg/s, which, for PKS 2155-304 is approximately unity), (ii) the mass-loading of the wind $\mu = L/\dot{M}c^2$ (Michel 1969), with $\dot{M}$ the mass flux in the jet, (iii) the magnetisation parameter

σ which is the ratio of the energy flux carried by electromagnetic fields to that carried by particles. For monoenergetic electrons and positrons of Lorentz factor γ,

$$\sigma = (\mu/\gamma) - 1 \qquad (2.2)$$

We specify its value at the "launching" radius, inside of which the ideal MHD approximation is assumed to hold, and denote this (constant) quantity by σ_0. The particle Lorentz factor at the launching point is then $\gamma_0 = \mu/(\sigma_0 + 1)$. An alternative, more intuitive measure of the mass-loading, is the pair multiplicity κ, defined as in pulsar physics (e.g., Lyubarsky & Kirk 2001). Here, we specify κ by its value $\kappa_{r_{\mathrm{g}}}$ at $r = r_{\mathrm{g}}$:

$$\kappa_{r_{\mathrm{g}}} \approx \frac{a_0}{4\mu} \left(\frac{c}{\omega r_{\mathrm{g}}} \right) \qquad (2.3)$$

3. The two-fluid model

We adopt the model of two cold, charged fluids $(e^{\pm})$ in a Kerr metric, following Khanna (1998). To keep the analysis tractable, only transverse waves are treated. We also assume radial propagation, circular polarisation and charge neutrality. It follows that the fluids have the same radial component of the four-velocity, which we write as a dimensionless momentum $p_{\parallel}$. The transverse components of the fluid momenta are equal and opposite and are denoted by $\pm p_{\perp}$. The evolution is described by the continuity equation, the equations of motion of the fluids and the two relevant Maxwell equations (Faraday's law and Ampère's law) in the small-wavelength approximation, $c/\omega r \sim \epsilon \ll 1$. Although smaller, we assume that the wavelength is of the same order in ϵ as the gravitational radius r_{g}. Under these conditions general relativistic effects do not appear in the governing equations. In the lowest-order system of equations, we search for large-amplitude plane-wave solutions using the approach introduced by Akhiezer & Polovin (1956).

3.1. *Plane waves*

Expressing all wave quantities in terms of the phase ϕ enables the fluid equations to be converted into a set of ordinary differential equations, which can be integrated to find nonlinear wave solutions. Concentrating on waves with subluminal phase velocity, $\beta_{\mathrm{w}} < 1$, one finds that the particles are in resonance with the wave in the sense that the radial components of the fluid velocities equal the phase velocity of the radially propagating wave. In this case the plasma current is directed along the magnetic field and, to lowest order, the forces exerted on the fluids by the fields vanish. Viewed from a frame that moves radially with speed β_{w}, the electric field vanishes, and the wave is simply a static magnetic field of constant magnitude whose direction rotates through 2π radians over one wavelength. The rate at which the B-vector rotates is arbitrary, being determined by the dependence of the fluid proper density n on phase. In the following, we select the simplest case, where n, $|B|^2$ and $|p_{\perp}|^2$ are all constant and the wave is a monochromatic magnetic shear, in rotating (complex) coordinates: $B \propto p_{\perp} \propto \mathrm{e}^{\pm i\phi}$.

3.2. *Radial evolution*

The radial dependence of the wave goes through three phases. Defining the dimensionless radius $R = \mu\omega r/(a_0 c)$, on finds that $|p_{\perp}| \ll 1$ when $R \ll \mu/\sigma_0$, and the wave is essentially a cold MHD structure in which the inertia associated with the current is negligible. There is no acceleration of either the wave speed or the fluids in this regime, and the magnetisation parameter σ remains constant at its initial value. At intermediate radii,

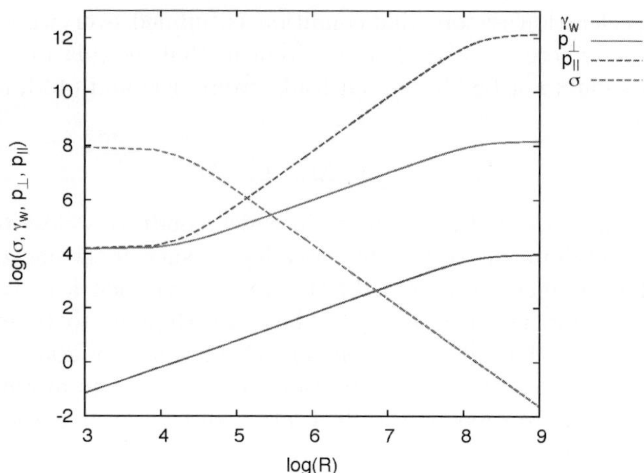

Figure 1. The magnetisation parameter σ, radial and transverse fluid momenta $p_\parallel$ and $|p_\perp|$ and Lorentz factor of the wave, $\gamma_{\rm w}$, as functions of the dimensionless radius $R = \mu r \omega / (a_0 c)$, for $\sigma_0 = 10^8$ and $\mu = 1.4 \times 10^{12}$.

where $|p_\perp| \gg 1$, and $\sigma \gg 1$, one readily finds an approximate solution, provided the wave is launched not too close to the sound speed (i.e., for $\gamma_0 \gg \sigma_0^{1/2} \gg 1$):

$$\gamma_{\rm w} \approx R \qquad\qquad \gamma \approx R^2 \sigma_0 / \mu$$
$$\sigma \approx \mu^2 / \left(R^2 \sigma_0\right) \quad |p_\perp| \approx R \sigma_0 / \mu \ . \tag{3.1}$$

This solution is valid until σ becomes of the order of unity, which occurs at $R \approx \mu/\sqrt{\sigma_0}$ Finally, at large radius, $R \gg \mu/\sqrt{\sigma_0}$, only kinetic energy remains: $\gamma \approx \mu$, $\gamma_{\rm w} \approx \mu/\sqrt{\sigma_0}$, $|p_\perp| \approx \sqrt{\sigma_0}$. This behaviour is illustrated in Fig. 1, which plots the solutions for values of interest in the case of PKS 2155-304: $\sigma_0 = 10^8$, $\mu = 1.4 \times 10^{12}$.

4. Application to blazar variability

The radius $r_{\rm acc}$ at which acceleration begins, corresponding to $R \approx \mu/\sigma_0$, is

$$r_{\rm acc} \quad\approx\quad r_{\rm g} a_0^{1/3} \kappa_{r_{\rm g}}^{2/3} \left(\omega r_{\rm g}/c\right)^{-1/3} = 1.2\, \Delta t_{100}^{1/3}\, \kappa_{r_{\rm g}}^{2/3}\, L_{46}^{1/6}\, M_9^{2/3}\ {\rm pc} \ , \tag{4.1}$$

where the variation timescale in units of $100\,{\rm s}$ is $\Delta t_{100} = (2\pi/\omega)/(100\,{\rm s})$ and $M_9 = M/10^9 {\rm M}_\odot$. We analyse the case of mildly supersonic launch: $\sigma_0 \approx \mu^{2/3}$, so that

$$\gamma_0 \quad=\quad 6.5 \times 10^3\, \Delta t_{100}^{1/3}\, \kappa_{r_{\rm g}}^{-1/3}\, L_{46}^{1/6}\, M_9^{-1/3} \ . \tag{4.2}$$

The solutions presented in section 3 eventually convert all of the Poynting flux to kinetic energy flux at large radius. However, in the case of blazars, this is unlikely to be realised, since the resulting Lorentz factor $(= \mu)$ is very large. Instead, dissipative processes, which we have so far neglected, are likely to intervene. These include (i) instabilities in the wave-solution, which can lead to an effective friction between the fluids, equivalent to a finite resistivity in MHD language, (ii) interaction of the jet with the external medium, causing collimation, or matter entrainment, and (iii) interaction with ambient photons, causing jet deceleration and giving rise to inverse Compton photons.

If jet radiation is modulated at the wave frequency, it is straightforward to derive a criterion on the size of the emitting region (assumed $\sim r$) and the Lorentz factor of the jet such that the modulations are not washed out by the travel time of the signal across the source (Michel 1971, Arons 1979, Kirk *et al.* 2002): $\gamma_{\rm w}^2\, 2\pi c/\omega > r$. Since

$\gamma_{\rm w} \propto r$ in the acceleration region, this condition is fulfilled everywhere, provided it is satisfied at $r = r_{\rm acc}$, where $\gamma_{\rm w} = \gamma_0$. The requirement that modulation on a timescale of $100\Delta t_{100}$ seconds should not be filtered out leads, from (4.1) and (4.2), to an upper limit on the multiplicity:

$$\kappa_{r_{\rm g}} \quad < \quad 14\,\Delta t_{100}\,L_{46}^{1/8}\,M_9^{-1} \tag{4.3}$$

Equation (4.3) implies that those sources in which conditions close to the black hole might favour the development of prolific cascades — such as strong disk emission — should not exhibit extreme variability. In the absence of such a cascade, the magnetosphere is able to support vacuum gaps. These are thought to produce highly non-stationary discharges in the pulsar environment (Levinson et al 2005, Timokhin 2010). Thus, if our model is correct, the rapid variations of the TeV gamma-ray flux of a blazar could be a direct manifestation of the fluctuating discharge of a vacuum gap in the black-hole magnetosphere.

5. Summary and conclusions

Circularly polarized, large-amplitude waves are likely to be generated in the polar-cap regions of a rotating black hole endowed with an external magnetic field. These regions are believed to be charge-starved, so that the ideal MHD description might fail because the inertia of the current carriers cannot be neglected. In the case of the subluminal magnetic shear which we analyse above, this happens when $p_\perp$ becomes relativistic, and it causes the wave and jet to accelerate. Additional effects, such as collimation or interaction with an external radiation field may enhance the acceleration process. However, a detailed investigation remains to be done.

We do not address the question of the mechanism of γ-ray production in the accelerating jet. However, if, as seems inevitable, the emission is modulated at the wave frequency, we show that the large spatial extent of the emission region fails to wash out this signal, provided that the density of pairs injected in the magnetosphere by an electromagnetic cascade is sufficiently small.

References

Aharonian, F. *et al.* 2007, ApJ, 664, L71
Akhiezer, A. I. & Polovin, R. V. 1956, Sov. Phys. JETP, 3, 696
Arons, J. 1979, Space Sci. Rev., 24, 437
Begelman, M. C., Fabian, A. C., & Rees, M. J. 2008, MNRAS, 384, L19
Blandford, R. D. & Znajek, R. L. 1977, MNRAS, 179, 433
De Villiers, J. & Hawley, J. F. 2003, ApJ, 592, 1060
HESS Collaboration, Abramowski, A. *et al.* 2010, arXiv:1005.3702
Khanna, R. 1998, MNRAS, 294, 673
Kirk, J. G., Skjæraasen, O., & Gallant, Y. A. 2002, A&A, 388, L29
Levinson, A., Melrose, Judge, D. A., & Luo, Q. 2005, ApJ, 631, 456
Lyubarsky, Y. & Kirk, J. G. 2001, ApJ, 547, 437
McKinney, J. C. & Gammie, C. F. 2004, ApJ, 611, 977
Medin, Z. & Lai, D. 2010, MNRAS, 406, 1379
Michel, F. C. 1969, ApJ, 158, 727
Michel, F. C. 1971, Comments on Astrophysics and Space Physics, 3, 80
Spitkovsky, A. 2006, ApJ, 648:L51–L54
Timokhin, A. N. 2010, arXiv:1006.2384

Discussion

PE'ER:

(*a*) Is there any evidence of an annihilation line?

(*b*) Are you considering any protons that can carry kinetic energy and momentum?

KIRK:

(*a*) Not to my knowledge. I also would not expect a signature, because the pair density in the jet is very low.

(*b*) I have assumed a pair plasma. If protons are present (perhaps introduced into the funnel by decaying neutrons) the structure of the nonlinear wave modes is more complicated, but could, in principle, be analysed using the same approach.

LEVINSON: How is it possible to have variations on a time scale shorter than $r_{\rm g}/c$?

KIRK: I don't think there is any fundamental reason for excluding structure in a non-axisymmetric magnetic field on scales less than $r_{\rm g}$, at a position close to $r = r_{\rm g}$. I just assume these are transported radially to large distance. The transverse length scale is stretched, but the radial length scale remains fixed in the lab. frame for a relativistic flow. Provided the Lorentz factor is large, the photons that are produced at large radius carry the information on the radial length scale to the observer, and this is not smeared out by the large transverse size of the source.

Jets at all Scales
Proceedings IAU Symposium No. 275, 2011
G. E. Romero, R. A. Sunyaev & T. Belloni, eds.

© International Astronomical Union 2011
doi:10.1017/S174392131001567X

Jets at lowest mass accretion rates

Dipankar Maitra[1], Andrew Cantrell[2], Sera Markoff[3], Heino Falcke[4], Jon Miller[1], and Charles Bailyn[2]

[1] Dept. of Astronomy, University of Michigan,
Ann Arbor, MI, USA 48109
email: **dmaitra,jonmm@umich.edu**

[2] Dept. of Astronomy, Yale University
New Haven, CT, USA 06511
email: **andrew.cantrell,charles.bailyn@yale.edu**

[3] Astronomical Institute "Anton Pannekoek", University of Amsterdam
1098 XH Amsterdam, The Netherlands
email: **s.b.markoff@uva.nl**

[4] Dept. of Astronomy, Radboud University
6500 GL Nijmegen, The Netherlands
email: **H.Falcke@astro.ru.nl**

Abstract. We present results of recent observations and theoretical modeling of data from black holes accreting at very low luminosities ($L/L_{Edd} \lesssim 10^{-8}$). We discuss our newly developed time-dependent model for episodic ejection of relativistic plasma within a jet framework, and a successful application of this model to describe the origin of radio flares seen in Sgr A*, the Galactic center black hole. Both the observed time lags and size-frequency relationships are reproduced well by the model. We also discuss results from new Spitzer data of the stellar black hole X-ray binary system *A0620–00*. Complemented by long term SMARTS monitoring, these observations indicate that once the contribution from the accretion disk and the donor star are properly included, the residual mid-IR spectral energy distribution of *A0620–00* is quite flat and consistent with a non-thermal origin. The results above suggest that a significant fraction of the observed spectral energy distribution originating near black holes accreting at low luminosities could result from a mildly relativistic outflow. The fact that these outflows are seen in both stellar-mass black holes as well as in supermassive black holes at the heart of AGNs strengthens our expectation that accretion and jet physics scales with mass.

Keywords. black hole physics, accretion, accretion disks, acceleration of particles, Galaxy: nucleus, radiation mechanisms: general

1. Introduction

Collimated relativistic outflows or "jets" are observed to be very closely associated with compact accretors like black holes and neutron stars where $GM/Rc^2 \lesssim 1$. Such jets are known to emit across a broad range of the electromagnetic spectrum, from radio to X-rays, and perhaps even γ-rays. Sometimes jets can be imaged directly, but often their presence is inferred, such as from flat to slightly inverted spectral energy distribution (SED) in radio through IR (see, e.g., Markoff *et al.* 2001; Maitra *et al.* 2009a; Vila & Romero 2010), correlation between fluxes (see, e.g., Gallo *et al.* 2003; Russell *et al.* 2010), and scaling relations connecting radio and X-ray luminosities with the black hole mass (Merloni *et al.* 2003; Falcke et al 2004; Gültekin *et al.* 2009).

In X-ray binaries (XRB), compact steady jets appear to turn off (on) as the source makes a transition from nonthermal to thermal (thermal to nonthermal) X-ray state, at luminosities typically a few percent of the Eddington luminosity. See e.g. Homan &

Belloni (2005), Remillard & McClintock (2006) for details of X-ray states, and Fender (2006) for a review on XRB jets. However, there is no clear consensus as to any lower luminosity limit for jets, or even about the mode of accretion at low mass accretion rates ($\dot{M}$). While it is generally agreed that the radiative efficiency of the emitting plasma is low (e.g. Narayan & Yi 1994; Blandford & Begelman 1999) at low $\dot{M}$, it is not even clear whether the observed emission originates in an inflow or an outflow.

Obviously instrumental limitations become important at the lowest luminosities. Nevertheless, ever-improving technological advances are making it possible to detect sources at luminosities as low as 10^{-9} L_{Edd}, a limit which is constantly decreasing. Results of these fascinating experiments point strongly to the presence of outflows even at the lowest luminosities, for stellar as well as supermassive black holes. Here we present a case study of two such sources: Sagittarius A*, the supermassive black hole at the center of our galaxy, and the stellar mass X-ray binary A0620–00, both of which suggest the presence of a jet at luminosities $\lesssim 10^{-8}$ L_{Edd}.

2. Sagittarius A*

Located 8.4 ± 0.6 kpc (Reid *et al.* 2009) away at the Galactic Center, Sagittarius A* (hereafter Sgr A*) is the nearest "supermassive" black hole with a mass of $\sim 4 \times 10^6$ $M_\odot$ (Ghez *et al.* 2008). Despite its relative proximity, the source of emission from Sgr A* is still unknown. This is partly due to the extremely high extinction in the direction of the Galactic Center, and partly to the fact that with a bolometric luminosity $\sim 10^{-9}L_{Edd}$, Sgr A* is extremely under-luminous compared to other active galactic nuclei (AGN). However, flares and data from multiwavelength campaigns provide important clues about the nature of emission from Sgr A*. We have developed a time-dependent jet model which for the first time allows comparing the model predictions with radio flare data from Sgr A*. Taking into account relevant cooling mechanisms, we calculate the frequency-dependent time lags and photosphere size and compare with recent observations (see Fig. 1). Details of the modeling is presented in Maitra *et al.* (2009b).

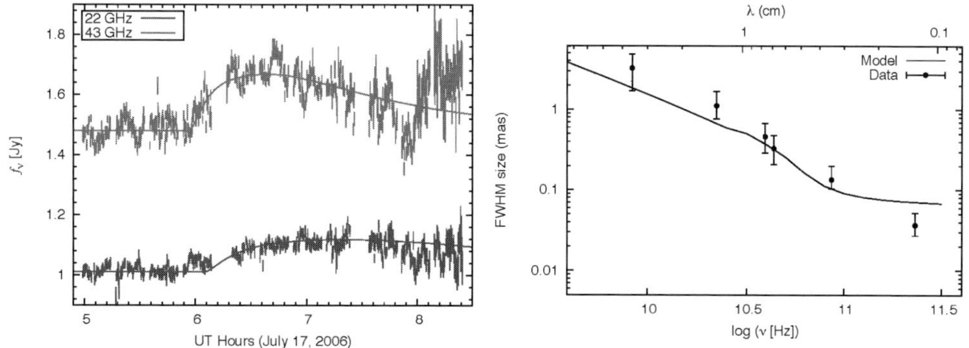

Figure 1. *Left:* Comparison of model with Sgr A* data for the flare on 2006 July 17 at 43 and 22 GHz. The 43 GHz data with error bars from Yusef-Zadeh *et al.* (2008) are shown in red, and 22 GHz data+model in blue. We model the flare that created a peak in the 43 GHz light curve close to 6.5 hours UT. *Right:* Comparison of model-predicted frequency-size relationship (solid line) with observations. The data are from Bower *et al.* (2004), Shen *et al.* (2005), and Doeleman *et al.* (2008). Figures taken from Maitra *et al.* (2009b).

3. A0620-00

Since its massive outburst in 1975, A0620–00 has been in quiescence, with luminosities around 10^{-8} L_{Edd} . It has been detected in quiescence in X-rays, radio, IR, near-IR and optical using various space-based as well as ground-based instruments.

Despite easy detection in the optical and near-IR, estimating the mass of the compact accretor has been quite a challenge. An accurate (dynamical) measurement of the accretor mass requires correct knowledge of the orbital inclination. Historically attempts to measure the inclination from light curves have given widely varying results with $38° < i < 75°$. However, recently Cantrell *et al.* (2008, 2010) have analyzed ~ 30 years of A0620–00 optical and near-IR (OIR) data and showed that previous disagreements on the inclination of A0620–00 were caused by (1) improper estimation of the disk flux (which can be significant even in quiescence), (2) nightly variability in light curve shape, and (3) not recognizing that even during quiescence, the OIR light curve shows three distinct optical states (named *passive, loop,* and *active,* and characterized by magnitude, color, and aperiodic variability). Of these three states the passive state is the faintest, shows consistent periodicity and is perhaps the closest to true quiescence. Consistent estimates of the binary parameters are obtained when the above effects are properly accounted for, and the best estimates of (i, M) are $(51.0° \pm 0.9°, 6.6 \pm 0.25$ M$_\odot$). Accurate knowledge of the stellar contribution also allows determining any nonstellar residuals in the photometric data.

Here we present results of SMARTS long-term photometric monitoring of A0620–00 in OIR and a *Spitzer*/IRAC observation in 2006 November 26. Both the long-term monitoring as well as the *Spitzer* snapshot suggest the presence of nonthermal emission, presumably from a jet (see Fig. 2).

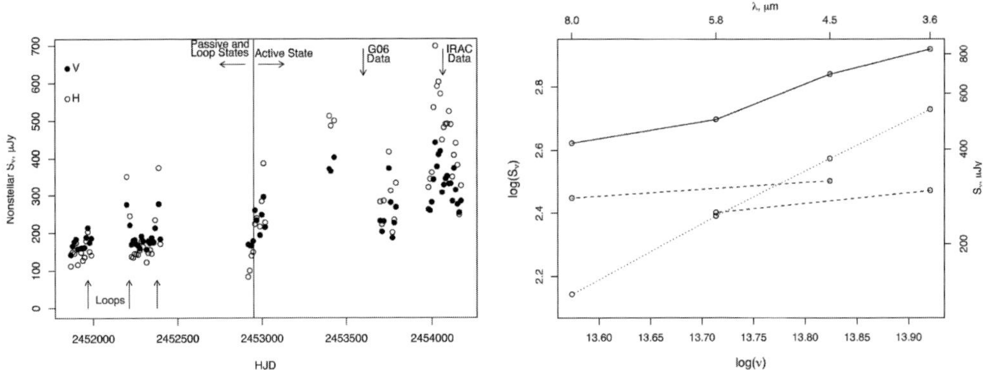

Figure 2. *Left:* Dereddened, nonstellar V- (filled points) and H-band (open points) flux densities from SMARTS monitoring of A0620–00. OIR state transitions as described in Cantrell *et al.* 2010 are marked, as are the dates of the *Spitzer*/IRAC data and the Gallo *et al.* (2006) data. A0620–00 is consistently redder in the active state than in the passive state, possibly due to a nonthermal contribution in the active state. *Right:* The solid line shows the IRAC SED of A0620–00. Also shown are the expected contributions from the donor + accretion disk (dotted line), and the residual "excess" (observed SED minus the donor+disk; dashed lines). The dashed lines connect the two pairs of bands which were obtained simultaneously. The excess has a relatively flat spectrum with slope ~ 0.3, suggesting that it could be at least partially nonthermal.

4. Discussion and Conclusions

Recent multiwavelength data show that relativistic outflows or jets can be important at luminosities as low as $\sim 10^{-9} L_{\mathrm{Edd}}$, and hence affect the evolution and energetics of compact accretors. Here we have presented a case study of two low-luminosity accreting black hole sources, Sgr A* and A0620–00. The two sources differ by five orders of magnitude in mass, but still the data point toward the presence of an outflow in both.

In particular, for Sgr A* we show that a time-dependent relativistic jet model can successfully reproduce:

- The quiescent broadband spectral energy distribution of Sgr A*,
- The observed 22 and 43 GHz light curve morphologies and time lags,
- The frequency-size relationship.

These results suggest that the observed radio emission from Sgr A* is most easily explained by a stratified, optically thick, mildly relativistic jet. Frequency-dependent measurements of time-lags and intrinsic source size provide strong constraints on the bulk motion of the jet plasma.

For A0620–00, radio, *Spitzer*, and SMARTS data obtained during quiescence suggest:

- Significant excess above the expected thermal emission from donor + accretion disk; spectral index of the excess -0.9$<\alpha<$-0.5 ($f_\nu \sim \nu^\alpha$), i.e. consistency with optically thin jet synchrotron emission in the NIR,
- Nearly flat IR spectral index from *Spitzer* observations is suggestive of optically thick synchrotron emission in IR.

Techniques like Doppler tomograms of quiescent systems will lead to better understanding of gas flow in the accretion disc of A0620–00 and also that of the gas stream-disc impact point. Future VLBI, submm and time-lag measurements over a broader range of wavelengths will reveal greater details about ongoing physical processes within a few gravitational radii for sources like Sgr A* and M87.

References

Blandford, R. & Begelman M. 1999, *MNRAS*, 303, L1

Bower, G. *et al.* 2004, *Science*, 304, 704

Cantrell, A. *et al.* 2008, *ApJ*, 673, L159

Cantrell, A. *et al.* 2010, *ApJ*, 710, 1127

Doeleman, S. *et al.* 2008, *Nature*, 455, 78

Falcke, H. *et al.* 2004, *A&A*, 414, 895

Fender, R. 2006, In "Compact Stellar X-ray Sources", Cambridge Astrophysics Series, 39, 381

Gallo, E. *et al.* 2003, *MNRAS*, 344, 60

Ghez, A. *et al.* 2008, *ApJ*, 689, 1044

Gültekin, K. *et al.* 2009, *ApJ*, 706, 404

Homan, J. & Belloni, T. 2005, *Ap&SS*, 300, 107

Maitra, D. *et al.* 2009a, *MNRAS*, 398, 1638

Maitra, D. *et al.* 2009b, *A&A*, 508, L13

Markoff, S. *et al.* 2001, *A&A*, 372, L25

Merloni, A. *et al.* 2003, *MNRAS*, 345, 1057

Narayan, R. & Yi I. 1994, *ApJ*, 428, L13

Reid, M. *et al.* 2009, *ApJ*, 700, 137

Remillard, R. & McClintock J. 2006, *ARA&A*, 44, 49

Russell, D. *et al.* 2010, *MNRAS*, 405, 1759

Shen, Z.-Q. *et al.* 2005, *Nature*, 438, 62

Vila, G. & Romero G. 2010, *MNRAS*, 403, 1457

Yusef-Zadeh, F. *et al.* 2008, *ApJ*, 682, 361

Discussion

BENAGLIA: Could you think on radio polarization observations to confirm the non-thermal emission?

MAITRA: Not with instruments available at the time: for the radio source associated with A0620-00 is a $\sim 50\,\mu$Jy source, and the expected polarization level is of a few %. It might be feasible with the EVLA.

MIRABEL: Some time delays at different wavelengths are observed in the ejecta of GRS 1915+105, which implies that the time delays are determined by the adiabatic expansion of the plasma irrespective of the mass of the black hole. How do you get rid of the scattering to determine the frequency-size relation?

MAITRA: Yes, the VLT observations of Sgr A* have shown the existence of NIR flares. The NIR/X-ray flares trace particle (re)organization and cooling very close to the black hole, while marginally affecting the optically thick radio flux (that we modelled here). The presence of tge scattering screen in the direction of Sgr A* had indeed been the biggest obstacle towards obtaining the intrinsic size of Sgr A*. However, this problem has recently been overcome with the innovation of VLBI closure amplitude analysis (e.g. Doeleman *et al.* 2001). Closure amplitude method gives an accurate measurement of the source size. The scatter broadening follows a λ^2 law and this has been measured fairly accurately for Sgr A*, in turn giving good estimates of the intrinsic size of Sgr A*.

RUSSEL: What is the spectral index of the optical-IR flares in quiscence of A0620?

MAITRA: According to the preliminary done so far, equally good fits to all data can be obtained assuming any jet slope in the range $-0.9 < \alpha < -0.5$, where $f_\nu \sim \nu^\alpha$.

MILLER-JONES: What is the time resolution of the A0620-00 light curves and how finely can you bin them?

MAITRA: The minimum exposure time required to obtain good signal-to-noise for A0620-00 using the CTIO+SMARTS telescope s about 10 minutes. Right now we observe A0620 approximately once every night, but the number of observations per night could be increased (at least temporarily) if needed.

Jets at all Scales
Proceedings IAU Symposium No. 275, 2011
G. E. Romero, R. A. Sunyaev & T. Belloni, eds.

© International Astronomical Union 2011
doi:10.1017/S1743921310015681

Modelling magnetically dominated and radiatively cooling jets

Martín Huarte-Espinosa[1,2], **Adam Frank**[1] and **Eric Blackman**[1]

[1]Department of Physics and Astronomy, University of Rochester,
600 Wilson Boulevard, Rochester, NY, 14627-0171
emails: `martinhe; afrank; blackman @pas.rochester.edu`
[2]Kavli Institute for Cosmology Cambridge, Madingley Road,
Cambridge CB3 0HA, UK

Abstract. Using 3D-MHD Eulerian-grid numerical simulations, we study the formation and evolution of rising magnetic towers propagating into an ambient medium. The towers are generated from a localized injection of pure magnetic energy. No rotation is imposed on the plasma. We compare the evolution of a radiatively cooling tower with an adiabatic one, and find that both bend due to pinch instabilities. Collimation is stronger in the radiative cooling case; the adiabatic tower tends to expand radially. Structural similarities are found between these towers and the millimeter scale magnetic towers produced in laboratory experiments.

Keywords. stars: winds, outflows, ISM: jets and outflows, methods: numerical, MHD

---

1. Introduction

Non-relativistic jets are observed in the vicinities of several Protostellar Objects, Young Stellar Objects (YSOs) and post-AGB stars. Plausible models suggest that jets are launched and collimated by accretion, rotation and magnetic mechanisms in the "central engine" (see Pudritz *et al.* 2007, for a review). The relative extent over which magnetic energy dominates the outflow kinetic energy in a propagating jet has traditionally divided them into two classes: magnetocentrifugal (Blandford & Payne 1982; Ouyed & Pudritz 1997; Blackman et al. 2001; Mohamed & Podsiadlowski 2007), in which magnetic fields only dominate out to the Alfvén radius, or Poynting flux dominated (PFD, Lynden-Bell 1996; Ustyugova *et al.* 2000; Lovelace *et al.* 2002; Nakamura & Meier 2004) in which magnetic fields dominate the jet structure, acting as a magnetic piston over very large distances from the engine. PFD jets carry large electric currents along which generate strong tightly wound helical magnetic fields around the jet axis. Simulations of such jets have found that these magnetic fields play a role in the formation of current-driven kink instabilities and the stabilization of Kelvin-Helmholtz (KH) modes in jets (e.g. see Nakamura & Meier 2004).

The effects of plasma radiative-cooling have been followed only in few simulations of magnetized, kinetic energy-dominated jets. These, have been found to form both thin cocoons and nose cones (e.g. Blondin et al. 1990; Frank et al. 1998), and to be more susceptible to KH instabilities relative to adiabatic jets (Hardee & Stone 1997, and references therein). Correlations between the structure of kinetic energy-dominated jets and their power have been extensively explored. There has been less work on PFD jets.

Magnetic fields with initially primarily poloidal (radial and vertical) geometries anchored to accretion discs were shown to form tall, highly wound and helical magnetic structures, or magnetic towers, that expand vertically when laterally supported in pressure equilibrium with the ambient gas (Lynden-Bell 1996, 2003). The local injection of pure toroidal magnetic energy, without imposing any rotation on the plasma, has been shown to form magnetic towers in laboratory experiments (Lebedev *et al.* 2005. These experiments were modeled with numerical simulations by Ciardi *et al.* 2007) and are analogous to 3D-MHD numerical simulations of AGN jets by Li et al. (2006, and subsequent papers). Here we use a modified version of the implementation of Li et al., in order to study magnetically driven and radiatively-cooling jets, or magnetic towers, motivated by the contexts of Protostellar Objects, YSOs and Planetary Nebulae.

2. Model

We form magnetic towers using 3D-MHD Eulerian-grid numerical simulations by locally injecting pure magnetic energy and compare the evolution of these towers as they propagate into an ambient "interstellar" medium (ISM) for radiative cooling vs. adiabatic cases. No rotation is imposed on the plasma at the base. The ISM gas is modelled with an ideal gas equation of state, a ratio of specific heats of $\gamma = 5/3$, a uniform number density of $100\,\mathrm{cm}^{-3}$, a constant temperature of $10000\,\mathrm{K}$ and null velocity. We start the simulations with a helical magnetic field inside a central cylinder with both radius and height of $50\pi\,\mathrm{AU}$. The helical geometry of the initial magnetic field is described from its vector potential (i.e. $\mathbf{B} = \nabla \times \mathbf{A}$)

$$\mathbf{A}(r,z) = \begin{cases} \frac{r}{4}(\cos 2\,r + 1)(\cos 2\,z + 1)\hat{\phi} + \frac{\alpha}{8}(\cos 2\,r + 1)(\cos 2\,z + 1)\hat{k}, & \text{for } r, z < \pi/2; \\ 0, & \text{for } r, z \geqslant \pi/2, \end{cases}$$

$$(2.1)$$

in cylindrical coordinates. The parameter α is an integer with units of length and determines the ratio of toroidal to poloidal magnetic fluxes, and electric currents. For the initial conditions $\alpha = 3$, but $\alpha = 15$ at all later times.

A cubic computational domain with 128^3 fixed cells is used. Boundary conditions are set to periodic at both $x = \pm 800\,\mathrm{AU}$ and $y = \pm 800\,\mathrm{AU}$, to reflective at $z = 0$, and to outflow at $z = 1600\,\mathrm{AU}$. We use BlueHive†, an IBM parallel cluster of the Center for Research Computing of the University of Rochester, to run simulations for about two weeks, using 32 processors.

Using the above initial and boundary conditions, we solve the equations of radiative-magnetohydrodynamics in three-dimensions with the AMR parallel code *AstroBEAR*‡ (Cunningham *et al.* 2009). A source term is implemented to the induction equation to continuously inject magnetic fields via (2.1), into the computational domain. For numerical stabilization, static gas is also injected into regions where $\mathbf{A}(r, z)$ is not null, but its contribution to both mass and thermal energy is negligible. The ionization of both H and He, the chemistry of H2 and optically thin cooling are considered (in one of our simulations) using tables of Dalgarno & McCray (1972). To focus on the minimalist physics of magnetic tower expansion, no gravitational, viscous or general-relativistic processes are considered. This implementation extends the one of Li et al. (2006) to stellar scales and to the radiative-MHD regime.

3. Results and discussion

Here we discuss some of our preliminary results. Higher resolution simulations, further details and analyses will be presented in Huarte Espinosa, Frank & Blackman 2011 (in prep.).

† https://www.rochester.edu/its/web/wiki/crc/index.php/BlueHive_Cluster
‡ http://www.pas.rochester.edu/~bearclaw/

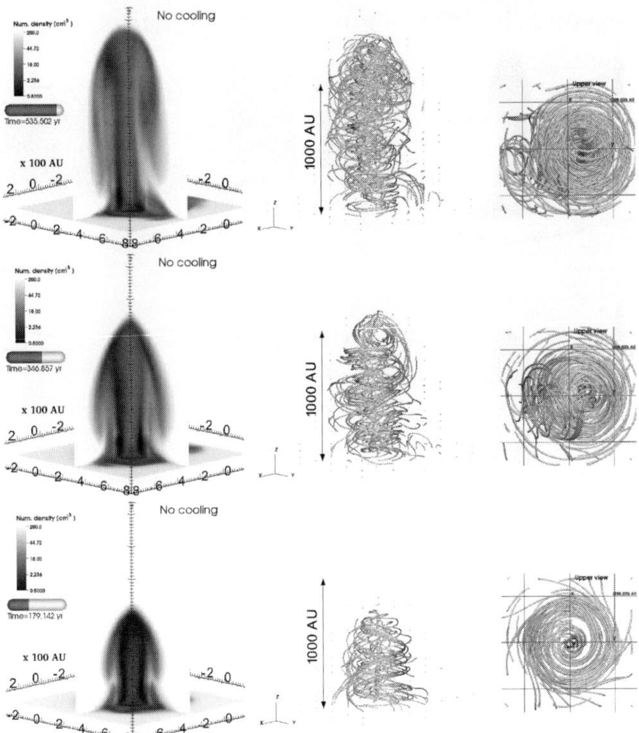

Figure 1. Evolution of the adiabatic magnetic tower. Left: logarithmic density grayscale maps. Middle: magnetic field lines. Right: field lines upper view. Time is the same row-wise. Open field lines are a visualization effect.

The adiabatic and radiatively cooling magnetic towers are shown in Figures 1 and 2, respectively. The ratio of thermal to magnetic pressures takes values within (0.1,1) at regions where $r, z \leqslant 50\pi$ AU, throughout the simulations. Vertical pressure gradients primarily due to the toroidal magnetic field in the tower, cause the plasma to accelerate along the z direction, and a shock is driven in the plasma. A self-pinched cavity of ~ 150 AU is formed in about a few 10 yr by a collimated outflow, the vertical speed of which is $\sim 100\,\mathrm{km\,s^{-1}}$. This "cocoon" is filled with tightly wound helical magnetic fields and gas that is about 0.01 times less dense than that of the ambient. On average, the cocoon thermal pressure is larger than the magnetic pressure by factors that increase radially from 1 to 4, approximately, at $z \sim 300$ AU. This is different than the magnetically dominated jets of Li et al. (2006). We note that, as opposed to other magnetized jet launch simulations (e.g. Shibata & Uchida 1986), no rotation has been imposed on the plasma.

In the cocoon, the toroidal magnetic flux dominates the poloidal flux. A poloidal electric current develops and the radial pressure gradient from the toroidal magnetic field causes radial expansion of the adiabatic tower (see left column of Figure 1) that exceeds its collimating hoop stress. However, Figure 2 shows that cooling suppresses this effect significantly. The upper view of the magnetic field lines (rightmost column in Figures 1 and 2) shows two nested toroidal magnetic surfaces. The ambient pressure collimates the outer magnetic surface, whereas its magnetic tension collimates the field lines of the inner magnetic surface. Such structures resemble the ones seen in the millimetric-scale magnetic towers produced in laboratory experiments by Ciardi *et al.* (2007). Pinch

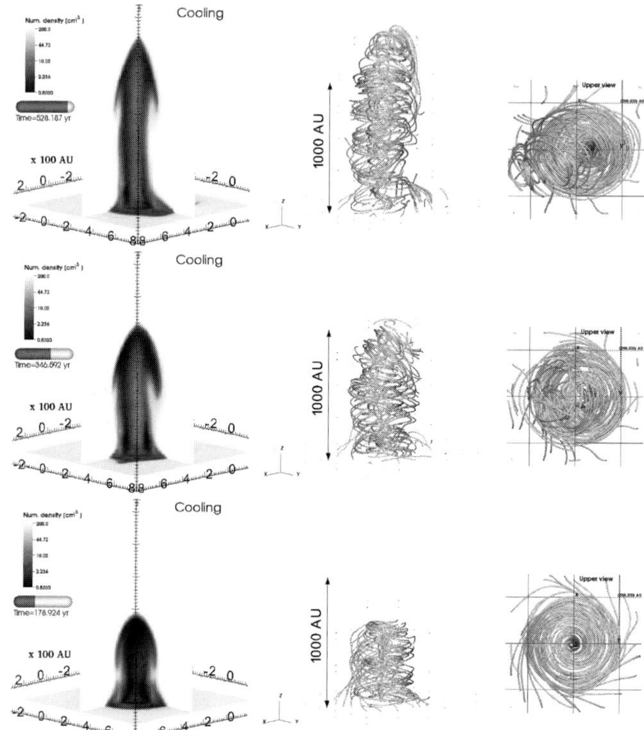

Figure 2. Evolution of the cooling magnetic tower. Panel structure is as in Figure 1.

instabilities cause the towers to bend, and poloidal field lines seem to pile up, particularly in the cooling tower.

References

Blackman, E. G., Frank, A., & Welch, C. 2001, *ApJ*, 546, 288

Blandford, R. D., & Payne, D. G. 1982, *MNRAS*, 199, 883

Blondin, J. M., Fryxell, B. A., & Konigl, A. 1990, *ApJ*, 360, 370

Lebedev, S. V., *et al.* 2005, *MNRAS*, 361, 97

Ciardi, A., *et al.* 2007, *Phys. of Plasmas*, 14, 056501

Cunningham A. J., Frank A., Varnière P., Mitran S., & Jones, T. W. 2009, *ApJS*, 182, 519

Dalgarno A. & McCray R. A. 1972, *ARA&A*, 10, 375

Frank, A., Ryu, D., Jones, T. W., & Noriega-Crespo, A. 1998, *ApJL*, 494, L79

Hardee, P. E. & Stone, J. M. 1997, *ApJ*, 483, 121

Li, H., Lapenta, G., Finn, J. M., Li, S., & Colgate, S. A. 2006, *ApJ*, 643, 92

Lovelace, R. V. E., Li, H., Koldoba, A. V., Ustyugova, G. V., & Romanova, M. M. 2002, *ApJ*, 572, 445

Lynden-Bell, D. 1996, *MNRAS*, 279, 389

Lynden-Bell, D. 2003, *MNRAS*, 341, 1360

Mohamed S. & Podsiadlowski P. 2007, *ASPC*, 372, 397

Nakamura, M. & Meier, D. L. 2004, *ApJ*, 617, 123

Ouyed, R. & Pudritz, R. E. 1997, *ApJ*, 482, 712

Pudritz, R. E., Ouyed, R., Fendt, C., & Brandenburg, A. 2007, *Protostars and Planets V*, 277

Shibata, K. & Uchida, Y. 1986, *PASJ*, 38, 631

Ustyugova, G. V., Lovelace, R. V. E., Romanova, M. M., Li, H., & Colgate, S. A. 2000, *ApJ*, 541, L21

Discussion

ROMERO: In the case you have as a compact object a neutron star, can you estimate the maximum field on the surace of the star necessary to allow the formation of a tower? How high is this field?

HUARTE-ESPINOSA: I don't know. I suppose I could because I can control the injection of magnetic energy.

FERREIRA: YSO observations show that there is not enough pressure to confine jets. So, could magnetic towers be formed or even maintained?

HUARTE-ESPINOSA: The fields should expand. It could be interesting to do simulations with other, weaker density profiles to see what happens to the internal magnetic surface or jet. See the comment by David Meier.

MEIER: Concerning the confinement question, manetic towers do not need large external pressure to confine them- certainly nothing like the pressure in the visible jet plasma. The main jet is confined in the foward current part of the jet. Then moving radially outward, the jet is self-confined in the current-free region. Finally, along the return current surface (from the core of the jet) the magnetic pressure is weak, and it only requires a weak ambient pressure to confine the outer magnetic tower.

YUAN: If the jet is magnetically dominated, why cooling can play an important dynamical role?

HUARTE-ESPINOSA: Cooling can play an important role in the shock region at large distances from the central magnetized region.

FENDT: I expect reconnection being much more important than cooling as the whole field structure will be reconfigured. What do you think?

HUARTE-ESPINOSA: Reconnection plays a role in reality indeed. It would be interesting to redo the simulations with magnetic resistivity to see this. I honestly don't know how important reconnection would be relative to cooling. Yet, the towers in my simulations do not expand due to totation. So, line reconfiguration would not be as important as it would for a rotation-formed tower.

Jets at all Scales
Proceedings IAU Symposium No. 275, 2011
G. E. Romero, R. A. Sunyaev & T. Belloni, eds.

© International Astronomical Union 2011
doi:10.1017/S1743921310015693

The astrophysical jets

Wolfgang Kundt

Argelander Institute of Bonn University,
Auf dem Hügel 71,
D-53121 Bonn, Germany
email: wkundt@astro.uni-bonn.de

Abstract. In this published note I attempt to sketch my understanding of the universal working scheme of all the astrophysical jet sources, or 'bipolar flows', on both stellar and galactic scales, also called 'microquasars', and 'quasars'. A crucial building block will be their medium: extremely relativistic $e^{\pm}$-pair plasma performing quasi loss-free $\mathbf{E} \times \mathbf{B}$-drifts through self-rammed channels, whose guiding equi-partition $\mathbf{E}$- and $\mathbf{B}$-fields convect the electric potential necessary for eventual single-step post-acceleration, at their terminating 'knots', or 'hotspots'. The indispensable pair plasma is generated in magnetospheric reconnections of the central rotator. Already for this reason, black holes cannot serve as jet engines.

Keywords. black hole physics, plasmas, ISM: jets and outflows, galaxies: jets

---

1. How to make jets?

Astrophysical jet sources involve powerful engines of the Universe: The engines can eject charges at very-high energies (VHE), continually at almost luminal speeds to astronomical distances of order $\lesssim$Mpc, for $\gtrsim 10^7$yr, in two antipodal directions, in a beamed manner, aimed within an angle of $\approx 1\%$, whereby particles arrive with $\lesssim$PeV energies at their termination sites. Militaries would appreciate availing of such guns. How does non-animated matter achieve such a difficult goal, time and again, on various size and mass scales? After years of intense deliberation, it is my distinct understanding that all the jet engines follow a uniform working pattern.

For its multiple tasks, biology uses reliable molecular engines with stable working patterns that can be easily reproduced; but here we deal with anorganic matter! How to achieve reliability? No jet without abundant jet substance, or rather jet 'plasma', which should be much lighter than its surrounding medium – for confinement reasons – i.e., much smaller in rest mass, and much hotter. This constraint is optimised by relativistic pair plasma, the lightest plasma in the Universe; and the lighter the higher its Lorentz factors.

This jet plasma must be supplied continually, throughout the lifetime of a jet engine. We know that our Sun can generate relativistic $e^{\pm}$-pairs, at a low rate, even nowadays, in connection with its establishing a hot corona. When the Sun was young, born at the inner edge of its (planetary) accretion disk with a spin period of 3.6 hours (instead of the present 27.3 days), its pair-formation rate has been estimated to have well sufficed to power its young-stellar-object (YSO) stage (Blome & Kundt 1988, Kundt 2005). Whereby the necessary magnetic reconnections are thought to result from 'magnetic spanking' of the inner edge of its accretion disk, which serves as a heavy and sharp-edged obstacle. Correspondingly, forming white dwarfs, as well as neutron stars surrounded by accretion disks are expected to be even stronger generators of relativistic pair plasma.

Continual supply of abundant (relativistic) pair plasma is a necessary though not a sufficient condition on a jet engine to function: The freshly generated plasma must be post-accelerated, and funneled into two antipodal channels. We maintain that for the above-listed engines, the funneling of their twin-jets is achieved by buoyancy, in the quasi-spherical gravitational potential of their central rotator, and that the bunching in momentum space, and post-acceleration of the charges take place via (i) their scattering on the ambient photon gas, via (ii) phase-riding on the outgoing low-frequency magnetic waves (LFW) of the central rotator, like in the Crab nebula (Kulsrud et al 1972), and via (iii) blowing its ambient medium into the shape of Blandford & Rees's funneling deLaval nozzles (Kundt & Krishna 1980, 2004).

And what is the working pattern of the central engines (CEs) of active galactic nuclei (AGN), during their – quite similar-looking – jet formations? As is known since the 70s, a supermassive black hole (SBH) cannot anchor a corotating magnetosphere; it can only swallow. No pair formation, no buoyant escape, no LFWs. Fortunately, ever since 1979, I have convinced myself (and even a few of my referees) that the CEs of AGN are not SBHs, rather (nuclear-) burning disks (BDs), which act analogously to heavy, magnetised rotators, via their strongly shearing inner-galactic diskal coronas (Kundt 1979, 1990, 1996, 2001, 2005, 2009a,b; Kundt & Krishna 1980, 2004). They share the magnetic reconnections, the buoyant escape, the LFW post-acceleration, and self-blown deLaval nozzles with the stellar CEs, thus creating quite similar-looking twin jets.

A complete description of the functioning of the jet engines, their dynamics, stabilities, and various morphologies, takes more space than is available here. The reader can find it in the quoted literature, or in the forthcoming proceedings of the XIVth Brazilian School of Cosmology and Gravitation (held in 2010). Suffice it to mention that pair formation via magnetic reconnections have been recently assessed by Dal Pino *et al.* (2010), that pair annihilations in neutron-star jet sources have been first reported by Kaiser & Hannikainen (2002), and that functional regularities in micro-quasars and quasars are described in Körding *et al.* (2006, 2008). Finally, if the CEs of the AGN are BDs rather than BHs, the rare, gigantic, supersoft X-ray outbursts reported by Komossa may have to be re-interpreted as caused by innermost stars on perturbed orbits colliding with the heavy innermost disk of their galaxy.

References

Blome, H.-J., Kundt, W. 1988, *Astrophys. & Space Sci.*, 148, 343-361

Dal Pino, E. M. de G., Piovezan, P. P., Kadowaki, L. H. S. 2010, *Astron. Astrophys.*, 518; arXiv:1005.3067v1

Kaiser, C. R. & Hannikainen, D. C. 2002, *MNRAS*, 330, 225-231

Körding, E. G., & Jester, S., & Fender, R. 2006, *MNRAS*, 372, 1366-1378

Körding, E. G., Jester, S., & Fender, R. 2008, *MNRAS*, 383, 277-288

Komossa, S. 2005, in *Growing Black Holes, ESO Symposia, Springer*, 159-163

Kulsrud, R. M., Ostriker, J. P., & Gunn, J. E. 1972, *Phys. Rev. Lett.*, 28, 636-639

Kundt, W. 1979, *Astrophys. & Space Science*, 62, 335-345

Kundt, W. 1990, *Astrophys. & Space Sci.*, 172, 109-134

Kundt, W. 1996, in *Jets from Stars and Galactic Nuclei, Lecture Notes in Physics 471, Springer*, W. Kundt (ed.), 140-144

Kundt, W. 2001, in *MICROQUASARS, Proceedings of the Third Miroquasar Workshop*, A.J. Castro-Tirado, J. Greiner & J.M. Paredes (eds.), Kluwer, 273-277

Kundt, W. 2005, *Astrophysics, A New Approach, Springer*, Chapter 11

Kundt, W. 2009a, in *XIIIth Brazilian School of Cosmology & Gravitation, AIP Conf. Proc. 1132*, M. Novello & S.E.P. Bergliaffa (eds.), Melville, N.Y., 288-302

Kundt, W. 2009b, in *General Relativity and Gravitation, 1967-1980*, 9, 41

Kundt, W. & Gopal-Krishna 1980, *Nature*, 288, 149-150.

Kundt, W. & Krishna, G. 2004, *J. Astrophys. Astr.*, 25, 115-127

Jets at all scales
Proceedings IAU Symposium No. 275, 2011
G. E. Romero, R. A. Sunyaev & T. Belloni, eds.

© International Astronomical Union 2011
doi:10.1017/S174392131001570X

Probing the accretion disk - jet connection via instabilities in the inner accretion flow. From microquasars to quasars

Agnieszka Janiuk[1], Bożena Czerny[2], Monika Mościbrodzka[3], and Aneta Siemiginowska[4]

[1]Center for Theoretical Physics, PAN, Al. Lotnikow 32/46, 02-668 Warsaw, Poland
email: `agnieszka.janiuk@gmail.com`

[2]N. Copernicus Astronomical Center, Bartycka 18, 00-716 Warsaw, Poland

[3]Department of Physics, University of Illinois, 1110 West Green Street, Urbana, IL 61801, USA

[4]Harvard Smithsonian Center for Astrophysics, 60 Garden St, Cambridge, MA 02138, USA

Abstract. We present various instability mechanisms in the accreting black hole systems which might indicate at the connection between the accretion disk and jet. The jets observed in microquasars can have a peristent or blobby morphology. Correlated with the accretion luminosity, this might provide a link to the cyclic outbursts of the disk. Such duty-cycle type of behaviour on short timescales results from the thermal instability caused by the radiation pressure domination. The same type of instability may explain the cyclic radioactivity of the supermassive black hole systems. The somewhat longer timescales are characteristic for the instability caused by the partial hydrogen ionization. The distortions of the jet direction and complex morphology of the sources can be caused by precession of the disk-jet axis.

Keywords. physical data and processes: accretion, X-rays:binaries, galaxies: active

1. Overview

Radiation pressure instability. The black hole accretion disk is subject to the thermal and viscous instability if the radiation pressure dominates over the gas pressure. If the accretion rate outside the unstable region (i.e. the mean accretion rate) is such that the

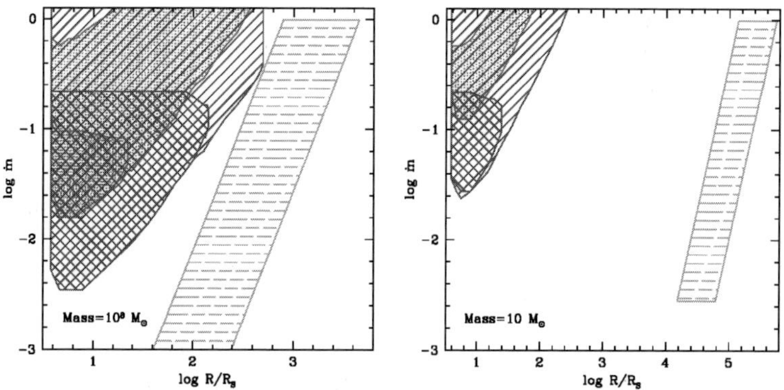

Figure 1. Extension of the radiation pressure (solid and dotted lines) and hydrogen ionization (dashed lines) unstable zones, depending on mean accretion rates (Eddington units). The results are for two heating prescriptions, as well as a zero and non-zero fraction of jet power.

disk is unstable, the source enters a cycle of bright, hot states, separated by the cold, low luminosity states. The outburst amplitudes and durations are sensitive to black hole mass, viscosity and the mean accretion rate. If the viscous heating is proportional to the total pressure, the outburst amplitudes are very large - they can be reduced if the heating is proportional to the square root of the gas times the total pressure.

Partial Hydrogen ionization instability. Another type of the thermal - viscous instability is due to the partial hydrogen ionization. The disk also cycles between the two states: a hot and mostly ionized state with a large local accretion rate and a cold, neutral state with a low accretion rate. This instability has originally been proposed to explain the luminosity variations observed in cataclysmic variables . The same mechanism is responsible for the eruptions in soft X-ray transients (see e.g., Lasota (2001) for review) as well as may operate in active galactic nuclei (e.g., Janiuk *et al.* (2004)).

Disk precession. As studied by Janiuk *et al.*(2009), the acoustic instabilities of the Papaloizou-Pringle type can arise in the supersonic inner parts of the accretion flow. The azimuthal modes of such instabilities will lead to the tilt of the disk and its precession.

2. Implications

We speculate that in the hot state, the luminous core will power a radio jet, while during the cold state the radio activity ceases. In case of the regular periodic outbursts of GRS 1915+105 (see e.g. Fender & Belloni (2004)), lasting from ~ 100 to ~ 2000 s, this approach is successful (e.g. Janiuk *et al.* (2002)). Only the limit cycle mechanism (likely driven by the radiation pressure instability) explains the absence of the direct transitions from its state C to the state B, as well as the observed QPO oscillations. Scaling the timescale with the black hole mass by a factor 10^8 gives the outbursts durations of $10^2 - 10^4$ yrs, and amplitudes are sensitive to the energy fraction deposited in the jet (see Fig. 1). The radiation pressure instability operating above $\dot{m} = 0.025$ may be responsible for very short ages of Compact Symmetric Objects (Czerny *et al.* (2009)). The radiostructures may be influenced by jet precession (Kunert-Bajraszewska *et al.*(2010)).

In the ionization instability, the derived separations between outbursts are on order of 10^6 yrs for a $10^8 M_\odot$ black hole, while the outburst duration is an order of magnitude shorter. The location of the unstable zone is sensitive to the black hole mass (see Fig. 1). For an X-ray binary, the disk maximum radius is limited by the Roche-lobe size and might be too small for an appropriate temperature range to appear. In AGN, the partly ionized zone is much closer to the black hole. If the two unstable zones are very close to each other, the rate of supply of material to the radiation pressure dominated region may be modulated on long timescales, independently of the environment in the host galaxy.

Acknowledgements

This work was supported in part by grant NN 203 512638 from the Polish Ministry of Science.

References

Czerny, B., *et al.* 2009, *ApJ*, 698, 840
Fender, R. & Belloni, T. 2004, *ARAA*, 42, 317
Janiuk, A., Czerny, B., & Siemiginowska, A. 2002, *ApJ*, 576, 908
Janiuk, A., Czerny, B., Siemiginowska, A., & Szczerba, R., 2004,*ApJ*, 602, 595
Janiuk, A., Sznajder, M., Moscibrodzka, M., & Proga, D. 2009, *ApJ*, 705, 1503
Kunert-Bajraszewska, M., Janiuk, A., Gawronski, M., & Siemiginowska, A. 2010, *ApJ*, 718, 1345
Lasota, J.-P., 2001, *New Astron. Revs.*, 45, 449

Jets at all Scales
Proceedings IAU Symposium No. 275, 2011
G. E. Romero, R. A. Sunyaev & T. Belloni, eds.

© International Astronomical Union 2011
doi:10.1017/S1743921310015711

The optimal locations for shock acceleration in MHD jets

Peter Polko[1], **David L. Meier**[2] **and Sera Markoff**[1]

[1] Astronomical Department "Anton Pannekoek", University of Amsterdam,
Postbus 94249, NL-1090GE, Amsterdam, the Netherlands
email: `P.Polko@uva.nl` and `S.B.Markoff@uva.nl`

[2] Jet Propulsion Laboratory, California Institute of Technology, Pasadena, CA 91109, USA
email: `David.L.Meier@jpl.nasa.gov`

Abstract. Jets can contribute to the spectra of X-ray binaries (XRBs) and active galactic nuclei (AGN) from the radio through the γ-ray bands; thus understanding their physics is key for interpreting the data. Recent VLBI observations suggest that jets begin to accelerate particles into power-law distributions at a point offset from the black hole by $\sim 10^4 \ r_{\rm g}$, possibly via a collimation shock. Spectral fitting of simultaneous, broadband data from both XRBs and AGN in jet-dominated states corroborates this picture. From a magnetohydrodynamical (MHD) point of view, it is natural to associate the onset of particle acceleration with the final MHD critical point in the flow, the modified fast point (MFP), where causal contact with the upstream flow is broken. In this way a standing disruption like a shock can form, and this location might vary with the physical parameters of the jet. In order to study this issue, we have used the self-similar formalism of Vlahakis & Königl (2003, hereafter VK03) to simplify the MHD equations and to derive solutions that cross the critical points. We have found a new parameter space of solutions that cross the MFP at a finite height above the disc and are relativistic, spanning a range of Lorentz factors $\Gamma \leqslant 10$ (Polko *et al.* 2010). We present these results, as well as preliminary work connecting the relativistic formalism to the non-relativistic conditions with gravity near the base of the jets.

Keywords. acceleration of particles, ISM: jets and outflows, MHD, methods: analytical

1. Introduction

The assumptions in VK03 to make the MHD equations tractable are self-similarity and axisymmetry, which are also seen in numerical simulations (see figure 2 and 11 from McKinney, 2006), idealised MHD and time-independence. Due to self-similarity, VK03 had to ignore the effect of gravity. Because we want to fit our solutions to a corona model where gravity certainly plays a role, we compared the relativistic VK03 with the non-relativistic Vlahakis *et al.* (2000) including gravity and found a relativistic expression for the gravity term.

2. Results

We found a solution that smoothly crosses all three critical points (Figure 1). Initially the jet is accelerated thermally (with a corresponding decrease in specific relativistic enthalpy), at later times magnetic acceleration takes over (as shown by a decrease in the Poynting-to-mass ratio). All solutions so far are in the return-current regime ($F = 0.75$) as we have not changed this parameter yet. The solutions span a wide range of properties such as initial temperature, final velocity and magnetisation of the jet allowing a multitude of systems to be described, including but not limited to XRBs and AGN.

Figure 1. The top panel shows an example solution where the value of the numerator and denominator of the "wind equation" cross zero at the MFP and the modified slow point (MSP) (the Alfvén point, indicated by the vertical line, has been divided out for clarity). The variable θ is the angle in radians between the radial vector crossing the field line and the axis of symmetry. The y-axis is logarithmic at large values and linear near zero. The accretion disc is located at $\theta = \pi/2$. The ratio $(\mathrm{d}M^2/\mathrm{d}\theta)$ stays finite, showing it is indeed a smooth crossing. The lower panel shows the Poynting–to-mass flux ratio (S), the specific relativistic enthalpy (ξ) and the Lorentz factor (Γ) for this same solution. It can be seen very clearly that the initial acceleration is thermal, leaving the jet cold $(\xi \approx 1)$, and that later acceleration is magnetic. The lines curve back at the right side because the field lines overcollimate, leading to a possible collimation shock. The MFP occurs after overcollimation has begun. The cylindrical radius is scaled to the Alfvén point.

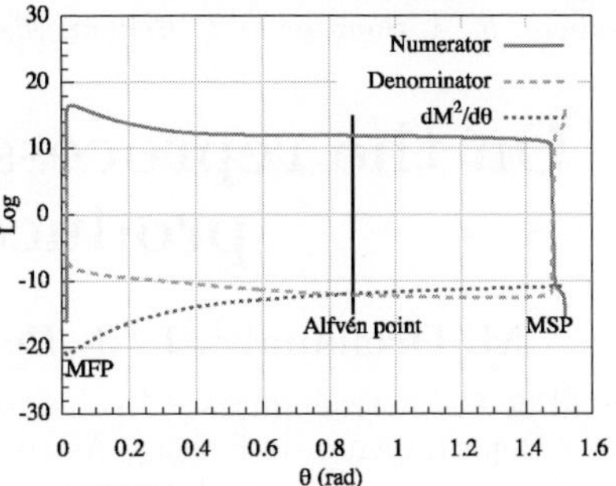

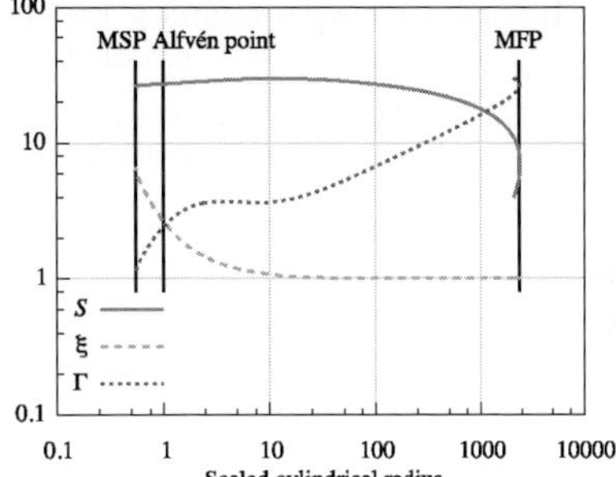

3. Conclusions & Future work

We have shown for the first time that it is possible to obtain solutions to the ideal MHD equations assuming an axisymmetric, self-similar jet that smoothly cross all three critical points. Every solution found overcollimates shortly before the MFP, allowing a shock to develop, that, because it occurs beyond the MFP, cannot disrupt the flow upstream causing the shock, allowing it to be a stable feature in the jet. With the identification of the MFP as the start of the acceleration region, which fits to observations have determined to also be a steady location (e.g., Markoff *et al.* 2005; Migliari *et al.* 2007; Markoff *et al.* 2008; Maitra *et al.* 2009) but is currently a free parameter, it is therefore possible to independently determine where particles first get accelerated into a power-law distribution. With the wide range in solutions it should be possible to fit the boundary conditions imposed by observations for both XRBs and AGN. After connecting our solutions to a corona model and determining the extent of the possible solutions, we can replace the hydrodynamical acceleration in the spectral fitting routine of Markoff *et al.* 2005 with the MHD formalism presented here, allowing us to fit the spectra of XRBs more self-consistently and to obtain better estimates for the parameters describing the XRB system.

References

Maitra, D., Markoff, S., Brocksopp, C., Noble, M., Nowak, M., & Wilms, J. 2009, *MNRAS*, 398, 1638
Markoff, S., Nowak, M. A., & Wilms, J. 2005, *ApJ*, 635, 1203
Markoff, S., Nowak, M., Young, A., *et al.* 2008, *ApJ*, 681, 905
McKinney, J. C. 2006, *MNRAS*, 368, 1561
Migliari, S., *et al.* 2007, *ApJ*, 670, 610
Polko, P., Meier, D. L., & Markoff, S. 2010, *ApJ*, in press
Vlahakis, N. & Königl, A. 2003, *ApJ*, 596, 1080
Vlahakis, N., Tsinganos, K., Sauty, C., & Trussoni, E. 2000, *MNRAS*, 318, 417

Jets at all Scales
Proceedings IAU Symposium No. 275, 2011
G. E. Romero, R. A. Sunyaev & T. Belloni, eds.

© International Astronomical Union 2011
doi:10.1017/S1743921310015723

On the reprocessing of gamma-rays produced by jets

M. Orellana[1,2], L. J. Pellizza[3] and G. E. Romero[4,1]

[1]Facultad de Cs. Astronómicas y Geofísicas - Universidad Nacional de La Plata, Argentina
[2]Departamento de Física y Astronomía, Universidad de Valparaíso, Chile
email: morellana@fcaglp.unlp.edu.ar

[3]Instituto de Astronomía y Física del Espacio, UBA - CONICET, Argentina
email: pellizza@iafe.uba.ar

[4]Instituto Argentino de Radioastronomía (IAR), CCT La Plata - CONICET, Argentina
email: romero@iar.unlp.edu.ar

Abstract. Systems of two very different sizescales are known to produce very high-energy (VHE) radiation in their jets: AGNs and microquasars. The produced VHE photons ($E_\gamma \sim 1$ TeV) can be absorbed by the intense environmental soft photon fields, coming from the companion star (in high mass binaries) or from the accreting material (disk+corona in AGNs), as these are the dominant sources at energies around $\sim (m_e c^2)^2/E_\gamma$. Energetic pairs are created by the photon-photon annihilation, and, depending on how efficient are the competing cooling channels, the absorption can lead to a reprocessing by Inverse Compton pair-cascade development. A self-consistent modeling of these systems as gamma-ray sources should then include, along with the emission and absorption processes, a thorough treatment of the pair cascades. We discuss here on this issue, focusing on our (preliminary) results of numerical simulations devoted to a study case similar to the high-mass microquasar candidate LS 5039.

Keywords. radiation mechanisms: nonthermal, gamma rays: theory

---

1. Introduction

Many efforts have been devoted to the calculation of electromagnetic (EM) cascades using both numerical (e.g. Protheroe 1986, Bednarek 1997 and subsequent works, Orellana *et al.* 2007) or semi-analytical techniques (Aharonian & Plyasheshnikov 2003). At high energies the angles of emission of secondary electrons and photons are very small, thus the shower develops essentially in the direction of the incident particle (radially away from the source) when one deals with EM cascades in the absence of magnetic fields.

In a more realistic case, the presence of magnetic fields changes the problem dramatically, which become 3D as the electrons are deflected from their original direction†. More importantly, the radiative synchrotron channel take out energy of the system and can

---

† Four different regimes can in principle be distinguished, in similar as Gabici & Aharonian (2007) pointed out concerning different values of the intergalactic magnetic field: Regime I: $B \ll B_{\rm iso}$. The EM cascade is not affected at all. The electrons in the cascade do not suffer synchrotron losses, nor are they deflected. Thus, the EM cascade develops along a straight line. Regime II: $B_{\rm iso} \lesssim B \ll B_{\rm syn}$. No energy is subtracted to the EM cascade due to synchrotron losses, but low energy electrons are effectively isotropized by the magnetic field. Regime III: $B \gtrsim B_{\rm syn}$. The development of the EM cascade is strongly suppressed since its very first steps due to strong synchrotron losses. Regime IV: $B \gg B_{\rm syn}$. The magnetic field is strong enough as to maintain a pure synchrotron/curvature cascade, as in the polar regions of pulsars. The inequalities in these items leave a gray area to be explored by detailed calculations.

take the leading of the energy losses of the charged particles changing the character of the cascade development.

We have devised a scheme for the simulation of electromagnetic cascades in the presence of magnetic fields from an *ab initio* point of view, and versatile enough as to be applied to many different systems (Pellizza, Orellana & Romero, 2010). There are three main advantages of our scheme. First, the code can be applied to any γ-ray production model for which the density, energy and momentum of the primary photons are given, with arbitrary target photon fields and magnetic field configurations. Second, it computes three dimensional cascades that give information on the spectral energy distribution of photons leaving the system, and its dependence with time, position and propagation direction. Finally, it provides a self-consistent treatment of synchrotron emission, and its dependence with position within the physical system and time.

2. Aplication to LS 5039: preliminary results

We have computed the cascade produced in the high-mass γ-ray loud binary LS 5039. Though its nature is not clear, it remains as an excellent microquasar candidate for which the VHE photons produced at the base of the jets triger EM cascades during their propagation (see e.g. Bosch-Ramon *et al.* 2006, for an emission model, and Cerutti *et al.* 2010 for alternative cascade treatment). Here we have considered power-law injected photons (photon index 2), through the background stellar radiation field (taken as a black body with $T = 3 \times 10^4$ K, and $R_\star = 10^{12}$ cm), and for simplicity, a constant and isotropic injected luminosity was assumed. Note the low eccentricity and inclination of the system ($i \sim 30°$) are responsible for a moderate variation in the opacity as a function of the orbital phase.

The ambient (ordered) magnetic field, B_0, is assumed uniform in a simplifying approach. We take it to be $B_0 = 0.001$ G, a value that is expected to give EM cascades in the anisotropic regime, which haven't been explored by other authors so far. See e.g. the plot for the cascades domains as estimated by Cerutti *et al.* (2010), for LS 5039. The pairs are confined by the magnetic field inside the binary system during a certain (non-negligible) time interval while they radiate by Synchrotron and IC processes. These can lead to modifications in the light curve (delayed from injection) that we are planning to explore in a forthcoming work. The resulting observable spectra, normalized to the total energy injected by primary particles for different orbital phases (phase zero corresponds to the periastron), were obtained and will also be presented elsewhere.

References

Aharonian, F. A. & Plyasheshnikov, A. V. 2003, *Astropart. Phys.* 19, 525

Bednarek, W. 1997, *A&A* 322, 523

Bosch-Ramon, V., Romero, G. E., & Paredes, J. M. 2006, *A&A* 447, 26

Cerutti, B. *et al.* 2010, *A&A*, in press [arXiv:1006.2683]

Gabici, S. & Aharonian, F. A. 2007, *Ap&SS* 309, 465

Orellana, M., Bordas, P., Bosch-Ramon, V., Romero, G. E., & Paredes, J. M. 2007, *A&A* 476, 9

Protheroe, R. J. 1986, *MNRAS* 221, 769

Pellizza, L. J., Orellana, M., & Romero G. E. 2010, *IJMPD*, Vol. 19, 671–676

Jets at all Scales
Proceedings IAU Symposium No. 275, 2011
G. E. Romero, R. A. Sunyaev & T. Belloni, eds.

© International Astronomical Union 2011
doi:10.1017/S1743921310015735

Broad emission lines for a negatively spinning black hole

T. Dauser[1], **J. Wilms**[1], **C. S. Reynolds**[2], **and L. W. Brenneman**[3]

[1]Dr. Karl Remeis-Observatory and ECAP, Sternwartstr. 7, 96049 Bamberg, Germany
email: `thomas.dauser@sternwarte.uni-erlangen.de`
[2]Department of Astronomy and Maryland Astronomy Center for Theory and Computation,
University of Maryland, College Park, MD 20742, USA
[3]Harvard-Smithsonian C. for Astrophysics, 60 Garden Street, Cambridge, MA 02138, USA

Abstract. We present an extended scheme for the calculation of the profiles of emission lines from accretion disks around rotating black holes. The scheme includes disks with angular momenta which are parallel and antiparallel with respect to the black hole's angular momentum, as both configurations are assumed to be stable (King *et al.* 2005). Based on a Green's function approach, an arbitrary radius dependence of the disk emissivity and arbitrary limb darkening laws can be easily taken into account, while the amount of precomputed data is significantly reduced with respect to other available models. We discuss line shapes for such disks and present a code for modelling observational data with this scheme in X-ray data analysis programs. A detailed discussion will soon be presented in a forthcoming paper (Dauser *et al.* 2010).

Keywords. accretion, accretion disks, black hole physics, galaxies: nuclei, galaxies: active, lines: profiles

1. How does a black hole get negative spin?

Misalignments between the angular momenta of a black hole and its accretion disk are a prediction of stochastic evolution models for AGN (e.g., Volonteri *et al.* 2005). In galactic black holes, misalignments can be produced by the initial supernova kick (Brandt & Podsiadlowski 1995). As shown by King *et al.* (2005), both parallel and antiparallel alignments of the disk and black hole angular momenta are stable configurations; misaligned discs will evolve to one of them. In fact, accretion onto rapidly-spinning retrograde black holes, where the angle between the angular momenta is greater than $90°$, may be of some importance for understanding the properties of powerful radio-loud AGN. Garofalo (2009) argues that an accretion disk around a retrograde black hole is a particularly potent configuration for generating powerful jets, which might also explain the lack of radio-loud AGN (Garofalo *et al.* 2010).

2. Line Profiles

Since the shape of the observed emission line (e.g. at $6.4\,\mathrm{keV}$ for a neutral Fe Kα transition) depends on the spin of the black hole, a, and the emissivity (i.e., the intensity dependence on the radius) and inclination of the surrounding accretion disk, the diagnostic power of relativistic lines is very high, as they provide one of the most direct ways to probe the physics of the region of strong gravity close to the black hole.

In order to allow for a comparison of the pure frame-dragging effects, the inner edge of the accretion disk in the left part of Fig. 1 was set to $9\,r_\mathrm{g}$ for all profiles, which is the smallest inner radius possible for an accretion disk around a maximally negatively rotating black hole. The major difference between the different spins is the relative strength

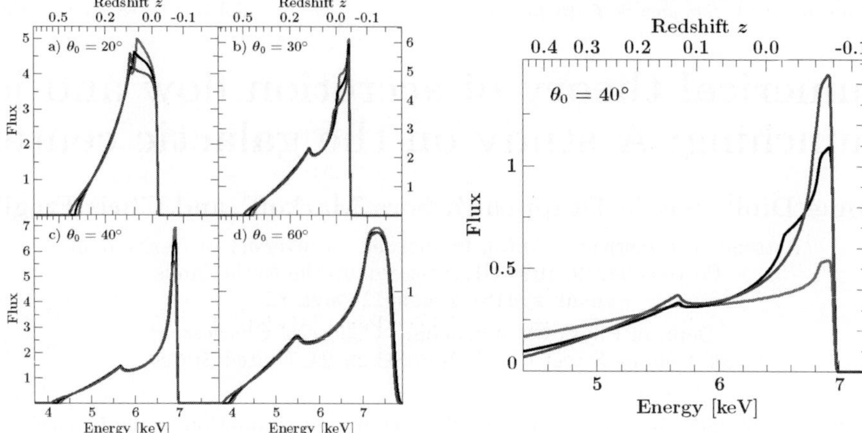

Figure 1. Line profiles for a relativistic iron line at 6.4 keV, an emissivity of r^{-3} and under an inclination angle of θ_o. The maximal spinning black hole ($a = +0.998$) is drawn in red, the non-rotating ($a = 0$) in black and the line for maximal negative spin ($a = -0.998$) in blue. Left: Accretion disk with an inner radius of $9\,r_g$. Right: Line for a disk extending down to the marginal stable orbit, which varies from $1.23\,r_g$ ($a = +0.998$) to $9\,r_g$ ($a = -0.998$).

of the core of the line to the red wing, which decreases with decreasing a. For this case of a large inner radius, the most significant differences in line shape are seen for low values of θ_0 while the red tails are virtually indistinguishable.

The more realistic case, where the disk extends down to the marginally stable orbit, is shown in the right part of Fig. 1. Since the inner edge of the disk is closer to the black hole for positively spinning black holes, more strongly redshifted photons emerge and therefore broader lines (especially for emissivities strongly peaked inwards).

3. Data Analysis

As none of the currently available line models like the `kerrdisk` (Brenneman & Reynolds 2006) or the `ky`-family of models (Dovčiak *et al.* 2004) are valid for retrograde accretion disks, we implemented a model function, called `relline`, that can be added to data analysis software such as ISIS or XSPEC for spin $-0.998 \leqslant a \leqslant 0.998$ and inclination $0° \leqslant \theta \leqslant 89°$. Both, an additive and a convolution model are provided at `www.sternwarte.uni-erlangen.de/research/relline/`.

Acknowledgements: The research presented in these contributions was partially financed by the European Commission under grant ISN 215212.

References

Brandt N., & Podsiadlowski P., 1995, *MNRAS* 274, 461

Brenneman L. W., & Reynolds C. S., 2006, *ApJ* 652, 1028

Dauser T., Wilms J., Reynolds C. S., & Brenneman L. W., 2010, *MNRAS in press, arxive:1007.4937*

Dovčiak M., Karas V., & Yaqoob T., 2004, *ApJ Suppl.* 153, 205

Garofalo D., 2009, *ApJ* 699, 400

Garofalo D., Evans D. A., & Sambruna R. M., 2010, *MNRAS* 406, 975

King A. R., Lubow S. H., Ogilvie G. I., & Pringle J. E., 2005, *MNRAS* 363, 49

Volonteri M., Madau P., Quataert E., & Rees M. J., 2005, *ApJ* 620, 69

Jets at all Scales
Proceedings IAU Symposium No. 275, 2011
G. E. Romero, R. A. Sunyaev & T. Belloni, eds.

© International Astronomical Union 2011
doi:10.1017/S1743921310015747

Numerical theory of accretion flow and jet launching: A study on the galactic center

Salomé Dibi[1], Samia Drappeau[1], Sera Markoff[1] and Chris Fragile[2]

[1]Astronomical Institute "Anton Pannekoek", University of Amsterdam,
Postbus 94249, 1090 GE Amsterdam, the Netherlands
email: s.dibi-rousselle@uva.nl

[2]Dept. of Physics & Astronomy, College of Charleston,
58 Coming Street Charleston, SC 29424, United States

Abstract. We obtained the first spectral predictions from a simulation of the Galactic Center to include radiative processes internally. We performed simulations with and without cooling, with and without spin, and for different initial configurations of the magnetic field, in order to test the effect on jet launching and inner accretion disk characteristics. By exploring parameter space, we will attempt to place new constraints on the controversial question about the presence or not of a jet from Sgr A*, as well as study jet launching in general. We have shown that, as expected, the spin of the BH affects the structure of the jet. The presence of cooling also strongly influences the inner structure of the accretion disk and therefore affects jet launching. These results show that radiative cooling is not negligible, as is usually assumed for the very underluminous supermassive BH, Sgr A*. On the contrary, the inclusion of cooling has a very visible influence on the accretion disk. Furthermore it creates an important difference in the resulting spectra.

Keywords. Sgr A*, MHD, radiation mechanisms: general, plasmas, radiative transfer, diffusion, acceleration of particles.

1. Modeling Sgr A* with Cosmos^{++}

We present preliminary results of a series of fully relativistic 2D magneto-hydrodynamic simulations (using Cosmos^{++}; Anninos *et al.* 2005) of the accretion flow around a supermassive black hole (BH). The initial state of the simulation is shown in Figure 1. Our

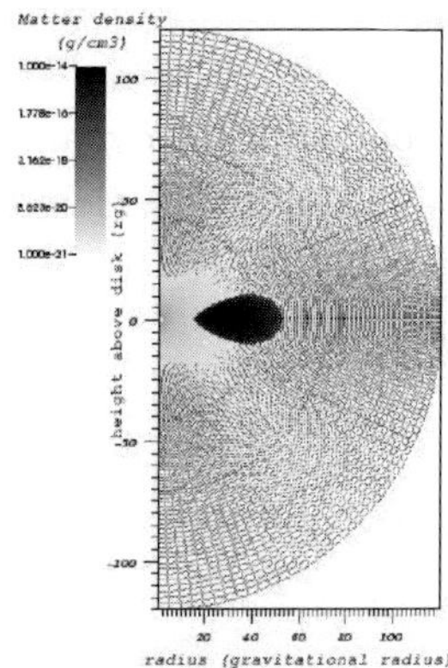

Figure 1. This figure shows the initial 2D state of the simulation (black and white color scale in density) for a $4.3 \times 10^6 M_\odot$ BH at (0,0), with an edge-on view of the accretion disk. The black shows the initial concentration of the material inside the accretion disk, set to values that capture the measured accretion rate onto Sgr A*. Together with the matter density, the mesh grid domain used for the computation is represented. Symmetry about the rotation axis of the accretion flow is assumed, as necessary for 2D.

simulations are some of the first that self-consistently include radiative cooling which allows us to compare our predictions to actual observations. In the present work we choose Sgr A* for our first tests as its accretion flow is relatively well constrained compared with other sources. We performed simulations with and without cooling, with and without spin, and for different initial configurations of the magnetic field. S. Drappeau (see proceedings article, this volume) has used our output to generate spectra to compare with observational data of the Galactic center. The simulations were all run in 2D, and for 7 complete orbits in order to approximate a steady state.

2. Results

For the models with normalized spin parameter of $a = 0.99$, the jet is clearly more relativistic (Lorentz factor 3). Even without spin, plasma is ejected along the axes, but it is less collimated and less relativistic (Fig. 2). For a non-rotating BH ($a = 0$), the outflow

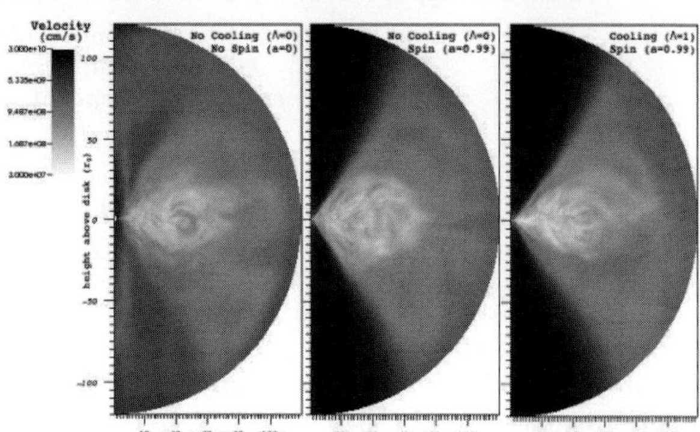

Figure 2. These panels show the time averaged velocity magnitude of the plasma over the last two orbits for three different simulations of Sgr A*. The models were run with the same initial magnetic field configuration (one poloidal loop crossing the disk, parameter $B_p = 1$) to test the spin (parameter a) and the radiative cooling (parameter Λ). Black and white color scale in velocity from $3 * 10^7$ to $3 * 10^{10} cm/s$.

is an order of magnitude slower which is not surprising as the magnetic field strength is also ten times weaker. The presence of cooling strongly influences the inner structure of the accretion disk and therefore affects jet launching. The resulting inner accretion disk is flatter, because in the models without cooling the disk and jet are puffed up by gas pressure due to higher temperatures. These results show that radiative cooling is not negligible, as is usually assumed for the very underluminous supermassive BH, Sgr A*. On the contrary, the inclusion of cooling has a very visible influence on the accretion disk. Furthermore, Samia Drappeau developed a code to generate spectra from our simulations output, taking into account the radiative loss terms included in Cosmos[++] (bremsstrahlung, synchrotron and inverse Compton). The initial results show significant differences in the spectrum for the models with cooling versus without. Including the radiative cooling in the dynamics is essential, because it creates an important difference in the resulting spectra. Indeed, in the infrared, which probes radiation from the accretion flow very close to the BH, the predicted emission is two orders of magnitude higher compared to that without cooling.

3. Conclusion

These preliminary results are thus very interesting and exciting. We are finishing an exploration of parameter space in 2D, and then will choose some of those cases for 3D followup.

References

Peter Anninos, P. Chris Fragile, Jay D Salmonson 2005, *The Astrophysical Journal*, 635, 723

Jets at all Scales
Proceedings IAU Symposium No. 275, 2011
G. E. Romero, R. A. Sunyaev & T. Belloni, eds.

© International Astronomical Union 2011
doi:10.1017/S1743921310015759

Self-consistent spectra from GRMHD simulations with radiative cooling
A link to reality for Sgr A*

Samia Drappeau[1], Salomé Dibi[1], Sera Markoff[1] and Chris Fragile[2]

[1]Sterrenkundig Instituut 'Anton Pannekoek', Universiteit van Amsterdam,
Postbus 94249, 1090GE Amsterdam, the Netherlands
email: `s.drappeau@uva.nl, s.dibi-rousselle@uva.nl, s.markoff@uva.nl`

[2]Dept. of Physics & Astronomy, College of Charleston,
Charleston, USA
email: `fragilep@cofc.edu`

Abstract. Cosmos++ (Anninos *et al* 2005) is one of the first fully relativistic magneto-hydro-dynamical (MHD) codes that can self-consistently account for radiative cooling, in the optically thin regime. As the code combines a total energy conservation formulation with a radiative cooling function, we have now the possibility to produce spectra energy density from these simulations and compare them to data. In this paper, we present preliminary results of spectra calculated using the same cooling functions from 2D Cosmos++ simulations of the accretion flow around Sgr A*. The simulation parameters were designed to roughly reproduce Sgr A*'s behavior at very low (10^{-8}-10^{-7} M$_\odot$/yr) accretion rate, but only via spectra can we test that this has been achieved.

Keywords. Sgr A*, MHD, radiation mechanisms: general, plasmas, radiative transfer, diffusion, acceleration of particles.

1. Methodology

Given some characteristics like temperature, mass density and magnetic pressure from the simulation, we can generate broad-band spectra with emission coming from Bremsstrahlung, Synchrotron and Compton effects. These signatures can then be compared to data (Fig. 1).

Simulation characteristics are time-averaged over the last two stable orbits before generating a spectrum. We then obtain a steady-state emission spectrum to compare to data. Five models have been simulated, given these initial conditions:

1. No cooling, no spin, 1-loop magnetic field
2. No cooling, spin = 0.99, 1-loop magnetic field
3. Cooling, spin = 0.99, 1-loop magnetic field
4. No cooling, no spin, 4-loop magnetic field
5. No cooling, spin = 0.99, 4-loop magnetic field

2. Preliminary results

This work is still in development. Further simulations of Sgr A* are ongoing. But from our first results we can already draw some interesting conclusions:

1. Synchrotron emissions are higher than expected. This divergence between models and data is due to temperature and magnetic field simulations values, 2 orders of magnitudes higher than typical Sgr A* values.

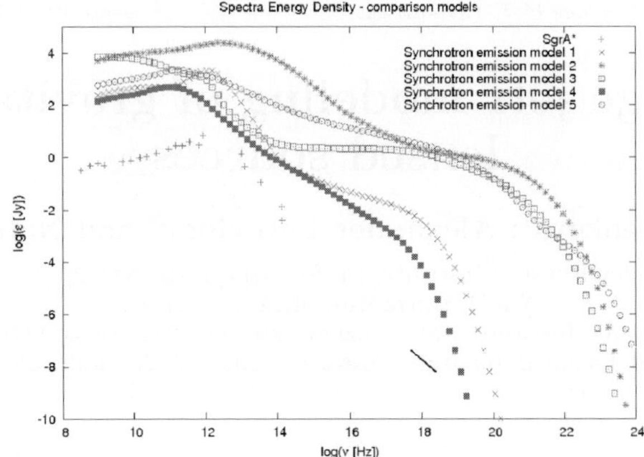

Figure 1. Sgr A* data (*plus symbol and line*) from Melia & Falcke (2001). We show here synchrotron emission from no cooling, no spin, 1-loop magn. field model (*cross*), from no cooling, spin = 0.99, 1-loop magn. field model (*star*), cooling, spin = 0.99, 1-loop magn. field model (*dot square*), no cooling, no spin, 4-loop magn. field model (*filled square*) and no cooling, no spin, 4-loop magn. field model (*dot circle*).

2. Bremsstrahlung emission is too faint, as expected, most of the X-rays are dominated by cooler bremsstrahlung emission from the outer radii (e.g. Quataert 2002), a part of the accretion disk which is not simulated presently.

3. We did not discriminate parts of simulation known to be not realistic which results in none-physical emission, e.g. the tail in synchrotron emission might be coming from the base of the "jet", an area not well handled by the simulation.

4. Turning on cooling routines makes the emission to dramatically drop in frequency range where the accretion disk is radiating.

5. Tuning black hole spin to maximum generates too much magnetic field, hence too much emission.

3. Conclusion

These spectra show Dibi *et al.* (see proceedings article, this volume) first set of runs are a good start in modeling Sgr A* even though they are generating too much power. We have great expectations in models 1 and 4 and it will be interesting to see how their spectra behave when cooling routines are turned on. Future works also include being able to isolate emission coming from parts of the simulation we know can not be trust. And in general we want to be able to designate regions where originate each part of the emission.

References

Anninos, P., Fragile, P. C., & Salmonson, J. D. 2005, *ApJ*, 635, 723
Blumenthal, G. R. & Gould, R. J. 1970, *Review of Modern Physics*, 42, 237
Esin, A. A., Narayan, R., Ostriker, E., & Yi, I. 1996, *ApJ*, 465, 312
Fragile, P. C. & Meier, D. L. 2009, *ApJ*, 693, 771
Mahadevan, R., Narayan, R., & Yi, I. 2009, *ApJ*, 465, 327
Melia, F. & Falcke, H. 2001, *ARAA*, 39, 309
Quataert, E. 2002, *ApJ*, 575, 855
Stepney, S. & Guilbert, P. W. 1983, *MNRAS*, 204, 1269

Jets at all Scales
Proceedings IAU Symposium No. 275, 2011
G. E. Romero, R. A. Sunyaev & T. Belloni, eds.

© International Astronomical Union 2011
doi:10.1017/S1743921310015760

The image jets modeling of gravitationally lensed sources

Tatiana Larchenkova[1], Alexander Lutovinov[2] and Natalya Lyskova[3]

[1] ASC of P. N. Lebedev Physical Institute, Profsoyuznaya str. 84/32, Moscow 117997, Russia
email: `tanya@lukash.asc.rssi.ru`
[2] Space Research Institute, Profsoyuznaya str. 84/32, Moscow 117997, Russia
[3] Moscow Physical-Technical Institute, Institutskyi per., 9, Dolgoprudnyi 141700, Russia

Abstract. The jets image modelling of gravitationally lensed sources have been performed. Several basic models of the lens mass distribution were considered, in particular, a singular isothermal ellipsoid, an isothermal ellipsoid with the core, different multi-components models with the galactic disk, halo and bulge. The obtained jet images were compared as with each other as with results of observations. A significant dependence of the Hubble constant on the model parameters was revealed for B0218+357, when the circular structure was took into account.

Keywords. jets, gravitational lensing, Hubble constant, B0218+357

It is known, that due to the gravitational lensing of a compact object with a relativistic jet the multiple images of the object itself and of the jet can arise. Such gravitationally lensing systems are observed and the most bright from them are PKS 1830-211 (Nair *et al.*, 1993) and B0218+357 (Patnaik *et al.*, 1993). The modelling of images of such sources and study their temporal behavior are began actual due to forthcoming launches of cosmic radiointerferometers.

For the case of the lensing on the spiral galaxy we considered several multicomponents models: Model I – disk and softened halo, approximated by the isothermal ellipsoid located in the singular isothermal halo of a dark matter; Model II – disk, approximated by the Kuzmin disk (Kuzmin, 1956) in the isothermal halo; Model III – disk and bulge, approximated by the Kuzmin disk, in the isothermal halo (Keeton & Kochanek 1998).

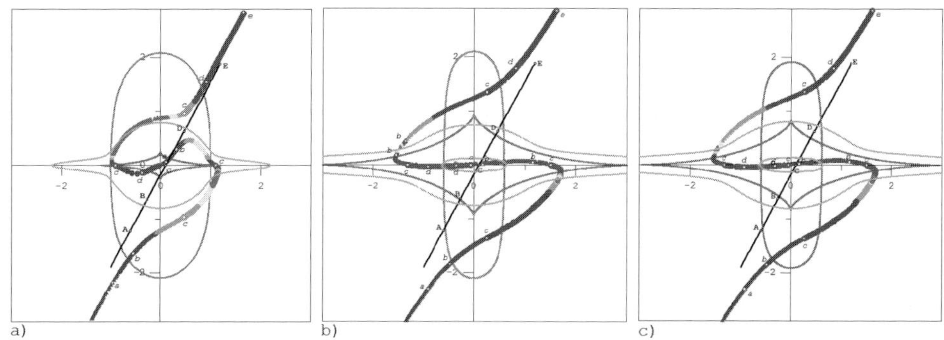

Figure 1. Mapping of a gravitationally lensed jet for three models (I – a, II – b, III – c). Designations are: gray – caustics, light-gray – critical curves, black line – jet. Jets images depending on the magnification M are shown by magenta ($0 < |M| < 1$), blue ($1 < |M| < 3$), green ($3 < |M| < 7$), yellow ($7 < |M| < 10$) and red ($|M| > 10$). The lower case letters on the jet images correspond to the same capital letters on the jet itself (for details see Larchenkova *et al.* 2011).

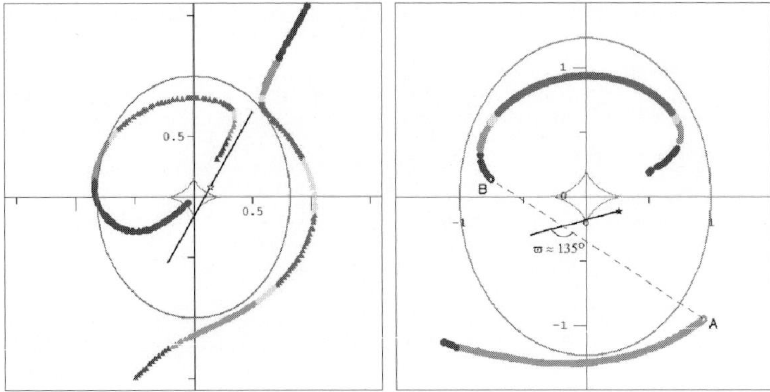

Figure 2. (*left*) The nearly circular image of the lensed jet in the Model I with parameters: $q_{3d} = 0.05$, $a_d = 1.0$, jet slope 60, inclination of all components $i = 30$. (*right*) Results of the B0218+357 jet modelling with the Model I. sThe jet is directed at the angle of $\varpi \simeq 135^o$ to the line connected images A and B, and cross the tangential caustic near its bottom cusp.

Depending on the disk axes ratio q_{3d}, its characteristics size a_d and jet parameters, the different geometry of caustics and jet images was obtained. Some typical results are presented in Fig. 1 for different lens models. It is clearly seen that the inclusion of the low-mass bulge (similar to the Milky Way one) does not change significantly the picture of the jet lensing (see Fig. 1 b, c).

There is a general question: is it possible to obtain a circular-like structure by the jet lensing? The one of answers is that: to obtain a circular-like structure it is necessary to cross the tangential caustics by the jet near tangentially to their cusps. An example of the appearance of such a structure are shown in Fig. 2 (left).

The system B0218+357 is a very interesting and important for astrophysical and cosmological investigations due to several reasons: the presence of a large scale jet and the circular-like structure in radio images; the time delay between images of the compact core, which was measured with the high accuracy; the significant distance from other extragalactic objects. For this system there were obtained several set of parameters of model I and II reproducing results of observations. An example of the mutual arrangement of the lens, source, jet and circular-like structure is presented in Fig. 2 (right). In a combination with the measured time delay the model parameters can give us estimations of the Hubble constant. It was revealed that the considered models lead to a wide variety of its value $H \simeq 35 - 90$ km s^{-1} Mpc^{-1} (for details see Larchenkova *et al.* 2011).

This work was supported by the program of Russian Academy of Sciences "Origin, Structure and Evolution of Objects in the Universe" and government contract P1336.

References

Keeton, C. R. & Kochanek, C. S., 1998, *ApJ*, 495, 157
Kuzmin, G., 1956, *Sov. Astron*, 33, 27
Larchenkova, T., Lutovinov, A., & Lyskova, N. 2011, *Astron. Letters*, in press
Nair, S., Narasimha, D., & Rao, A. P., 1993, *ApJ*, 407, 46
Patnaik, A. *et al.*, 1993, *MNRAS*, 261, 435

Part 2. Active galactic nuclei

Jets at all Scales
Proceedings IAU Symposium No. 275, 2011
G. E. Romero, R. A. Sunyaev & T. Belloni, eds.

© International Astronomical Union 2011
doi:10.1017/S1743921310015772

Blazar jet physics in the age of Fermi

Charles D. Dermer[1], on behalf of the Fermi LAT Collaboration

[1]Naval Research Laboratory, Code 7653,
4555 Overlook Ave., SW, Washington, DC 20375-5352 USA
email: `charles.dermer@nrl.navy.mil`

Abstract. The impact of the Fermi Gamma-ray Space Telescope on blazar research is reviewed. This includes a brief description of the Fermi Large Area Telescope, a summary of the various classes of extragalactic sources found in the First Large Area Telescope AGN Catalog, and more detailed discussion of the flat spectrum radio quasar 3C454.3 and the BL Lac object PKS 2155-304. Some theoretical studies related to ongoing blazar research with Fermi are mentioned, including implications of γ-ray observations of radio galaxies on blazar unification scenarios, variability in colliding shells, and whether blazars are sources of ultra-high energy cosmic rays.

Keywords. galaxies: jets, gamma rays: observations, gamma rays: theory

1. Introduction

Even without the Fermi Gamma-ray Space Telescope, the study of blazar jets would be rapidly advancing from discoveries made with ground-based γ-ray Cherenkov telescopes and AGILE. With Fermi, we have a telescope that is giving the most complete census yet of the $\gtrsim 100$ MeV sky. Fermi makes sensitive measurements of the γ-ray spectra of blazars where the energy output often dominates. Unexpected puzzles have arisen. This includes the origin of the multi-GeV cutoffs in 3C 454.3 and other FSRQs. Another is the relation of blazars to radio galaxies detected at γ-ray energies. The blazar sequence (Fossati *et al.* 1998), if it is that, remains sketchily understood, even with our new knowledge from Fermi.

Here we briefly review the results of blazar studies in the age of Fermi, including the 1st Fermi LAT AGN Catalog (1LAC; Abdo *et al.* 2010a) and the misaligned AGN (MAGN) sample of radio galaxies detected with Fermi. An overview of various classes of galaxies and AGNs detected with Fermi is given, including some detail on individual sources. 3C 454.3 and PKS 2155-304 Theoretical questions arising from these observations are raised, and some speculations in terms of colliding shells ejected from the central supermassive black hole are made. The likelihood that blazars accelerate ultra-high energy cosmic rays is considered.

2. The Large Area Telescope on Fermi

The Large Area Telescope (LAT) on the Fermi Gamma-ray Space Telescope is a pair conversion telescope that tracks the direction of an electron-positron pair formed by an incident γ ray. The converter-tracker is made of a 4×4 array of 16 modules. Each module has 18 tracker planes that uses Si strips to follow the direction of the converted pair. The upper 16 tracker planes have tungsten converters, with the uppermost 12 "front" planes using thin layers of tungsten to minimize multiple scattering and provide a narrow point spread function (psf), and the lower 4 "back" planes using thick tungsten layers to increase the effective area at the expense of the psf. The bottom two planes have no converters, only Si tracking planes. The energy is deposited in the calorimeter, which is made of 96 CsI(Tl) crystals. The entire assembly is surrounded by a segmented anti-coincidence detector used to reject charged particles. See Atwood *et al.* (2009) for more detail.

The LAT has a field-of-view, roughly defined by the solid angle within which the effective area is more than one-half of the on-axis effective area, of 2.4 sr, which corresponds to $\approx 1/5^{\text{th}}$ of the full sky. For the most stringent *diffuse* event class used to identify γ rays, the on-axis effective area increases from ≈ 1500 cm^2 at 100 MeV to ≈ 8000 cm^2 at 1 GeV, and is roughly constant at higher energies. The energy resolution is better than 10% between ≈ 50 MeV and ≈ 50 GeV. For front-converting events, the 68% acceptance cone is $\approx 3.5°$ and $\approx 0.6°$ at photon energies $E \approx 100$ MeV and $E \approx 1$ GeV, respectively. Updated instrument performance that improves on pre-flight Monte Carlo and muon calibrations by using inflight corrections to the instrument response functions are found at http://www-glast.slac.stanford.edu/software/IS/glast_lat_performance.htm.

Fermi nominally operates in a zenith-pointing rocking mode that allows the LAT to survey the entire sky every two orbits, or $\approx$ every 3 hours. It typically rocks 35° north and south of the zenith on alternate orbits. Autonomous repoints to gamma-ray bursts and two dedicated pointings, one to 3C 454.3 and a second to the flaring Crab, have been executed by Fermi.

3. LBAS and 1LAC

Two lists of AGNs detected with Fermi have now been published by the Fermi Collaboration. After 3 months of science observations between 2008 August 4 and 2008 October 30, the LAT Bright AGN Sample (LBAS) was released (Abdo *et al.* 2009a). It consists of 106 high-latitude ($|b| > 10°$) sources associated with AGNs. These sources have a test statistic $TS > 100$, corresponding to $\gtrsim 10\sigma$ significance, and are a subset of the 205 sources listed in the Fermi LAT bright source list (Abdo *et al.* 2009b). By comparison, the 3rd Egret Catalog (3EG; Hartman *et al.* 1999) of γ-ray sources and the revised EGRET catalog (EGR; Casandjian & Grenier 2008) list 31 sources with significance $> 10\sigma$, of which 10 are at high latitude. Remarkably, 5 of the $> 10\sigma$ EGRET sources are not found in the LAT bright source list. These are the flaring blazars NRAO 190, NRAO 530, 1611+343, 1406-076, and PKS 1622-297, the most luminous EGRET blazar (Mattox *et al.* 1997).

The 1LAC (Abdo *et al.* 2010a) is a subset of the 1451 sources in the First LAT Fermi Source Catalog (1FGL; Abdo *et al.* 2010b) derived from analysis of data taken during 11 months of observation between 2008 August 4 and 2009 July 4. There are 1043 1FGL sources at high latitudes, of which 671 are associated with 709 AGNs, with the larger number of AGNs than 1LAC sources due to multiple associations. Associations are made by comparing the localization contours with counterparts in various source catalogs, for example, the flat-spectrum 8.4 GHz CRATES (Combined Radio All-Sky Targeted Eight GHz Survey; Healey *et al.* 2007) and the Roma BZCAT blazar catalog (Massaro *et al.* 2009). The probability of association is calculated by comparing the likelihood of chance associations with catalog sources if randomly distributed. Positional coincidence only affords an association; correlated flux variability between different wavebands is required for a firm identification.

Of the 671 associations, 663 are considered "high-confidence" associations due to more secure positional coincidences. The "clean" sample is a subset of the high-confidence associations consisting of 599 AGNs with no multiple associations or other peculiarities. As listed in Table 1, these subdivide into 275 BL Lac objects, 248 flat spectrum radio quasars, 26 other AGNs, and 50 AGNs of unknown types. The "New Classes" category

Table 1. Classes of γ-ray emitting AGNs and galaxies in the 1LAC "clean" sample

Class	Number	Characteristics	Prominent Members	Other
All	599			
BL Lac objects	275	weak emission lines	AO 0235+164	
...LSP	64	$\nu_{pk}^{syn} < 10^{14}$ Hz	BL Lacertae	
...ISP	44	10^{14} Hz $< \nu_{pk}^{syn} < 10^{15}$ Hz	3C 66A, W Comae	
...HSP	114	$\nu_{pk}^{syn} > 10^{15}$ Hz	PKS 2155-304, Mrk 501	
FSRQs	248	strong emission lines	3C 279, 3C 354.3	
...LSP	171		PKS 1510-089	
...ISP	1			
...HSP	1			
New Classes[1]	26			
...Starburst	3	active star formation	M82, NGC 253	
...MAGN	7	steep radio spectrum AGNs	M87, Cen A, NGC 6251	
...RL-NLS1s	4	strong FeII, narrow permitted lines	PMN J0948+0022	
...NLRGs	4	narrow line radio galaxy	4C+15.05	
...other sources[2]	9			
Unknown	50			

[1] Total adds to 27, because the RL-NLS1 source PMN J0948+0022 is also classified as an FSRQ in the 1LAC
[2] Includes PKS 0336-177, BZU J0645+6024, B3 0920+416, CRATES J1203+6031, CRATES J1640+1144, CGRaBS J1647+4950, B2 1722+40, 3C 407, and 4C +04.77

contains non-blazar AGNs, including starburst galaxies and radio galaxies of various types (e.g., narrow line and broad line). An AGN is classified as an "unknown" type either because it lacks an optical spectrum, or the optical spectrum has insufficient statistics to determine if it is a BL Lac objects or a FSRQ. In comparison with the 671 AGNs in the 1LAC taken with 11 months of Fermi data, EGRET found 66 high-confidence ($> 5\sigma$) detections of blazars out of total of 271 sources in the 3EG, with another 27 lower-confidence detections with significance between 4σ and 5σ. Thus the 1LAC already represents an order-of-magnitude increase in the number of AGNs over EGRET. There are $\approx 300 - 400$ unassociated and therefore unidentified high-latitude Fermi sources in the LBAS.

4. Classification of radio-emitting AGNs and unification

Different classes of extragalactic AGNs are defined according to observing frequency. We already noted the association of Fermi sources with BL Lac objects and flat spectrum radio quasars (FSRQs). This represents an optical classification. The precise definition used by the Fermi team is that an AGN is a BL Lac object if the equivalent width of the strongest optical emission line is < 5Å, and the optical spectrum shows a Ca II H/K break ratio < 0.4 in order to ensure that the radiation is predominantly nonthermal (the Ca II break arises from old stars in elliptical galaxies). The wavelength coverage of the spectrum must satisfy $(\lambda_{max} - \lambda_{min})/\lambda_{max} > 1.7$ in order that at least one strong emission line would have been detected if present. This helps guard against biasing the classification for AGNs at different redshifts where the emission lines could be redshifted out of the relevant wavelength range. For sources exhibiting BL Lac or FSRQ characteristics at different times, the criterion adopted is that if the optical spectrum conforms to BL Lac properties at any time, then it is classified as a BL Lac object.

The criterion for classification of radio galaxies according to their radio properties stems from the remarkable correlation between radio morphology and radio luminosity (Fanaroff & Riley 1974). The twin-jet morphology of radio galaxies is seen in low-power radio galaxies, whereas the lobe and edge-brightened morphology is found in high-power radio galaxies, with a dividing line at $\approx 2 \times 10^{25}$ W/Hz at 178 MHz, or at a bolometric radio luminosity of $\approx 2 \times 10^{40}$ erg s^{-1}. Besides a radio-morphology/radio-power classification, radio spectral hardness can also be used to characterize sources as flat-spectrum and steep-spectrum sources. The misaligned AGNs (MAGNs) are 1LAC sources associated with steep ($F_\nu \propto \nu^{-\alpha_r}$, with $\alpha_r > 0.5$) radio-spectrum objects typically at 178 MHz in the Third Cambridge and Molonglo radio catalogs that show extended radio structures in radio maps. By contrast, some radio galaxies show hard radio spectra in the 1 – 10 GeV range, and can be subdivided according to the widths of the optical emission lines into broad- and narrow-line radio galaxies (Perlman *et al.* 1998). Because of the close relation between core dominance and γ-ray spectral properties, core dominance can be used to infer the alignment of a radio-emitting AGN (e.g., Lister & Homan 2005).

Blazars and radio galaxies can also be classified according to their broadband spectral energy distribution (SED) when there is sufficient multiwavelength coverage to reconstruct a spectrum from the radio through the optical and X-ray bands. When the peak frequency $\nu_{pk}^{\rm syn}$ of the synchrotron component of the spectrum is $< 10^{14}$ Hz, then a source is called low synchrotron-peaked (LSP), whereas if the SED has $\nu_{pk}^{\rm syn} > 10^{15}$ Hz, then it is referred to as high synchrotron-peaked (HSP). Intermediate synchrotron-peaked (ISP) objects have 10^{14} Hz $< \nu_{pk}^{\rm syn} < 10^{15}$ Hz. SEDs of the bright Fermi LBAS sources are constructed in Abdo *et al.* (2010c). Essentially all FSRQs are LSP blazars, whereas BL Lac objects sample the LSP, ISP, and HSP range.

According to the standard unification scenario for radio-loud AGNs (Urry & Padovani 1995), radio galaxies are misaligned blazars, and FR1 and FR2 radio galaxies are the parent populations of BL Lac objects and FSRQs, respectively. To establish this relationship requires a census of the various classes of sources that takes into account the different beaming properties for the Doppler-boosted radiation of blazars. Even if analysis of data of radio galaxies and blazars supports the unification hypothesis, this paradigm still does not explain the reasons for the differences between radio-quiet and radio-loud AGNs, or between BL Lac objects and FSRQs.

Other classes of extragalactic Fermi sources found in the 1LAC include starburst galaxies, narrow line radio galaxies (NLRGs), radio-loud narrow line Sy 1s (RL-NLS1s), and radio-quiet AGNs. The recent γ-ray detections of the starburst galaxies M82, NGC 253, and NGC 4945 (though the latter source has a Sy 2 nucleus that might contribute to the γ-ray emission) with Fermi, VERITAS, and HESS opens up a new avenue of research in cosmic-ray studies. Five NLRGs are reported in the 1LAC. These objects have narrow emission lines in their optical spectrum, suggesting that they are observed at large angles with respect to the jet direction, with the surrounding dust torus obscuring the broad line region (BLR).

RL-NLS1s have also been recently established as a γ-ray source class (Abdo *et al.* 2009c). These objects show narrow Hβ lines with FWHM line widths $\lesssim 2000$ km s^{-1}, weak forbidden lines ($[OIII]/H\beta < 3$) and a strong Fe II bump, and are therefore classified as narrow-line type I Seyferts (Pogge 2000). By comparison with the $\sim 10^9 M_\odot$ black holes in blazars, the host galaxies of RL-NLS1s are spirals that host nuclear black holes with relatively small ($\sim 10^6 - 10^8 M_\odot$) mass that accrete at a high Eddington rate. The detection of these objects challenges scenarios where radio-loud AGNs are hosted by elliptical galaxies that form as a consequence of galaxy mergers.

The 1LAC includes 10 associations with radio-quiet AGNs. Radio-quiet AGNs have not yet been established as a γ-ray source class. In 8 of these cases, at least one blazar, radio galaxy, or CRATES source is also found close to the γ-ray source. In the remaining two cases, the association probabilities are weak. Thus none appear in the 1LAC "clean" sample. It remains unclear whether any radio-quiet sources, including Sy 2 galaxies such as NGC 4945 or NGC 1068 (Lenain *et al.* 2010), produce γ rays from Sy nuclei rather than cosmic-ray processes.

5. Properties of Fermi AGNs

Various correlations are found by comparing γ-ray properties of Fermi AGNs according to their radio, optical, or SED classification. Probably the most pronounced correlation is between the γ-ray spectral index Γ_γ measured between 100 MeV and 100 GeV, and optical AGN type. FSRQs have significantly softer spectra than BL Lac objects, with $\langle \Gamma_\gamma \rangle \cong 2.40 \pm 0.17$ for FSRQs and $\langle \Gamma_\gamma \rangle \cong 1.99 \pm 0.22$ for BL Lac objects in the LBAS (Abdo *et al.* 2009a). The SED classification shows that the γ-ray spectral index progressively hardens from $\langle \Gamma_\gamma \rangle \cong 2.48, 2.28, 2.13$, and 1.96 when the class varies from FSRQs to LSP-BL Lacs, ISP-BL Lacs, and HSP-BL Lacs, respectively.

Fermi data reveal complex γ-ray blazar spectra. All FSRQs and LSP-BL Lac objects, and most ISP blazars show breaks in the $\approx 1 - 10$ GeV range (Abdo *et al.* 2010d). Such a break was apparent from the first observation of the bright blazar 3C 454.3 (Abdo *et al.* 2009d), and will be discussed in more detail below. This has the unfortunate consequence, however, to reduce the utility of the FSRQs for EBL studies. The HSP blazars, though, are generally well-described by a flat or rising νF_ν SED in the GeV range, with spectral breaks between $\approx 10 - 100$ GeV implied from $\gtrsim 100$ GeV measurements with air Cherenkov telescopes.

There is a much higher ratio, close to unity, of BL Lacs to FSRQs in the 1LAC compared to the 3EG, where the ratio was $\approx 25\%$. This is in large part due to the better multi-GeV sensitivity of Fermi over EGRET which finds far more HSP BL Lac objects than with EGRET, though stronger redshifting effects on the higher-redshift FSRQs than BL Lacs could also contribute. Only 121 out of 291 BL Lac objects had measured redshifts at the time of publication of the 1LAC. For sources with measured redshift z, BL Lac objects are mostly found at low ($z \lesssim 0.4$) redshifts, with only a few HSP BL Lac objects at higher redshifts. By contrast, the FSRQs span a wide range from $z \approx 0.2$ to the highest redshift 1LAC blazar with $z = 3.10$.

This significant redshift incompleteness hampers interpretation of AGN properties, and the overall behavior and explanation of the blazar sequence (see discussion in Abdo *et al.* 2010a), and the role of RL NLS1s in understanding the sequence Foschini *et al.*(2009), if it is not due to selection biases (Padovani *et al.* 2003)). Nevertheless, a plot of Γ_γ as a function of apparent isotropic γ-ray luminosity L_γ in the 100 MeV – 10 GeV band for those sources with measured redshift can be constructed (Fig. 1; from Abdo *et al.* 2010e). It shows that the hard-spectrum BL Lac objects typically have much lower L_γ than the FSRQs. This divide has been interpreted as a change in the accretion regime at approximately 1% of the Eddington luminosity (Ghisellini *et al.* 2009). In addition, the nearby radio galaxies with $z \lesssim 0.1$ inhabit a separate portion of the Γ_γ vs. L_γ plane, and are characterized by lower γ-ray luminosities than their parent populations. The values of L_γ and Γ_γ of the two more distant steep spectrum radio sources, 3C 207 ($z = 0.681$) and 3C 380 ($z = 0.692$), and the FR2 radio galaxy PKS 0943-76 ($z = 0.27$) fall, however, within the range of γ-ray luminosities measured from FSRQs.

Both core and lobes can significantly contribute to the measured γ-ray luminosities. In the case of Centaurus A, the values of L_γ of the core and lobes are comparable, with the lobe emission primarily attributed to Compton-scattered CMBR (Abdo *et al.* 2010f). The significant or dominant lobe component means that the core luminosity of misaligned AGNs can be less than the measured L_γ unless the lobe emission is resolved.

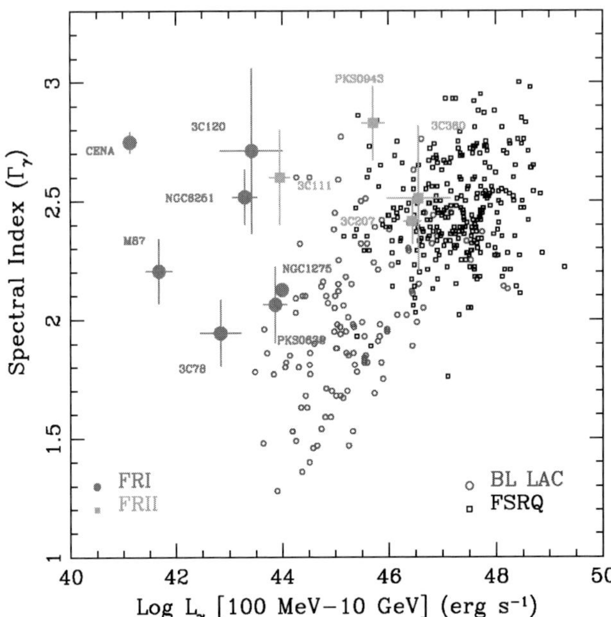

Figure 1. Gamma-ray spectral slopes of FR1 radio galaxies (red circles), FR2 radio sources (green squares), BL Lac objects (open blue circles) and FSRQs (open black squares), are plotted as a function of their 100 MeV - 10 GeV γ-ray luminosity (Abdo *et al.* 2010e). The γ-ray emitting misaligned AGNs (MAGNs) are the red and green points.

5.1. *3C 454.3*

The FSRQ 3C 454.3, at $z = 0.859$, underwent giant flares and became the brightest γ-ray source in the sky for a week in 2009 December and again in 2010 April (Ackermann *et al.* 2010). The latter outburst triggered a pointed-mode observation by Fermi. During the December outburst, its daily flux reached $F = 22(\pm1) \times 10^{-6}$ ph($>$ 100 MeV) cm^{-2} s^{-1}, corresponding to an apparent isotropic luminosity of $\approx 3 \times 10^{49}$ erg s^{-1}, making it the most luminous blazar yet detected with Fermi. Using the measured flux and a one-day variability timescale at the time that the most energetic photon (with energy $E \approx 20$ GeV) was detected, implies a minimum Doppler factor of $\delta_{\rm D,min} \approx 13$. Assuming that the outflow Lorentz factor $\Gamma \approx 20$, consistent with the inferred value of $\delta_{\rm D,min}$ and with radio observations at a different epoch (Jorstad *et al.* 2005), then simple arguments suggests a location $R \lesssim c\Gamma^2 t_{\rm var}/(1 + z) \approx 0.2(\Gamma/20)^2 (t_{\rm var}/{\rm day})$ pc, which is at the outer boundary of the BLR. Flux variability on time scales as short as 3 hr was measured at another bright flux state, which suggests that the γ-ray emission site would be even deeper in the BLR. This stands in contrast to inferences based on coherent optical polarization changes over timescales of weeks prior to a γ-ray flare in 3C 279 (Abdo *et al.* 2010g), which seems to place the emission site at much larger distances.

Figure 2 shows the light curve of 3C 454.3 taken from the public website on Fermi monitored sources†, measured in durations of one day over the course of the Fermi mission. Intense flaring occurs during periods of enhanced activity, as if the engine is being fueled prior to reaching a state coinciding with maximum energy output. Indeed, γ-ray flux enhancements reach a plateau preceding a major flare, and the 2008 July outburst shows strong resemblance to those in 2009 August and 2010 December.

† fermi.gsfc.nasa.gov/ssc/data/access/lat/msl_lc/

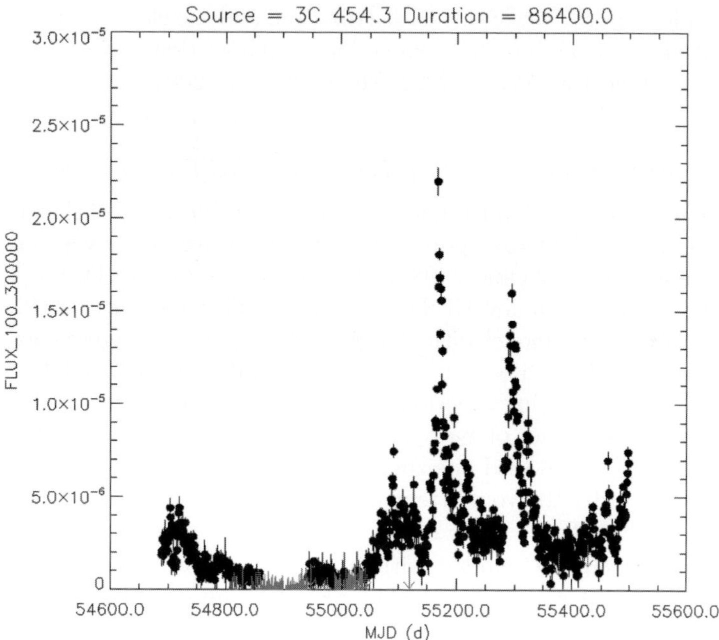

Figure 2. Fermi LAT 1-day light curve of 3C 454.3, showing the giant flares in December 2009, and April 2010 (MJD 55200 corresponds to 4 January 2010).

As noted already in the initial Fermi report (Abdo *et al.* 2009d), the spectrum of 3C 454.3 breaks strongly by 1.2($\pm$0.3) units at $E_{br} \approx 2$ GeV. Such a break is inconsistent with simple radiative cooling scenarios, which predict a break by 0.5 units. The more recent analysis of 3C 454.3 data (Ackermann *et al.* 2010), including the major outbursts, confirms the strong spectral break and finds that E_{br} is very weakly dependent on the flux state, even when the flux changes by more than an order of magnitude. No obvious pattern is found in the spectral index/flux plane, as might be expected in simple acceleration and cooling scenarios (e.g., Kirk *et al.* 1998).

The origin of the spectral break in 3C 454.3 bears on several important issues in FSRQs: the location of the γ-ray emission site; the source of soft target photons in Compton-scattering models; and the relation of FSRQs and BL Lacs in view of the disappearance of such breaks in ISP and HSP blazars. Such a break would be readily understood if the target field was sufficiently intense to attenuate the blazar radiation by $\gamma\gamma$ absorption processes, but the intense line Lyα radiation field at 10.2 eV observed in GALEX measurements of 3C 454.3 (Bonnoli *et al.* 2010) implies $E_{br} \gtrsim 30$ GeV (Reimer 2007). Photon attenuation deep within the BLR by He II recombination and Lyα radiation with $E > 54.4$ eV has been proposed (Poutanen & Stern 2010), but the model lacks a consistent treatment of synchrotron and Compton-scattered blazar flare emission. A full spectral model for the SED of 3C 454.3 can fit the break with a distribution of nonthermal electrons that scatters accretion-disk and BLR radiation (Finke & Dermer 2010), but a solution insensitive to changes in the location of the emission site requires a BLR with a wind-like density.

The spectral break could also be due to Klein-Nishina effects in scattering, as has been proposed to explain the SED of PKS 1510-089 (Abdo *et al.* 2010h). The KN break due to upscattered Lyα radiation occurs at a few GeV, and the observed break energy is insensitive to the Doppler factor. But the break is not sharp enough to fit the spectrum of 3C

454.3 spectrum for a power-law electron distribution, and the electron spectrum resulting from cooling is likely to exhibit a hardening due to competition between synchrotron and Compton losses (Dermer & Atoyan 2002, Moderski *et al.* 2005).

5.2. *PKS 2155-304, HSP BL Lac objects, and FR1 radio galaxies*

PKS 2155-304, an X-ray selected BL Lac object at $z = 0.116$, is one of the most prominent representatives of the HSP blazar population. Its two-peaked SED was measured for 11 days between 2008 August 25 and 2008 September 6 at optical (ATOM), X-ray (RXTE and Swift), and γ-ray (Fermi and HESS) frequencies. This low-state SED is well fit by a one-zone synchrotron/SSC model with Doppler factor $\delta_{\rm D} = 32$, magnetic field $B' = 0.018$ G, and comoving size $R' = 1.5 \times 10^{17}$ cm (corresponding to a variability time of 2 d; Aharonian *et al.* 2009). During a period of extraordinary flaring on 2006 July 28, PKS 2155-304 exhibited a sequence of γ-ray flares varying on time scales as short as ≈ 5 min (Aharonian *et al.* 2007). Detailed one-zone synchrotron SSC model fits using a low EBL require $\delta_{\rm D} \gtrsim 100$ in order to avoid attenuation of the TeV γ rays and provide a good fit to the broadband SED (Finke *et al.* 2008).

One-zone synchrotron/SSC models with $\delta_{\rm D} \gtrsim 10$ give good fits to other HSP BL Lac objects such as Mrk 421 and Mrk 501. The SEDs of radio galaxies, in contrast, are fit with much smaller Doppler factors. The SED of the core of Cen A, for instance, can be fit with $\delta_{\rm D} \approx 1$ and bulk Lorentz factors $\Gamma \approx$ few (Abdo *et al.* 2010i). Likewise, the SEDs of the FR1 radio galaxies NGC 1275 (Abdo *et al.* 2009e) and M87 (Abdo *et al.* 2009f) are well fit with $\delta_{\rm D} \approx 2$ and $\Gamma \sim 4$.

The much larger values of Γ for BL Lac objects than for their putative parent population, the misaligned FR1 radio galaxies, is contrary to simple unification expectations. Moreover, even though the γ-ray luminosities from FR1 radio galaxies are much smaller than that of BL Lac objects (Fig. 1), they are still larger than expected by debeaming the radiation of BL Lac objects with $\Gamma \gtrsim 20$. Additional soft target photons that can be Compton scattered to high energies result in a reduction of the value of $\delta_{\rm D}$ compared to those implied by the one-zone synchrotron/SSC model. These target photons can be produced in a structured jet, as in the spine and sheath model (Chiaberge *et al.* 2000). Another soft photon source arises if blazar flows decelerate from the inner jet to the pc scale (Georganopoulos & Kazanas 2003), in accord with the mildly relativistic flows at the sub-pc scale found in radio observations of Mrk 421 and Mrk 501.

Given the results of synchrotron/SSC modeling, the Fermi observations suggest that the γ-ray emission from the core of a radio galaxy originates from a slower region than the emission of BL Lac objects. This can be understood in a colliding shell model if shells with large opening angles θ_j have lower Γ and are less energetic than shells with narrow opening angles. Such a circumstance might also explain the short variability timescale of PKS 2155-304, as we now show.

Consider a simple colliding shell model where the second shell is much more powerful and has a much larger Lorentz factor than the first shell. Denote the Lorentz factors of the first and second shells by $\Gamma_{a(b)} = 1/\sqrt{1 - \beta_{a(b)}^2}$, with $\zeta \equiv \Gamma_b/\Gamma_a \gg 1$. The collision radius takes place at $r_{\rm coll} \approx 2\Gamma_a^2 \Delta t_*$, where $\Delta t_* \gtrsim R_{\rm S}/c$, and ejection time scales are required to be greater than $R_{\rm S}/c$, where $R_{\rm S}$ is the Schwarzschild radius. If the second shell is much more powerful than the first, then a strong forward shock is formed that travels through shell a with Lorentz factor $\bar{\Gamma}_f$ given by the relative Lorentz factor Γ_{rel} of the two shells, so that $\bar{\Gamma}_f \cong \Gamma_{rel} = \Gamma_a \Gamma_b (1 - \beta_a \beta_b) \to (\zeta + \zeta^{-1})/2 \to \zeta/2$ (see Sari & Piran 1995).

For short timescale variability, both the radial and angular timescales must be much shorter than R_S/c. The radial timescale

$$t_{rad} = \frac{1+z}{\delta_D}\frac{\Gamma_a \Delta t_*}{\bar{\beta}_f \bar{\Gamma}_f} \approx \frac{\Gamma_a \Delta t_*}{\Gamma \bar{\Gamma}_f} \approx 2\Delta t_*/\zeta^2 \qquad (5.1)$$

for low-redshift sources, noting that $\Gamma_a \Delta t_*/\bar{\Gamma}_f$ is the width of shell a in the frame of shell b, and shell widths $\sim c\Delta t_*$ in the engine frame are required to be $\gtrsim R_S/c$. Here the Doppler factor $\delta_D \approx \Gamma \approx \Gamma_b$, where Γ is the shocked fluid Lorentz factor. The angular timescale

$$t_{ang} = \frac{(1+z)r_{\rm coll}}{\Gamma^2 c} \approx 2\frac{\Gamma_a^2 \Delta t_*}{\Gamma^2} \approx 2\Delta t_*/\zeta^2 \ . \qquad (5.2)$$

In both cases, the variability timescale can be much shorter than R_S/c when $\zeta \gg 1$.

The criterion for a strong forward shock is that the ratio of apparent luminosities $L_{*b}/L_{*a} \gtrsim \zeta^4$. This may seem an extreme requirement if the opening angles of the slow and fast shells are the same. But if $\theta_{a(b)} \propto 1/\Gamma_{a(b)}$, $\theta \sim 1/\Gamma$, a condition arising in simulations of relativistic jets (e.g., Komissarov *et al.* 2009, and references therein), then excessive energy requirements for the fast shell are mitigated. The maximum radiative efficiency is $\approx \zeta^2 L_{*a}/2$ (Dermer & Razzaque 2010), so this system is necessarily very inefficient. This suggestion may not only resolve the short variability measured in PKS 2155-304, but also, as noted above, possibly relieve the Doppler-factor conflict in unification schemes of blazars and radio galaxies. A more detailed treatment is in preparation.

6. Blazars and ultra-high energy cosmic rays

In Fermi acceleration scenarios, two conditions are required to accelerate particles to ultra-high energies $E \gtrsim 10^{20}$ eV. The first is that the isotropic power must exceed $\approx 10^{46}\Gamma^2/Z^2$ erg s^{-1}, where Γ is the shocked fluid Lorentz factor, and Ze is the charge. The second is that the sources have luminosity density $\gtrsim 10^{44}$ erg Mpc^{-3} yr^{-1} within the GZK radius to power UHECRs against photohadronic losses on the CMBR. Consider the FR1 radio galaxy NGC 1275, at ≈ 75 Mpc. Its γ-ray luminosity from the 1LAC is $L_\gamma \cong 1.2 \times 10^{44}$ erg s^{-1}. For the mildly relativistic outflow speeds from synchrotron/SSC models, it has adequate power to accelerate Fe nuclei. Moreover, it is radiating nonthermal γ-ray power within this volume with a luminosity density $\approx 20 \times 10^{45}$ erg Mpc^{-3} yr^{-1}. If a small fraction of this power is channeled into UHECRs, then FR1 radio galaxies and BL Lac objects are favored to be the sources of UHECRs, provided that UHECRs are high-Z ions (Dermer & Razzaque 2010).

7. Summary

The Fermi LAT is providing a uniform spectral and temporal GeV database with much better sensitivity than either the EGRET or AGILE missions. Besides BL Lac objects and Flat Spectrum Radio Quasars, several new classes of γ-ray galaxies have been established, including radio galaxies, radio-loud narrow line Sy 1 galaxies, and star-forming galaxies. The γ-ray spectral slope of blazars is strongly correlated with γ-ray luminosity and whether they are BL Lac objects or FSRQs, though a large fraction of BL Lac objects still lack redshift measurements. FSRQs exhibit spectral cutoffs between $1 - 10$ GeV, the reason for which is not well understood. Gamma-ray emitting FR1 radio galaxies have much lower γ-ray luminosities than BL Lac objects, but not as low as expected for one-zone synchrotron/SSC models. Colliding shells with a range of collision Lorentz factors might account for the different γ-ray luminosities of radio galaxies and blazars, and the short variability timescale measured in PKS 2155-304. BL Lacs and FR1 galaxies have sufficient power and luminosity density to account for the UHECRs.

Acknowledgements

This work is supported by the Office of Naval Research and NASA. I thank L. Foschini for comments and T. Piran for criticism. The *Fermi* LAT Collaboration acknowledges support from a number of agencies and institutes for both development and the operation of the LAT as well as scientific data analysis. These include NASA and DOE in the United States, CEA/Irfu and IN2P3/CNRS in France, ASI and INFN in Italy, MEXT, KEK, and JAXA in Japan, and the K. A. Wallenberg Foundation, the Swedish Research Council and the National Space Board in Sweden. Additional support from INAF in Italy and CNES in France for science analysis during the operations phase is also gratefully acknowledged.

References

Aharonian, F., *et al.* 2007, *ApJL*, 664, L71

Aharonian, F., *et al.* 2009, *ApJL*, 696, L150

Abdo, A. A., *et al.* 2009a, *ApJ*, 700, 597 (LBAS)

Abdo, A. A., *et al.* 2009b, *ApJS*, 183, 46 (0FGL)

Abdo, A. A., *et al.* 2009c, *ApJL*, 707, L142 (RL-NLS1s)

Abdo, A. A., *et al.* 2009d, *ApJ*, 699, 817 (3C 454.3)

Abdo, A. A., *et al.* 2009e, *ApJ*, 699, 31 (NGC 1275)

Abdo, A. A., *et al.* 2009f, *ApJ*, 707, 55 (M87)

Abdo, A. A., *et al.* 2010a, *ApJ*, 715, 429 (1LAC)

Abdo, A. A., *et al.* 2010b, *ApJS*, 188, 405 (1FGL)

Abdo, A. A., *et al.* 2010c, *ApJ*, 716, 30 (LBAS broadband SED)

Abdo, A. A., *et al.* 2010d, *ApJ*, 710, 1271 (LBAS GeV SED)

Abdo, A. A., *et al.* 2010e, *ApJ*, 720, 912 (MAGN)

Abdo, A. A., *et al.* 2010f, *Science*, 328, 725 (Cen A lobes)

Abdo, A. A., *et al.* 2010g, *Nature*, 463, 919 (3C 379)

Abdo, A. A., *et al.* 2010h, *ApJ*, 721, 1425 (PKS 1510-089)

Abdo, A. A., *et al.* 2010i, *ApJ*, 719, 1433 (Cen A core)

Ackermann, M., *et al.* 2010, *ApJ*, 721, 1383 (3C 454.3 flares)

Atwood, W. B., *et al.* 2009, *ApJ*, 697, 1071

Bonnoli, G., Ghisellini, G., Foschini, L., Tavecchio, F., & Ghirlanda, G. 2010, *MNRAS*, in press (arXiv:1003.3476)

Casandjian, J.-M. & Grenier, I. A. 2008, *A&A*, 489, 849

Chiaberge, M., Celotti, A., Capetti, A., & Ghisellini, G. 2000, *A&A*, 358, 104

Dermer, C. D. & Atoyan, A. M. 2002, *ApJL*, 568, L81

Dermer, C. D. & Razzaque, S. 2010, *ApJ*, 724, 1366

Fanaroff, B. L. & Riley, J. M. 1974, *MNRAS*, 167, 31P

Finke, J. D. & Dermer, C. D. 2010, *ApJL*, 714, L303

Finke, J. D., Dermer, C. D., Böttcher, M. 2008, *ApJ*, 686, 181

Foschini, L., for the Fermi/LAT Collaboration, Ghisellini, G., Maraschi, L., Tavecchio, F., & Angelakis, E. 2009, arXiv:0908.3313

Fossati, G., Maraschi, L., Celotti, A., Comastri, A., & Ghisellini, G. 1998, *MNRAS*, 299, 433

Georganopoulos, M. & Kazanas, D. 2003, *ApJL*, 594, L27

Ghisellini, G., Maraschi, L., & Tavecchio, F. 2009, *MNRAS*, 396, L105

Healey, S. E., Romani, R. W., Taylor, G. B., Sadler, E. M., Ricci, R., Murphy, T., Ulvestad, J. S., & Winn, J. N. 2007, *ApJS*, 171, 61

Hartman, R. C., *et al.* 1999, *ApJS*, 123, 79

Jorstad, S. G., *et al.* 2005, *AJ*, 130, 1418

Kirk, J. G., Rieger, F. M., & Mastichiadis, A. 1998, *A&A*, 333, 452

Komissarov, S. S., Vlahakis, N., Königl, A., & Barkov, M. V. 2009, *MNRAS*, 394, 1182

Lenain, J.-P., Ricci, C., Türler, M., Dorner, D., & Walter, R. 2010, arXiv:1008.5164

Lister, M. L. & Homan, D. C. 2005, *AJ*, 130, 1389

Massaro, E., Giommi, P., Leto, C., Marchegiani, P., Maselli, A., Perri, M., Piranomonte, S., & Sclavi, S. 2009, *A&A*, 495, 691

Mattox, J. R., Wagner, S. J., Malkan, M., McGlynn, T. A., Schachter, J. F., Grove, J. E., Johnson, W. N., & Kurfess, J. D. 1997, *ApJ*, 476, 692

Moderski, R., Sikora, M., Coppi, P. S., & Aharonian, F. 2005, *MNRAS*, 363, 954

Padovani, P., Perlman, E. S., Landt, H., Giommi, P., & Perri, M. 2003, *ApJ*, 588, 128

Perlman, E. S., Padovani, P., Giommi, P., Sambruna, R., Jones, L. R., Tzioumis, A., & Reynolds, J. 1998, *AJ*, 115, 1253

Pogge, R. W. 2000, *New Astronomy Reviews*, 44, 381

Poutanen, J. & Stern, B. 2010, *ApJL*, 717, L118

Reimer, A. 2007, *ApJ*, 665, 1023

Sari, R. & Piran, T. 1995, *ApJL*, 455, L143

Urry, C. M. & Padovani, P. 1995, *PASP*, 107, 803

Discussion

DAVID MEIER: Early in your talk you mentioned 10 possible RQ objects that have been detected, but I did not see them on your Γ_γ vs. luminosity plot. Where do they lie on that plot?

CHUCK DERMER: The association of radio-quiet AGN with γ-ray sources is very tentative, as we discuss in the 1LAC paper. Moreover, none of them survive in the "clean" sample shown in the plot. Because the detection of γ-ray emission from radio-quiet AGNs could also originate from cosmic-ray processes, as in the case of starburst galaxies, and so might require the detection of γ-ray flux variability to confirm, the Fermi Collaboration is not yet ready to make a definitive statement. The analysis is ongoing.

TSVI PIRAN: In internal shocks the time scale, because of angular spreading, is the variability time of the inner engine. I don't see how you can avoid this in your model.

CHUCK DERMER: The collision radius is set by the speed of the slower shell. The angular spreading is set by the speed of the shocked fluid which, for a strong forward shock, can be much greater than the speed of the slower shell. The smaller opening angle of the high speed, large luminosity shell also reduces the angular spreading time scale. I agree that this process cannot be very efficient, but if $\sim 10\%$ of the Eddington luminosity is channeled into a narrow opening angle jet $\sim 10^{-2}$ for $\Gamma \sim 100$ during bright flares, only $\sim 1\%$ need be converted to yield $\approx 10^{46}$ erg s^{-1} from PKS 2155-304.

ELISABETE M. DE GOUVEIA DAL PINO: Regarding the production of UHECRs by FR1/BL Lac sources, surely they have the "power" to produce them, but how do the CRs escape from these compact sources?

CHUCK DERMER: The escape of UHECRs without suffering photodisintegration in the AGN radiation fields poses an important constraint on the location of UHECR acceleration sites. In FR1 radio galaxies and BL Lac objects, the broad line region is absent, and the accretion-disk radiation is weak. Whether the host galaxy radiation field poses a challenge to this model will require more detailed calculations.

Jets at all Scales
Proceedings IAU Symposium No. 275, 2011
G. E. Romero, R. A. Sunyaev & T. Belloni, eds.

© International Astronomical Union 2011
doi:10.1017/S1743921310015784

The influence of collimation on the appearance of relativistic jets

Pierre-Olivier Petrucci, Timothe Boutelier and Gilles Henri

Laboratoire d'Astrophysique de GrenOble,
Université Joseph Fourier - Grenoble 1 / CNRS
UMR 5571, BP 53, 38041 Grenoble Cedex 09, France
email: pierre-olivier.petrucci@obs.ujf-grenoble.fr

Abstract. The question of the collimation of relativistic jets is the subject of a lively debate in the community. We numerically compute the apparent velocity and the Doppler factor of a non homokinetic jet using different velocity profile, to study the effect of collimation on the appearance of relativistic jets (apparent velocity and Doppler factor). We argue that if the motion is relativistic, the high superluminal velocity are possible only if the geometrical collimation is smaller than the relativistic beaming angle γ^{-1}. In the opposite case, the apparent image will be dominated by the part of the jet traveling directly towards the observer resulting in no apparent velocity. Furthermore, getting rid of the homokinetic hypothesis yields a complex relation between the observing angle and the Doppler factor, resulting in important consequences for the numerical computation of AGN population and unification scheme model.

Keywords. galaxies: active – BL Lacertae objects: individual: – galaxies: jets – gamma-rays: theory – radiation mechanisms: non-thermal

1. Introduction

Jet opening angles are observed in different type of objects like AGNs or YSO (Junor *et al.* 1999; Horiuchi *et al.* 2006).These observations show a decrease of the jet opening angle with distance from the central core, indication of collimation processes. Jet models also predict a variation of the jet opening angle (e.g. Ferreira 1997; Casse & Keppens 2002; McKinney 2006; Hawley & Krolik 2006; Zanni *et al.* 2007)and indeed some of them are enable to reproduce the observations (Dougados *et al.* 2004). However, the existence of a jet opening angle is generally omitted in radiative jet models. We investigate the importance of the jet opening angle using a simple formalism (Boutelier *et al.* 2010, B10 hereafter).

2. Formalism

We consider the simple case of a shell initially spherical, propagating with a relativistic speed characterized by the Lorentz factor γ_0 on the jet axis. In the jet rest frame, we assume that the surface emissivity is uniform with a flat spectrum. The geometrical collimation of the jet is characterized by θ_{jet}, and the velocity distribution is described by a function $\gamma(\theta)$, where θ is the angle to the jet axis. For a point of this surface referenced by the angle θ, the velocity vector is directed in the θ direction (see Fig. 1).

For a given observational angle θ_{obs} defined between the jet axis and the line of sight, we project on the sky plane the surface observed at a given observational time T_{obs} where:

$$T_{obs}(M) = \frac{r}{\beta(\alpha)c} - \frac{r\cos\alpha}{c} = \frac{r}{\beta(\alpha)c}(1 - \beta(\alpha)\cos\alpha) \qquad (2.1)$$

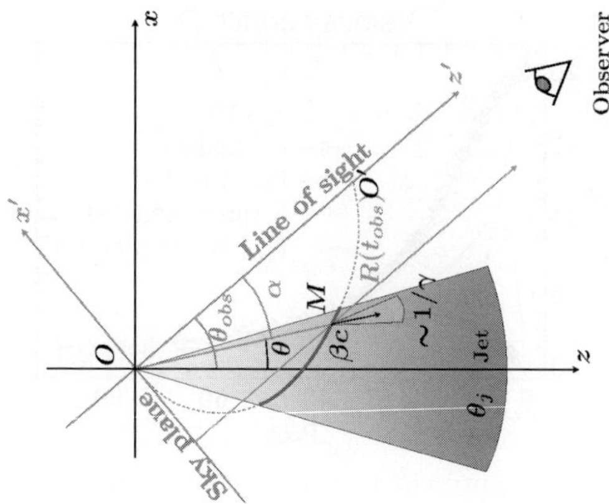

Figure 1. Sketch of the jet model in the case of velocity distribution D_1. See text for the signification of the different parameters

where $\beta(\alpha)$ is the velocity of M deduced from the velocity distribution $\gamma(\theta)$ in the jet frame and $\alpha = \theta_{obs} - \theta$. Hence, two points of the jet M_1 and M_2 will be seen by the observer at the same instant if the observational times reach the condition: $T_{obs}(M_1) = T_{obs}(M_2)$. Let's choose as a reference point, the intersection between the propagating shell and the jet axis. At a given instant t, this point is at distance $r_0(t)$ from the origin, and is characterized by the Lorentz factor γ_0. The parametric equation of the jet surface seen at a given observational time expressed in the observer's frame is then:

$$r(\alpha) = r_0(t) \left(\frac{\beta(\alpha)}{\beta_0} \right) \left[\frac{1 - \beta_0 \cos\theta_{obs}}{1 - \beta(\alpha)\cos\alpha} \right] \tag{2.2}$$

No characteristic scale is involved in this equation which is auto-similar.

The observed flux on the sky plane is related to the intrinsic flux in the source rest frame $S_{\nu,int}$ by the Doppler factor: $S_{\nu,obs} = S_{\nu,int}\delta^3$†. Due to the velocity distribution $\gamma(\theta)$, each point shell have an intrinsic velocity different in norm and direction, and then a different apparent speed as measured by the observer. We choose to define the apparent speed of the whole structure as the one of the brightest point of the sky plane. This is what is expected from VLBI observations for which the apparent speed of a component is computed by fitting the position of the maximum intensity on a temporal sequence of observations. This assumption differs however from the previous works done on the same subject (Gopal-Krishna et al. 2004; Gopal-Krishna *et al.* 2006, 2007). These authors estimate the apparent velocity from the average of the apparent speed of each point of the structure weighted by the Doppler factor boost (see discussion in Boutelier *et al.* 2010). These works also do not take into account the light travel effects that distort the emitting region as seen by the observer. Due to the axial symmetry hypothesis on the jet geometry, the problem of determining the direction of the brightest point of a tridimensional surface projected on the sky plane can be treated in a bi-dimensional approach. Indeed, the direction of the maximum of intensity is necessary in the plane defined by the jet axis and the observer line of sight. Hence, the following work solve only the two dimensions case, as represented on the Fig. 1.

3. Results

Knowing the shape of the observed surface, it is now possible to project it on the sky plane and to compute the intensity profile. It can be shown (B10) that the intensity

† We assume a flat spectrum

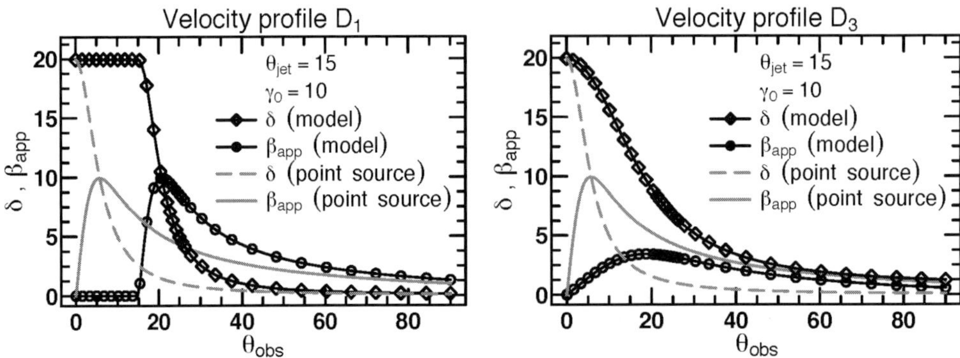

Figure 2. Jet apparent velocity in unit of c (black empty circle) and Doppler factor (black empty diamond), as a function of the observational angle θ_{obs}, computed for different jet velocity profiles; **left:** conical velocity distribution. **right:** gaussian velocity distribution. The jet opening angle is $\theta_{jet} = 15°$, and the Lorentz factor on the jet axis is $\gamma_0 = 10$. As a comparison, the theoretical expression for the apparent velocity (red line) and the Doppler factor (red dashed line) for a point like source of the same Lorenz factor is also shown.

profile is parametrized by the following two equations:

$$\begin{cases} x'(\alpha) = r_0 \sin\alpha \left(\dfrac{\beta(\alpha)}{\beta_0}\right) \left[\dfrac{1 - \beta_0 \cos\theta_{obs}}{1 - \beta(\alpha)\cos\alpha}\right] \\[2ex] I(\alpha) = I_0 \left[\dfrac{\sqrt{1 - \beta(\alpha)^2}}{1 - \beta(\alpha)\cos\alpha}\right]^3 \end{cases} \tag{3.1}$$

We compare two different types of velocity profile $\gamma(\theta)$ (cf. Fig 2): a conical profile, which assumes $\gamma = \gamma_0$ for $\theta < |\theta_{jet}|$, and a gaussian profile where $\gamma(\theta) = 1 + (\gamma_0 - 1)\exp\left[-\ln 2\left(\frac{\theta}{\theta_j}\right)^2\right]$ For the gaussian velocity profile for which the velocity is never constant in the jet, the apparent velocity is always positive for every observational angle but $\theta_{obs} = 0°$. The reason is that in that case, the brightest point of the jet is never on the line of sight, but slightly shifted. However, we emphasize that the apparent velocity is rather small compare to the point-like source approximation. This is due to the lower intrinsic Lorentz factor at this point of the jet, but also because the brightest point is very close to the line of sight ($\alpha < \gamma^{-1}$). This is confirmed by the high values of Doppler factor: $\delta > \beta_{app}$, $\forall\, \theta_{obs}$. We can observe on Fig. 2 that the maximal apparent speed is reached for an observational angle close to the jet opening angle $\theta_{obs} \approx \theta_{jet} \pm \epsilon$.

3.1. *Maximum apparent velocity*

In order to study together the effect of the jet opening angle and the Lorentz factor, we have computed with our model the maximum apparent velocity $\beta_{app,max}(\theta_{jet}, \gamma_0)$ and the associated Doppler factor, for each velocity profile. We show on Fig. 3 the result for the gaussian profile. It confirms the affirmations that we make in the previous sections, i.e. that the jet opening angle decrease dramatically the apparent velocity, even for high jet Lorentz factor. Together with the increase of the opening angle, $\beta_{app,max}$ get farther than the ideal case, that would be materialized by vertical lines on Fig. 3.

More precisely for high Lorentz factor ($\gamma_0 \gg 1$) and small observational angles ($\alpha < \gamma^{-1}$), the Doppler factor can be wrote as $\delta(\alpha, \gamma) \approx \frac{2\gamma}{1 + \gamma^2 \alpha^2}$. If $d\gamma/d\theta \ll \gamma^2$ (i.e. the characteristic angular scale $\Delta\theta$ on which the Lorentz factor varies $\gg 1/\gamma$, the relativistic

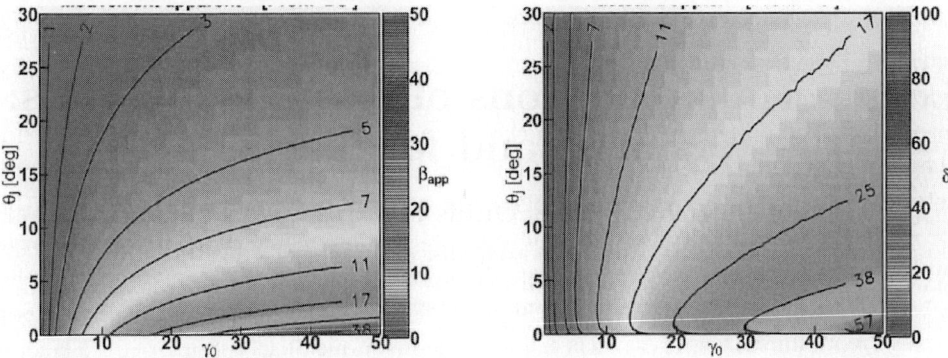

Figure 3. Contour of maximal apparent velocity ($\beta_{app,max}$ on the left) and the associated Doppler factor (on the right), as a function of the geometrical collimation of the jet θ_{jet} and of the Lorentz factor on the jet axis γ_0, for a gaussian velocity profile $\gamma(\theta)$.

beaming angle) it can be shown that the maximum Doppler factor is reached for $\alpha \approx \frac{\dot{\gamma}}{2\gamma^3}$. The corresponding apparent velocity is then of the order of $\beta_{app} \simeq \frac{d\gamma/d\theta}{\gamma} \simeq \frac{1}{\Delta\theta}$.

4. Conclusion

The jet angular aperture can have a significant effect on the jet appearance velocity. For a $\gamma(\theta)$ characterized by an angular scale $\Delta\theta$ the apparent velocity is upper limited by $1/\Delta\theta$. In consequence, large apparent velocities require highly collimated jets. Note also that small apparent velocity and high Doppler factor can be obtained with large opening angle. This has to be taken into account in beamed/unbeamed populations studies. Moreover, jet opening angles varying along the jet would result to different apparent velocity at different position in the jet (e.g. VLBI vs. VLA). Finally, as already discussed by Gopal-Krishna *et al.* (2007), TeV blazars, which require apparently large Doppler factor but show generally subluminal motion, could be characterized by large opening angles.

References

Boutelier, T., Henri, G., & Petrucci, P.-O., 2010, *A&A*, submitted
Casse, F. & Keppens, R. 2002, *ApJ*, 581, 988
Dougados, C., Cabrit, S., Ferreira, J., *et al.* 2004, *Astrophysics and Space Science*, 293, 45
Ferreira, J. 1997, *A&A*, 319, 340
Gopal-Krishna, Dhurde, S., & Wiita, P. J. 2004, *ApJ*, 615, L81
Gopal-Krishna, Sircar, P., & Dhurde, S. 2007, *A&A*, 28, 29
Gopal-Krishna, Wiita, P. J., & Dhurde, S. 2006, *MNRAS*, 369, 1287
Hawley, J. F. & Krolik, J. H. 2006, *ApJ*, 641, 103,
Horiuchi, S., Meier, D. L., Preston, R. A., & Tingay, S. J. 2006, *PASJ*, 58, 211
Junor, W., Biretta, J. A., & Livio, M. 1999, *Nature*, 401, 891
McKinney, J. C. 2006, *MNRAS*, 368, 1561
Zanni, C., Ferrari, A., Rosner, R., Bodo, G., & Massaglia, S. 2007, *A&A*, 469, 811

Jets at all Scales
Proceedings IAU Symposium No. 275, 2011
G. E. Romero, R. A. Sunyaev & T. Belloni, eds.

© International Astronomical Union 2011
doi:10.1017/S1743921310015796

Long-term simulations of extragalactic jets: cavities and feedback

Manel Perucho, Vicent Quilis and José María Martí

Departament d'Astronomia i Astrofísica, Universitat de València
Burjassot 46293
email: `manel.perucho@uv.es`, `vicent.quilis@uv.es`, `jose-maria.marti@uv.es`

Abstract. We present long-term numerical simulations of powerful extragalactic relativistic jets in two dimensions. The jets are injected in a realistic atmosphere with powers 10^{44}, 10^{45} and 10^{46} erg/s, during tens of Myrs. After this time, the jet injection is switched off. We follow the evolution of the jets and associated shocks from 1 kpc to hundreds of kiloparsecs during more than 100 Myrs. The 10^{45} erg/s jet was simulated with leptonic and baryonic composition. Our results show that, for powerful jets, the main heating mechanisms are the driving shock-wave and mixing. We discuss the implications that these results have in the frame of cooling flows in clusters.

Keywords. galaxies:jets, galaxies: cooling flows, galaxies: active, relativistic hydrodynamics

1. Introduction

The rates of cooling of the cluster medium among the galaxies should result in the fall of this gas onto the galaxies, due to the energy loss of this gas (Fabian 1994). This should occur via the so called cooling flows, with mass-fall rates of up to 1000 $M_\odot$/yr. However, these flows are not observed commonly. The present paradigm for explaining this fact relies on heating by Active Galactic Nuclei (see McNamara & Nulsen 2007 for a recent review on the topic). There has been debate about the way in which the heating may occur and whether it is possible at all that the energy brought by AGN jets from the core of the active galaxies to the cluster medium is enough to substantially reduce the cooling flow. The processes that have been invoked are heating by the bow-shock produced at the head of the jet, sound waves, turbulent heating in the wake of buoyant cavities or heat conduction, among others. Cavities have been observed in a number of clusters, despite the difficulty in the detection, as deficits in X-ray emission, normally filled by radio-emission from jet lobes in AGN. These cavities are thought to be close to pressure equilibrium with their environments because any bright rims or shells surrounding them appear to be cooler than the ambient, implying that they are not shocks.

Turbulent heating of the ambient in the wake of the buoyant cavities may become important when the pressure jump between the shocked material and the ambient becomes small. The assumption that buoyancy is the dominant cause of cavity motion comes from the idea that radio lobes reach pressure balance with the surrounding medium in a relative short time. After equilibrium is reached, the region inside the shocked ambient medium, including the shocked jet material and some mixed shocked ambient through the contact discontinuity, i.e., the cocoon, would turn into an X-ray cavity. Buoyancy, and further mixing favoured by smaller velocity gradients through the contact discontinuty would then be possible. In the last years shocks in powerful FRII jets have been reported, like those in Hercules A (Nulsen *et al.* 2005), Hydra A (Simionescu *et al.* 2009) or MS0735.6+7421 (McNamara *et al.* 2005) at distances of hundreds of kiloparsecs to the active nucleus. Some numerical studies have focused on the evolution of cavities by buoyancy (see, e.g., Quilis *et al.* 2001, Brüggen *et al.* 2009). Different works have addressed the problem from the evolution of jets, simulated as an injection of energy and momentum in central cells (Omma *et al.* 2004, Vernaleo & Reynolds 2007, O'Neill & Jones 2010), related to the accretion of cooling flows (Brighenti & Mathews 2006, Cattaneo & Teyssier 2007), or directly injected in one boundary (Zanni *et al.* 2005, Vernaleo & Reynolds 2006). These works have considered intermittency in the galactic activity and some included pressure and density profiles. Most of them report successful transfer of injected energy to the ambient particles, but Vernaleo & Reynolds (2006), who find a failure of the injected jets to stop the cooling flows. It is our aim to test this scenario via numerical simulations of relativistic jets evolving in a realistic environment from a close distance to the nucleus, 1 kpc, to hundreds of kiloparsecs, with the largest possible resolution. We present two-dimensional axisymmetric, RHD simulations following the evolution of jets with different properties, in which the injection of particles and energy at the inlet is gradually stopped after tens of Myrs.

2. Simulations

We used the finite-volume code *Ratpenat*, which solves the equations of relativistic hydrodynamics in conservation form using high-resolution-shock-capturing methods. *Ratpenat* was parallelized with a hybrid scheme with both parallel processes (MPI) and parallel threads (OpenMP) inside each process (see Perucho *et al.* 2010). The code also solves the equation of state of relativistic gas with two populations of particles, namely leptons and baryons.

The jets are injected in a grid including an ambient composed by neutral hydrogen with a King-like density profile in hydrostatic equilibrium. This profile has been taken from the fit to X-ray data of 3C 31 by Hardcastle *et al.* (2002) as typical for an elliptical galaxy. The density at 1 kpc from the galactic nucleus is 0.1 m_p/cm^3. The corresponding dark-matter distribution forcing this equilibrium consists of a halo of $10^{14}\,M_\odot$ within 1 Mpc. We present here results from 4 simulations (see Table 1 for specific properties of each jet). The resolution depends on the simulation with cell size ranging from 50×50 parsecs (J1) to 100×100 parsecs (J2, J3, J4). The jets are injected at 1 kpc, with radius of 100 pc, i.e., one or two cells depending on resolution in a grid that is extended axially and radially as the bow-shock grows. In the radial direction, a grid with the given resolution extends up to 50 kpc (J1) or 100 kpc (J2, J3, J4). Then a grid with increasing cell size is added up to 100 kpc (J1) or 150 kpc (J2, J3, J4). Beyond this distance, the radial cell size is then kept constant at its maximum and the grid increased following the evolution of the bow-shock. Reflection boundary conditions are set at the basis of the grid to mimic the effect of a counter-jet. With the given grid and jet properties, the time-step during the first part of the simulations is of 50 to 100 years. The total injected energy ranges from a 3×10^{59} to 10^{61} erg, depending on the simulation.

The injection of energy and particles through the jet is stopped after 50 Myrs (J1, J3, J4) or 16 Myrs (J2). Before this moment, the dynamics of the jets are dominated by the injection, following expected evolutions. During and after the switch-off, the decrease in the velocity of the head is immediate due to the short time-scales needed by the relativistic fluid in the jet to reach the terminal shock. Mach numbers of the bow-shocks fall from around 10 to values between 1 and 2, but remain in these values from the switch-off to times close to 200 Myrs in J1, J2 and J4. This represents a factor 3 of the active phase in the case of J1 and J4, and more than a factor 10 for J2, the most powerful jet. In all the simulations, the aspect ratio of the bow-shock approaches sphericity after

Table 1. Parameters of the simulated jets. Column 1 gives the model, column 2 the injection velocity, column 3 the injection density, column 4 the leptonic number, column 5 the jet power, column 6 the maximum resolution, and column 7 the switch-off time.

Model	Velocity [c]	Density [kg/m^3]	X_e	L_k [erg/s]	max. resol. [pc/cell]	t_{off} [Myrs]
J1	0.9	8.3×10^{-26}	1.0	10^{45}	50	50
J2	0.984	8.3×10^{-26}	1.0	10^{46}	100	16
J3	0.9	8.3×10^{-27}	1.0	10^{44}	100	50
J4	0.9	8.3×10^{-26}	0.5	10^{45}	100	50

the end of the active phase. The jet material mixes completely with the shocked ambient, forming a low-density region, precursor of the buoyant cavity (see Fig. 1). Although this process starts before the switch-off, it is enhanced after this time. The calculation of the integrated X-ray luminosity across a 3D-box, produced from the axisymmetric grid, shows that this region already appears as less bright than the surroundings. None of our simulations has reached the stage of cavity buoyancy, as the pressure jump persists.

This mixed region is, at the late stages, formed by shocked ambient with polution of jet particles, with maxima of around two per cent of the latter in a given cell, implying very efficient mixing. Croston *et al.* 2008 report that radio lobes of low-power radiogalaxies present lower pressure than expected if it is obtained from the emission of hot particles. They suggest that the missing pressure is provided by entrained ambient particles, what could be confirmed by our results.

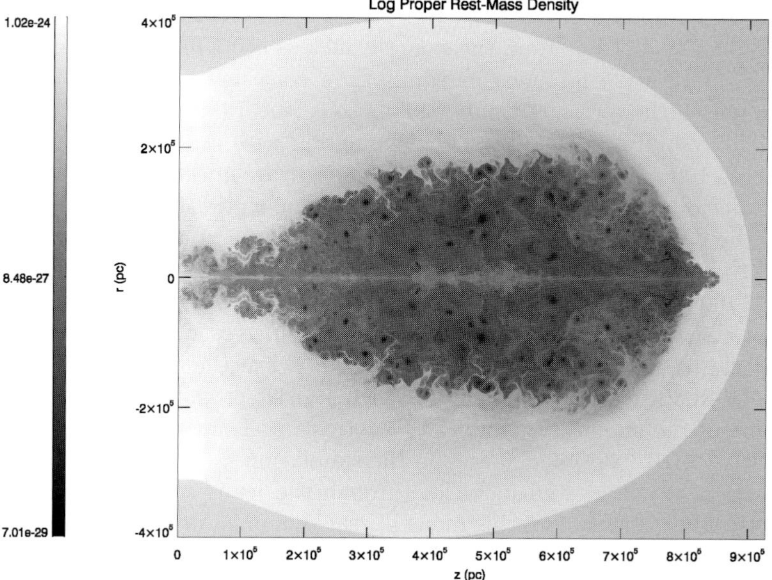

Figure 1. Maps of rest-mass density of simulation J2 at time 1.8 Myrs. The units are kg/m^3.

The energetic balance shows that around 95 % of the injected energy goes to the ambient medium through shock-heating, mixing and acceleration. The rest is kept by the jet particles, invested into the gain of potential energy or lost in radiation. The amount of mass of the ambient that undergoes shock heating ranges from 10^{11} to 10^{12} $M_\odot$ depending on the power injected during the active phase, in agreement with observational results (see McNamara & Nulsen 2007). From this mass, typically 0.1 to 1.0% is mixed by instabilities arising in the contact discontinuity between the shocked ambient and the shocked jet fluid.

3. Conclusions

Our simulations show that the injection of a collimated, relativistic flow in the galactic and intra-cluster medium brings a large amount of energy to this gas, mainly through shock-heating and turbulent mixing with the injected, hot jet material. Contrary to what previously assumed, it seems that the shocks are persistent even during times longer than three times the duration of the active phase, and up to ten times in the case of J2. This result shows that buoyant cavities could represent the fate of radio-galactic relics, but are not needed in the process of stopping cooling flows. The amount of displaced and entrained gas is consistent with observations. In addition, we show that entrainment is very important and this could give an answer to reported missing pressure in radio-lobes if only hot particles are taken into account.

Acknowledgements

The authors want to acknowledge the use of supercomputational resources from the Spannish Supercomputational Network (RES), technical support and assistance from the staff of the nodes "Tirant", "Mare Nostrum" and "Magerit". The authors acknowledge financial support from the Spannish "Ministerio de Ciencia e Innovación" (MICINN) grants AYA2010-21322-C03-01, AYA2010-21097-C03-01 and CONSOLIDER2007-00050, and from the "Generalitat Valenciana" grant "PROMETEO-2009-103". MP acknowledges support from MICINN through a "Juan de la Cierva" contract.

References

Brighenti, F. & Mathews, W. G., 2006, *ApJ*, 643, 120

Brğgen, M., Scannapieco, E., & Heinz, S., 2009, *MNRAS*, 395, 2210

Cattaneo, A. & Teyssier, R., 2007, *MNRAS*, 376, 1547

Croston, J. H., Kraft R. P., & Hardcastle M. J., 2007, *ApJ*, 660, 191

Croston, J. H. *et al.*, 2008, *MNRAS*, 386, 1709

Fabian, A. C., 1994, *ARAA*, 32, 277

Hardcastle, M. J. *et al.*, 2002, *MNRAS*, 334, 182

Kraft, R. P., Nulsen, P. E. J., & Birkinshaw, M., 2007, *ApJ*, 665, 1129

McNamara, B. R. & Nulsen, P. E. J. 2007, *ARAA*, 45, 117

McNamara, B. R., Nulsen, P. E. J., Wise, M. W. *et al.*, 2005, *Nature*, 433, 45

Nulsen, P. E. J., Hambrick, D. C., McNamara, B. R. *et al.*, 2005, *ApJ* (Letters), 625, 9

Omma, H., Binney, J., Bryan, G., & Slyz, A., 2004, *MNRAS*, 348, 1105

O'Neill, S. M. & Jones, T. W., 2010, *ApJ*, 710, 180

Perucho, M. & Martí, J. M., 2007, *MNRAS*, 382, 526

Perucho, M., Martí, J. M., Cela, J. M., Hanasz, M., de la Cruz, R., & Rubio, F., 2010, *A&A*, 519, A41

Quilis, V., Bower, R. G., & Balogh, M. L., 2001, *MNRAS*, 328, 1091

Simionescu, A., Roediger, E., Nulsen, P. E. J. *et al.*, 2009, *A&A*, 495, 721

Vernaleo, J. C. & Reyynolds, C. S., 2006, *ApJ*, 645, 83

Vernaleo, J. C. & Reyynolds, C. S., 2007, *ApJ*, 671, 171

Zanni, C., Murante, G., Bodo, G., Massaglia, S., Rossi, P., & Ferrari, A., 2005, *A&A*, 429, 399

Discussion

YUAN: The jet is well collimated, but the observed cavity is spherical-like. Some people argue we must require a B-field.

PERUCHO: I disagree. Observations seem to indicate that the lobes are dominated by thermal gas. In our simulations, the cavities form after the jets are switched off, without any need of magnetic fields.

Title of your IAU Symposium
Proceedings IAU Symposium No. 275, 2011
G. E. Romero, R. A. Sunyaev & T. Belloni, eds.

© International Astronomical Union 2011
doi:10.1017/S1743921310015802

Radiation from matter entrainment in astrophysical jets: the AGN case

A. T. Araudo[1,2], V. Bosch-Ramon[3] and G. E. Romero[1,2]

[1]Instituto Argentino de Radioastronomía (CCT La Plata, CONICET),
C.C.5, 1894 Villa Elisa, Buenos Aires, Argentina
email: aaraudo@fcaglp.unlp.edu.ar, romero@fcaglp.unlp.edu.ar

[2]Facultad de Ciencias Astronómicas y Geofísicas, Universidad Nacional de La Plata,
Paseo del Bosque, 1900 La Plata, Argentina

[3]Dublin Institute for Advanced Studies, 31 Fitzwilliam Place, Dublin 2, Ireland
email: valenti@cp.dias.ie

Abstract. Jets are found in a variety of astrophysical sources. In all the cases the jet propagates with a supersonic velocity through the external medium, which can be inhomogeneous, and inhomogeneities could penetrate into the jet. The interaction of the jet material with an obstacle produces a bow-like shock within the jet in which particles can be accelerated up to relativistic energies and emit high-energy photons. In this work, we explore the active galactic nuclei scenario, focusing on the dynamical and radiative consequences of the interaction at different jet heights. We find that the produced high-energy emission could be detectable by the current γ-ray telescopes. In general, the jet-clump interactions are a possible mechanism to produce (steady or flaring) high-energy emission in many astrophysical sources in which jets are present.

Keywords. galaxies: active, radiation mechanisms: nonthermal

1. Introduction

Jets at different scales are present in astrophysical sources such as protostars, microquasars and active galactic nuclei (AGN). The medium that surrounds the jets can be inhomogeneous and clumps from this external medium can interact with the jets.

In young stellar objects, the interaction of matter from the external medium with protostar outflows has been proposed to explain the formation of Herbig-Haro (HH) objects. The dynamical properties of this interaction have been numerically simulated by Raga *et al.* (2003), and they conclude that clumps with mass between 10^{-4} and 0.1 times the mass of a giant planet can significantly contribute to the molecular mass of HH outflows. However, the velocity of HH jets is not relativistic ($\sim$ hundred km s^{-1}), and particle acceleration in the bow shock formed in the jet might be inefficient or the possible non-thermal emission produced by accelerated particles might be hidden by thermal radiation.

In high-mass microquasars, the interaction of clumps from the companion stellar wind (Owocki & Cohen, 2006) with the jets of the compact object (black hole or neutron star) can produce strong bow-like shocks within the jet. Electrons and protons accelerated in those shocks can radiate significant amount of γ rays in the form of flaring and steady emission (Araudo *et al.* 2009). In the case where only few clumps interact with the jet, the produced emission is sporadic and can explain the GeV flares detected from some γ-ray binaries (e.g. Abdo *et al.* 2009).

At extragalactic scales, AGN are composed by an accreting supermassive black hole (SMBH) at the center of the galaxy and relativistic jets. Surrounding the SMBH there is a population of clouds (Krolik *et al.* 1981) that move at velocities > 1000 km s^{-1} forming the called broad line region (BLR). In the present contribution, we illustrate the most important dynamical and radiative consequences of the interaction of clouds from the BLR with the base of AGN jets. We found that γ rays from jet-cloud interactions should be detectable by present and future instrumentation in nearby low-luminous AGN at high energy (HE) and very high energy (VHE), and in powerful and nearby quasars only at HE because the VHE radiation is absorbed by the dense nuclear photon fields. In the case of sources exhibiting boosted γ rays (blazars), the isotropic radiation from jet-cloud interactions will be masked by the jet beamed emission, which will not be the case in non-blazar sources.

2. The jet-cloud interaction

We adopt clouds with density $n_c = 10^{10}$ cm^{-3}, size $R_c = 10^{13}$ cm, and velocity $v_c = 10^9$ cm s^{-1}. The jet Lorentz factor is fixed to $\Gamma = 10$, implying a jet velocity $v_j \approx c$, and the radius/height relation is fixed to $R_j = 0.1\,z$. The jet density n_j in the laboratory reference frame (RF) can be estimated as $n_j = L_j/((\Gamma-1)\,m_p\,c^3\sigma_j)$, where $\sigma_j = \pi R_j^2$ and L_j is the kinetic power of the matter-dominated jet.

The completely penetration of a cloud into the jet, the ram pressure of the latter should not destroy the former before the cloud has fully entered into the jet. This means that the time required by the cloud to penetrate into the jet, $t_c \sim 2R_c/v_c = 2 \times 10^4$ s, should be shorter than the cloud lifetime inside the jet. To estimate this cloud lifetime, we first compute the time required by the shock in the cloud to cross it (t_{cs}). The velocity of this shock is $v_{cs} \sim \chi^{-1/2}\,c$, where $\chi = n_c/n_j(\Gamma-1)$, is derived by ensuring that the jet and the cloud shock ram pressures are equal. This yields a cloud shocking time of

$$t_{cs} \sim \frac{2R_c}{v_{cs}} \simeq 7\times10^4 \left(\frac{R_c}{10^{13}\text{ cm}}\right) \left(\frac{n_c}{10^{10}\text{ cm}^{-3}}\right)^{1/2} \left(\frac{z}{10^{16}\text{ cm}}\right) \left(\frac{L_j}{10^{44}\text{ erg s}^{-1}}\right)^{-1/2} \text{ s.} \quad (2.1)$$

Rayleigh-Taylor (RT) and Kelvin-Helmholtz (KH) instabilities produced in the cloud surface by the interaction with the jet material will affect the obstacle. The timescale in which these instabilities grow up to a scale length $\sim R_c$ is $t_{\rm RT/KH} \sim t_{cs}$. For this reason, we take t_{cs} as the characteristic timescale of our study. Therefore, for a penetration time (t_c) at least as long as $\sim t_{cs}$, the cloud will remain an effective obstacle for the jet flow. Setting $t_c \sim t_{cs}$, we obtain the minimum value for the interaction height $z_{\rm int}$, giving

$$z_{\rm int}^{\rm min} \approx 2.5 \times 10^{15} \left(\frac{v_c}{10^9\text{ cm s}^{-1}}\right)^{-1} \left(\frac{n_c}{10^{10}\text{ cm}^{-3}}\right)^{-1/2} \left(\frac{L_j}{10^{44}\text{ erg s}^{-1}}\right)^{1/2} \text{ cm.} \quad (2.2)$$

For $z_{\rm int} > z_{\rm int}^{\rm min}$, the jet crossing time $t_j \sim 2R_j/v_c$ results larger than t_{cs}. Once the cloud is inside the jet, a bow shock around the cloud is formed on a time $t_{bs} \sim Z/v_j \sim 10^2$ s at a distance $Z \sim 0.3 R_c$ cm from the cloud. In this bow shock, particles can be accelerated up to relativistic energies more efficiently than in the cloud shock (because the bow-shock velocity is $v_{bs} \gg v_{cs}$).

3. Non-thermal emission

The non-thermal luminosity of particles accelerated in the bow shock can be estimated as a fraction $\eta_{\rm nt}$ of the bow shock luminosity: $L_{\rm nt} \sim \eta_{\rm nt} L_{bs}$, where $L_{bs} \sim (\sigma_c/\sigma_j)\,L_j$ and $\sigma_c = \pi R_c^2$. The accelerator/emitter magnetic field in the bow-shock RF (B) can be determined by relating $U_B = \eta_B U_{\rm nt}$, where $U_B = B^2/8\pi$ and $U_{\rm nt} = L_{\rm nt}/(\sigma_c c)$ are the magnetic and

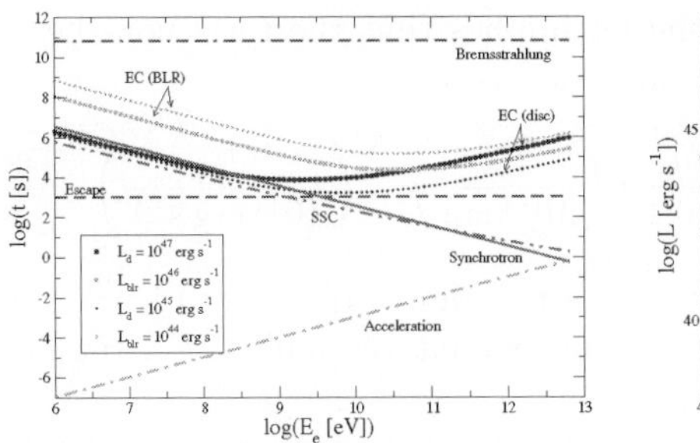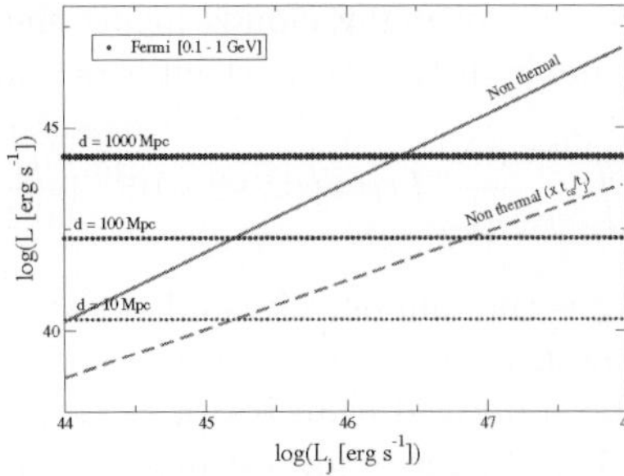

Figure 1. Left: acceleration gain, escape, and cooling lepton timescales are plotted. SSC is plotted when the steady state is reached and EC for both the BLR and disc photon fields are shown for the conditions of faint (BLR: 10^{44}; disc: 10^{45} erg s^{-1}) and bright sources (BLR: 10^{46}; disc: 10^{47} erg s^{-1}). Synchrotron and relativistic bremsstrahlung are also plotted. Right: Upper limits to the γ-ray luminosity produced by N_c^j clouds inside the jet as a function of L_j in FR II sources. Two cases are plotted, one assuming that clouds cross the jet without disruption (green solid lines), and one in which the clouds are destroyed in a time as short as t_{cs} (green dashed lines). In addition, the sensitivity levels of *Fermi* in the range 0.1–1 GeV (maroon dotted lines) are plotted for three different distances $d = 10$, 100, and 1000 Mpc.

the non-thermal energy densities, respectively. Fixing $\eta_{nt} = 0.1$ and $\eta_B = 0.01$, we obtain $L_{nt} \sim 4 \times 10^{39}(L_j/10^{44}$ erg s$^{-1})$ erg s^{-1} and $B \sim 10$ G. Regarding the acceleration mechanism, we adopt the following prescription for the acceleration rate: $\dot{E}_{acc} = 0.1\,q\,B\,c$, where q is the electron charge.

We assume that electrons and protons are injected into the bow-shock region following a power law in energy of index 2.2 and with an exponential cutoff at the maximum energy. The injection luminosity is L_{nt}. Particles are affected by different losses that balance the energy gain from acceleration. The escape time downstream from the relativistic bow shock considers advection ($t_{adv} \sim 3\,R_c/c \sim 10^3$ s) and diffusion ($t_{diff} = 3\,Z^2/2\,r_g\,c$) timescales. In the latter expression, r_g is the particle gyroradius and the Bohm regime has been considered. The most important radiative losses that affect the lepton injection are synchrotron and SSC, determining a maximum energy E_e^{max} of several TeV, as is shown in Figure 1 (left). In the case of protons, pp cooling is negligible in the bow-shock region and the maximum energy is constrained by equating the acceleration and diffusion timescales, given $E_p^{max} \sim 5 \times 10^3\,(B/10\,{\rm G})$ TeV. Then, protons with energies $> 0.4 E_p^{max}$ can reach the cloud by diffusion and radiate there via pp more efficiently than in the jet, since $n_c \gg n_j(z_{int})$.

4. Many clouds interacting with the jets

Many of the BLR clouds can be simultaneously inside the jet at different z, each of them producing non-thermal radiation. Therefore, the total luminosity can be much larger than that produced by just one interaction, which is $\sim L_{nt}$. The number of clouds within the jets, at $z_{int}^{min} \leqslant z \leqslant R_{blr}$, can be computed from the jet (V_j) and cloud (V_c) volumes, resulting in $N_c^j = 2\,f\,V_j/V_c \sim 9(L_j/10^{44}$ erg s$^{-1})^2$, where the factor 2 accounts for the two jets and $f \sim 10^{-6}$ is the filling factor of clouds in the whole BLR. In reality, N_c^j is correct if one neglects the cloud disruption inside the jet. However, even under cloud fragmentation, strong bow shocks can form around the cloud fragments before these have accelerated up to close v_j. Then, the real number of interacting clouds inside the jet is difficult to estimate, but it could be between $(t_{cs}/t_j)\,N_c^j$ and N_c^j.

The presence of many clouds inside the jet implies that the total non-thermal luminosity available in the BLR-jet intersection region is

$$L_{\mathrm{nt}}^{\mathrm{tot}} \sim 2 \int_{z_{\mathrm{int}}^{\mathrm{min}}}^{R_{\mathrm{blr}}} \frac{\mathrm{d}N_{\mathrm{c}}^{\mathrm{j}}}{\mathrm{d}z} L_{\mathrm{nt}}(z)\,\mathrm{d}z \sim 2\times10^{40} \left(\frac{\eta_{\mathrm{nt}}}{0.1}\right) \left(\frac{R_{\mathrm{c}}}{10^{13}\,\mathrm{cm}}\right)^{-1} \left(\frac{L_{\mathrm{j}}}{10^{44}\,\mathrm{erg\,s^{-1}}}\right)^{1.7} \frac{\mathrm{erg}}{\mathrm{s}}, \quad (4.1)$$

where $\mathrm{d}N_{\mathrm{c}}^{\mathrm{j}}$ is the number of clouds located in a jet volume $\mathrm{d}V_{\mathrm{j}} = \pi\,(0.1z)^2\,\mathrm{d}z$. In all the calculations L_{blr} has been fixed to $0.1\,L_{\mathrm{j}}$, as approximately found in FR II galaxies, and R_{blr} has been derived from Kaspi *et al.* (2005).

In Fig. 1 (right), we show estimates of the γ-ray luminosity when many clouds interact simultaneously with the jet. For this, we have followed a simple approach assuming that most of the non-thermal luminosity is converted into γ rays. This will be the case as long as the escape and synchrotron cooling time are longer than the IC cooling time (EC+SSC) at the highest electron energies. In Araudo *et al.* (2010), we present more detailed calculations by applying the model presented in this contribution to two characteristic sources, Cen A and 3C 273.

5. Discussion

We have studied the interaction of clouds from the BLR with the base of jets in AGNs. For very nearby sources, such as Cen A, the interaction of large clouds ($R_{\mathrm{c}} > 10^{13}$ cm) with jets may be detectable as a flaring event, although the number of these large clouds and thereby the duty cycle of the flares are difficult to estimate. Given the weak external photon fields in these sources, VHE photons can escape without experiencing significant absorption. Therefore, jet-cloud interactions in nearby FR I may be detectable in both the HE and the VHE range as flares with timescales of about one day.

In FR II sources, many BLR clouds could interact simultaneously with the jet. The number of clouds depends strongly on the cloud lifetime inside the jet, which could be of the order of several t_{cs}. Nevertheless, we note that after cloud fragmentation many bow shocks may still form and efficiently accelerate particles if these fragments move more slowly than the jet. Since FR II sources are expected to exhibit high accretion rates, radiation above 1 GeV produced in the jet base can be strongly attenuated by the dense disc and the BLR photon fields, although γ rays below 1 GeV should not be affected significantly. Since jet-cloud emission should be rather isotropic, it would be masked by jet beamed emission in blazar sources, although since powerful/nearby FR II jets do not display significant beaming, these objects may produce detectable γ rays through jet-cloud interactions. As shown in Fig. 1 (right), close and powerful sources could be detectable by deep enough observations of *Fermi*.

References

Abdo, A. A. *et al.* 2009, *ApJS*, 183, 46

Araudo, A. T., Bosch-Ramon, V., & Romero, G. E. 2009, *A&A*, 503, 673

Araudo, A. T., Bosch-Ramon, V., & Romero, G. E. 2010, *A&A* (in press) [arXiv:1007.2199]

Begelman, M. C., Blandford, R. D., & Rees, M. J. 1984, *Rev. Mod. Phys.*, 56, 255

Kaspi, S., Maoz, D., Netzer, H., Peterson, B. M., Vestergaard, M., & Jannuzi, B. T. 2005, *ApJ*, 629, 61

Krolik, J. H., McKee, C. F., & Tarter, C. B. 1981, *ApJ*, 249, 422

Owocki, S. P., & Cohen D. H. 2006, *ApJ*, 648, 5650

Raga, A. C., Velzquez, P. F., de Gouveia dal Pino, E. M., Noriega-Crespo, A., & Mininni, P. 2003, *RMxAC*, 15, 115

Discussion

MIRABEL: Can you apply the same model to the MQSO Cyg X-3 that has a WRayet donor and LGRBs that usualy have WRayet progenitors?

ARAUDO: We have applied the model to high-mass microquasars (HMMQ), studying the interaction between clumps from the wind of the companion star with the jet of the compact object (Araudo, Bosch-Ramon & Romero 2009). In this work, we have considered a source with similar parameters to the system Cygnus X-1. Now, we are working on the application of the model to the HMMQ Cygnus X-3. We have not applied the model to gamma-ray bursts.

PERUCHO: Stellar winds shocked by the jet flow could also be an interesting scenario to check. Have you considered it?

ARAUDO: We have not considered the interaction between stars and jets in AGNs, but we are going to do this in the near future. A similar scenario has been studied by Bednarek & Protheroe (1997).

DRAPPEAU: How do calculate the number of clouds at any given time? i.e. What is the destruction rate and the entrainment rate of clouds in jets?

ARAUDO: To estimate the number of clouds into the jet, we calculate $N_c \sim 2 x f x V_j / V_c$, where $f \sim 10^{-6}$ is the filling factor of clouds in the broad line region (BLR), V_c is the volume of the clouds, and V_j is the volume of the jet intesected by the BLR. In this estimation, we not consider the destruction of clouds into the jet. The previous equation for N_c takes into account that the entrainment is equal to the escape rate of the jet, and is independent of time.

Jets at all Scales
Proceedings IAU Symposium No. 275, 2011
G. E. Romero, R. A. Sunyaev & T. Belloni, eds.

© International Astronomical Union 2011
doi:10.1017/S1743921310015814

Time-dependent multi-zone radiation transfer modeling of fast blazar variability

Giovanni Fossati[1] and Xuhui Chen[1]

[1]Department of Physics and Astronomy, Rice University
6100 Main St., Houston, TX 77005, USA
email: gfossati@rice.edu, Xuhui.Chen@rice.edu

Abstract. We present the first applications of a new time-dependent multi-zone jet radiation transfer code to the study the multiwavelength emission of the TeV Blazar Mrk 421. The code couples Fokker-Planck and Monte Carlo methods. For the first time all light travel time effects are fully considered as well as proper self-consistent treatment of Compton cooling, which depends on them. The first tests focus on the March 2001 observations of Mrk 421, still one of the best datasets available for phenomenology and X-ray/TeV data coverage. We summarize the results of scenarios of variability induced by injection of relativistic electrons in a blob encountering a shock, and with different combinations with a second component, either co-spatial or independent from the active region.

Keywords. radiation mechanisms: nonthermal, radiative transfer, galaxies: active, BL Lacertae objects: individual (Mrk 421), galaxies: jets

1. Blazars multiwavelength spectra and variability

Blazars emit strongly from radio through γ-ray energies. Their spectral energy distribution (SED) comprises two major continuum, non-thermal, components attributed to synchrotron and inverse Compton (IC) radiation by ultrarelativistic electrons (Maraschi *et al.* 1992, Marscher & Travis 1996, Dermer *et al.* 1992, Sikora *et al.* 1994). Rapid and large-amplitude multiwavelength variability is a defining observational characteristic of blazars (Ulrich *et al.* 1997).

In a subclass of blazars, commonly referred to as high-peaked BL Lacs (HBL) the synchrotron component peaks (in νF_ν) in the X-ray band, the higher energy component (IC) reaches up to TeV γ-rays, a combination accessible observationally thanks to ground based Cherenkov telescopes and the availability of several X-ray observatories. Hence, the brightest HBLs have been studied extensively. Simultaneous X-ray/γ-ray observations showed that variations around the two peaks are well correlated (Fossati *et al.* 2000a,b, 2008, Sambruna *et al.* 2000, Krawczynski *et al.* 2004, Błażejowski *et al.* 2005, Aharonian *et al.* 2009). The variability observed at/above $\nu_{\rm peak}$ is likely the result of the rapid change of the electron distribution, in the regime where acceleration and cooling are approximately balanced: by studying the variability around $\nu_{\rm peak}$, we are investigating the behavior of the electrons that are the most direct probe of the physical conditions of the emitting plasma (Inoue & Takahara 1996, Kirk *et al.* 1998, Kusunose *et al.* 2000).

2. Time-dependent modeling

The theoretical interpretation of the multiwavelength observations has remained relatively basic (Sikora *et al.* 2001, Krawczynski *et al.* 2002, Böttcher & Chiang 2002). One of the biggest challenges is the treatment of light crossing time effect (LCTE), rarely fully modeled, which affect the observer's "perception" of the variability (e.g. because of delayed times) and the physical evolution within the source, mainly because it affects the correct computation of the IC cooling of the emitting particles. Previous work dealt with the effect of time delays from the observer point of view (Chiaberge & Ghisellini 1999, Kataoka *et al.* 2000, Katarzyński *et al.* 2008), and in some cases included the internal effect to calculate the IC emission but without properly accounting for it when calculating IC energy losses (Sokolov & Marscher 2005, Graff *et al.* 2008).

2.1. *Our code*

We have developed a code in which for the first time all the LCTE are fully considered, internal and external, as well as proper self-consistent treatment of Compton cooling, which depends on them. This code affords us the freedom to "play" with (inhomogeneous and varying) physical conditions, internal and external to the active region, and simulate a broad range of scenarios for blazar variability. The code and its first application are discussed in detail by Chen *et al.* (2010). It couples Fokker-Planck and Monte Carlo (MC) in a 2 dimensional (cylindrical) geometry. It is built on the MC radiative transfer code developed by Böttcher and collaborators (2003). MC is ideal for multizone 2D/3D radiative transfer problems because it tracks the trajectory of every photon, thus automatically accounting for LCTE, regardless of the geometry.

3. Observational open questions

We focused on observational findings that seem to be common to various sources:

- (quasi-)symmetry of flares, in the sense that the rise and decay timescales are often very similar. This would suggest that the flare evolution is governed by a factor that is energy independent, such as the geometry of the active region via its crossing time.

- Amplitude and phase correlations of variations in different bands, typically X-ray and TeV γ-rays for the best observed blazars so far. Particularly challenging is that the TeV emission has been seen to vary super-quadratically with respect to variations in X-ray, in the case of Mrk 421 and PKS 2155$-$304, the two best studied HBLs (Fossati *et al.* 2008, Aharonian *et al.* 2009).

- Intra-X-ray band lags, which can be soft or hard, without (so far) a clear observational understanding of what determines the sign. Physically, at least qualitatively, there are good reasons to expect lags of both signs depending on the combination of the relevant timescales (e.g. cooling, acceleration).

- Limited variability in the optical band. The case for two components, a flaring and a steady one, to interpret some of the observations has become compelling with better of multiwavelength observations.

4. Summary of results

The first tests focus on the March 19 2001 flare of Mrk 421 an ideal test-bench because of its isolation, large amplitude and good data coverage.

We studied simple scenarios generally intended to simulate variability caused by the encounter of a blob with a shock, i.e. produced by injection of relativistic electrons as a "shock front" crosses the emission region. We consider emission from two components, with the second component either being pre-existing and co-spatial and participating in the evolution of the active region (background), or being spatially independent, only diluting our observation (foreground) (Chen *et al.* 2010).

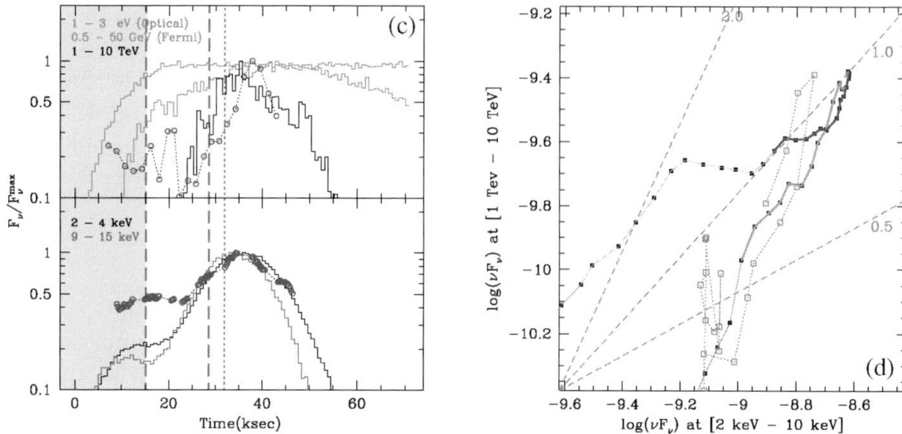

Figure 1. Summary of results of the case with pre-existing background electron population. Left: Light curves, normalized to their peak values. In blue the *RossiXTE*/PCA 2–4 keV and TeV data. The grey shaded area marks the initial phase not meaningful because it corresponds to the source "setup". The long dashed vertical grey lines mark the injection period. The dotted red line marks the time when the largest slice of the active region becomes visible, i.e. when one could expect the flare peak accounting for LCTE. Right: The flux vs. flux plot for X-ray and γ-rays. Colors highlight different time intervals (red end at t=25 ks, and each color spans 10 ks). The grey points and dotted line show the March 19 2001 data (shifted for plotting purposes).

A scorecard of the results of the three main cases is reported in Table 1. Figures 1 show an example of the simulations results for case #1, light curves and flux–flux correlation.

While each scenario seems to reproduce adequately some of the observed features, none of them was able to reproduce all the characteristics of the 2001 March 19 flare. Features particularly challenging to match are: 1) The symmetry of the light curves, in particular for the TeV band; 2) The intra-band X-ray time lag, showing a systematic soft lag; 3) Reproducing the (super)quadratic relationship between TeV/X-ray fluxes.

The first two points are among those more affected by the spatial extent and geometry of the source, whose influence varies with observed energy band because of the relative importance of geometrical and physical time-scales. The impact of the spatial extent of the source on the observed phenomenology is indeed quite significant. It affects not only the shape of the light curve (e.g. its symmetry), but also other less obvious observables such as time lags. Differences in physical time-scales for particles of different energy effectively adds a further geometric effect by inducing inhomogeneities (e.g. stratification) in the source. The impact of the geometry effects, due to the source intrinsic structure and to the stratification of properties due to the physical processes, emphasizes the necessity of a code like the one we introduce here for modeling the variable high energy emission from blazar jets.

The difficulty of producing a quadratic relationship between X-ray and γ-ray fluxes during the declining phase of the flare indicates that radiative cooling cannot fully explain electron cooling, and that there is need for an energy-independent mechanism. One possibility is adiabatic cooling, which could be associated with expansion of the blob.

One of the most interesting aspects of our analysis was the comparison between two possible hypotheses for the presence of an additional SED component. Disentangling variable and steady components is necessary to see more clearly the properties of the transient one and in turn understand its nature.

Table 1. Summary of Simulations Results

Feature	Obs.	Case #1		Case #2		Case #3	
Flare symmetry							
soft X-ray	Y	Y	+	N	–	N	–
hard X-ray	Y	Y	+	Y	+	Y	+
TeV γ-ray	Y	N	–	N	–	N	–
Flux-Flux Correlation							
trend up	2	2	+	2	+	2	+
trend down	2	2, 1	$\simeq$	1	–	1	–
paths overlap?	Y	Y/N	$\simeq$	N	–	N	–
Time Lags							
X-ray$-$X-ray	hard (2 ks)	soft	–	soft	–	soft	–
X-ray$-\gamma$-ray	γ-ray (2 ks)	N	–	Y (2 ks)	+	Y (8 ks)	–

Case #1: injection of new electrons in a region filled with a pre-existing population, then evolving together.
Case #2: injection in an empty volume, whose radiation is diluted by that of a separate emission component.
Case #3: like #1, with parameters adjusted to match better the TeV spectrum.

Our simulations would seem to favor the scenario with a simply diluting component. One important difference between the two alternatives concerns the IC emission. If the observed SED consists of the sum of two independent contributions, then the only seed photons for variable IC component will be those produced by the injected electrons themselves. Starting from an empty blob, the energy density of synchrotron seed photons needs some time to build up, which naturally results in a delay in the variation of the IC scattered γ-rays. This delay is present in the simulations for case #2, as in the observations.

References

Aharonian, F. *et al.* 2009, *Astronomy & Astrophysics*, 502, 749

Błażejowski, M. *et al.* 2005, *Astrophysical Journal*, 630, 130

Böttcher, M. & Chiang, J. 2002, *Astrophysical Journal*, 581, 127

Böttcher, M., Jackson, D. R., & Liang, E. P. 2003, *Astrophysical Journal*, 586, 389

Chen, X., Fossati, G., Liang, E., & Böttcher, M. 2010, *Monthly Notices of the RAS*, submitted

Chiaberge, M. & Ghisellini, G. 1999, *Monthly Notices of the RAS*, 306, 551

Dermer, C. D., Schlickeiser, R., & Mastichiadis, A. 1992, *Astronomy & Astrophysics*, 256, L27

Fossati, G. *et al.* 2008, *Astrophysical Journal*, 677, 906

Fossati, G. *et al.* 2000a, *Astrophysical Journal*, 541, 153

Fossati, G. *et al.* 2000b, *Astrophysical Journal*, 541, 166

Graff, P. B., *et al.* 2008, *Astrophysical Journal*, 689, 68

Inoue, S. & Takahara, F. 1996, *Astrophysical Journal*, 463, 555

Kataoka, J. *et al.* 2000, *Astrophysical Journal*, 528, 243

Katarzyński, K., *et al.* 2008, *Monthly Notices of the RAS*, 390, 371

Kirk, J. G., Rieger, F. M., & Mastichiadis, A. 1998, *Astronomy & Astrophysics*, 333, 452

Krawczynski, H., Coppi, P. S., & Aharonian, F. 2002, *Monthly Notices of the RAS*, 336, 721

Krawczynski, H. *et al.* 2004, *Astrophysical Journal*, 601, 151

Kusunose, M., Takahara, F., & Li, H. 2000, *Astrophysical Journal*, 536, 299

Maraschi, L., Ghisellini, G., & Celotti, A. 1992, *Astrophysical Journal*, 397, L5

Marscher, A. P. & Travis, J. P. 1996, *Astronomy & Astrophysics Supplement Series*, 120, 537

Sambruna, R. M. *et al.* 2000, *Astrophysical Journal*, 538, 127

Sikora, M., Begelman, M. C., & Rees, M. J. 1994, *Astrophysical Journal*, 421, 153

Sikora, M., *et al.* 2001, *Astrophysical Journal*, 554, 1

Sokolov, A. & Marscher, A. P. 2005, *Astrophysical Journal*, 629, 52

Ulrich, M.-H., *et al.* 1997, *Annual Reviews of Astronomy & Astrophysics*, 35, 445

Jets at all Scales
Proceedings IAU Symposium No. 275, 2011
G. E. Romero, R. A. Sunyaev & T. Belloni, eds.

© International Astronomical Union 2011
doi:10.1017/S1743921310015826

Variability studies in blazar jets with SF analysis: caveats and problems

Dimitrios Emmanoulopoulos[1], Ian M. McHardy[1] and Phil Uttley[1]

[1]School of Physics & Astronomy, University of Southampton
SO17 1BJ, Southampton, United Kingdom
email: D.Emmanoulopoulos@soton.ac.uk

Abstract. Blazars are radio-loud active galactic nuclei (AGN) dominated by relativistic jets seen at small angles to the line-of-sight. They exhibit dramatic flux variations across the electromagnetic spectrum. The fastest variations are observed in the X-ray and γ-ray bands on time-scales of hours or even minutes. Currently, a substantial part of the blazar literature has been based on the study of these temporal variations through the use of structure function (SF) analysis, the results of which are believed to put great constrains on the jet-physics.

The SF is often invoked in the framework of shot-noise models to determine the temporal properties of individual shots within the jet as well as their geometrical sizes. We argue, that for blazar variability studies, the SF-results are sometimes erroneously interpreted leading to misconceptions about the actual source properties. Based on extensive simulations we caution that spurious breaks will appear in the SFs of almost all light-curves, even though these light-curves may contain no intrinsic characteristic time-scale.

Finally, it is also commonly thought that SFs are immune to the sampling problems, such as data-gaps, which affects the estimators of the underlying power spectra density function such as the periodogram. However, we show that SFs are also troubled by gaps which can induce artefacts.

Keywords. (galaxies:) BL Lacertae objects: general, galaxies: jets, X-rays: general, methods: statistical

---

1. Intorduction

One of the most extensively used tools in the field of blazar variability is the structure function (SF) (e.g. Hughes *et al.* 1992) which measures the mean value of the flux-variance for measurements, $x(t)$, that are separated by a given time interval, τ, where $SF(\tau) = \left\langle [x(t) - x(t + \tau)]^2 \right\rangle$. The SF is commonly characterized in terms of its slope β, where $SF(\tau) \propto \tau^{\beta}$.

We will show, through extensive simulations, that much of the existing literature on blazar SFs tends to misinterpret observed SF-characteristics, such as breaks, as being real or physically meaningful. Often the SF-breaks are either artefacts intrinsic to SFs, or subject to much greater statistical variation than inferred from the commonly-used fitting procedures

We create artificial light-curves, lacking any sort of characteristic time-scales, and we study the SF-behaviour of MRK 501, derived from both the thoroughly studied *ASCA* data-set Tanihata *et al.* (2001) (TAN01), and the long-look light-curve of *All-Sky Monitor* (*ASM*) (onboard (*Rossi X-ray Timing Explorer*)), *RXTE*). Finally, we test the statistical robustness of the most commonly used fitting-procedures which are employed in order to derive astrophysically interesting quantities from the SF, as well as the sensitivity of SF to the presence of data-gaps. A complete description of our simulations and our results can be found in Emmanoulopoulos *et al.* (2010) (EM10).

2. SF and data-length: MRK 501

In this section we examine the SF-behaviour of MRK 501 by employing the *ASCA* data-set of TAN01 in 2–10 keV ($\sim$10 days), with a sample interval of 5678.3 sec, and the ASM data-set ($\sim$2000 days), in bins of 15 days.

We simulate stationary light-curves based on the procedure described by Timmer & Koenig (1995), which uses as an input the power spectral density (PSD) function of the observed light-curve. The output of the method is an ensemble of stochastic time-series produced having the same statistical properties with the observed data-set. Moreover, this methodology allows us to take into account the *red-noise leak* and *aliasing* effects.

Initially, we estimate the periodogram of the ASCA data-set which can be very well fitted by a simple featureless power-law with index $\alpha = -1.80 \pm 0.09$ ($\chi^2 = 3.80$ for 6 degrees of freedom with a null hypothesis probability of 0.70). Next we produce 2000 artificial light-curves having the same power-law PSD of index of $\alpha = -1.80$ and the same length as the studied *ASCA* data-set. For each artificial light-curve we estimate the SF and we localize the position of the break by producing an interpolated version of each SF. The distribution of the position of the SF-breaks is shown in the left panel of Fig.1 (grey area) having a mean value of 0.92 and and a standard deviation of 0.09 days. These simulations show us clearly that stochastic data-sets having the same length as the *ASCA* data-set of MRK 501 and the same featureless PSD, exhibit breaks in the SF around a day. Thus, the apparent break seen in the SF for MRK 501 from TAN01 should not be associated with any sort of physically meaningful time-scale.

By extending our simulations to even longer time-scales (2000 days long), we can check the effects of the data-length in the SF-estimates. From a total of 2000 long-term light-curves 1893 exhibit SF-breaks whose distribution is well represented by a Gaussian distribution having a mean 399.35 days and a standard deviation of 74.73 days (Fig.2, right panel in EM10).

In order to compare our predicted break with real data we employ the long-term *RXTE-ASM* light-curve of MRK 501 (Emmanoulopoulos 2008) and we estimate its SF. The SF-break occurs around 402 days (Fig.3, right panel in EM10), something which is absolutely in accordance with the results from our simulated light-curves of the same length which do not include any sort of characteristic time-scale.

The aforementioned example reflects in a clear way that the SF deals only with the properties of the observed light-curve, ignoring the properties of the true underlying variability process taking place within the jet.

3. SF and gappy data-sets

The SF method is considered by several researchers (e.g. Kataoka *et al.* 2001, Kataoka *et al.* 2002, Zhang *et al.* 2002) to be the ideal method of studying the time properties of gappy data-sets as it is believed to be less distorted by gaps than frequency-domain methods.

We produce an ensemble of 2000 artificially produced light-curves, having PSD slope of -1.5 and being 2000 t.u. long, with a given gappy pattern (EM10). Then we apply the bootstrap method (e.g. Czerny *et al.* 2003) 1000 times to each one of the afore-mentioned light-curves, we estimate for each SF bin i the quantity $|\log_{10}[SF^i_{\mathrm{gappy}}(\tau)] - \log_{10}[SF^i_{\mathrm{conti}}(\tau)]| \cdot err_{\mathrm{boot},i}(\tau)^{-1}$, and finally we estimate its mean value. As we can see

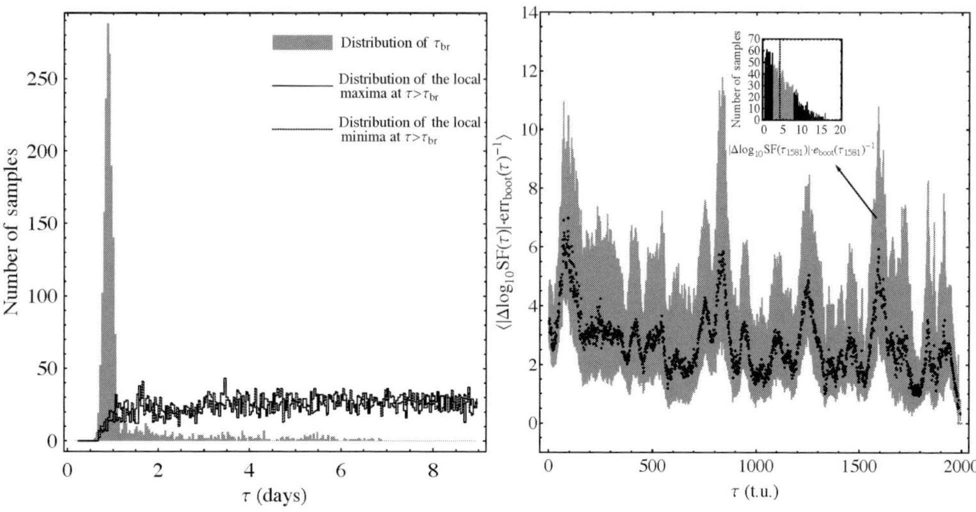

Figure 1. [Left panel] The grey area depicts the distribution of the SF-breaks coming from the 2000 simulated light-curves (the histogram-bins have a length of 0.04 days). The solid and dashed lines represent the distribution of the local maxima and local minima for $\tau > \tau_{\mathrm{br}}$ mapping the positions of the wiggling features. [Right panel] The mean difference between gappy and continuous SFs coming from an ensemble of 2000 artificial light-curves.

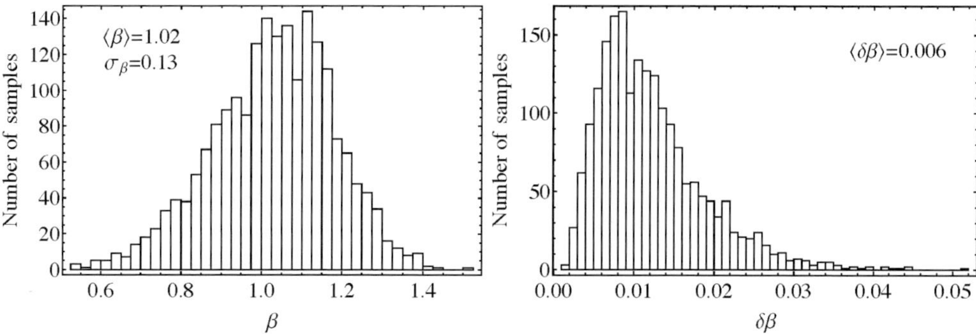

Figure 2. [Left panel] The distribution of the SF slopes β has a mean of 1.02 and a standard deviation 0.13 (the histogram-bins have a length of 0.025). [Right panel] The distribution of the errors coming from the fit $\delta\beta$, having a mean of 0.006 (the histogram-bins have a length of 0.001).

from the right panel of Fig.1, the various estimates differ significantly from unity, having a mean value of 2.79 ± 0.91. That means that data-gaps do affect the SF-results in an erratic way by introducing systematic deviations which depend on the light-curve realization. Moreover, the bootstrap method does not yield statistically meaningful errors that reflect the true deviations between the continuous and the gappy SFs.

4. SF and fitting procedures

One of the major problems that affects the results of the SF-analysis is that the various estimates, $SF(\tau_i)$, are not statistically independent of each other. This problem affects severely the fitting routines e.g. least-squares, maximum likelihood, that are commonly used in the published blazar-SF-literature to derive the SF-breaks and the -slopes, yielding very small estimates of uncertainty in the fitting parameters (EM10).

Having produced 2000 light-curves, 500 time-units (t.u.) long from a power-law PSD having an index -2, we calculate for each one of them the logarithm of the SF-estimates. We attribute to every estimate a standard error based on the error of the sample mean for each time bin, which is one of the most commonly used methods (e.g. Zhang *et al.* 2002, Czerny *et al.* 2003). From Fig.2 we can see that the errors $\delta\beta$ derived from the fit for the individual SF-slopes β (right panel) are very small in comparison to the actual scatter of the fitted β (left panel).

Since the SF-slope β is associated with the nature of the variability process within the jet (e.g. red-noise, white-noise), different realizations of the same variability process may appear having significantly different SF-slopes due to error underestimation.

5. Conclusions

Our results can be summarized as follows:

• Strong SF-breaks frequently occur in data-sets lacking any sort of characteristic time-scales. The position of these physically uninteresting breaks depends on the length of the observations and the shape of the underlying PSD.

• Data-gaps affect severely the SF-estimates in an unpredictable way, introducing systematic deviations. The bootstrap method can not yield statistically meaningful errors that depict the true deviations between the gappy and the continuous SFs.

• Non-independence and non-Gaussianity make impossible the estimation of a meaningful goodness-of-fit from based on normal fitting procedures. We see, for example, that the derived uncertainties on the SF quantities i.e. positions of breaks and slopes, are always much smaller than the actual scatter of these variables during multiple realizations of the same variability process.

Acknowledgements

DE and IMM acknowledge the Science and Technology Facilities Council (STFC) for support under grant ST/G003084/1. PU is supported by an STFC Advanced Fellowship. This research has made use of NASA's Astrophysics Data System Bibliographic Services.

References

Czerny B., Doroshenko V. T., Nikołajuk M., Schwarzenberg-Czerny A., Loska Z., & Madejski G. 2003, *MNRAS*, 342, 1222

Emmanoulopoulos, D. 2008, *Blazar Variability across the Electromagnetic Spectrum*, *PoS*, (BLAZARS2008)038, Palaiseau, France

Emmanoulopoulos, D., McHardy, I. M., & Uttley, P. 2010, *MNRAS*, 404, 931

Hughes P. A., Aller H. D., & Aller M. F. 1992, *ApJ*, 396, 469

Kataoka, J., Takahashi, T., Wagner, S. J., Iyomoto, N., Edwards, P. G., Hayashida, K., Inoue, S., Madejski, G. M., Takahara, F., Tanihata, C., & Kawai, N. 2001, *ApJ*, 560, 659

Kataoka, J., Tanihata, C., Kawai, N., Takahara, F., Takahashi, T., Edwards, P. G., & Makino, F. 2002, *MNRAS*, 336, 932

Tanihata C., Urry C. M., Takahashi T., Kataoka J., Wagner S. J., Madejski G. M., Tashiro M., & Kouda M. 2001, *ApJ*, 563, 569

Timmer J. & Koenig M. 1995, *A&A*, 300, 707

Zhang, Y. H., Treves, A., Celotti, A., Chiappetti, L., Fossati, G., Ghisellini, G., Maraschi, L., Pian, E., Tagliaferri, G., & Tavecchio, F. 2002, *ApJ*, 572, 762

Discussion

KAWAI: What is your suggestion for estimating the characteristic time scale from data with irregular gaps?

EMMANOULOPOULOS: The best way is to use the Fourier domain, i.e.m power spectral density, even in the case of gappy irregular data sets. Through simulations you can take into account aliasing and sampling effects in a robust and statistically correct way (e.g. Uffley *et al.* 2009)

MEIER: So, should we then conclude that blazars have no preferred time scale, i.e., that their PSD is simply a power law?

EMMANOULOPOULOS: Yes, the PSD of blazars (up to now) is a simple power law, having no breaks or other features.

Jets at all scales
Proceedings IAU Symposium No. 275, 2011
G. E. Romero, R. A. Sunyaev & T. Belloni, eds.

© International Astronomical Union 2011
doi:10.1017/S1743921310015838

The far-infrared view of M87
as seen by the Herschel Space Observatory

M. Baes[1], M. Clemens[2], E. M. Xilouris[3], J. Fritz[1], W. D. Cotton[4],
J. I. Davies[5], G. J. Bendo[6], S. Bianchi[7], L. Cortese[5], I. De Looze[1],
M. Pohlen[5], J. Verstappen[1], H. Böhringer[8], D. J. Bomans[9],
A. Boselli[10], E. Corbelli[7], A. Dariush[5], S. di Serego Alighieri[7],
D. Fadda[11], D. A. Garcia-Appadoo[12], G. Gavazzi[13], C. Giovanardi[7],
M. Grossi[14], T. M. Hughes[5], L. K. Hunt[7], A. P. Jones[15], S. Madden[16],
D. Pierini[8], S. Sabatini[17], M. W. L. Smith[5], C. Vlahakis[18], S. Zibetti[19]

[1]Sterrenkundig Observatorium, Universiteit Gent, Belgium

[2]INAF-Osservatorio Astronomico di Padova, Italy

[3]National Observatory of Athens, Greece

[4]National Radio Astronomy Observatory, USA

[5]Department of Physics and Astronomy, Cardiff University, UK

[6]Astrophysics Group, Imperial College London, UK

[7]INAF-Osservatorio Astrofisico di Arcetri, Firenze, Italy

[8]Max-Planck-Institut für Extraterrestrische Physik, Garching, Germany

[9]Astronomical Institute, Ruhr-University Bochum, Germany

[10]Laboratoire d'Astrophysique de Marseille, France

[11]NASA Herschel Science Center, California Institute of Technology, USA

[12]European Southern Observatory, Santiago, Chile

[13]Università di Milano-Bicocca, Italy

[14]CAAUL, Observatório Astronómico de Lisboa, Portugal

[15]Institut d'Astrophysique Spatiale (IAS), Université Paris-Sud 11, France

[16]Laboratoire AIM, CEA/DSM, Université Paris Diderot, France

[17]INAF-Istituto di Astrofisica Spaziale e Fisica Cosmica, Roma, Italy

[18]Leiden Observatory, The Netherlands

[19]Max-Planck-Institut für Astronomie, Heidelberg, Germany

Abstract. The origin of the far-infrared emission from the nearby radio galaxy M87 remains a matter of debate. Some studies find evidence of a far-infrared excess due to thermal dust emission, whereas others propose that the far-infrared emission can be explained by synchrotron emission without the need for an additional dust emission component. We observed M87 with PACS and SPIRE as part of the Herschel Virgo Cluster Survey (HeViCS). We compare the new Herschel data with a synchrotron model based on infrared, submm and radio data to investigate the origin of the far-infrared emission. We find that both the integrated SED and the Herschel surface brightness maps are adequately explained by synchrotron emission. At odds with previous claims, we find no evidence of a diffuse dust component in M87.

---

1. Introduction

At a distance of 16.7 Mpc, M87 is the dominant galaxy of the Virgo Cluster. It is one of the nearest radio galaxies and was the first extragalactic X-ray source to be identified. Because of its proximity, many interesting astrophysical phenomena can be

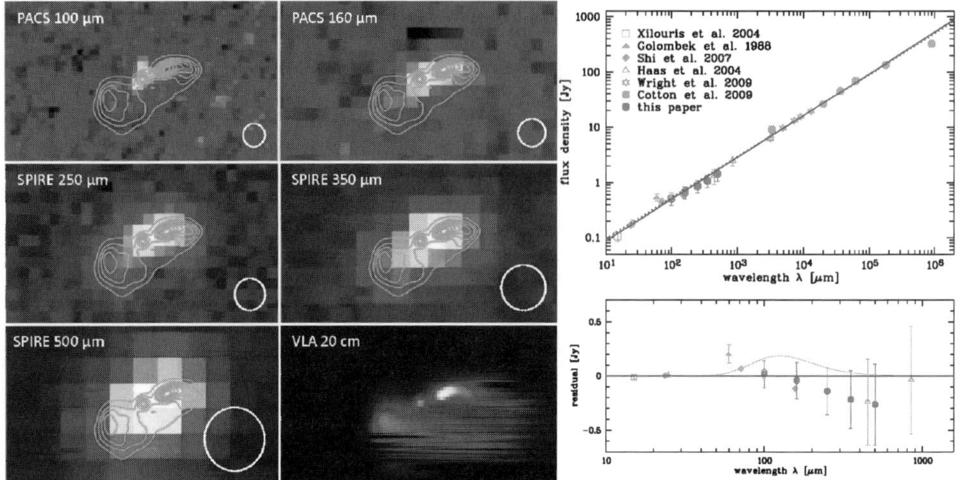

Figure 1. Left: the Herschel view of the central regions of M87. The bottom right image is a VLA 20 cm image from the FIRST survey. The 20 cm radio contours have been overlaid on the Herschel images. The field of view of all images is 160" × 90", beam sizes are indicated in the bottom right corner. Top right: the global SED of M87 from mid-infrared to radio wavelengths. Where no error bars are seen, they are smaller than the symbol size. The solid line in the plot is the best-fit power law of the ISOCAM, IRAS, MIPS, SCUBA, GBT, WMAP, and VLA data; the dotted line has only been fitted to the SCUBA, GBT, WMAP, and VLA data. Bottom right: residual between data and the best-fit synchrotron model in the infrared-submm wavelength range. The cyan line is a modified black-body model with $T = 23$ K and $M_{\rm d} = 7 \times 10^4\ M_\odot$.

studied in more detail in M87 than in other comparable objects. Particularly remarkable is the prominent jet extending from the nucleus, visible throughout the electromagnetic spectrum. The central regions of M87, in particular the structure of the jet, have been studied and compared intensively at radio, optical, and X-ray wavelengths. (e.g. Biretta *et al.* 1991; Meisenheimer *et al.* 1996; Böhringer *et al.* 2001; Perlman *et al.* 2001; Sparks *et al.* 2004; Perlman & Wilson 2005; Kovalev *et al.* 2007; Simionescu *et al.* 2008).

Compared to the available information at these wavelengths, our knowledge of M87 at far-infrared (FIR) wavelengths is rather poor. A controversial issue is the origin of the FIR emission in M87, i.e., the question of whether the FIR emission is caused entirely by synchrotron emission or whether there is an additional contribution from dust associated with either the global interstellar medium or a nuclear dust component. Several papers, based on IRAS, ISO and Spitzer observations, arrive at different conclusions (Perlman *et al.* 2007; Buson *et al.* 2009; Xilouris *et al.* 2004; Shi *et al.* 2007; Tan *et al.* 2008).

We investigate the nature of the FIR emission of M87 using new FIR data from the Herschel Space Observatory (Pilbratt *et al.* 2010), obtained as part of the science demonstration phase (SDP) observations of the Herschel Virgo Cluster Survey (HeViCS, Davies *et al.* 2010). HeViCS is an approved open time key program, which has been awarded 286 h of observing time in parallel mode with the PACS and SPIRE instruments. We will ultimately map four 4×4 square degree regions of the cluster at 100, 160, 250, 350 and 500 μm, down to the 250 μm confusion limit of about 1 MJy sr^{-1}. The HeViCS SDP observations consisted of a single cross-scan of one 4×4 deg^2 field at the centre of the Virgo Cluster. While these SDP observations comprise only 6% of the total HeViCS observations, the analysis based on these observations already gives a prelude to the primary HeViCS science goals including the effects of the environment on the dust medium of galaxies, the FIR luminosity function, the complete SEDs of galaxies and a detailed

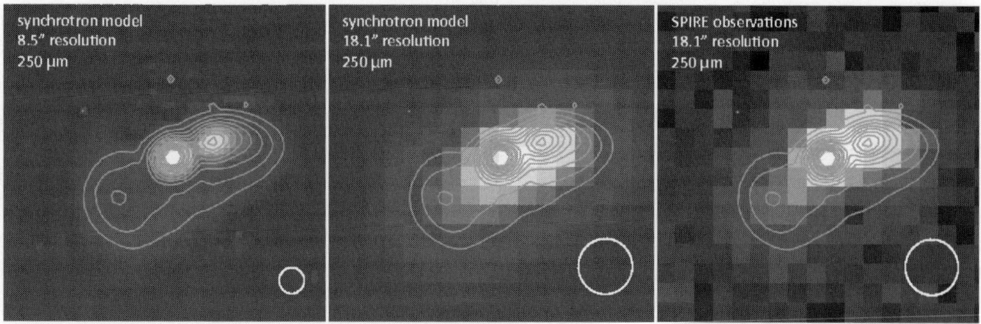

Figure 2. A comparison between the synchrotron model image and the observed image at 250 μm. The left panel shows the synchrotron image at the model resolution, the central panel shows the same model convolved to the SPIRE 250 μm beam and pixel size. The right panel shows the observed SPIRE 250 μm image. In all panels, the green lines are the contours of the synchrotron model at the model resolution.

analysis of the dust content of dwarf and early-type galaxies (Davies *et al.* 2010; Cortese *et al.* 2010; Clemens *et al.* 2010; Smith *et al.* 2010; Grossi *et al.* 2010; De Looze *et al.* 2010; Boselli *et al.* 2010). The observations also enabled us to study the intensity and nature of the FIR emission of M87 (Baes *et al.* 2010).

2. Analysis and conclusion

The left part of Figure 1 shows the PACS and SPIRE images of the central 160" × 90" region of M87, which is clearly detected in all five bands. The top right panel in Figure 1 shows the integrated SED in the infrared-submm-radio region between 15 μm and 100 cm, with the new Herschel data as well as ISOCAM, IRAS, MIPS, and SCUBA, GBT, WMAP, and VLA data gathered from the literature (Xilouris *et al.* 2004; Golombek *et al.* 1988; Shi *et al.* 2007; Haas *et al.* 2004; Cotton *et al.* 2009; Wright *et al.* 2009); the actual flux densities can be found in Baes *et al.* (2010). The solid line is the best-fit power law for all literature data and has a slope $\alpha = -0.76$; the dotted line fits only the submm and radio data and has a slope $\alpha = -0.74$. The bottom right panel in Figure 1 shows the residual from the best-fit power law in the infrared-submm wavelength region; clearly, the integrated Herschel fluxes are in full agreement with synchrotron radiation. The cyan line in this figure is a modified black-body fitted with $T = 23$ K and $M_d = 7 \times 10^4\ M_\odot$. This temperature is the mean dust equilibrium temperature in the interstellar radiation field of M87, determined using the SKIRT radiative transfer code (Baes *et al.* 2003, 2005) and based on the photometry from Kormendy *et al.* (2009). The dust mass was adjusted to fit the upper limits of the residuals. It is clear that the SED of M87 is incompatible with dust masses higher than $10^5\ M_\odot$.

Although indicative, the analysis of the integrated SED does not definitively identify the origin of the FIR emission in M87. Approximating the global SED as a single power-law synchrotron model is indeed an oversimplification of the complicated structure of M87. Several studies have shown that M87 contains three distinct regions of significant synchrotron emission, each with their own spectral indices: the nucleus, the jet and associated lobes in the NW region, and the SE lobes (e.g., Biretta *et al.* 1991; Meisenheimer *et al.* 1996; Perlman *et al.* 2001; Shi *et al.* 2007). We have constructed a synchrotron model for the central regions of M87, based on a newly reduced MIPS 24 μm map and archival MUSTANG 90 GHz and 15, 8.2, 4.9, 1.6, and 0.3 GHz maps (Cotton *et al.* 2009). We fitted a second-order polynomial synchrotron model to each pixel of the

MIPS + radio data cube and used this synchrotron model to predict the emission of M87 at 250 μm (the SPIRE 250 μm image provides the optimal compromise between S/N and spatial resolution). Figure 2 shows the comparison between the synchrotron model prediction at 250 μm and the SPIRE observations. At the model resolution, the three distinct components are visible, but when we convolve this synchrotron model image with the SPIRE 250 μm beam, the three different components merge into a single extended structure with one elongated peak slightly west of the nucleus. Comparing the central and right panels of Figure 2, we see that the synchrotron model is capable of explaining the observed SPIRE 250 μm image satisfactorily.

We conclude that for both the integrated SED and the SPIRE 250 μm map, we have found that synchrotron emission is an adequate explanation of the FIR emission. We do not detect a FIR excess that cannot be explained by the synchrotron model. In particular, we have no reason to invoke the presence of smooth dust emission associated with the galaxy interstellar medium, as advocated by Shi *et al.* (2007). For a dust temperature of 23 K, which is the expected equilibrium temperature in the interstellar radiation field of M87, we find an upper limit to the dust mass of 7×10^4 $M_\odot$. Our conclusion is that, seen from the FIR point of view, M87 is a passive object with a central radio source emitting synchrotron emission, without a substantial diffuse dust component.

References

Baes, M., *et al.* 2003, *MNRAS*, 343, 1081

Baes, M., Dejonghe, H., & Davies, J. I. 2005, *AIP Conf. Series*, 761, 27

Baes, M., *et al.* 2010, *A&A*, 518, L53

Biretta, J. A., Stern, C. P., & Harris, D. E. 1991, *AJ*, 101, 1632

Böhringer, H., *et al.* 2001, *A&A*, 365, L181

Boselli, A., *et al.* 2010, *A&A*, 518, L61

Buson, L., *et al.* 2009, *ApJ*, 705, 356

Clemens, M. S., *et al.* 2010, *A&A*, 518, L50

Cortese, L., *et al.* 2010, *A&A*, 518, L49

Cotton, W. D., *et al.* 2009, *ApJ*, 701, 1872

Davies, J. I., *et al.* 2010, *A&A*, 518, L48

De Looze, I., *et al.* 2010, *A&A*, 518, L54

Golombek, D., Miley, G. K., & Neugebauer, G. 1988, *AJ*, 95, 26

Grossi, M., *et al.* 2010, *A&A*, 518, L52

Haas, M., *et al.* 2004, *A&A*, 424, 531

Kormendy, J., Fisher, D. B., Cornell, M. E., & Bender, R. 2009, *ApJS*, 182, 216

Kovalev, Y. Y., *et al.* 2007, *ApJ*, 668, L27

Meisenheimer, K., Roeser, H.-J., & Schloetelburg, M. 1996, *A&A*, 307, 61

Perlman, E. S., *et al.* 2001, *ApJ*, 551, 206

Perlman, E. S. & Wilson, A. S. 2005, *ApJ*, 627, 140

Perlman, E. S., *et al.* 2007, *ApJ*, 663, 808

Pilbratt, G. L., *et al.* 2010, *A&A*, 518, L1

Shi, Y., *et al.* 2007, *ApJ*, 655, 781

Simionescu, A., *et al.* 2008, *A&A*, 482, 97

Smith, M. W. L., *et al.* 2010, *A&A*, 518, L51

Sparks, W. B., *et al.* 2004, *ApJ*, 607, 294

Tan, J. C., Beuther, H., Walter, F., & Blackman, E. G. 2008, *ApJ*, 689, 775

Xilouris, E. M., *et al.* 2004, *A&A*, 416, 41

Wright, E. L., *et al.* 2009, *ApJS*, 180, 283

Discussion

KYLAFIS: The complete absence of dust is for me surprising. Is it specific to M87 or is it general?

BAES: So far, we have observed ellipticals in the Virgo Cluster and found that all truly passive ellipticals, were not detected, which implied a stringente limit on the dust. Further observations with Herschel will confirm these preliminary findings

Jets at all Scales
Proceedings IAU Symposium No. 275, 2011
G. E. Romero, R. A. Sunyaev & T. Belloni, eds.

© International Astronomical Union 2011
doi:10.1017/S174392131001584X

The jet in M87 from e-EVN observations

G. Giovannini[1,2], C. Casadio[1], M. Giroletti[1], M. Beilicke[3],
A. Cesarini[4] & H. Krawczynski[3]

[1]Istituto di Radioastronomia-INAF,
via Gobetti 101, 40129 Bologna, Italy
email: ggiovann@ira.inaf.it

[2]Dipartimento di Astronomia,
via Ranzani 1, 40127 Bologna, Italy

[3]Department of Physics, Washington University,
St. Louis, MO 63130, USA

[4]School of Physics, National University of Ireland
Galway, University Road, Galway, Republic of Ireland

Abstract. One of the most intriguing open questions of today's astrophysics is the jet physical properties and the location and the mechanisms for the production of MeV, GeV, and TeV gamma-rays in AGN jets. M87 is a privileged laboratory for a detailed study of the properties of jets, owing to its proximity, its massive black hole, and its conspicuous emission at radio wavelengths and above. We started on November 2009 a monitoring program with the e-EVN at 5 GHz. We present here results of these multi-epoch observations and discuss the two episodes of activity at energy E>100 GeV that occured in this period. One of these observations was obtained at the same day of the first high energy flare. We added to our results literature data obtained with the VLBI and VLA. A clear change in the proper motion velocity of HST-1 is present at the epoch ∼2005.5. In the time range 1998 – 2005.5 the apparent velocity is subluminal, and superluminal (∼2.7c) after 2005.5.

Keywords. Galaxies: jets, Galaxies: M87, Galaxies: active

1. Introduction

The giant radio galaxy Messier 87 (M87), also known as 3C 274 or Virgo A, is one of the best studied radio sources and a known γ-ray-emitting AGN. It is located at the center of the Virgo cluster of galaxies at a distance = 16.7 Mpc, corresponding to an angular conversion 1 mas = 0.081 pc. The massive black hole at the M87 center has an estimated mass = 6×10^9 solar masses, with a scale of 1 mas = 140 R$_S$. The bright jet is well resolved in the X-ray, optical, and radio wave bands.

The jet is characterized by many substructures and knots. In 1999 HST observations revealed a bright knot at about 1" from the core, named HST-1. This feature is active in the radio, optical, and X-ray regimes. It was discussed by Perlman *et al.* 1999, who compared optical and radio images. Biretta *et al.* 1999 measured in the range 1994–1998 a subluminal speed = 0.84c for the brightest structure (HST-1 East), which appears to emit superluminal features moving at 6c. However this motion was measured in regions on a larger scale with respect to the VLBI structures discussed here. In this time range HST-1 in the radio band was a faint jet structure (a few mJy/beam, see Fig. 1), but starting from 2000 it increased by more than a factor 50 and it reached a flux density ∼100 mJy in 2005 (See Fig. 2 and Harris *et al.* 2009).

VLBI observations of the M87 inner region show a well resolved, edge-brightened jet structure. At very high resolution (43 and 86 GHz) near to the brightest region the jet has a wide opening angle, and we refer to the many published papers which discuss the possible presence of a counter-jet and the location of the radio core; see e.g. Junor *et al.* 1999, Krichbaum *et al.* 2005. After a few milliarcsec (mas) the jet appears well collimated and limb-brightened.

Very High Energy (VHE) γ-ray emission was reported by the High Energy Gamma-Ray Astronomy (HEGRA) collaboration in 1998/99 (Aharonian *et al.* 2003), confirmed by the High Energy Stereoscopic System (HESS) in 2003–2006 (Aharonian *et al.* 2006), and by VERITAS in 2007 (Acciari *et al.* 2008). Coordinated intensive campaigns have permitted to detect the source again in 2008 (Acciari *et al.* 2009) and as recently as February and April 2010 (Mariotti *et al.* 2010). Steady emission at MeV/GeV energies has also been detected by *Fermi*/LAT (Abdo *et al.* 2009).

Various models have been proposed to explain the multi-wavelength emission and in particular to constrain the site of the VHE emission in M87. The inner jet region was favoured by the observed short TeV variability timescales according to Aharonian *et al.* 2006. The VHE emission could then be produced in the BH magnetosphere (Neronov *et al.* 2007) or in the slower jet layer (Tavecchio *et al.* 2008), with the spine accounting for the emission from the radio to the GeV band; this would lead to a complex correlation between the TeV and radio components.

However, VLBA observations at 1.7 GHz by Cheung *et al.* 2007 resolved HST-1 in substructures with superluminal components. Aharonian *et al.* 2006 discussed that HST-1 cannot be excluded as a source of TeV γ rays, however they conclude that the more promising possibility is that the site of TeV γ-ray production is the nucleus of M87 itself. Comparing multifrequency data Harris *et al.* 2008 suggested that the TeV emission from M87 was originated in HST-1.

Finally, Acciari *et al.* 2009 reported rapid TeV flares from M87 in February 2008, associated by an in increase of the radio flux from the nucleus, while HST-1 was in a low state, thus concluding that the TeV flares originate in the core region.

In this context we started at the end of 2009 a program to observe with the e-EVN M87 at 5 GHz to study the properties of the M87 core, jet, and HST-1 structure.

2. Observations and Data Reduction

The observations have been carried out in e-VLBI mode, with data acquired by EVN radio telescopes, directly streamed to the central data processor at JIVE, and correlated

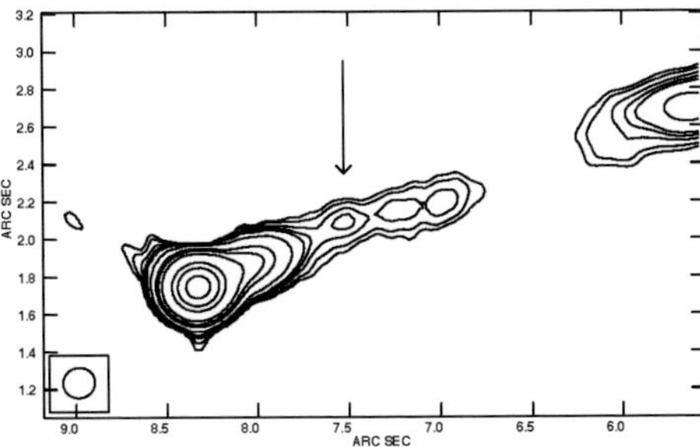

Figure 1. M87 jet obtained on March 1998 at 15 GHz with the VLA in A configuration. An arrow indicates the HST-1 position. Levs are: -3 1.5 2 3 4 5 10 30 50 100 500 1000 2000 mJy/beam. The HPBW is 0.16"

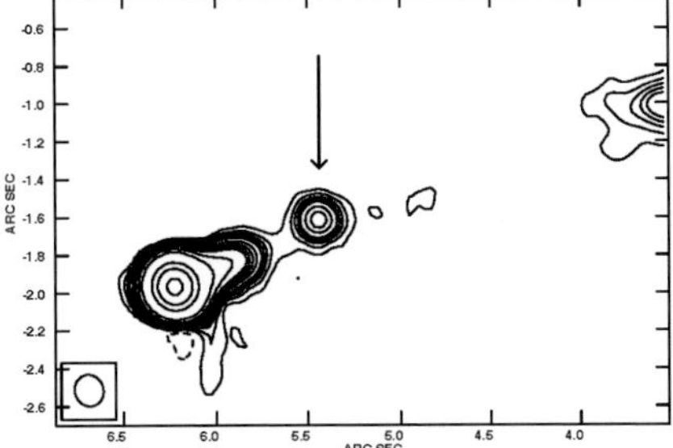

Figure 2. M87 jet obtained on June 2003 at 15 GHz with the VLA in A configuration. An arrow indicates the HST-1 position. Levs are: -3 2 4 6 8 10 15 20 30 50 100 500 1000 2000 mJy/beam. The HPBW is 0.16"

in real-time. The observing frequency of 5 GHz was chosen to simultaneously grant a large field of view and a high angular resolution. For observations taking advantage of the long baselines provided by the Arecibo and Shanghai telescopes, our clean beam with uniform weights is about 2.0×0.9 mas in PA $-25°$.

We obtained 6 epochs at 5 GHz, namely on 2009 November 19, 2010 January 27, February 10, and March 28, and as Target of Opportunity on 2010 March 6, May 18, and June 9.

As a result of the large bandwidth (a rate of 1 Gbps was sustained by most stations), long exposure (up to 6 hours per epoch), and extended collecting area, the rms noise in our images is mostly dynamic range limited. As an average value, we can quote $0.5 - 0.8$ mJy beam^{-1} in the nuclear region and $0.1 - 0.2$ mJy beam^{-1} in the HST-1 region. We present here preliminary results. Data reduction was carried out in the standard mode using the AIPS and CalTech package.

3. Results

3.1. *The inner jet region*

The jet orientation and velocity has been discussed in many papers comparing observational data on the jet brightness and proper motion. Recently Acciari *et al.* 2009 assumed as a likely range $\theta = 15 - 25$ deg.

Because of the very similar uv-coverage, we used our images to search for evidence of a possible proper motion, comparing different epoch position of jet substructures and subtracting images at different epochs (with the same grid, angular resolution and similar uv-coverage) to look for possible systematic trends. No evidence was found in anycase.

We find a marginal evidence of a nuclear flux density increasing in the last three epochs.

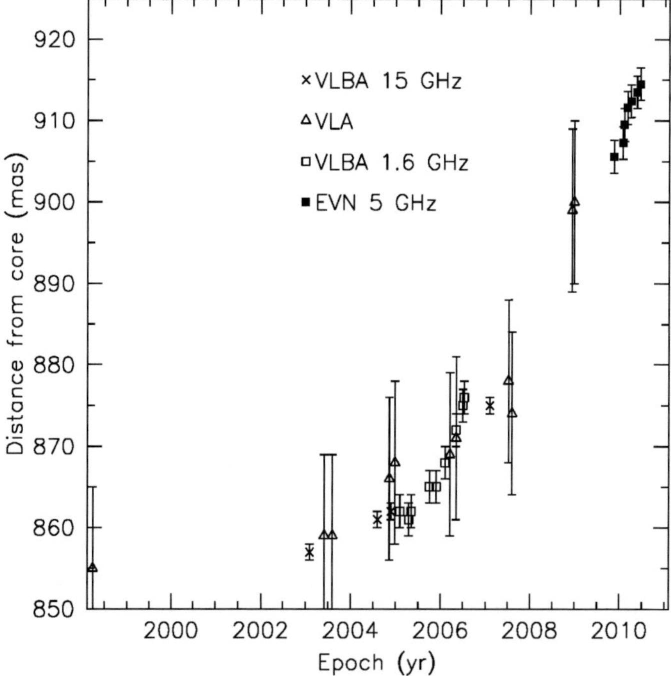

Figure 3. Distance of HST-1 brightest peak from the M87 core at different epochs

In these epochs we have an increase of the core flux density and of the inner jet (within $\sim$8 mas) flux density. The data analysis is still in progress since it is not easy to separate the core and jet flux density because of the source structure.

3.2. *HST-1*

In our observations HST-1 is clearly resolved. It is oriented in E-W direction forming an angle of $\sim$20° with the jet axis. The HST-1 size is in agreement with a very small ($\sim$0°) jet opening angle confirming the high jet collimation in the sub-arcsecond region.

To better study the dynamic of this structure we searched archive VLA data at high resolution (A configuration) and high frequency (U, and Q bands). We refer to Harris *et al.* 2009 for a discussion of the flux density variability. Here we only want to compare different epochs to derive the HST-1 dynamic.

We started to analyze data from 1998, even if HST-1 is very faint before of 2003. Starting from 2003.6 the HST-1 structure is well evident (see e.g. Fig. 2 obtained on June 2003) and well separated by the jet structure near the core.

We estimated from e-EVN and VLA data the distance of HST-1 from the core. In e-EVN data we measured the distance between the core and the brightest knot in HST-1, in VLA images we used the HST-1 peak, being this structure unresolved. Adding the values obtained at 1.5 and 15 GHz by Cheung *et al.* 2007 and Chang *et al.* 2010, respectively, we can study the HST-1 proper motion with a good statistic from 2003 to present epoch. The apparent proper motion of HST-1 is shown in Fig. 3.

A clear change in the proper motion velocity is present at the epoch $\sim$2005.5, coincident with the TeV γ-ray activity and the maximum radio/X-ray flux density of HST-1. In the time range 2003 – 2005.5 the apparent velocity is 0.5c – 0.6c; in the time range 2005.5 – 2010.25 the apparent velocity is $\sim$2.7c. We note also a possible decrease in the apparent velocity in 2007 with a restarted high velocity motion from 2008 (near the time of the high energy flare) up to now. Assuming a jet orientation angle = 25° a proper motion of 2.7c corresponds to an intrinsic velocity = 0.94c.

4. Summary

With our new e-EVN data we have obtained images of the nuclear region of M87 and of the jet substructure HST-1.

The radio core flux density is constant in the first three epochs with an average flux density $\sim$1805 mJy and slightly increasing in the last three epochs: 2013 mJy in 2010.25.

The HST-1 structure is well resolved in many substructures. A complex proper motion is clearly present. Comparing e-EVN data with archive VLA data and published VLBA data at 1.7 and 15 GHz we find a strong evidence that in 2005.5 HST-1 increased its velocity from an apparent velocity $\sim$0.5c to 2.7c. With present data it is not possible to discuss if this change in velocity is related to the M87 VHE activity and/or to the maximum radio/X-ray flux density of HST-1 at this epoch. A more regular and longer monitor and a multi-frequency comparison is necessary to clarify this point.

References

Abdo, A. A., Ackermann, M., Ajello, M., *et al.* 2009, ApJ, 707, 55
Acciari, V. A., Beilicke, M., Blaylock, G., *et al.* 2008, ApJ, 679, 397
Acciari, V. A., Aliu, E., Arlen, T., *et al.* 2009, Sci, 325, 444
Aharonian, F., Akhperjanian, A., Beilicke, M., *et al.* 2003, A&A, 403, L1
Aharonian, F., Akhperjanian, A. G., Bazer-Bachi, A. R., *et al.* 2006, Sci, 314, 1424

Biretta, J. A., Sparks, W. B., & Macchetto, F. 1999, ApJ, 520, 621
Chang, C. S., Ros, E., Kovalev, Y. Y., & Lister, M. L. 2010, A&A in press (arXiv:1002.2588)
Cheung, C. C., Harris, D. E., & Stawarz, L., 2007, ApJ, 663, L65
Harris, D. E., Cheung, C. C., Stawarz, L., & *et al.* 2008, in ASP Conf. Ser. 386, 80
Harris, D. E., Cheung, C. C., Stawarz, L., Biretta, J. A., Perlman, E. S. 2009, ApJ, 699, 305
Junor, W., Biretta, J. A., & Livio, M. 1999, Natur, 401, 891
Krichbaum, T. P., Zensus, J. A., & Witzel, A. 2005, AN, 326, 548
Mariotti, M. 2010, ATel, 2431, 1
Neronov, A. & Aharonian, F. A. 2007 ApJ 671, 85
Perlman, E. S., Biretta, J. A., Zhou, F., Sparks, W. B., & Macchetto, F. D. 1999, AJ, 117, 2185
Tavecchio, F. & Ghisellini, G. 2008, MNRAS, 385, 98

Discussion

MEIER: Cheung *et al.* (2007) claim that HST-1 is a complex of components, with one quasi-stationary and others moving up to $4.3\,c$. Please, comment on the differences between your data and Cheung *et al.*'s.

GIOVANNINI: We collected new anda archive data from 1998 to now. The data show a clear HST-1 proper motion with a shift in position larger than the HST-1 size, therefore, all the structure is moving. We agree with Cheung *et al.* (2007) that HST-1 is complex with many sub-structures, variable in flux density and position. However, in our data we do not see a stationary component. All the structure is moving at about the same velocity, the same velocity of the brightest sub-component.

Jets at all Scales
Proceedings IAU Symposium No. 275, 2011
G. E. Romero, R. A. Sunyaev & T. Belloni, eds.

© International Astronomical Union 2011
doi:10.1017/S1743921310015851

The picture of relativistic jet from *Fermi*-LAT and multi-band observations of blazar 3C 279

Masaaki Hayashida[1] and Greg Madejski[1] for the *Fermi*-LAT collaboration and members of 3C 279 multi-band campaign

[1]Kavli Institute for Particle Astrophysics and Cosmology, SLAC National Accelerator Laboratory, Stanford University, CA, 94025, USA
email: `mahaya@slac.stanford.edu`

Abstract. Strong and variable radiation detected over all accessible energy bands in blazar arises from a relativistic, Doppler-boosted jet pointing close to our line of sight. Flat Spectrum Radio Quasar 3C 279 was one of the brightest γ-ray blazars in the sky at the time of the discovery with EGRET. Since the successful launch of the *Fermi* Gamma-ray Space telescope in 2008, we have organized extensive multi-band observational campaign of 3C 279 from radio to γ-ray bands, also including optical polarimetric observations. The uninterrupted monitoring in the γ-ray band by *Fermi*-LAT together with the multi-band data provide us with new insights of the relativistic jet of blazar. Here, we present the results of the first-year multi-band campaign of 3C 279 including the discovery of a γ-ray flare event associated with a dramatic change of the optical polarization - as well as a discovery of an "orphan" X-ray flare, unassociated with prominent outbursts in other bands.

Keywords. galaxies: jets – quasars: individual (3C 279) – gamma rays: observations – X-rays: galaxies – polarization

1. Blazar 3C 279

The flat spectrum radio quasar 3C 279 was the first bright γ-ray blazar reported by the EGRET instrument aboard the Compton Gamma-Ray Observatory to show strong and rapidly variable γ-ray emission (Hartman *et al.* 1992, Kniffen *et al.* 1994, Wehrle *et al.* 1998); recently, it also has been detected at photon energies above 100 GeV by the MAGIC ground-based Cherenkov telescope (Albert *et al.* 2008). This blazar, at the redshift $z = 0.536$, harbors a black hole with mass $M \simeq (3 - 8) \times 10^8 M_\odot$ (Woo & Urry 2002, Nilsson *et al.* 2009); for specificity, we adopt $6 \times 10^8 M_\odot$. It shows superluminal expansion best described as the jet material propagating with the bulk Lorentz factor $\Gamma_{\rm jet} = 16 \pm 3$ at a small angle ($\theta \sim 2°$) to our line of sight (Jorstad *et al.* 2005). In this paper, we report the results of the first-year *Fermi*-LAT and multi-band campaign of 3C 279 including optical polarimetric observations.

2. Multi-band observational campaign with *Fermi*-LAT

The successful launch and deployment of the *Fermi* Gamma-ray Space Telescope inaugurated a new capability to monitor the γ-ray sky. The main instrument on-board *Fermi*, the Large Area Telescope (LAT: Atwood *et al.* 2009) features a wide (2.4 steradian) field of view and rocks by 35–50° north and south of the orbital plane every second orbit, it observes the entire sky every ~ 3 hours, allowing an essentially uninterrupted flux history of 3C 279 in the γ-ray band. The observations of the source were conducted in other

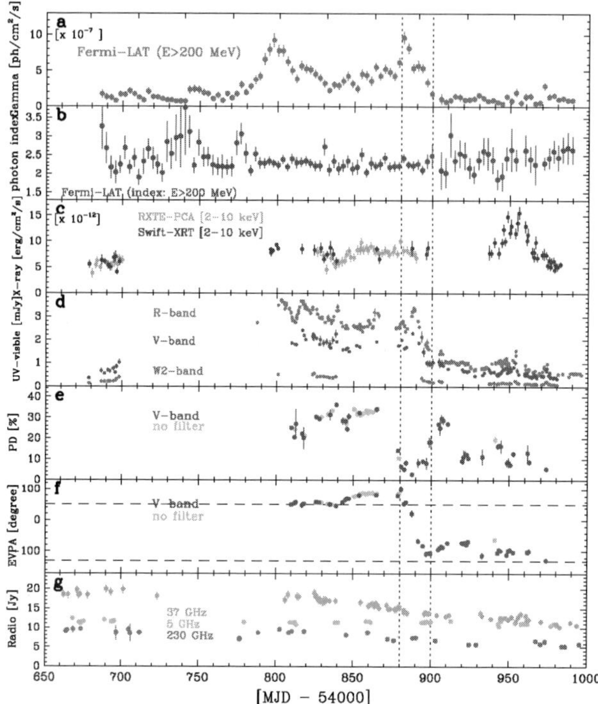

Figure 1. History of fluxes in γ-ray above 200 MeV (a), X-ray of 2-10 keV (c), optical-UV (d), radio (g) bands as well as γ-ray photon index (b), optical polarization degree (e) and angle (f) of 3C 279 between July 2008 and June 2009 (54650 and 55000 MJD). The two dashed vertical lines correspond to 54880 and 54900 MJD, respectively. The full version of the light curves of the campaign including some additional bands can be seen in Abdo *et al.* (2010).

bands as well, and the best coverage was obtained after the start of the routine scientific operation of the *Fermi*-LAT (on August 4th 2008, 54682 MJD). In Figure 1 we plot the flux history in the γ-ray band by *Fermi*-LAT, as well as in the X-ray band measured by *RXTE*-PCA and *Swift*-XRT, optical, infrared and radio bands by the GASP-WEBT observatories† and *Swift*-UVOT. Polarization information in the optical band are also provided by Kanata (V-band:Watanabe *et al.* 2005) and KVA (no filter) telescopes.

3. Gamma-ray flare with a change of optical polarization

Towards the end of the high state there is a sharp γ-ray flare at 54880 MJD. This sharp γ-ray flare coincides with a significant drop of the level of optical polarization (polarization degree: PD), from $\sim 30\%$ down to a few %, lasting for $\Delta t \sim 20$ days. Subsequently, both γ-ray and optical fluxes gradually decrease together and reach the quiescent level, followed by a temporary recovery of the high degree of polarization. This event is associated with a dramatic change of the electric vector position angle (EVPA) of the polarization, in contrast to being relatively constant before the event at $\sim 50°$ (parallel to the jet direction observed by Very Long Baseline Interferometry observations in radio bands; see, e.g., Jorstad *et al.* 2005). Since the EVPA has $\pm 180° \times n$ ambiguity, we selected values on the assumption of a smooth change of the EVPA, such that it would follow the overall trend. The polarization angle increases slightly at 54880 MJD – coincident with the γ-ray flare – then decreases by $208°$ with an average rate of $\sim 12°$ per day, and returns

† See details of the program in `http://www.to.astro.it/blazars/webt/gasp/homepage.html`

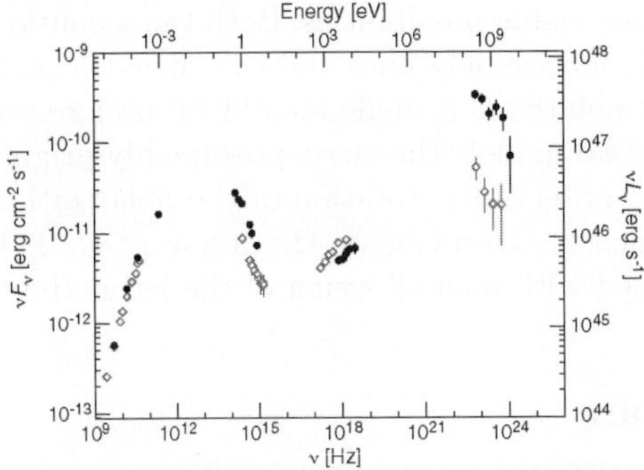

Figure 2. Energy spectrum from radio to γ-ray band of 3C 279 at two different epochs. The filled circle points were taken between 54880 and 54885 MJD, corresponding to the first five days of the sharp γ-ray flare accompanying the dramatic polarization change event while the open diamond points were taken between 54950 and 54960 MJD, around the peak of the isolated X-ray flare. The original version is taken from Abdo *et al.* (2010).

to a level nearly exactly $180°$ from the original level, resembling closely the behavior of optical polarization measured in BL Lacertae (Marscher *et al.* 2008), but at a 4 times slower rate. This clearly indicates that the sharp γ-ray flare is unambiguously correlated with the dramatic change of optical polarization due to a single, coherent event, rather than a superposition of multiple shorter duration events.

The gradual rotation of the polarization angle is unlikely to originate in an axially symmetric jet since any compression of the jet plasma by, e.g., a perpendicular shock moving along the jet and viewed at a small but constant angle to the jet axis would change the degree of polarization, but would not result in a gradual change of EVPA. Instead, it can reflect a non-axisymmetric magnetic field distribution (as in, e.g., Konigl & Choudhuri 1985). The first possibility may be due to propagation of an emission knot following a helical path in a magnetically dominated jet as was recently investigated in the context of the optical polarization event seen in BL Lacertae (Marscher *et al.* 2008) or may involve the "global" bending of a jet. Since the magnetic field in the emission region is anisotropic - presumably concentrated in the plane of a shock / disturbance propagating along the jet - the degree and angle of observed polarization then depends on the instantaneous angle θ of the direction of motion of the radiating material to the line of sight. The maximum rotation rate of the polarization angle would correspond to the minimum of θ. Similar geometry - albeit on larger scales - has been observed in another blazar, PKS 1510-089 (Homan *et al.* 2002). Nonetheless, in both scenarios, the coherent polarization event is produced by a density pattern co-moving along the jet, and therefore, it is possible to estimate the distance traveled by the emitting material during the flare $\Delta r_{\rm event}$; this in turn allows us to constrain the distance of the dissipation region from the black hole $r_{\rm event} \gtrsim \Delta r_{\rm event} \sim 10^{19} \, (\Delta t_{\rm event}/20 \, {\rm days}) \, (\Gamma_{\rm jet}/15)^2 \, {\rm cm}$, which is ~ 5 orders of magnitude larger than the gravitational radius of the black hole in 3C 279. The constraints on the distance of the dissipation region can be relaxed under "flow-through" scenarios – where the emission patterns may move much slower than the bulk speed of the jet or not propagate at all: one such example is the model involving swings ("wobbling") of the jet associated with jet instabilities such that its boundary moves relative to our line of sight. In this case the time scale for the observed variation is the time scale for the jet motion. Consequently, the emission region easily can be much closer (by a factor $\Gamma^2_{\rm jet}$) to the black hole than in the "helical" or "bent jet" scenarios, since the natural radial scale for $\Delta t_{\rm event} \sim 20$ days is $r_{\rm event} \sim c\Delta t_{\rm event} \sim 500 - 1000$ gravitational radii (see, e.g., Lyutikov *et al.* 2003).

This leaves us with three viable possibilities. Both the scenario involving a knot propagating along the helical magnetic field lines and the "flow-through" scenario above imply that the rotation of the polarization angle should be preferentially following the same direction, since in those two models the twist presumably originates in the inner accretion disk. In our case, we observe the rotation of the polarization angle that is opposite in direction than that measured previously (Larionov *et al.* 2008), leaving us with the "bent jet" model combined with a small swing of the jet as the most compelling.

4. Isolated X-ray flare

Concurrent X-ray observations indicate relatively steady flux during the γ-ray flare with the optical polarization change, but reveal a significant, symmetrical flare at 54950 MJD with duration of ~ 20 days, similar to the duration of the γ-ray flare. It suggests the X-ray photons are produced at a comparable distance from the black hole as the optical/γ-ray photons. Importantly, this X-ray flare is not accompanied by a prominent optical or γ-ray flare. The X-ray spectrum during the isolated flare remains much harder than the optical spectrum (see Figure 2), and therefore cannot be attributed to a temporary extension of the high-energy tail of the synchrotron emission, but instead, may be generated by inverse-Compton scattering of low-energy electrons. However, the similarity of profiles of the γ-ray and X-ray flares argues against the latter being just a delayed version of the former due to, e.g., particle cooling. Therefore, it must be produced independently by another mechanism involving primarily lower energy electrons.

More details of the campaign results and discussions are described in Abdo *et al.* (2010).

The *Fermi*-LAT Collaboration acknowledges support from a number of agencies and institutes for both development and the operation of the LAT as well as scientific data analysis. These include NASA and DOE in the United States, CEA/Irfu and IN2P3/CNRS in France, ASI and INFN in Italy, MEXT, KEK, and JAXA in Japan, and the K. A. Wallenberg Foundation, the Swedish Research Council and the National Space Board in Sweden. Additional support from INAF in Italy and CNES in France for science analysis during the operations phase is also gratefully acknowledged.

References

Abdo, A. A., *et al. Nature*, **463**, 919–923 (2010).
Albert, J., *et al. Science*, **320**, 1752–1754 (2008).
Atwood, W. B., *et al. Astrophys. J.*, **697**, 1071–1102 (2009).
Dermer, C., Schlickheiser, R., & Mastichiadis, A. *Astron. Astrophys.*, **256**, L27–L30 (1992).
Hartman, R. C., *et al. Astrophys. J. (Lett.)*, **385**, L1–L4 (1992).
Homan, D. C., *et al. Astrophys. J.*, **580**, 742–748 (2002).
Jorstad, S. G., *et al. Astron. J.*, **130**, 1418–1465 (2005).
Kniffen, D. A., *et al. Astrophys. J.*, **411**, 133–136 (1994).
Konigl, A. & Choudhuri, A. R. *Astrophys. J.* **289**, 188–192 (1985).
Larionov, V. M., *et al. Astron. Astrophys.* **492**, 389–400 (2008).
Lyutikov, M., Pariev, V. I., & Blandford, R. *Astrophys. J.* **597**, 998–1009 (2003)
Marscher, A. P., *et al. Nature*, **452**, 966–969 (2008).
Nilsson, K., *et al. Astron. Astrophys.*, **505**, 601–604 (2009)
Watanabe, M., *et al. Pub. Astron. Soc. Pacif.*, **117**, 870–884 (2005).
Wehrle, A. E., *et al. Astrophys. J.*, **497**, 178–187 (1998).
Woo, J.-H. & Urry, C. M. *Astrophys. J.* **579**, 530–544 (2002).

Discussion

BEDNAREK: In the LAT spectrum during the flare consistent with the MAGIC spectrum, you model only the LAT spectrum?

HAYASHIDA: The exptrapolation of our LAT spectrum to higher energies goes to below the MAGIC data points. But they are not simultaneous data and the LAT spectrum is still 4-5 times lower than the historical maximum flux among the EGRET observations.

ROMERO: Changes in the polarization angle were observed long before Fermi launching. Actually, Ileana Andruchow has published in A&A the very same model you have shown here. It would be nice, I think, if the Fermi team checked the relevant literature. Just a comment.

HAYASHIDA: Thank you for your pointing out. In fact, we have been aware of your work on optical polarization observations. However, our work focuses in the long term monitoring and the event of a gradual and large rotation (about 180°) in polarization degree, while that paper has results of short term (4 days) and discusses only microvariability. It is rather different from our work and we did not have room for listing all previous polarimetric studies. We appreciate your work and we would like to emphasize the important of polarimetric observations.

ANDRUCHOW: Did you perform any statistical tests in order to analyze the variability in an isolated curve?

HAYASHIDA: When we apply chi-squared tests, for example, fitting with a constant flux, probability is almost zero, as one can easily expect from light curves even by eyes. Then we also calculated the fractional variability index to quantify the variability. Although the data sampling rates are not the same along different bands, the gamma-ray band shows the largest value.

Jets at all Scales
Proceedings IAU Symposium No. 275, 2011
G. E. Romero, R. A. Sunyaev & T. Belloni, eds.

© International Astronomical Union 2011
doi:10.1017/S1743921310015863

Unveiling the nature of extragalactic jets with Chandra observations

F. Massaro[1], **C. C. Cheung**[2] **and D. E. Harris**[1]

[1]Harvard - Smithsonian Astrophysical Observatory, Center for Astrophysics
60 Garden Street, Cambridge, MA 02138, USA
email: fmassaro@head.cfa.harvard.edu

[2]NRL/NRC, 4555 Overlook Ave SW, Washington, DC, 20375 USA

Abstract. In 1974 Fanaroff & Riley divided the extended radio sources into two classes, on the basis of their radio morphology and power. For several years we have been collecting basic parameters for extragalactic jets detected in the X-rays, looking for an extension of the classification criterion, based on their radio and X-rays properties. The fact that different processes have been proposed to explain their X-ray radiation, (synchrotron vs inverse Compton emission) suggests the possibility of a new classification scheme. However, comparing the radio-to-X-ray properties of the extragalactic jets, several aspects on their nature became unexpectedly unclear.

Keywords. radiation mechanisms: non-thermal, X-rays: galaxies, galaxies: active, galaxies: jets.

1. Introduction

Radio galaxies and Quasars show a small number of features: jets, hotspots, and lobes. Following the definition of Bridle (1986), jets are highly collimated structures, distinguishable from other extended components either spatially or by brightness contrast, starting in the central region of the host active galactic nucleus (AGN). As suggested by Leahy *et al.* (1997), hotspots are brightness peaks which are neither a core nor part of a jet, usually lying where the jet terminates. Lobes are extended components on kpc scale, typically double, roughly symmetrical and ellipsoidal lying on both sides of the host radio galaxy or quasar.

In 1974, it was first noticed by Fanaroff and Riley that the relative positions of regions of high and low surface brightness in the extended regions of radio galaxies are correlated with their radio luminosity (central bright=class I vs. edge bright=class II). They found that nearly all sources with luminosity $L_{178MHz} \leqslant 2 \times 10^{25} \ h_{100}^{-2}$ W Hz^{-1} str^{-1} were of Class I (i.e. FR I) while the brighter sources were nearly all of Class II (i.e. FR II).

The radio galaxy morphology turns out to reflect the method of energy transport in the radio source. FR I radio galaxies typically have bright jets in the centre, while FR IIs have faint jets but bright hotspots at the ends of their lobes (e.g. Baum *et al.* 1995). In particular, FR Is show surface brightness higher toward their cores while FR IIs toward their edges at radio frequencies.

The radio to optical emission arising from extended structures of radio galaxies and quasars is widely interpreted as synchrotron radiation by relativistic particles whereas the origin of X-ray emission in extended structures (jets and hotspots) is still unclear, but is certainly non-thermal. The main dichotomy lies in which mechanism, synchrotron or inverse Compton (IC) scattering, dominates the X-ray emission of jets. The former describes emission from low power jets, typically FR I, while the latter provides a good explanation for high power radio galaxy (i.e. FR II) and quasars.

In the past ten years, we have collected a sample of radio galaxies and quasars for which the X-ray emission associated with their radio jets or hotspots has been detected by Chandra. The main goal of this project (hereinafter XJET project, Harris *et al.* 2010) is the development of a new classification criterion based not only on the radio morphology and power but also on the X-ray properties of these extended structures. We aim to build a method to distinguish between knots in jets and hotspots, both in radio galaxies and quasars, linked with the radiative process responsible for their radio and X-ray emission.

The list and all the basic informations on the sources in our sample (e.g. coordinates, redshift, radio classification etc.) can be found on the XJET webpage†. Our sample consists of 96 sources for a total of 194 extended components.

2. X-ray Data reduction

The X-ray data reduction has been performed following the standard reduction procedure described in the Chandra Interactive Analysis of Observations (CIAO) threads ‡, using CIAO v4.2 and the Chandra Calibration Database (CALDB) version 4.2.

Level 2 event files were generated using the *acis_process_events* task, after removing the hot pixels with *acis_run_hotpix*. Events were filtered for grades 0,2,3,4,6 and we removed pixel randomization.

Lightcurves for every dataset were extracted and checked for high background intervals, that have been eventually removed according to the criteria described in Chandra Interactive Analysis of Observations (CIAO) threads.

Astrometric registration was achieved by changing the appropriate keywords in the fits header so as to align the nuclear X-ray position with that of the radio.

We created 3 different fluxmaps (soft, medium, hard, in the ranges 0.5 – 1, 1 – 2, 2 – 7 keV, respectively) by dividing the data with monochromatic exposure maps (with nominal energies of soft=0.75 keV, medium=1.4 keV, and hard=4 keV). The exposure maps and the flux maps were regridded to a common pixel size which was usually 1/4 the size of a native ACIS pixel (native=0.492″). For sources of large angular extent we used 1/2 or no regridding. To obtain maps with brightness units of ergs cm^{-2} s^{-1} pixel^{-1}, we multiplied each event by the nominal energy of its respective band.

To measure observed fluxes we construct appropriate regions (usually circular) as well as two adjacent background regions of the same size. We then measure the net flux in each region and in each energy band. After applying a small correction which is the ratio of the mean energy of the events within the 'on' aperture to the nominal energy applied earlier to all events, a one σ error is assigned based on the usual $\sqrt{number - of - counts}$ in the on and background regions.

3. Data analysis

We developed two main *tools* to investigate the radio and the X-ray emission of the selected jets and hotspots.

By measuring the radio fluxes, using the radio maps available in the public archives (e.g. NVSS, NED) or kindly provided by our colleagues, we calculated the X-ray-to-radio flux ratio ρ, where the X-ray fluxes have been derived directly from our fluxmaps analysis. On the other hand, both the X-ray and the radio luminosities, L_X and L_R, for the whole

† ($http://hea-www.harvard.edu/XJET/$)
‡ ($http://cxc.harvard.edu/ciao/guides/index.html$)

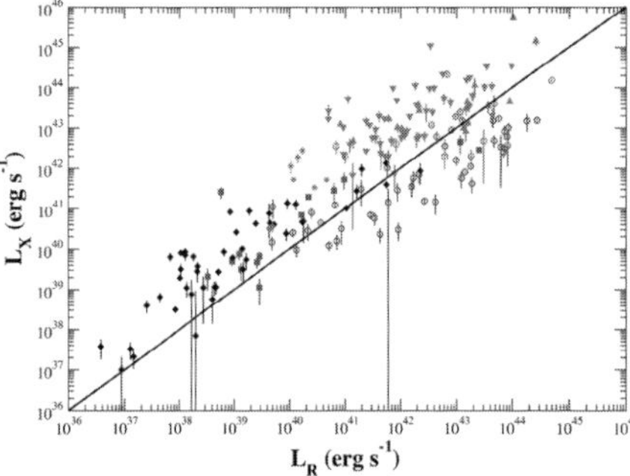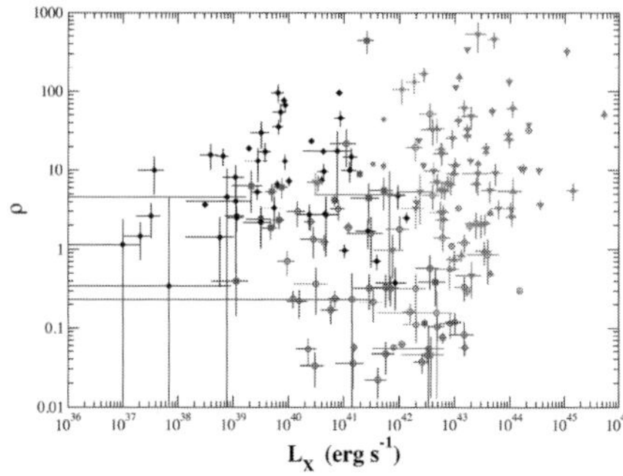

Figure 1. a - Left panel) The X-ray luminosity vs the radio luminosity of all the knots and hotspots in the selected radio galaxies and quasars. b - Right panel) The luminosity ratio vs the X-ray luminosity of the selected knots and hotspots. The nomenclature here reported has been adopted for all the plots in the paper, indicating: FR I knots (filled blue diamonds), FR II knots (filled blue squares), CDQ and LDQ knots (filled red triangles), BL Lac knots (green stars), FR II hotspots (i.e. hs2, open blue circles), LDQ hotspots (open red cricles).

sample of extended components (i.e. knots and hotspots), in radio galaxies and quasars, are plotted in Fig. 1a.

In addition, we also evaluated the X-ray hardness ratio HR using the hard, H, and the medium, M, X-ray fluxes by using the following relation: $(H-M)/(H+M)$. We neglected the soft X-ray flux in the estimate of HR because uncertain absorption corrections are much larger for the soft band than for the medium and hard bands. The hardness ratio HR has been used as a good indicator of the X-ray spectral index α_X of our extended components, because the short exposure of the available X-ray observations combined with their low intrinsic flux (i.e. $\sim 10^{-15}$ erg s^{-1} cm^{-2}) do not allow us to estimate α_X precisely for the whole sample. We also note that because HRs are evaluated using X-ray fluxes there is a 1-1 relationship between α_X and HR.

We compare all the values of ρ and HR dividing the knots and the hotspots of our sources in 5 categories as defined in the following. The nomenclature chosen has been adopted for all the figures. We considered knots in FR Is (i.e. k1, filled blue diamonds) and in FR IIs (i.e. k2, filled blue squares), knots in quasars (i.e. kq, filled red triangles) and in BL Lacs (i.e. kbl, green stars), while for the hotspots we defined those in FR IIs (i.e. hs2, open blue circles) and in LDQs (i.e. hsq, open red cricles).

All the main results derived form our analysis comparing the ρ and the HR values of the different extended structure here defined are reported in the next section.

4. Results

The main observational results arising from our analysis are here summarized.

• The X-ray luminosities of knots in the BL Lac jets lie between FR I and Quasars (e.g. Fig. 1, 2a).

  • Hotspots typically do not show luminosity ratios higher than 10 (e.g. Fig. 1, 2a).

  • All knots have luminosity ratios higher 0.1 (e.g. Fig. 1, 2a).

• Knots of radio galaxies and quasars have similar luminosity ratio distributions (e.g. Fig. 1, 2).

• The distribution of the luminosity ratios, both for hotspots and knots, are similar between radio galaxies and quasars. This is a surprising point because, following the general interpretation, the X-ray emission arising from these extended structures has a

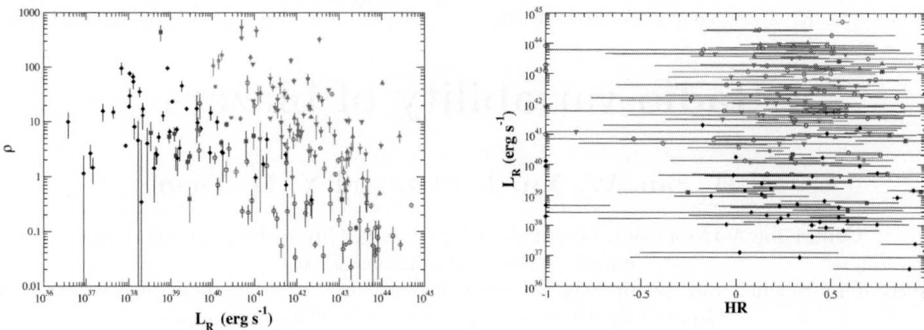

Figure 2. a - Left panel) The luminosity ratio vs the radio luminosity of the selected knots and hotspots. b - Right panel) The radio luminosity *vs* the X-ray hardness ratios.

different origin (synchrotron *vs* inverse Compton processes). So it was expected to find some differences between their luminosity ratio distributions.

• There are no differences between knots of quasars and radio galaxies in terms of HR (e.g. Fig. 2b).

• There are no differences between the HRs in hotspots of quasars and radio galaxies (e.g. Fig. 2b).

Consequently, the main conclusion arising from our analysis is that, despite the three radiative processes proposed to interpret the X-ray emission in knots and hotspots of both radio galaxies and quasars suggesting a different distributions of luminosity ratios and X-ray hardness ratios, no difference have been found.

Acknowledgements

F. Massaro acknowledges the Foundation BLANCEFLOR Boncompagni-Ludovisi, n'ee Bildt for the grant awarded him in 2010 to support his research. The work at SAO is supported by the NASA grant NNX10AD50G.

References

Fanaroff, B. L. & Riley J. M. 1974, *MNRAS*, 167, 31
Bridle, A. H. 1986. *Extragalactic jets - Trends and correlations. Can. J. Phys.*, 64, 353
Leahy, J. P. *et al.* 1997 *MNRAS*, 291, 20
Baum, S. A., Zirbel, E. L., & O'Dea, C. P. 1995 *ApJ*, 451, 88
Harris, D. E., Massaro, F., & Cheung, C. C. 2010, *AIPC.* 1248, 355

Discussion

SAMBRUNA: A synchrotron counterpart can be present in a QSR knot contributing to X-ray emission, that HR may miss.

MASSARO: Unfortunately, the low statistic, due to short exposure of many of the considered Chandra observation do not allow us to distinguish between single and multicomponent models. So, it is certainly true that a second component may affect the estimates of the hardness ratios in the X-rays.

Jets at all Scales
Proceedings IAU Symposium No. 275, 2011
G. E. Romero, R. A. Sunyaev & T. Belloni, eds.

© International Astronomical Union 2011
doi:10.1017/S1743921310015875

Radio variability of blazars

J. H. Fan, W. Xu, J. Pan, and Y. H. Yuan

Center for Astrophysics, Guangzhou University, Guangzhou 510006, China
email: jhfan_cn@yahoo.com.cn
Astronomy Science and Technology Research Laboratory of Department of Education of
Guangdong Province, Guangzhou 510006, China

Abstract. In this work, we present the analysis results using UMRAO preliminary data base. We used the light curves 1) to get the shortest timescales and then to get the brightness temperature so that we can estimate the Doppler factors; 2) to investigate the periodicity and discuss the variability index. We also used the data base to discuss the polarization properties of blazars. We found that the periodicity distribution in BL Lacs and that in the flat spectrum radio quasars should be from the same distribution. The Doppler factor in FSRQs is higher than that in BL. The polarization in BLs are higher than that in the flat spectrum radio quasars

Keywords. galaxies: nuclei-galaxies: jets-radio continuum: galaxies

1. Introduction

Blazar is a special subclass of active galactic nuclei (AGNs). It consists of two sub-classes, namely BL Lacertae objects–BLs and flat spectrum radio quasars-FSRQs. In this sense, the properties of BLs and FSRQs can be concluded as the properties of a blazar. Therefore, a blazar shows high and variable luminosity, high and variable polarization, superluminal motions in their radio components, high gamma-ray emission, has strong emission lines or has no emission line feature at all (e.g. Fan *et al.* 2005, 2009, 2010 and references therein). Variability is one of the most interesting properties, that is studied in the whole wavebands, particularly in the optical band (see Bai *et al.* 1998, Boettcher *et al.* 2009, Cellone *et al.* 2007, Dai 2010, de Vries *et al.* 2006, Efimov 2010, Fan *et al.* 1998, 2002, 2007, 2009a, 2009b, Fidelis *et al.* 2009, Gupta *et al.* 2008a, 2008b, 2008c, Kurtanidze *et al.* 2008, Liu 2010, Poon *et al.* 2009, Pyatunina *et al.* 2004, Qian *et al.* 2003, Qian *et al.* 2004, Raiteri *et al.* 2003, Romero *et al.* 2000, Romero *et al.* 2002, Sillanpaa *et al.* 1996, Takalo *et al.* 2010, Villata *et al.* 2004, Wagner *et al.* 1995, Webb *et al.* 1998, Wu *et al.* 2010, Xie *et al.* 1999, Xie *et al.* 2004, Zhang *et al.* 2010 etc. and reference therein).

In the radio bands, some monitoring programs provide data base for one to analyze the radio variability properties, from which we can discuss the emission mechanism and even the relationship between BLs and FSRQs. Based on the UMRAO data base, Aller *et al.* (1992) and Aller *et al.* (2003) investigated the statistical behavior of the flux and the linear polarization of AGNs, and the relation between different radio bands for the Pearson-Readhead sample. In the present paper, we will show our work in the radio variabilities based on the preliminary data base. In the 2nd section, we will show the Doppler factors estimated for the sources, in the 3rd section, we will show the periodicity and the variability index analysis, in the 4th section, we will show the polarization properties.

2. Time scale and Doppler factor

Variability is one of the most extremely properties of blazars. The variabilities were detected on different time scales. Different time scales bring us different information of the emitting source. The short-time scale perhaps sheds some lights on the emission size. If we assume the time scale as the measure of the size of the source, then the observed flux variation may be converted into a brightness temperature (Wagner *et al.* 1995)

$$T_B = (4.5 \times 10^{10} \text{K}) \Delta F [\frac{\lambda d_L}{t_{obs}(1+z)}]^2, \qquad (2.1)$$

here ΔF is the variability of flux density in Jy, λ the wavelength in cm, d_L the distance in Mpc, and t_{obs} the time scale in days, z is redshift of the source.

The time scales can be determined using the shortest time scale at the three frequencies (4.8 GHz, 8 GHz, and 14.5 GHz). The detailed process is as follows. For each frequency, we calculated the time scales corresponding to large variation. If the variation is larger than 5 times of the uncertainty, $\Delta S \geqslant 5\sigma$, the timescale can be taken as a true one, and then we take the shortest time scale as the timescale at the frequency, so we have 3 times scales. Finally, we take the shortest time scale of the three timescales as the timescale of the source (Fan *et al.* 2009c). When the timescale, the variation, and the corresponding wavelength ($\lambda = c/\nu$) are used to relation (2.1), we can get the brightness temperature. When the brightness temperature is higher than 10^{12}K, then the Doppler factor is can be estimated using $\delta = (T_B/10^{12})^{1/3}$, or $\delta = (T_B/10^{12})^{1/5}$.

From the UMRAO data base, we get the Doppler factors, δ's are in the range of 1.02 to 25.82 for all considered blazars. If we considered the BLs and FSRQs separately, we have $\delta = 1.02$ to 19.89 with an average value of $< \delta >= 8.5 \pm 5.3$ for BLs, and $\delta = 1.34$ to 25.82 with an average value of $< \delta >= 12.29 \pm 6.45$ for FSRQs. Doppler factor in FSRQs is, on average, higher than that in BLs(Fan *et al.* 2009c).

3. Variability and periodicity

Blazars are variable in the whole electromagnetic wavebands. It is interesting to investigate the periodicity and the variability violence.

3.1. *Variability Parameter*

The variability violence can be indicated using a variability parameter(see Romero *et al.* 1999). In radio bands, there are three kinds of variability parameters (variability index (VI), the normalized variability amplitude(NVA), and the root mean square dispersion($RMSD$)) (Fan *et al.* 2007).

Variability index measuring the peak-to-trough variations, can be calculated

$$VI = \frac{(S_{max} - \sigma_{S_{max}}) - (S_{min} + \sigma_{S_{min}})}{(S_{max} - \sigma_{S_{max}}) + (S_{min} + \sigma_{S_{min}})},$$

here, S_{max} and S_{min} are the highest and the lowest flux densities with $\sigma_{S_{max}}$ and $\sigma_{S_{min}}$ being the corresponding uncertainties of the fluxes.

Normalized variability amplitude (NVA) can be calculated using the mean $\langle S \rangle$ and standard deviation σ_{tot} of the flux points and the mean error level σ_{err} (see Edelson *et al.* 1996).

$$NVA = \sqrt{\frac{\sigma_{tot}^2 - \sigma_{err}^2}{\langle S \rangle^2}}$$

Root mean square dispersion (RMSD) can be determined by

$$\sigma_{RMSD} = \frac{1}{\langle S \rangle} \sqrt{\frac{1}{N-1} \sum_{i=1}^{n} (S_i - \langle S \rangle)^2},$$

where S_i's are the measured fluxes, $\langle S \rangle = \frac{1}{N} \sum_{i=1}^{n} S_i$ is the mean flux (see Kembhavi & Narlika 1999).

From the relevant data, we found that NVA is correlated with σ_{RMSD}, however the two parameters are not correlated with V.I.. That perhaps suggests NVA and σ_{RMSD} be good variability violence indicators.

3.2. *Periodicity*

Periodic variations are claimed in the optical bands (see Sillanpaa *et al.* 1988, Fan *et al.* 1997, 1998, 2002, Fan 2005). In the radio bands, there is no much work. For the periodicity analysis, the unevenly sampled data make periodicity investigation hard since this sampling will cause false periods. Recently there are some methods to deal with the period determination, such as Jurkevich method, DCF method, and Power spectral (Fourier) analysis (see Fan *et al.* 2002, Fan *et al.* 2007).

When the methods are used to the preliminary data base of UMRAO, we have got the periods at 8GHz for a sample of blazars. The period obtained is in the range of 2.2 to 20.8 years. If we consider FSRQs and BLs separately, we found that the physically significant periodicity at 8GHz are in the range of 2.2 to 20.8 years with an averaged value of 8.9±4.0 years for FSRQs, and 2.5 to 18.0 years with an averaged value of 8.1±3.4 years for BLs. A K-S test shows that the probability for the two distributions to be from the same distribution is 68.2% (Fan *et al.* 2007).

4. Polarization

From the UMRAO data base, we found that on average, the polarization in higher frequency tends to be higher in the radio range from 4.8GHz to 14.5GHz for all blazars. When we only considered the maximum polarization, similar behavior has been shown clearly. If we considered BLs and FSRQs separately, we found that BLs have higher maximum polarization than FSRQs on average. $\langle P^{max}_{4.80GHz}(\%) \rangle = 17\pm13$, $\langle P^{max}_{8.00GHz}(\%) \rangle = 26 \pm 21$, $\langle P^{max}_{14.5GHz}(\%) \rangle = 29 \pm 21$, for BLs, and $\langle P^{max}_{4.80GHz}(\%) \rangle = 9 \pm 10$ $\langle P^{max}_{8.00GHz}(\%) \rangle = 15 \pm 14$, and $\langle P^{max}_{14.5GHz}(\%) \rangle = 15 \pm 13$ for FSRQS (Fan *et al.* 2008). In fact, in the optical band, the polarization in BLs is also higher than that in FSRQs(see Wills *et al.* 1992, Fan 2002).

Acknowledgement: The work is partially supported by the National Natural Science Foundation of China (NSFC 10633010), the 973 programme (2007CB815405), the Bureau of Education of Guangzhou Municipality(No.11 Sui-Jiao-Ke[2009]), and GDUPS(2009). We thank Dr. Margo Aller for the mail of allowing us to use the preliminary data from the University of Michigan Radio Astronomy Observatory, which has been supported by the University of Michigan and the National Science Foundation.

References

Aller, M. F., Aller, H. D., & Hughes, P. A., 1992, *ApJ*, 399, 16
Aller, M. F., Aller, H. D., & Hughes, P. A., 2003, *ApJ*, 586, 33
Bai J. M., *et al.*, 1998, ApJSS, 132, 83.
Boettcher, M., Fultz, K., Aller, H. D., *et al.*, 2009, *ApJ*, 694, 174.

Cellone, S. A., *et al.*, 2007, *MNRAS*, 381, 60

Dai, B., 2010 *Multiwavelength Variability of Blazars, Sept. 22-24, 2010, Guangzhou, China*

de Vries, W. H., Becker, R. H., & White, R. L., 2006, *ASPC*, 360, 29.

Edelson, R. A., Alexander, T., Crenshaw, D. M., *et al.*, 1996, *ApJ*, 470, 364

Efimov, Yu., 2010 *Multiwavelength Variability of Blazars, Sept. 22-24, 2010, Guangzhou, China*

Fan, J. H., *et al.*, 1997, *A&A*, 125, 525.

Fan, J. H., Xie, G. Z., Pecontal, E. *et al.*, 1998, *ApJ*, 507, 173

Fan, J. H., Yuan Yu-Hai, Liu Yi, *et al.*, 2002, *PASJ*, 60, 1217

Fan, J. H., 2002, *PASJ*, 54, 55

Fan, J. H., *et al.*, 2002, *A&A*, 381, 1

Fan, J. H., Romero, Gustavo E., Wang, Yong-Xiang, *et al.* 2005, *ChJAA*, 5, 457

Fan, Jun-Hui, 2005, *ChJAS*, 5, 213

Fan, J. H., Liu, Y., Yuan, Y. H.. *et al.*, 2007, *A&A*, 462, 547

Fan, J. H., Peng, Q. S., Tao, J., Qian, B. C., & Shen, Z. Q., 2009a, *AJ*, 138, 1428.

Fan, J. H., Zhang, Y. W., Qian, B. C., *et al.*, 2009b, *ApJSS*, 181, 466.

Fan, J. H., Huang, Y., He, T. M., *et al.*, 2009c, *PASJ*, 61, 639.

Fan, J. H., Lu, Y., Qian, B. C., *et al.*, 2010, *RAA*, 10, 1100.

Fidelis, V. V., Yakubovskyi, D. A., & Voytkova, Yu. V., 2009, *Astron. Let.*, 35, 579.

Gupta, A. C., Deng, W. G., Joshi, U. C., *et al.*, 2008a, *New Astron.*, 13, 3759.

Gupta, A. C., Cha, S.-M., Lee, S., *et al.*, 2008b, *AJ*, 136, 2359.

Gupta, A. C., Acharya, B. S., Bose, D. *et al.*, 2008c, *CJA&A*, 8, 395.

Kembhavi A. K. & Narlika J. V., 1999, Quasars and Active Galactic Nuclei, Cambridge Uni.
Press

Liu, X., 2010, *Multiwavelength Variability of Blazars, Sept. 22-24, 2010, Guangzhou, China*

Poon, H., Fan, J. H., & Fu, J. N., 2009, *ApJSS*, 185, 511.

Pyatunina, T. B., Rakhimov, I. A., & Zborovskii, A. A., 2004, *Astron. Rep.*, 48, 439.

Qian, B. C. & Tao, J., 2003, *PASP*, 115, 490

Qian, B. C. & Tao, J., 2004, *PASP*, 116, 161

Raiteri C. M., Villata M., Tosti G., *et al.*, 2003, *A&A*, 402, 151

Romero G. E., *et al.*, 1999, A&AS, 135, 477

Romero, G. E., Cellone, S. A., & Combi, J. A., 2000, *AJ*, 120, 1192

Romero, G. E., Cellone, S. A., Combi, J. A., *et al.*, 2002, *A&A*, 390, 431

Sillanpaa, A., Haarala, S., Valtonen, M. J., *et al.*, 1988, *ApJ*, 325, 628.

Sillanpaa, A., Takalo, L. O., Pursimo, T. *et al.*, 1996, *A&A*, 305, L17

Takalo, L. O., *et al.*, 2010, *A&A*, 517, 63.

Villata M., Raiteri C. M., Kurtanidze O. M., *et al.*, 2004, *A&A*, 421, 103

Wanger, S. J. & Witzel, A., 1995, *ARA&A*, 33, 163

Webb J. R., *et al.*, 1998, *AJ*, 115, 2244

Wills, Beverley J., Wills, D., Breger, Michel, Antonucci, R. R. J., *et al.*, 1992, *ApJ*, 398, 454

Wu, J. H., 2010, *Multiwavelength Variability of Blazars, Sept. 22-24, 2010, Guangzhou, China*

Xie G. Z., *et al.*, 1999, *ApJ*, 522, 846

Xie, G. Z., *et al.*, 2004, *MNRAS*, 348, 831

Zhang, Haojing, Zhao, Gang, Zhang, Xiong, Bai, & Jinming, 2010, *Sc. Ch. G*, 53, 252

Jets at all Scales
Proceedings IAU Symposium No. 275, 2011
G. E. Romero, R. A. Sunyaev & T. Belloni, eds.

© International Astronomical Union 2011
doi:10.1017/S1743921310015887

A lepto-hadronic model for the high energy emission from the jets of FR I radiogalaxies

Matías M. Reynoso[1], María C. Medina[2], and Gustavo E. Romero[3]

[1]Instituto de Investigaciones Físicas de Mar del Plata
(Universidad Nacional de Mar del Plata - CONICET), Mar del Plata, Argentina
email: mreynoso@mdp.edu.ar

[2]Institute de Recherche sur les Lois Foundamentales de l'Univers (IRFU), Service de Physique
de Particules, Commisariat l'Energie Atomique, Saclay, France
email: clementina.medina@cea.fr

[3]Instituto Argentino de Radioastronomía (CCT La Plata-CONICET), Villa Elisa, Argentina
email: romero@fcaglp.unlp.edu.ar

Abstract. We present a lepto-hadronic model for the VHE emission from the relativistic jets of
FR I radiogalaxies. We assume that protons and electrons are accelerated in a compact region
near the base of the jet, and they cool emitting multiwavelength radiation as they propagate
along the jet. The particle distributions are obtained using an inhomogeneous steady-state transport equation that accounts for the cooling processes as well as the convection of particles in
the jet. The dominant processes that contribute to the photon SED are electron and proton
synchrotron radiation, inverse Compton interactions, and the inelastic collisions pp and $p\gamma$. The
accompanying neutrino output is obtained and the possibility of detection with Km3Net and
IceCube is discussed for the cases of Cen A and M87.

Keywords. galaxies: active, radiation mechanisms: nonthermal, gamma rays: theory, neutrinos

---

Outline of the model. FR I radiogalaxies are the AGNs in which the jet makes a large
angle with the line of sight. In this work we present a model based of relativistic leptons
and hadrons to explain the high energy emission from the jets of these objects. We show
the results obtained for the radiogalaxies Cen A and M87, which have been detected
by HESS and Fermi. We now enumerate the assumptions made to develop the present
model. Matter is captured by the central black hole though a dissipationless accretion
disk Bogovalov & Kelner (2010). A fraction of this accreted material is expelled in two
oppositely directed jets. Equipartition between jet kinetic energy and magnetic energy
takes place at $z_0 = 50R_g$ from the black hole. The jet has a small half-opening angle ξ, a
mildly relativistic bulk Lorentz factor Γ_j at z_0 &, and a middle viewing angle i_j. The bulk
kinetic power of the jet at z_0 is a fraction of the Eddington power: $L_j^{(\mathrm{kin})} = 0.5q_j L_{\mathrm{Edd}}$.
The magnetic field varies along the jet according to $B(z) = B_0 \left(\frac{z_0}{z}\right)^m$ with $m \in (1,2)$. Γ_j
increases slowly as magnetic energy decreases along the jet. Acceleration of relativistic
protons and electrons takes place at $z_{\mathrm{acc}} > z_0$, where the magnetic energy is in sub-
partition with the kinetic energy. The power carried by primary relativistic particles,
$L_{\mathrm{rel}} = L_p + L_e$, is a fraction of the bulk kinetic power of the jet, $L_{\mathrm{rel}} = q_{\mathrm{rel}} L_j^{(\mathrm{kin})}$. The
relation between the proton and electron power is given by the parameter a: $L_p = a\, L_e$.

Method of calculation. The maximum energies are obtained from the balance of the energy losses and gains. The main cooling for electrons is synchrotron radiation, and for protons we consider: $p\gamma$ collisions, synchrotron emission, adiabatic cooling, and pp collisions.
Synchrotron photons emitted by electrons, are the main target for IC and $p\gamma$ interactions.
They also cause internal absorption of gamma-rays. External absorption by the surround-
ing gas and dust is taken into account with an optical depth $\tau_{\gamma H}(E_\gamma) = N_H \sigma_{\gamma H}(E_\gamma)$.

Table 1. Parameters used in this work.

Parameter	Cen A	M87
$M_{\rm bh}$: Black Hole mass	$10^8\,M_\odot$	$6 \times 10^9\,M_\odot$
$L_{\rm j}^{\rm (kin)}$: jet power	$6 \times 10^{44}\,{\rm erg\ s}^{-1}$	$6.7 \times 10^{46}\,{\rm erg\ s}^{-1}$
$\Gamma_{\rm j}$: jet's bulk Lorentz factor at z_0	2.5	3.5
ξ: jet's opening angle	5°	1.5°
$i_{\rm j}$: viewing angle	25°	20°
$q_{\rm rel}$: jet's content of relativistic particles	0.1	0.1
a: hadron-to-lepton ratio	1	175
m: magnetic field dependance with z	1.5	1.5
z_0: jet's launching point	$7,4 \times 10^{14}\,{\rm cm}$	$4.4 \times 10^{16}\,{\rm cm}$
$z_{\rm acc}$: position of acceleration zone	$167 R_g$	$200 R_g$
$\Delta z_{\rm acc}$: extent of injection zone	$z_{\rm acc}\tan\xi$	$z_{\rm acc}\tan\xi$
η: acceleration efficiency	10^{-2}	10^{-4}
s : spectral index of injection Q	1.8	2.4
$E_{min,p}$: minimum energy protons	3.5 GeV	4.5 GeV
$E_{min,e}$: minimum energy electrons	51 MeV	77 MeV
N_H: gas and dust column density	$10^{23}\,{\rm cm}^{-2}$	$2.5 \times 10^{20}\,{\rm cm}^{-2}$

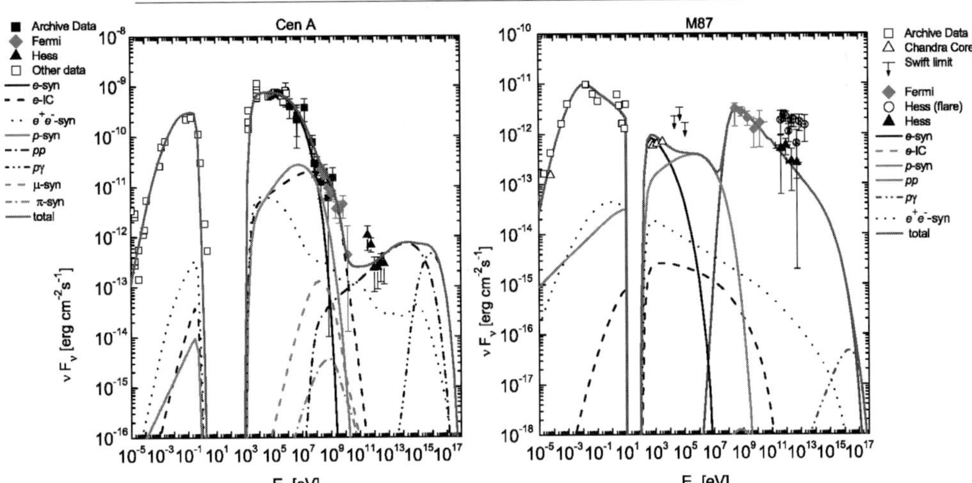

Figure 1. SEDs obtained for Cen A and M87.

The cross section $\sigma_{\gamma H}(E_\gamma)$ is taken as in Ryter (1996). The distribution of particles are obtained solving a steady-state transport equation with cooling and convection:

$$v(z)\frac{\partial N(E,z)}{\partial z} + \frac{\partial b(E,z)N(E,z)}{\partial E} + \frac{N(E,z)}{T_{\rm dec}(E)} = Q(E,z),$$

where $b(E,z) = -\frac{dE}{dt}$. The production of secondary particles $(\pi,\ \mu,\ e^\pm,\ \nu_i)$ in the jet by $p\gamma$, pp, and $\gamma\gamma$ interactions is taken into account.

Results for Cen A and M87. Table 1 presents the set of parameters used for applying our model to Cen A and M87. The resulting SED is shown in Fig 1. The computed SEDs are basically consistent with the multiwavelength data for both sources. The main processes are synchrotron emission of electrons and protons, IC interactions, pp and $p\gamma$ collisions. The accompanying neutrino output obtained is found to be stronger for Cen than for M87 for energies above 1 TeV (see Reynoso *et al.* (2010) for details).

References

Bogovalov, S. V. & Kelner, S. R. 2010, *IJMPD*, 19, 339.
Morganti, R., Oosterloo, T., Struve, C., & Saripalli, L. *A&A*, 485, L5.
Reynoso, M. M., Medina, M. C., & Romero, G. E. 2010, submitted, arXiv:1005.3025.
Ryter, C. E., 1996 *ApSS*, 236, 285.

Jets at all Scales
Proceedings IAU Symposium No. 275, 2011
G. E. Romero, R. A. Sunyaev & T. Belloni, eds.

© International Astronomical Union 2011
doi:10.1017/S1743921310015899

3D-MHD simulations of the evolution of magnetic fields in FR II radio sources

Martín Huarte-Espinosa[1,3], Martin Krause[4,5] and Paul Alexander[2,3]

[1]Department of Physics and Astronomy, University of Rochester 600 Wilson Boulevard,
Rochester, NY, 14627-0171; [2]Astrophysics Group, Cavendish Laboratory, 19 J. J.
Thomson Ave., Cambridge CB3 0HE, UK
emails: `martinhe@pas.rochester.edu, pa@mrao.cam.ac.uk`
[3]Kavli Institute for Cosmology Cambridge, Madingley Road, Cambridge CB3 0HA, UK;
[4]Max-Planck-Institut für Extraterrestrische Physik, Giessenbachstrasse, 85748
Garching, Germany. email: `krause@mpe.mpg.de`
[5]Universitätssternwarte München, Scheinerstr. 1, 81679 München, Germany

Abstract. 3D-MHD numerical simulations of bipolar, hypersonic, weakly magnetized jets and
synthetic synchrotron observations are presented to study the structure and evolution of magnetic fields in FR II radio sources. The magnetic field setup in the jet is initially random. The
power of the jets as well as the observational viewing angle are investigated. We find that synthetic polarization maps agree with observations and show that magnetic fields inside the sources
are shaped by the jets' backflow. Polarimetry statistics correlates with time, the viewing angle
and the jet-to-ambient density contrast. The magnetic structure inside thin elongated sources is
more uniform than for ones with fatter cocoons. Jets increase the magnetic energy in cocoons,
in proportion to the jet velocity. Both, filaments in synthetic emission maps and 3D magnetic
power spectra suggest that turbulence develops in evolved sources.

Keywords. galaxies: active, jets, methods: numerical, polarization, MHD

1. Introduction

Fanaroff-Riley class II radio sources (FRIIs, Fanaroff & Riley 1974) are extragalactic,
synchrotron in nature and show linear polarization fractions within 10–50% (Bridle &
Perley 1984). Stokes parameters are used to infer the magnetic structure in these sources.
Observed magnetic polarization vectors are generally parallel to the jets and to the lobe
boundaries, and follow both flux intensity gradients perpendicularly and lines between
multiple lobe hot spots (Bridle & Perley 1984). The linear polarization fraction in FRIIs
is typically higher at jet edges than in their beams, and also at source edges than in the
cocoons (Saikia & Salter 1988). The magnetic structure in FRIIs, as well as the way it
evolves and relates to AGN jet properties, is not clear.

2. Model and methodology

The equations of ideal MHD are solved in 3D using the code Flash 3.1 (Fryxell B.
et al. 2000), inside a cubic Cartesian domain with 200^3 fixed cells. The intra-cluster
medium is implemented as a monoatomic ideal gas ($\gamma = 5/3$), a stratified King density
profile (King 1972), magnetohydrostatic equilibrium with a central gravitational field
and magnetic fields with a Kolmogorov turbulent structure, with a thermal-to-magnetic
pressure ratio $\gtrsim 10$. Source terms are implemented in the equations to inject mass and
x-momentum in a central cylinder which takes weak and random magnetic fields from

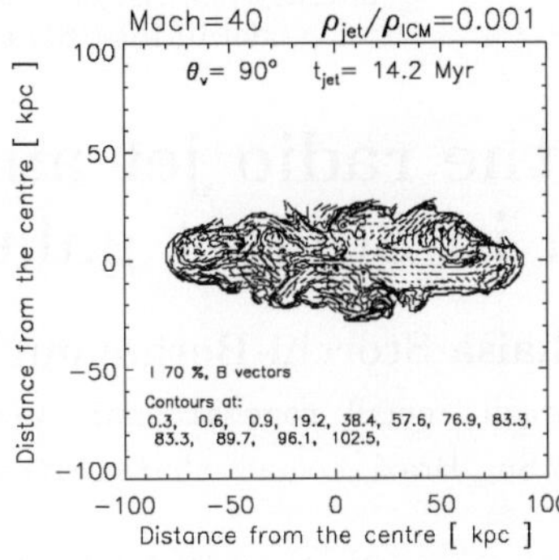
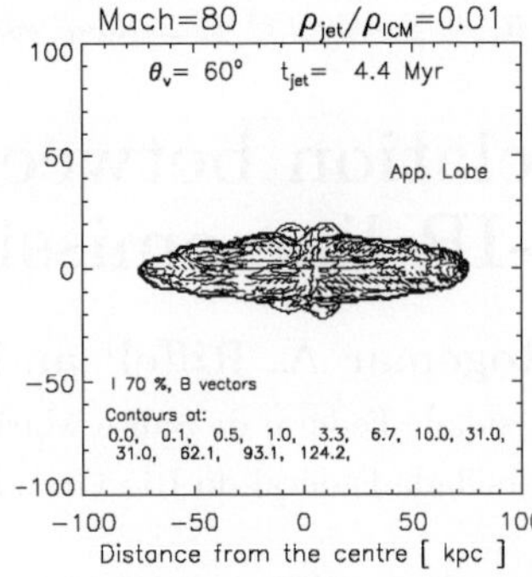

Figure 1. Synthetic polarization maps. Vectors follow the magnetic position angle and their length is proportional to the degree of linear polarization. Vectors are superimposed on contours of synchrotron emission (at 8 GHz) normalized to the mean emissivity. The initially random magnetic fields have been ordered by MHD processes. The left lobe in the right panel source is receding, yet beaming and light-travel effects are assumed to be negligible.

the innermost ambient medium. We investigate jet velocities with Mach=$\{40, 80, 130\}$ as well as $\eta = \rho_{\mathrm{jet}}/\rho_{\mathrm{amb}} = \{0.01, 0.001\}$.

3. Synthetic synchrotron emission

Synchrotron emission and Stokes parameters are calculated and integrated through the inflated model sources, along the line of sight. The density distribution of ultra-relativistic electrons is the product of the cocoon pressure and an incompressible tracer field injected with the jets. Synthetic polarization maps are produced for five model sources at different source expansion times, t_{jet}, and for viewing angles $\theta_v = \{30°, 60°, 90°\}$ (e.g. Figure 1).

4. Conclusions

Jets injected with initial random magnetic fields develop ordered fields by MHD processes within the radio source. Filaments suggest that turbulence develops in evolved sources. Polarimetry statistics correlates with time, θ_v and η, but not so with v_{jet}. Lighter jets show linear polarization degrees $\sim 39\%$ at the end of the simulations, independently of θ_v. This agrees with observations better than for the heavier sources which show better, and more realistic, field alignment with the jets than lighter sources. Some initial order in the magnetic fields may be required to meet all the constraints at once. See Huarte-Espinosa, Krause & Alexander 2011a (in prep.) for details.

The software used in these investigations was in part developed by the DOE-supported ASC / Alliance Center for Astrophysical Thermonuclear Flashes at the University of Chicago. MHE acknowledges funding from CONACyT México 196898/217314, and Dongwook Lee for the 3D-USM-MHD solver of Flash 3.1.

References

Bridle A. H. & Perley R. A. 1984, *ARA&A*, 22, 319
Fanaroff B. L. & Riley J. M. 1974, *MNRAS*, 167, 31
Fryxell *et al.* 2000, *ApJS*, 131, 273
King, I. R. 1972, *ApJL*, 174, L123
Matthews, A. P. & Scheuer, P. A. G. 1990, *MNRAS*, 242, 616.
Saikia, D. J. & Salter, C. J. 1988, *ARA&A*, 26, 93

Jets at all Scales
Proceedings IAU Symposium No. 275, 2011
G. E. Romero, R. A. Sunyaev & T. Belloni, eds.

© International Astronomical Union 2011
doi:10.1017/S1743921310015905

The relation between the radio jet and the near-IR line emission in Seyfert galaxies

Rogemar A. Riffel[1] and Thaisa Storchi-Bergmann[2]

[1] Universidade Federal de Santa Maria, Brazil – email: `rogemar@smail.ufsm.br`

[2] Universidade Federal do Rio Grande do Sul, Brazi – email: `thaisa@ufrgs.br`

Abstract. We used near-IR integral field spectroscopy, obtained with Gemini NIFS and GNIRS integral field units (IFUs), to map the ionized and molecular flux distributions and kinematics in the central few hundreds of parsecs of Seyfert galaxies.We conclude that the molecular gas emission can be considered a tracer of the feeding of the AGN, while the emission of the ionized gas a tracer of its feedback.

Keywords. galaxies: Seyfert, galaxies: jets, galaxies: kinematics and dynamics

1. Observations and Data Reduction

In this work we present two-dimensional maps for fluxes and velocity dispersion (σ) covering the inner hundreds of parsecs of the Seyfert galaxies ESO 428-G14, Mrk 1066 and NGC 591. These maps were constructed from J and K-band observations obtained using GNIRS-IFU (for ESO 428-G14) and NIFS (for Mrk 1066 and NGC 591) on Gemini telescopes. The data reduction was done using the software IRAF and followed the standard procedure. See Riffel *et al.* (2006, 2010) for more details.

2. Results and Conclusions

In Fig. 1 we present two-dimensional maps for the flux and σ obtained from the fitting of the H_2 2.12 μm, Paβ and [Fe II] 1.25 μm emission-line profiles. These maps show a tight relation between the ionized gas emission-line flux distributions and kinematics and the radio structure (thick contours), revealing that the radio jet plays a fundamental role not only in shaping the narrow-line region but also in the imprint of its kinematics. Moreover, the H_2 maps show only a weak correlation with the radio structure, suggesting that the radio jet is less important for the warm H_2 emission than for the emission of ionized gas.

We conclude that the molecular and ionized gas present distinct flux distributions and kinematics, with the former more restricted to the plane of the galaxies and the latter extending to high latitudes above it, being more associated with the radio emission. This conclusion is in good agreement with those found for other Seyfert galaxies and suggest that the molecular gas is a tracer of the feeding of the AGN and the ionized gas a tracer of its feedback.

References

Falcke H., Wilson A. S., Simpson C., & Bower G. A. 1996, *ApJ*, 470, L31

Nagar N. M., Wilson A. S., Mulchaey J. S., & Gallimore J. F. 1999, ApJS, 120, 209.

Riffel, R. A., Storchi-Bergmann, T., Winge, C., & Barbosa, F. K. B. 2006, *MNRAS*, 373, 2

Riffel, R. A., Storchi-Bergmann, & Nagar, N. M 2010, *MNRAS*, 404, 166

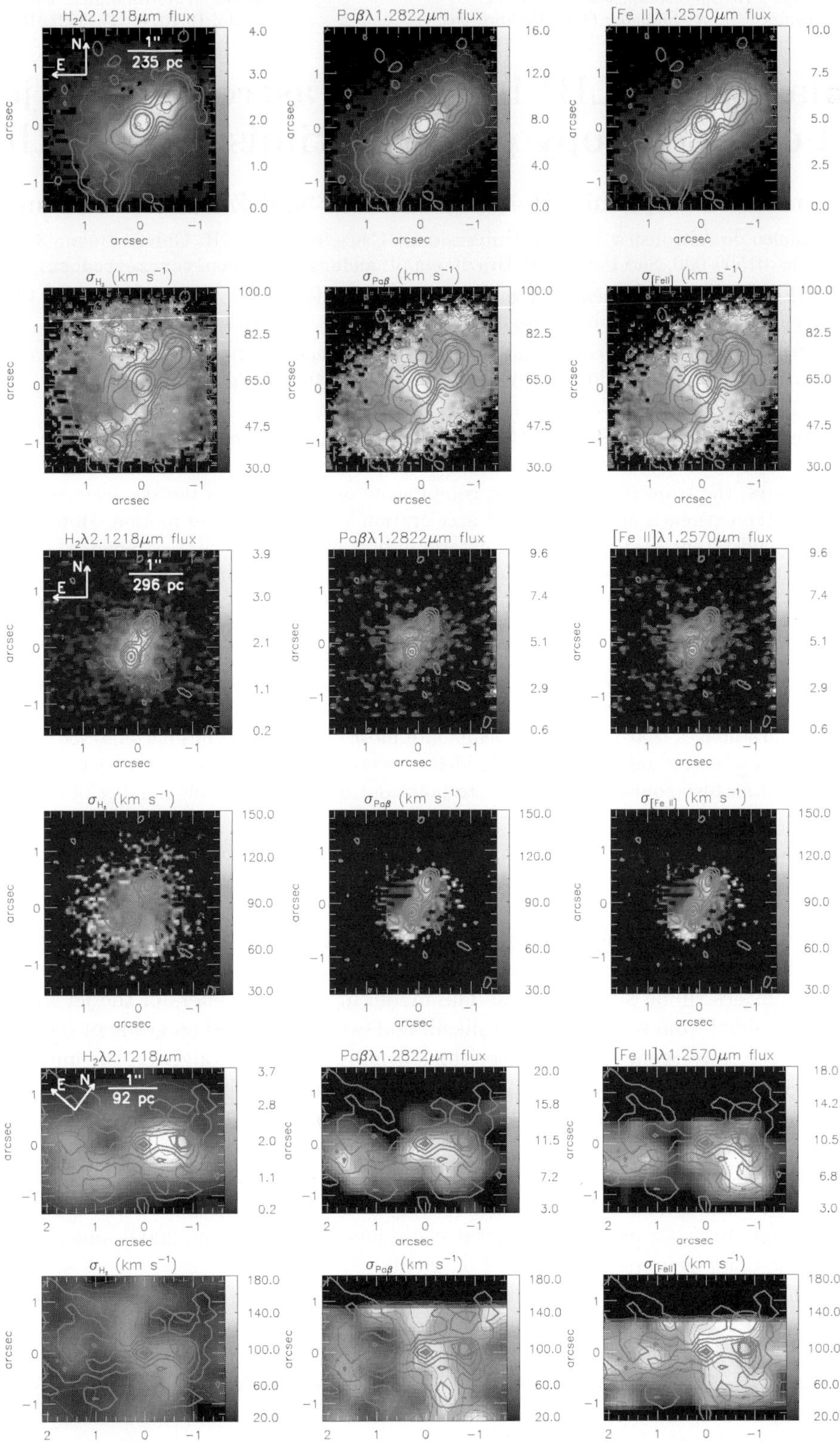

Figure 1. Resulting flux and σ maps for the $H_2\,2.12\,\mu m$, Paβ and [Fe II] $1.25\,\mu m$ (from left to right) for Mrk 1066 (1st and 2nd rows), NGC 591 (3rd and 4th rows) and ESO 428-G14 (5th and 6th rows). The thick contours are from radio continuum images from (Nagar *et al.* 1999) for Mrk 1066 and NGC 591 and from (Falcke *et al.* 1996) for ESO 428-G14.

Jets at all Scales
Proceedings IAU Symposium No. 275, 2011
G. E. Romero, R. A. Sunyaev & T. Belloni, eds.

© International Astronomical Union 2011
doi:10.1017/S1743921310015917

Analysing VLBI images of astrophysical jets via cross-entropy global optimisation method

Anderson Caproni[1], Hektor Monteiro[2] and Zulema Abraham[3]

[1]Núcleo de Astrofísica Teórica, Universidade Cruzeiro do Sul, R. Galvão Bueno 868, Liberdade, 01506-000, São Paulo, SP, Brazil. email: `anderson.caproni@cruzeirodosul.edu.br`

[2]UNIFEI, Instituto de Ciências Exatas, Universidade Federal de Itajubá, Av. BPS 1303, Pinheirinho, 37500-903, Itajubá, MG, Brazil

[3]Instituto de Astronomia, Geofísica e Ciências Atmosféricas, Universidade de São Paulo, R. do Matão 1226, Cidade Universitária, 05508-900, São Paulo, SP, Brazil

Abstract. Most of jets detected in AGN blazar sources exhibit a morphological structure usually composed by a spatially unresolved core and jet knots receding from it at relativistic velocities. In some cases, the trajectories of the jet components on the plane of the sky seem to be bent, indicating the existence of some kind of acceleration in the respective motion. However, such claims depend strongly on the correct determination of the structural parameters of the jet components, usually obtained from model fitting procedures performed either in the (u, v) or in the image planes. In this work we introduce a new model fitting technique to obtain structural parameters of knots present in VLBI jet images. Our method that is based on the cross-entropy technique minimises an performance function that depends on the sum of the squared residuals obtained from the comparison of an VLBI image and a model image, constructed by summing N_s elliptical Gaussian synthetic sources. We present in this work the cross-entropy model fittings of benchmark images that were built to simulate most of the conditions encountered in typical VLBI images of active galactic nuclei. Besides recovering the parameters of the jet components in all validation tests, our method is able to point out quantitatively the number of the sources present in the image.

Keywords. Methods: data analysis, techniques: interferometric, galaxies: jets, galaxies: active.

The determination of the structural parameters of AGN jet components detected from VLBI experiments is crucial to study jet kinematic, as well as to understand the underlying physics ruling such motions. The model fitting parameters of the jet knots are usually obtained from some maximum-likelihood estimator based on gradient techniques. However, the convergence of such algorithms usually depend strongly on the initial guess of the parameters and they are prone to find non-global minimum solutions, especially in the case of too complex images. To cope with those issues, we have developed an alternative method to obtain model fittings to interferometric radio images of astrophysical jets using a global optimisation algorithm known as cross-entropy. This statistical method was introduced by Rubinstein (1997), and it has been used in astrophysics in two different contexts: precession of relativistic jets (Caproni *et al.* 2009) and isochrones of Galactic open clusters (Monteiro *et al.* 2010).

Assuming that jet components can be modelled by N_s elliptical Gaussian functions characterised by six parameters (two-dimensional peak position in the image, peak intensity, eccentricity, amplitude and orientation angle of the major axis), our cross-entropy optimisation technique starts generating randomly a set of N individual tentative solutions composed by $6N_s$ parameters. Each tentative solution generates an image that is compared quantitatively with the observed image through an objective function that depends on the sum of the squared residuals obtained from the difference between them.

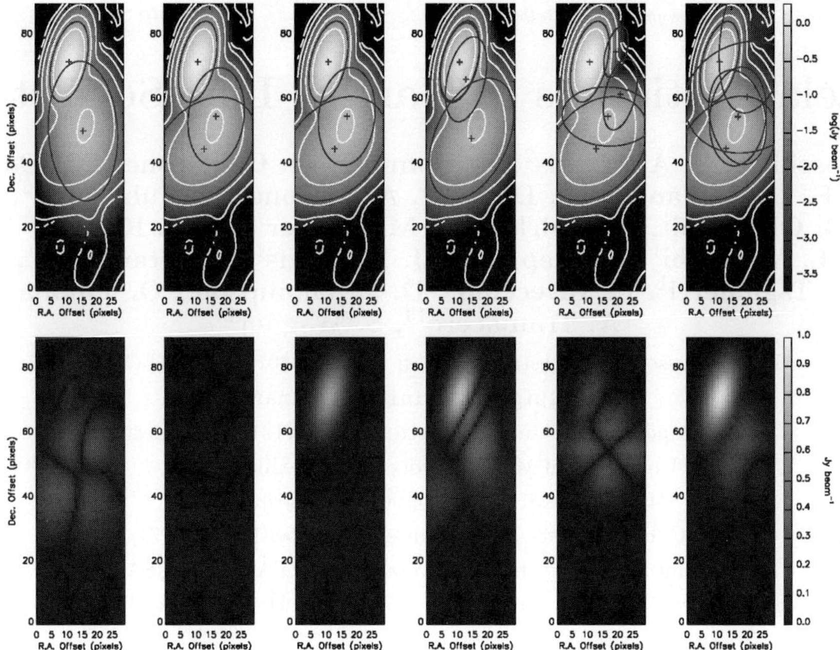

Figure 1. Results of model fitting for a synthetic image composed by 3 elliptical Gaussian sources superposed to a noise background extracted from an image obtained by the MOJAVE project. From the left to right, the number of the sources was varied from 2 to 7. Top: The contour lines are of the benchmark image, the grey-scale image is constructed from the fitted source parameters and the dark ellipses are the contours of the individual fitted sources (respective centres marked with crosses) at the FWHM. Bottom: The respective residual maps obtained from the absolute difference between the noise and the optimised model images (in linear scale).

The N_{elite}-best candidates (those that better minimises the objective function) are selected and used to construct the next set of tentative solutions. The optimisation process is halted after a pre-specified stopping criterion is fulfilled.

To validate our model fitting optimised procedure, we built benchmark tests that employed synthetic images created to simulate as realistic as possible typical interferometric radio maps. Those images are analysed by our model-fitting technique in order to check its capability of recovering the original parameters of the Gaussian sources. We show in Fig. 1 the model-fitting results from one of this benchmark test. It can also be noted a minimum in the distribution of the residuals for $N_{\mathrm{s}} = 3$, as expected in this particular case. A more complete discussion of the benchmark tests and their consequences will be presented in Caproni *et al.* (2011).

The authors acknowledge the Brazilian agencies FAPESP (Procs. 2010/090006-3 and 2006/57824-1) and CNPq for financial support. This research has made use of data from the MOJAVE database that is maintained by the MOJAVE team (Lister *et al.*, 2009, AJ, 137, 3718).

References

Caproni, A., Monteiro, H., & Abraham, Z. 2009, *MNRAS*, 399, 1415
Caproni, A., Monteiro, H., & Abraham, Z. 2011, *in preparation*
Monteiro, H., Dias, W. S., & Caetano, T. C. 2010, *A&A*, 516, 2
Rubinstein, R. Y. 1997, *European Journal of Operational Research*, 99, 89

Jets at all scales
Proceedings IAU Symposium No. 275, 2011
G. E. Romero, R. A. Sunyaev & T. Belloni, eds.

© International Astronomical Union 2011
doi:10.1017/S1743921310015929

Relativistic jets in Narrow-Line Seyfert 1

L. Foschini[1], E. Angelakis[2], G. Bonnoli[1], G. Calderone[3], M. Colpi[3],
F. D'Ammando[4], D. Donato[5], A. Falcone[6], L. Fuhrmann[2],
G. Ghisellini[1], G. Ghirlanda[1], M. Hauser[7], Y. Y. Kovalev[2,8],
L. Maraschi[1], E. Nieppola[9], J. Richards[10], A. Stamerra[11],
G. Tagliaferri[1], F. Tavecchio[1], D. J. Thompson[5], O. Tibolla[12],
A. Tramacere[13], S. Wagner[7]

[1]INAF – Osservatorio Astronomico di Brera, 23807 Merate (LC), Italy

email: luigi.foschini@brera.inaf.it

[2]Max-Planck-Institut für Radioastronomie, 53121 Bonn, Germany

[3]University of Milano Bicocca, 20100 Milano, Italy

[4]INAF – IASF-Palermo, 90146, Palermo, Italy

[5]NASA Goddard Space Flight Center, Greenbelt, MD 20771, USA

[6]Penn State University, University Park, PA 16802, USA

[7]Landessternwarte, Universität Heidelberg, Königstuhl, D 69117 Heidelberg, Germany

[8] Astro Space Center of the Lebedev Physical Institute, 117997 Moscow, Russia

[9]Metsähovi Radio Observatory, FIN-02540 Kylmala, Finland

[10]California Institute of Technology, Pasadena, CA 91125, USA

[11]University of Siena, 53100, Siena, Italy

[12]University of Würzburg, 97074, Würzburg, Germany

[13]INTEGRAL Science Data Centre, CH-1290, Versoix, Switzerland

Abstract. Narrow-Line Seyfert 1 (NLS1) class of active galactic nuclei (AGNs) is generally radio-quiet, but a small percent of them are radio-loud. The recent discovery by *Fermi*/LAT of high-energy $\gamma-$ray emission from 4 NLS1s proved the existence of relativistic jets in these systems. It is therefore important to study this new class of $\gamma-$ray emitting AGNs. Here we report preliminary results about the observations of the July 2010 $\gamma-$ray outburst of PMN J0948+0022, when the source flux exceeded for the first time 10^{-6} ph cm^{-2} s^{-1} ($E > 100$ MeV).

Keywords. Galaxies: jets – Galaxies: Seyfert – Gamma-rays: observations

---

The recent discovery of variable $\gamma-$ray emission from 4 NLS1s revealed the presence of a third class of $\gamma-$ray AGNs (Abdo *et al.* 2009a). This poses intriguing questions to the current knowledge of relativistic jet systems and on how these structures are generated. One of these sources, PMN J0948 + 0022 ($z = 0.5846$) is classified a typical NLS1, but it displays also strong, compact and variable radio emission, with inverted spectrum, suggesting the possibility of the presence of a relativistic jet (Zhou *et al.* 2003). The confirmation came with the detection of high-energy variable γ rays by *Fermi*/LAT (Abdo *et al.* 2009b, Foschini *et al.* 2010a). A multiwavelength campaign performed in March-July 2009 displayed coordinated variability at all frequencies, thus confirming that the source detected by *Fermi* is indeed the high-energy counterpart of PMN J0948+0022 (Abdo *et al.* 2009c).

PMN J0948 + 0022, the first NLS1 detected at γ rays, was soon followed by three more (Abdo *et al.* 2009a). Their main differences with respect to blazars and radio galaxies are the optical spectrum and the radio morphology, which is quite compact and

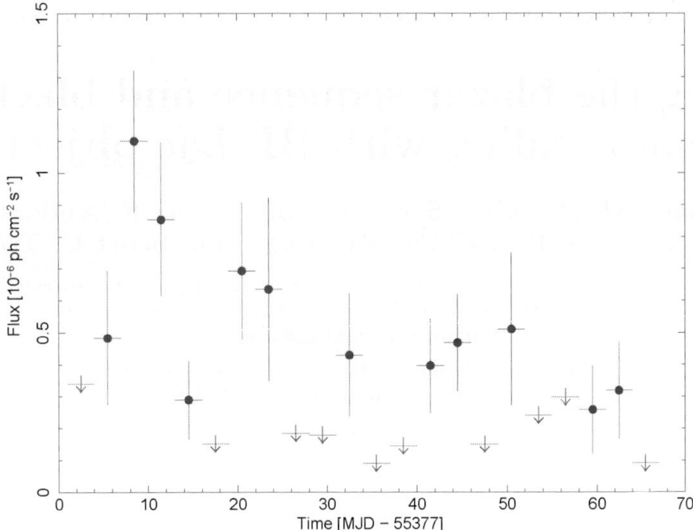

Figure 1. *Fermi*/LAT lightcurve $(0.1 - 100$ GeV) of PMN J0948 + 0022 in July-August 2010, with 3 days time bin. Blue points are detection with $TS > 10$ ($\sim 3\sigma$), while the others are 2σ upper limits. LAT data have been analyzed as described in Foschini *et al.* (2010b).

without extended structures. These characteristics point to systems with relatively low masses of the central black hole $(10^{6-8} M_\odot)$ and high accretion rates (up to 90% of the Eddington value), common for NLS1s, but not for blazars or radio galaxies. In addition, since the $\gamma-$ray NLS1s seem similar to blazars, i.e. small viewing angles, there should be a parent population with the jet viewed at large angles (as blazars vs radio galaxies). The first source of this type has been recently found (PKS $0558 - 504$, Gliozzi *et al.* 2010). Therefore, it seems that NLS1s could be a low mass set of systems "parallel" to blazars and radio galaxies. A key question was the power released by jets of NLS1s. The early observations and the 2009 MW campaign have shown that the maximum luminosity reached by PMN J0948 + 0022, the most powerful of these NLS1s, is $\sim 10^{47}$ erg s^{-1} $(0.1-100$ GeV). On the other hand, blazars can reach greater luminosities ($\sim 10^{49}$ erg s^{-1} in the case of 3C 454.3, e.g. Foschini *et al.* 2010b). The question was answered in July 2010, when PMN J0948+0022 underwent a strong outburst (Donato *et al.* 2010, Foschini 2010c) with a peak flux of $\sim 10^{-6}$ ph cm^{-2} s^{-1} $(0.1 - 100$ GeV), corresponding to a luminosity of $\sim 10^{48}$ erg s^{-1} (Fig. 1). Even if the source position was too close to the Sun for a full MW campaign, some coverage was obtained. Further details will be available in a forthcoming paper.

References

Abdo A. A. *et al.*, 2009a, *ApJ* 707, L142.
Abdo A. A. *et al.*, 2009b, *ApJ* 699, 976.
Abdo A. A. *et al.*, 2009c, *ApJ* 707, 727.
Donato D. *et al.*, 2010, *ATel* 2733, 12 July 2010.
Foschini L. *et al.*, 2010a, in: *Proceedings of Accretion and ejection in AGN: a global view.* L. Maraschi, G. Ghisellini, R. Della Ceca & F. Tavecchio eds, ASP Conference Series 427, San Francisco, in press (arXiv:0908.3313).
Foschini L. *et al.*, 2010b, *MNRAS*, in press (arXiv:1004.4518).
Foschini L., 2010c, *ATel* 2752, 22 July 2010.
Gliozzi M. *et al.*, 2010, *ApJ* 717, 1243.
Zhou H. Y. *et al.*, 2003, *ApJ* 584, 147.

Jets at All Scales
Proceedings IAU Symposium No. 275, 2011
G. E. Romero, R. A. Sunyaev & T. Belloni, eds.

© International Astronomical Union 2011
doi:10.1017/S1743921310015930

Testing the blazar sequence and black hole mass scaling with BL Lac objects

Richard M. Plotkin[1], Sera Markoff[1], Scott F. Anderson[2], Brandon C. Kelly[3], Elmar Körding[4], and Scott C. Trager[5]

[1]Astronomical Institute "Anton Pannekoek", University of Amstserdam,
Science Park 904, 1098 XH, Amsterdam, the Netherlands
email: r.m.potkin@uva.nl

[2]Dept. of Astronomy, University of Washington,
Box 351580, Seattle, WA 98195, USA

[3]Harvard-Smithsonian Center for Astrophysics,
60 Garden St., Cambridge, MA 02138, USA

[4]Université Paris Diderot and Service d'Astrophysique, UMR AIM,
CEA Saclay, F-91191 Gif-sur-Yvette, France

[5]Kapteyn Astronomical Institute, University of Groningen,
Postbus 800, NL-9700 AV Groningen, the Netherlands

Abstract. Jets from accreting black holes appear remarkably similar over eight orders of magnitude in black hole mass, with more massive black holes generally launching more powerful jets. For example, there is an observed correlation, termed the fundamental plane of black hole accretion, between black hole mass, radio luminosity, and X-ray luminosity. Here, we probe the high-mass tail (10^8–10^9 $M_\odot$) of the accreting black hole distribution with BL Lac objects. We build SEDs for hundreds of SDSS BL Lacs, and we use these SEDs to test the blazar sequence, a proposed anti-correlation between jet power and peak frequency. We then show our BL Lacs fit on the fundamental plane, supporting the non-linear scaling of jet radiation with black hole mass. The subset of BL Lacs considered here compose the largest sample yet used in the above types of studies, reducing potential selection effects and biases.

Keywords. galaxies: active, BL Lacertae objects: general, galaxies: jets

---

BL Lac objects are a subset of blazars unified with low-luminosity radio galaxies viewed nearly along the axis of a relativistic jet. Their spectral energy distributions (SEDs) are dominated by beamed jet emission, yielding synchrotron radiation peaking anywhere from the near-infrared to the soft X-ray, and inverse Compton emission at higher frequencies. Here, we use BL Lacs to shed light on two properties of relativistic jets. First, we investigate if their SED shape depends on luminosity (the blazar sequence), and then we test the scaling of jet radiation with black hole mass. We use a large uniformly selected BL Lac sample from the SDSS (Plotkin *et al.* 2010, hereafter P10), and we build SEDs using FIRST, NVSS, WENSS, and GB6 in the radio, 2MASS in the near-infrared, GALEX in the ultraviolet, and RASS and the XMM-Newton Slew Survey in the X-ray.

The Blazar Sequence: The blazar sequence was discovered by Fossati *et al.* (1998), and it has so far been supported by the *Fermi* Gamma-Ray Observatory. We test the blazar sequence by fitting parabolas to 409 SDSS BL Lacs with amply populated SEDs, and we then compare radio luminosity with peak frequency (Figure 1a). Only a very weak anti-correlation is observed (green solid line), with significance p=0.00425. However, if we only consider the 30 brightest radio sources and the 30 brightest X-ray sources to simulate the BL Lacs included in Fossati *et al.* (1998), then the anti-correlation becomes quite significant ($p<10^{-5}$, orange dashed line). We thus conclude deep blazar surveys

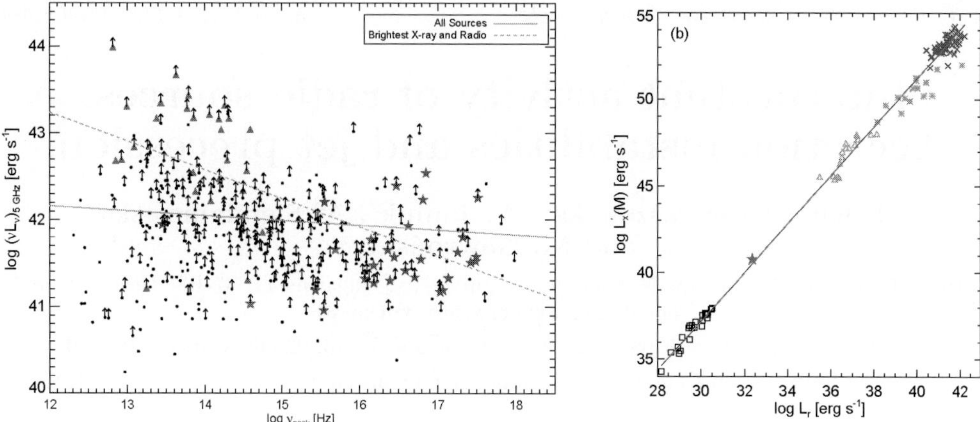

Figure 1. (a) Logarithm of rest-frame 5 GHz radio luminosity vs. logarithm of synchrotron peak frequency for 409 BL Lacs. The 30 brightest radio and 30 brightest X-ray sources are shown as red triangles and blue stars, respectively. Redshift (and therefore luminosity) lower limits are estimated if necessary from the fact we do not detect host galaxy emission (see P10). (b) The fundamental plane: log X-ray luminosity (normalized by black hole mass) vs. log radio luminosity. Included are X-ray binaries ($\sim 10 M_\odot$, black squares), Sgr A* ($\sim 10^6 M_\odot$, red star), LLAGN ($\sim 10^{7-8} M_\odot$, orange triangles), FR I ($\sim 10^{8-9} M_\odot$, green asterisks), and BL Lacs ($\sim 10^{8-9} M_\odot$, blue crosses). The FR I and BL Lac X-ray luminosities are extrapolated from the luminosity where the jet becomes optically thin (assuming $f_\nu \sim \nu^{-0.6}$), otherwise radiative losses from synchrotron cooling become important. We find $\log L_X = (1.40 \pm 0.04) \log L_r - (0.87 \pm 0.07) \log M_{bh} - (4.89 \pm 1.10)$ for our 113 objects.

reveal low-luminosity low-energy peaked BL Lacs that weaken the blazar sequence. We will explore this further by fitting our BL Lac SEDs with models more physical than a parabola (including multi-zone synchrotron and synchrotron self-compton).

Black Hole Mass Scaling: The fundamental plane is one of the strongest arguments supporting mass scaling (Merloni *et al.* 2003, Falcke *et al.* 2004). We use 58 data points from the KFC sample in Körding *et al.* (2006), and we replace their BL Lacs with 55 from P10 with central black hole mass estimates (Plotkin *et al.* 2010, subm.). Our motivation for including BL Lacs is we know their X-ray emission comes from a jet and not a corona (the objects in the KFC sample are also likely jet dominated). Using the Bayesian regression technique of Kelly (2007), we find $L_X \sim L_r^{1.40} M_{bh}^{-0.87}$ (Figure 1b, the regression is similar if we debeam the BL Lacs). This is consistent with the theoretical prediction of Falcke *et al.* (2004) for jet dominated systems. Merloni *et al.* (2003), however, find a different regression ($L_X \sim L_r^{1.64} M_{bh}^{-1.30}$) because their sample includes luminous X-ray binaries and AGN (i.e., hard X-rays are rather emitted by a corona). This study thus supports the notion that mass scaling investigations must compare black holes in similar accreting "states." *This work was supported by a Netherlands Organization for Scientific Research (NWO) Vidi Fellowship and NASA grant NNX09AF89G.*

References

Falcke, H., Körding, E., & Markoff, S. 2004, *A&A*, 414, 895
Fossati, G., Maraschi, L., Celotti, A., Comastri, A., & Ghisellini, G. 1998, *MNRAS*, 299, 433
Kelly, B. C. 2007, *ApJ*, 665, 1489
Körding, E., Falcke, H., & Corbel, S. 2006 *A&A*, 456, 439
Merloni, A., Heinz, S., & di Matteo, T. 2003, *MNRAS*, 345, 1057
Plotkin, R. M., *et al.* 2010 *AJ*, 139, 390 (P10)

Jets at all Scales
Proceedings IAU Symposium No. 275, 2011
G. E. Romero, R. A. Sunyaev & T. Belloni, eds.

© International Astronomical Union 2011
doi:10.1017/S1743921310015942

Intermittent activity of radio sources. Accretion instabilities and jet precession.

M. Kunert-Bajraszewska[1], A. Janiuk[2], A. Siemiginowska[3] and M. Gawroński[1]

[1] Toruń Centre for Astronomy, N. Copernicus University, Gagarina 11, 87-100 Toruń, Poland
email: magda@astro.uni.torun.pl

[2] Center for Theoretical Physics, PAN, Al.Lotnikow 32/46, 02-668 Warsaw, Poland

[3] Harvard Smithsonian Center for Astrophysics, 60 Garden St, Cambridge, MA 02138

Abstract. We consider the radiation pressure instability operating on short timescales ($10^3 - 10^6$ years) in the accretion disk around a supermassive black hole as the origin of the intermittent activity of radio sources. We test whether this instability can be responsible for short ages ($< 10^4$ years) of Compact Steep Spectrum sources measured by hot spots propagation velocities in VLBI observations and statistical overabundance of Gigahertz Peaked Spectrum sources. The implied timescales are consistent with the observed ages of the sources. We aslo discuss possible implications of the intermittent activity on the complex morphology of radio sources, such as the quasar 1045+352, dominated by a knotty jet showing several bends. It is possible that we are whitnessing an ongoing jet precession in this source due to internal instabilities within the jet flow.

Keywords. physical data and processes: accretion, galaxies: evolution, quasars: individual (1045+352)

1. Introduction

The compact radio sources consist of two population of objects: the gigahertz-peaked spectrum (GPS) and compact steep spectrum (CSS) sources which are considered to be young and evolve into large radio objects, during their lifetimes (O'Dea, 1998). However, it has already been pointed out by some authors (Gugliucci *et al.* 2005, Kunert-Bajraszewska *et al.* 2006) that there exists a group of GPS/CSS sources that will never evolve to become large scale objects, at least in a given cycle of activity if it is recurrent. These sources can be named short-lived radio objects.

We associated the existence of short-lived compact radio sources with the intermittent activity of the central engine caused by a radiation pressure instability within an accretion disk, which we briefly mention below.

2. Discussion and Results

According to the accretion disk instability model, for a given black hole mass, the larger the mean accretion rate, the longer the duration of a cycle episode, both in hot and cold states. In addition the viscosity parameter affects the results and for smaller viscosity, the cycle duration is longer. This relation was calibrated by Czerny *et al.* (2009) using a grid of models for various black hole masses and Eddington ratios. The outbursts are associated with the ejections of radio jets. The jets are then turned-off between the outbursts and each radio structure will represent a new outburst. In case of an apparently young, compact source we can suspect that in fact it is an old, reactivated object, in which

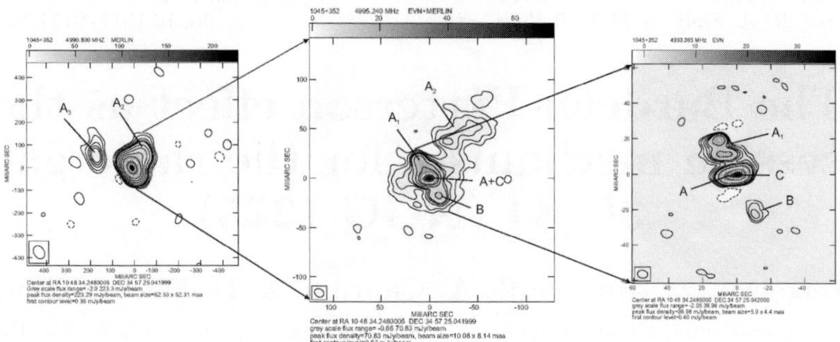

Figure 1. Radio images of 1045+352 at 5 GHz made with (from left): MERLIN, EVN+MERLIN, EVN. Contours increase by a factor 2 and the first contour level corresponds to 3σ. Indications: C - radio core, A-A_3 - radio jet, B - probable counter-jet.

the vast radio structures have already faded away and are not visible. We have applied i.a. the accretion disk instability model and jet precession model to explain the potential reactivation of the 1045+352 core (Kunert-Bajraszewska *et al.* 2010).

1045+352 is a CSS object and a HiBAL quasar at a medium redshift. Its linear size ($\sim$4 kpc) indcate it is a young object in the early phase of quasar evolution. The radio morphology of 1045+352 is dominated by the strong radio jet resolved into many sub-components and changing the orientation during propagation in the central regions of the host galaxy. As a consequence we observe at least three phases of jet activity indicate different directions of the jet outflow: components A_2-A_3 as the oldest one, structure A_1-B as the younger one, and the jet A as the current activity direction (Fig.1).

The results of the applied accretion disk instability model (Janiuk *et al.* 2002, Czerny *et al.* 2009) is not in agreement with the 1045+352 structure and size: (a) the calculated duration of the activity phase of the quasar is too short to enable the source grow to the observed size, (b) it cannot reproduce misalignment between the young and old radio structures. Such misalignment could be the result of the changed direction of the jet axis between the activity episodes - precession. In the case of 1045+352 we considered the precession of the innermost accretion disk due to internal instabilities within the accretion flow (Janiuk *et al.* 2008). This precession model reproduced well the observed complex structure of 1045+352. However, as discussed by Czerny *et al.* (2009), the mechanism of the accretion disk instability can be considered as the one that could explain the apparent statistical excess of the compact radio sources with respect to the galaxies with extended radio structures.

Acknowledgements

This work was supported by the Polish Ministry of Science and Higher Education under grant N N203 303635.

References

Czerny, B., Siemiginowska, A., Janiuk, A., Nikiel-Wroczyński, B., & Stawarz, L., 2009, *ApJ*, 698, 840

Gugliucci, N. E., Taylor, G. B., Peck, A. B., & Giroletti, M., 2005, *ApJ*, 622, 136

Janiuk A., Czerny B., & Siemiginowska A., 2002, *ApJ*, 576, 908

Janiuk A., Proga D., & Kurosawa R., 2008, *ApJ*, 681, 58

Kunert-Bajraszewska, M., Marecki, A., & Thomasson, P., 2006, *A&A*, 450, 945

Kunert-Bajraszewska, M., Janiuk, A., Gawroński, M. P., & Siemiginowska, A., 2010, *ApJ*, 718, 1345

O'Dea, C. P., 1998, *PASP*, 110, 493

Jets at all Scales
Proceedings IAU Symposium No. 275, 2011
G. E. Romero, R. A. Sunyaev & T. Belloni, eds.

© International Astronomical Union 2011
doi:10.1017/S1743921310015954

The Bardeen-Petterson effect as the precession mechanism for the radio galaxy 3C 84 (NGC 1275)

D. M. Teixeira[1], Z. Abraham[1], A. Caproni[2] & D. Falceta-Gonçalves[2,3]

[1]Instituto de Astronomia, Geofísica e Ciências Atmosféricas, Universidade de São Paulo, R. do Matão 1226, Cidade Universitária, CEP 05508-900, São Paulo, SP, Brazil

[2]Núcleo de Astrofísica Teórica, CETEC, Universidade Cruzeiro do Sul, R. Galvão Bueno 868, 01506-000, São Paulo, SP, Brazil

[3]Escola de Artes, Ciências e Humanidades, Universidade de São Paulo, Rua Arlindo Béttio, 1000, 03828-000, São Paulo, SP, Brazil
email: danilo@astro.iag.usp.br

Abstract. In this work we propose the Bardeen-Petterson effect as the precession mechanism of the jet precession in NGC 1275. To check if this is true we have estimated the angular momentum ratio and the aligment timescale predict by the theory and compared with the numerical results presented in the literature. We were able to explain the precession period assuming an accretion disk with column surface density in the form of a power law with exponent $0.6 < s < 0.7$ and a black hole rotation with a spin of $0.23 < a_* < 0.4$.

1. Introduction

The radio galaxy 3C 84, also known as NGC 1275, is located at a distance of 75 Mpc in the center of Perseus Cluster. Its optical morphology and the existence two systems of hydrogen lines seem to suggest that this galaxy is the result of an ongoing merge of two galaxies. X-ray emission has been observed in the nuclear region as well as in the halo. Bubble-like structures with different position angles with respect to the center of the cluster are seen in the X-ray maps; it has been suggested that these bubbles are inflated by a precessing jet. Recently, 3D numerical simulations (Falceta-Gonçalves *et al.* 2010) showed that, under certain conditions, a precessing jet can inflate multiple pair of bubbles. Assuming that the Bardeen-Petterson effect (Bardeen & Petterson 1975) can be the responsible for the jet precession in NGC 1275, they found that a precession period of $T_{\rm prec} = 5 \times 10^7$ years and a ratio between the angular momenta of the accretion disc and of the black hole of 1.1 could reproduce the observed X-ray maps.

2. Bardeen-Petterson effect

Frame dragging produced by a black hole with angular momentum J_{BH} causes precession of a particle if its orbital plane is inclined relative to the equatorial plane of the black hole; this is knows as Lense-Thirring effect (Lense & Thirring 1918).

The combined action of the Lense-Thirring effect with the inner viscosity of the disc causes the alignment of the angular momenta of the disc and the Kerr black hole; this is knows as Bardeen-Petterson effect (Bardeen & Petterson 1975) and tends to affect only the innermost parts of the disc due to the short range of the Lense-Thirring effect, while the outer parts tend to remain in its original configuration. The transition radius between

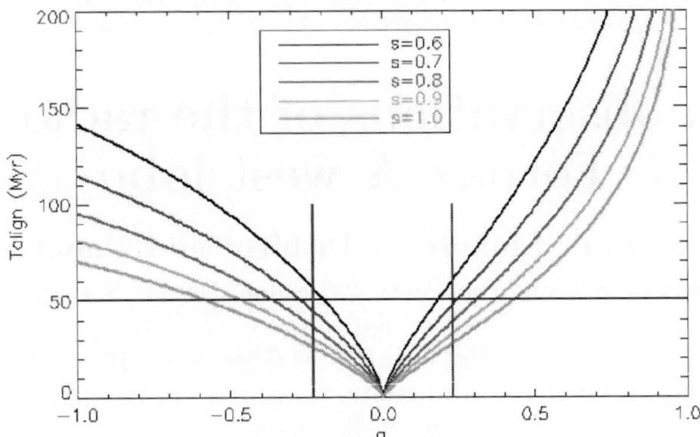

Figure 1. Alignment timescale predicted by the Bardeen-Petterson effect for different values of the exponent of the surface density equation. The dark green line corresponds to the minimum value of the black hole spin given by Daly (2009).

these two regions is known by Bardeen-Petterson radius and its location depends of the accretion disc configuration (Bardeen & Petterson 1975; Kumar & Pringle 1985; Nelson & Papaloizou 2000).

3. Results

Solving the equations for the Bardeen-Petterson effect given by Martin *et al.* (2007) we were able to find the precession period of NGC 1275 and put constrains on the black hole spin and the exponent of the surface density equation. The predicted values from the theory can be seen in the Fig. 1.

4. Conclusions

In this work we were able to explain the precession period found in the numerical simulations. We found that the accretion disk has an exponent of $0.6<s<0.7$ and that the black hole spin has a value of $0.23<a_*<0.4$.

References

Bardenn, J. M. & Petterson, J. A. 1975, *ApJ*, 195, L65
Daly, R. A. 1975, *ApJ*, 696, 32
Falceta-Gonçalves, D., Caproni, A., Abraham, Z., Teixeira, D. M., & dal Pino E. M. G. 2010, *ApJ*, 213, L74
Kumar, S. & Pringle, J. E. 1985, *MNRAS*, 213, 435
Lense, J. & Thirring, H. 1918, *Phys Z.*, 19, 156
Martin R. G., Pringle J. E., & Tout C. A. 2007, *MNRAS*, 381, 1617
Nelson, R. P. & Papaloizou, J. C. B. 2000, *MNRAS*, 315, 570

Jets at all Scales
Proceedings IAU Symposium No. 275, 2011
G. E. Romero, R. A. Sunyaev & T. Belloni, eds.

© International Astronomical Union 2011
doi:10.1017/S1743921310015966

Suzaku observations of the radio galaxy Fornax A west lobe

Hiromi Seta[1], Makoto S. Tashiro[1] and Naoki Isobe[2]

[1]Department of Physics, Saitama University, 255 Shimo-Okubo, Sakura, Saitama, 338-8570,
Japan
email: seta@heal.phy.saitama-u.ac.jp

[2]Department of Astronomy, Kyoto University, Kitashirakawa-Oiwake-cho, Skyo-ku, Kyoto
606-8502, Japan

Abstract. We performed mapping observations of the Fornax A west lobe with Suzaku in order to measure X-ray brightness distribution. Thanks to the low and stable background of Suzaku, we succeeded in detecting the faint diffuse X-ray emission from the west lobe. Performing careful corrections to the obtained images, we finally measured the X-ray brightness profile extending over the lobe. By comparing the X-ray and radio profiles, the magnetic field found to be fairly constant at ~ 1 μG over the lobe, while the electron energy distribution is suggested to concentrate on the lobe center.

Keywords. galaxies: individual: Fornax A, radiation mechanisms: non-thermal

1. Introduction

In the last several years, detections of cosmic microwave background boosted inverses Compton (IC) X-ray emission have been successively reported from radio lobes (e.g., Isobe *et al.* 2002). Using the X-ray and radio fluxes, we are able to estimate physical parameters such as the electron and magnetic energy densities, u_e, u_m, and the magnetic field strength B in the lobes.

In this paper, we evaluate spatial distributions of these physical parameters precisely by means of obtaining a detailed IC X-ray brightness distribution, which has ever assumed to have the same spatial distribution of radio emission. Fornax A is one of the brightest and largest radio galaxies. The hard X-ray up to 20 keV was detected for the first time from the west lobe with Suzaku, which showed that the Lorentz factor of the electron range at least 300 – 90000 with a single power-law energy distribution (Tashiro *et al.* 2009).

2. Observations and results

The Suzaku mapping observations were conducted five times from 2007 to 2009, with a total exposure of ~ 300 ks. The X-ray Imaging Spectrometer (XIS; Koyama *et al.* 2007) onboard Suzaku covers the energy range of 0.5 – 10 keV. The IC X-ray from the lobe is reported to dominate in energy range 2.0 – 5.5 keV the surrounding thermal emission and backgrounds (Tashiro *et al.* 2009). We show the backgrounds subtracted XIS image in 2.0 – 5.5 keV in Fig. 1a, which was smoothed with a two-dimensional Gaussian of $\sigma = 2'$. The obtained X-ray image shows the diffuse X-ray distribution associated to the radio lobe, but we see no significant trend along the jet axis. This supports that the lobe is dormant or inactive as suggested by Iyomoto *et al.* (1998) and Kim & Fabbiano (2003). Considering these evidences of inactivity of jets, we assume a concentric energy distribution below.

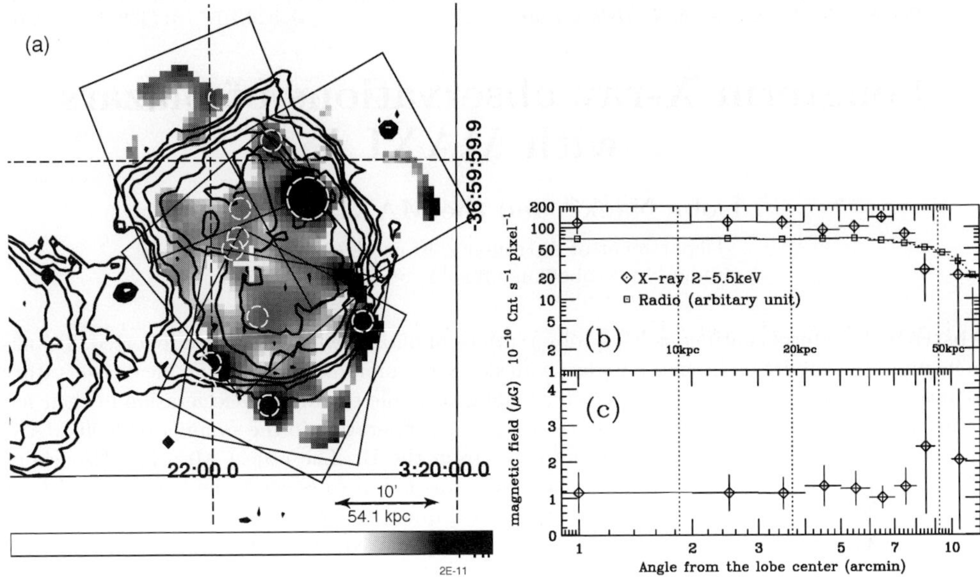

Figure 1. Panel (a) shows background subtracted XIS image in 2.0 – 5.5 keV. The 1.5 GHz VLA contour image (Fomalont *et al.* 1989) is overlaid. The squares are the fields of view for XIS. The dashed white circles are subtracted contaminant sources. The cross white point is the center of the radial profile as shown in the regions of panel (b) and (c). Panel (b) shows the obtained X-ray count rate radial profile. The radio radial profile is overlaid. Panel (c) shows the derived distribution of B from the ratio of X-ray and radio flux.

The panel (b) of Fig. 1 shows the obtained X-ray radial profile. The X-ray emission distributes up to $8'$ from the center, and decreases rapidly in the outer region. The X-ray distribution is similar to, or smaller than that of radio emission. This indicates that u_e distribution concentrates on the center of the lobe and is weak in the edge of lobe. By comparing the X-ray and radio fluxes in each radial region, the B is found to exhibit relatively uniform $\sim 1~\mu$G in the lobe as shown in Fig. 1. This result may indicate that the particles loss their energy in the ambient B filling the expanding lobe.

References

Fomalont, E. B., Ebneter, K. A., van Breugel, W. J. M., & Ekers, R. D. 1989, *ApJL*, 346, L17

Isobe, N., Tashiro, M., Makishima, K., Iyomoto, N., Suzuki, M., Murakami, M. M., Mori, M., & Abe, K. 2002, *ApJL*, 580, L111

Iyomoto, N., Makishima, K., Tashiro, M., Inoue, S., Kaneda, H., Matsumoto, Y., & Mizuno, T. 1998, *ApJL*, 503, L31

Kim, D.-W. & Fabbiano, G. 2003, *ApJL*, 586, 826

Koyama, K., *et al.* 2007, *PASJ*, 59, 23

Tashiro, M. S., Isobe, N., Seta, H., Matsuta, K., & Yaji, Y. 2009, *PASJ*, 61, 327

Jets at all Scales
Proceedings IAU Symposium No. 275, 2011
G. E. Romero, R. A. Sunyaev & T. Belloni, eds.

© International Astronomical Union 2011
doi:10.1017/S1743921310015978

Longterm X-ray observations of blazars with MAXI

Isobe Naoki[1] and the MAXI team.

[1]Department of Astronomy, Kyoto University
email: n-isobe@kusastro.kyoto-u.ac.jp

Abstract. Longterm continuous X-ray observations of blazars with MAXI are reported. Thanks to its unprecedentedly high sensitivity as an all sky X-ray monitor, MAXI is an ideal observatory to investigate variability of blazars, which should give a clue to particle acceleration in their jet, as well as the jet dynamics. Actually, since it started its operation in the summer of 2009, MAXI has successfully alerted two strong X-ray flares from the BL Lac object Mrk 421. Especially, in one of these flares, the X-ray flux of the object was found to become the highest in history. By closely examining the MAXI data, the physical quantities associated with the flares were estimated. These results clearly demonstrate the potential of MAXI for the variability of blazars.

Keywords. galaxies: jets, BL Lacertae objects: individual (Mrk 421), X-rays: galaxies

1. Introduction

One of the most outstanding properties of blazars is their rapid and high-amplitude intensity variation or flares, which is thought to be a useful probe for the flow dynamics, as well as to particle acceleration and cooling processes operating in their jet. Since the Ginga and ASCA era, X-ray observations have been a valuable tool for the blazar variability. Recently, based on the monitoring X-ray observations with the RXTE ASM and Swift, X-ray flares were successively detected from a number of blazars.

Since it was mounted on the Japanese Experiment Module "Kibo" of the International Space Station (ISS) in the summer of 2009, Monitor of All-sky X-ray Image (MAXI; Matsuoka *et al.* 2009) has been conducting successful all sky X-ray observations for more than one year. MAXI has the highest sensitivity as an all sky X-ray monitor, together with a high sky coverage ($\sim$95 % per day). These make this instrument the ideal observatory to investigate the longterm variability of blazars. Actually, MAXI has successfully made quick alerts of two strong X-ray flares from the High energy Peaked BL Lac object (HBL) Mrk 421 (Isobe *et al.* 2010a, 2010b). Currently, the MAXI lightcurve of more than 10 blazars are available from the MAXI web site (http://maxi.riken.jp/top/index.php).

2. Results

As shown in the 2 – 10 keV MAXI lightcurve of Mrk 421 in Figure 1 (1), the object was found to be highly variable. Especially, the MAXI has detected at least four X-ray flares with an X-ray flux exceeding the 100 mCrab level (2×10^{-9} ergs cm^{-2} s^{-1} in 2 – 10 keV). Due to relatively bad observational conditions for the first and second flares, we concentrate on the data associated with the third and forth ones on 2010 January 1 (MJD = 55136; Isobe *et al.* 2010a) and 2010 February 17 (MJD = 55146; Isobe *et al.* 2010b), respectively. These data are shown in Figure 1 (2) and (3), denoted as Epochs A and B. The maximum 2 – 10 keV flux in Epochs A and B was measured as 120 ± 10 mCrab at MJD = 55197.4 and 164 ± 17 mCrab at MJD = 55243.6, respectively. Especially,

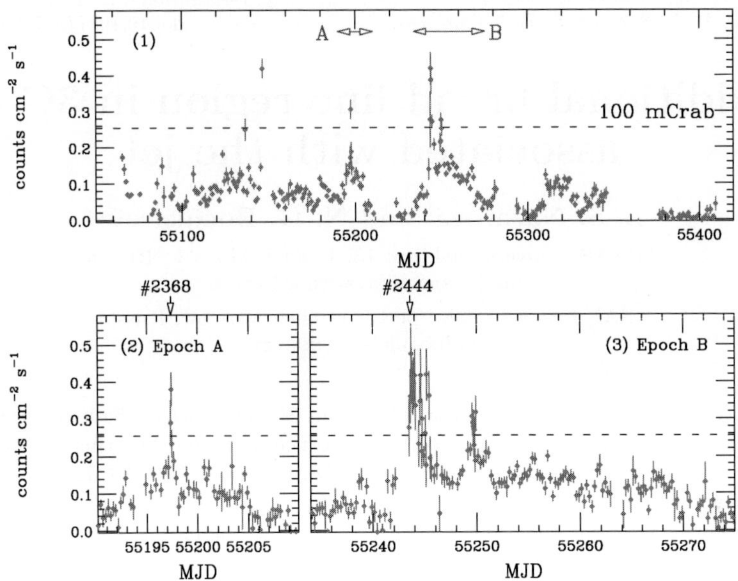

Figure 1. (1) Daily averaged 2 – 10 keV MAXI lightcurve of Mrk 421 from 2009 August 22 (MJD = 55062) to 2010 August 1 (MJD = 55409). The count rate for 100 mCrab is indicated by the dashed line. The data corresponding to the two arrows, denoted as A and B, are expanded to the panels (2) and (3), respectively. The two strong X-ray flares, alerted by MAXI (Isobe *et al.* 2010a, 2010b), are shown by the arrows on the top of the panels (2) and (3) with the article number on the Astronomer's Telegram.

the latter is found to be the highest among those ever reported from the objects (e.g., ~130 mCrab in 2008 June; Donnarumma *et al.* 2009). Therefore, we have concluded that MAXI discovered the strongest X-ray flare from Mrk 421. In addition, this event is probably concurrent with a strong very high energy γ-ray (> 100 GeV) flare with a flux of > 10 Crab, detected by the VERITAS Observatory (Ong *et al.* 2010).

Assuming that the decay time $t_{\rm d}$ of the flare is determined by the cooling timescale due to the synchrotron radiation, the dominant cooling process in typical HBLs, the magnetic field strength B is roughly estimated. For the flare in Epoch A, the decay time, $t_{\rm d} = 2.5 \times 10^4$ s, gives a magnetic field of $B = 0.045(\delta/10)^{-1/3}$ G with δ being the jet beaming factor (Isobe *et al.* 2010c). This is found to be consistent with the value determined derived in the previous studies (e.g., $B = 0.036 – 0.44$ G; Kino *et al.* 2002). In contrast, the decay time longer by a factor of 5 – 6 in the Epoch B flare, $t_{\rm d} = 1.3 \times 10^3$ s, corresponds to a weaker magnetic field of $B = 0.015(\delta/10)^{-1/3}$ G (Isobe *et al.* 2010c). We speculate that the Epoch B flare could be not a single event. These results, derived by the high quality continuous lightcurve with MAXI, clearly demonstrate the advantage of MAXI for the variability study of blazars.

References

Donnarumma, I. *et al.*, 2009, *ApJ*, 691, 13
Isobe N., *et al.*, 2010a, Astronomer's Telegram #2368
Isobe N., *et al.*, 2010b, Astronomer's Telegram #2444
Isobe N., *et al.*, 2010c, *PASJ*, in press (arXiv:1010.1003)
Kino, M., Takahara, F., & Kusunose, M., 2002, *ApJ*, 564, 97
Matsuoka, M., *et al.* 2009, *PASJ*, 61, 999
Ong, R. A., *et al.*, 2010, Astronomer's Telegram, 2443

Jets at all Scales
Proceedings IAU Symposium No. 275, 2011
G. E. Romero, R. A. Sunyaev & T. Belloni, eds.

© International Astronomical Union 2011
doi:10.1017/S174392131001598X

An additional broad line region in 3C 390.3 associated with the jet

L. S. Nazarova[1] and N. G. Bochkarev[2]

[1]EAAS, Universitetskij pr.13, 119992, Moscow, Russia
email:lsnazarova@rambler.ru

[2]SAI, Universitetskij pr 13, 119992, Moscow, Russia
email:boch@sai.msu.ru

Abstract. We suggest the existence of two BLRs in 3C 390.3 which have different locations. The BLR1 is located at a distance in accordance to the hydrogen line time lags of $\approx$20 days. This disk-like region emits predominantly low ionization lines. The BLR2 forms around the radio-jet and is located at the distance corresponding to time lags of $\approx$ 40-80 days. The BLR2 is responsible for most of the emission in UV lines. The Lα line partly forms in BLR2 (40-60%) and partly in BLR1.

Keywords. galaxies: Seyfert, double-peaked emission lines – modelling: broad line region

The double-peaked galaxy 3C 390.3 is the prototype of the class of AGNs showing complex broad-line profiles with displaced distributions and/or an anisotropic illumination of the broad-line region (BLR). The time lag in 3C 390.3 for optic lines small ($\approx$ 20 days)(Dietrich *et al.* 1998) compare to the lag in UV lines CIV and Lα ($\approx$ 37 and 60 days respectively)(O'Brien *et al.* 1998). This is differ from other AGNs.

The observed profiles of the CIV, Lα, Hβ and Hα lines have been divided into seven parts, the width of each part being equal to 2000 km s^{-1}. The core of the lines is measured between -1000 and +1000 km s^{-1}. Thus, the blue and red wings have a width of $\pm$7000 km s^{-1}. The difference of the line fluxes along of line profiles taken for Lα, CIV, Hβ and Hα lines show two clear components, one shortward which better seen in UV lines and one near to the line center. The CIV/Lα ratio is low in the low-velocity regions of the line profile, but becomes higher in the blue wing, particularly when 3C 390.3 is more active. The observed Lα/Hβ ratio is high at low velocity and decreases in the wings for both high and low states of the nuclear activity. The Hα/Hβ ratio is higher at low velocity in both the minimum and the maximum of nuclear activity.

The broad line spectrum is computed with the photoionization cod "CLOUDY"which consist in shell of gas (or cloud) in ionization and thermal equilibrium under the action of an external flux of radiation illuminating one face. The comparison of the observed and the computed CIV/Lα, Lα/Hβ and Hα/Hβ line ratios supports the suggestion that there are two different emission regions with a different densities :

– a low-density region with density: 10^{8-9}cm^{-3} located at the distance 40-80 days from the center (HIL zone);

– a high-density region with the density 10^{10-12}cm^{-3} or more located at the distances $\approx$20 days (LIL zone).

The contribution from the LIL zone with the high electron density N$_e$$\approx$$10^{10-12}$ cm^{-3} should be significant at the wings where Lα/Hβ $\approx$2. However the emission from HIL zone could predicts 40-60% contribution of the variable part Lα/Hβ $\approx$7,42.

Arshakian *et. al.* (2010) found a significant correlation between an ejected jet components and optical continuum flares. It is pointed out that the emission in the optical

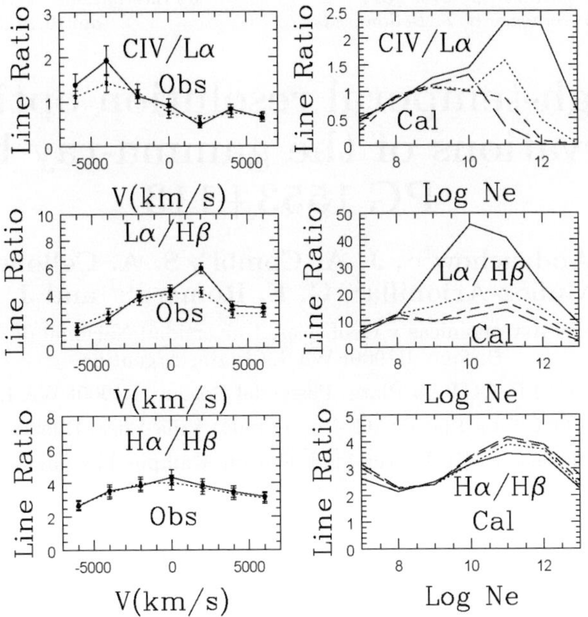

Figure 1. Line ratios CIV/Lα, Lα/Hβ and Hα/Hβ for the maximum states of nuclear activity. The observed line ratios - left side; the calculated line ratios - right side

lines might exist not only at the disk but also near to the radio-jet. According to the observed lags for Balmer and UV lines obtained of AGN Watch campaign and also taking into account our modeling we suggest the existing of two BLRs in 3C390.3 which have a different locations.

The BLR1 traditionally located at the disk at the distance according to the hydrogen lines ≈20 days. This disk-like region emits predominantly low ionization lines located in the equatorial plane where the BLR covering approaches 100%. The BLR2 forms around of the radio-jet and located at the distance ≈40–80 days from the center. The BLR2 responsible for the most emission in the UV lines (CIV line is varied at 84%). The Lα line partly forms in BLR2 – 40–60% and partly in BLR1.The bulk of the hydrogen lines forms at the disk and the wings – near to the radio-jet at the distance from the center ⩾80 days.

References

Arshakian T. G., Leon-Tavares J., Lobanov A. P., Chavushyan V. H., Shapovalova A. I., Burenkov, A. N., & Zensus J. A. 2010, *MNRAS* 401, 1231A

Deitrich, M., Peterson, B. M., Albrecht, P., Altmann, M., Barth, A. J., and 53 coauthors 1998, *ApJS* 115, 185D

O'Brien, P. T., Dietrich, M., Leight, K., Alloin, D., Clavel, J., Grenshaw, D. M., and 41 coauthors 1998, *ApJ* 509, 1630

Jets at all Scales
Proceedings IAU Symposium No. 275, 2011
G. E. Romero, R. A. Sunyaev & T. Belloni, eds.

© International Astronomical Union 2011
doi:10.1017/S1743921310015991

High-temporal resolution optical observations of the gamma-ray blazar PG 1553+113

I. Andruchow[1,2], J. A. Combi[3], S. A. Cellone[1,2],
A. J. Muñoz-Arjonilla[4], G. E. Romero[1,3] and J. Martí[4]

[1]Facultad de Ciencias Astronómicas y Geofísicas, Universidad Nacional de La Plata, Paseo del Bosque, B1900FWA La Plata, Argentina

[2]IALP, CONICET-UNLP, CCT La Plata, Paseo del Bosque, B1900FWA La Plata, Argentina

[3]IAR, CONICET, CCT La Plata, C.C. No. 5 (1894) Villa Elisa, Buenos Aires, Argentina

[4]Departamento de Física (EPS), Universidad de Jaén, Campus Las Lagunillas s/n, A3, 23071 Jaén, Spain

Abstract. We present here the results of an observational photo-polarimetry campaign at optical wavelengths of the blazar PG 1553+113, which was recently detected at very high energies ($> 100\,\mathrm{GeV}$) by the H.E.S.S and MAGIC γ-ray experiments.

Our high-temporal resolution data show significant variations in the linear polarization percentage and position angle at inter-night time-scales, while at shorter (intra-night) time-scales both parameters varied less significantly, if at all. Simultaneous differential photometry (at the B and R bands) shows no significant variability in the total optical flux.

Keywords. techniques: photometric - polarimetric, BL Lacertae objects: individual (PG 1553+113).

1. Introduction

PG 1553+113 is classified as a BL Lac object based on its featureless spectrum Miller & Green (1983) and significant optical variability Miller *et al.* (1988). It has been well studied from radio to very high energy (VHE), including several simultaneous multiwavelength campaigns Osterman *et al.* (2006), Mankuzhiyi *et al.* (2009). In X-rays, it has been detected at a number of different flux levels (e.g., Donato *et al.* (2001), Reimer *et al.* (2008)). No evidence for strong or fast flux variability has been detected in this energy range. At VHE, the object was detected with HESS Aharonian *et al.* (2006), and latter confirmed with MAGIC observations Albert *et al.* (2007). Recently, PG 1553+113 was detected in the GeV gamma-ray regime by the Fermi Gamma-ray Space Telescope Abdo *et al.* (2010). Combining archival radio, optical, X-ray an VHE γ-ray data, these authors modelled its SED, showing that it can be reasonably well fit with a simple, one-zone SSC model.

2. Observations and Results

We observed PG 1553+113 from April 21 to April 25, 2009 with the Calar Alto Faint Object Spectrograph (CAFOS) in imaging polarimetry mode, at the CAHA 2.2 m. telescope, Calar Alto, Spain. Exposure times were $150 - 240\,\mathrm{s}$ (B) and $50 - 120\,\mathrm{s}$ (R).

Polarimetry: Figure 2 presents the curves for the linear polarization (P) and position angle (θ) against time along the five nights. According to statistical criteria of Kersteven *et al.* (1976), the source resulted to be variable taking the five nights together in both, P

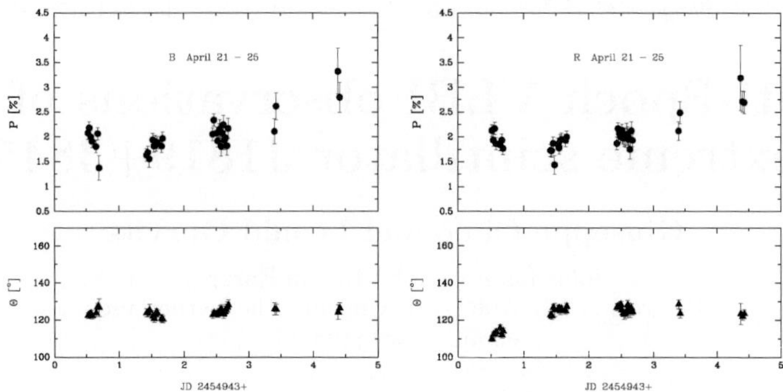

Figure 1. Polarization (top) and position angle (botton) curves for PG 1553+113 in filters B on the left and R on the right

and θ. The mean value for P was of 2.03% and for θ was of about 127.4 in both filters. Considering the microvaribilty, PG 1553+113 showed no important variations.

Photometry: The differential light-curves were obtained in the usual way. A statistical analysis of these light-curves reveals that PG 1553+113 showed no significant variability, neither in B nor in R. The dispersions of the curves ranged from 0.004 to 0.013 mag, and the mean apparent magnitudes for the whole campaign were $\langle B \rangle = 14.90 \pm 0.003$ mag and $\langle R \rangle = 14.18 \pm 0.003$ mag.

3. Discussion

The flux in both filters, B and R, remained steady during the observing run here reported. For the P and θ in the same period and in the same filters, the source showed small variations. Comparing our results with previous ones, ref. Osterman *et al.* (2006) found in their multiwavelenght campaing (April-May 2003) that PG 1553+113 was brighter in both optical bands by about 0.5 mag showing no signs of variation. In the same period, ref. Andruchow *et al.* (2005) followed the source in P (V filter) during two nights. P and θ showed then microvariations with a most likely flickering behaviour. We note that both, P and θ ere higher during 2003 and also, P seems to have rotated by $\sim 20°$ between both epochs.

The optical observations discussed in the previous paragraph bracket the VHE detections of PG 1553+113 by HESS Aharonian *et al.* (2006) and MAGIC Albert *et al.* (2007) in 2005 and 2006.

References

Abdo, A. A., Ackermann, M., Ajello, M., *et al.* 2010, *ApJ*, 708, 1310
Aharonian, F., Akhperjanian, A. G., Bazer-Bachi, A. R., *et al.* 2006, *A&A*, 448, 19
Albert, J., Aliu, E., Anderhub, H., *et al.*, 2007, *ApJ*, 654, L119
Andruchow, I., Romero, G. E., & Cellone, S. A. 2005, *A&A*, 442, 97
Donato, D., Ghisellini, G., Tagliaferri, G., & Fossati, G. 2001, *A&A*, 375, 739
Kesteven, M. J. L., Bridle, A. H., & Brandie, G. W. 1976, *AJ*, 81, 919
Mankuzhiyil, N., Dorner, D., Prandini, E., *et al.* 2009, *arXiv:0907.0740*
Miller, H. R. & Green, R. F. 1983, *BAAS*, 15, 957
Miller, H. R., Carini, M. T., Gaston, B. J., *et al.* 1988, *ESA Special Publication*, 281, 303
Osterman, M. A., Miller, H. R., Campbell, A. M., *et al.* 2006, *AJ*, 132, 873
Reimer, A., Costamante, L., Madejski, G., Reimer, O., & Dorner, D. 2008, *ApJ*, 682, 775
Schlegel, D. J., Finkbeiner, D. P., & Davis, M. 1998, *ApJ*, 500, 525

Jets at all Scales
Proceedings IAU Symposium No. 275, 2011
G. E. Romero, R. A. Sunyaev & T. Belloni, eds.

© International Astronomical Union 2011
doi:10.1017/S1743921310016005

Multi-Epoch VLBI observations of the extreme scintillator J1819+3845

Giuseppe Cimò and Leonid Gurvits

Joint Institute for VLBI in Europe,
Postbus 2, NL-9700TA, Dwingeloo, the Netherlands
email: `cimo@jive.nl`

Abstract. The quasar J1819+3845 has shown extreme variability with flux density variations in the radio regime up to 600% in less than one hour. In case of intrinsic high varibility, the short time scale sets a limit on the size of the emitting region and allows to estimate its brightness temperature. This would exceed 10^{21} K in the case of J1819+3845. Even an high relativistic jet beamed and doppler boosted in our line of sight cannot explain such an extreme violation of the Inverse Compton limit (10^{12} K). The variability of this source has been proven to be due to scattering in the Interstellar medium by a number of different experiments. Such an explanation requires a closeby scattering screen (few parsecs) and it results in a brightness temperature of about 10^{14} K. Many observing campaigns have been carried on to map the innermost jet structures of J1819+3845. Here we present the results of a number of VLBI observations, including space VLBI, to search for the missing jet in this puzzling source.

Keywords. Scintillation, VLBI, jets

1. Introduction

Intra-Day Variable sources (IDV's) are flat spectrum extragalactic synchrotron sources, often associated with optical quasars, BL Lac's or radio galaxies, that show flux density variability on time scales shorter than a day (Witzel *et al.* 1986 and Heeschen *et al.* 1987). These sources show also intrinsic variations on much longer time scales (months) that sometimes are associated with milli-arc-second structure that evolves on time scales of years. On the other hand, the intra-day variations are easily explained as a propagation effect due to scintillation in an interstellar screen of ionized gas (Rickett *et al.* 2001 and references therein).

The quasar J1819+3845 has shown extreme variability with flux density variations in the radio regime up to 600% in less than one hour (Dennett-Thorpe & De Bruyn 2000). In case of intrinsic high variability, the short time scale sets a limit on the size of the emitting region and allows one to estimate its brightness temperature. The latter would exceed 10^{21} K. Even a highly relativistic jet beamed and doppler boosted toward the line of sight cannot explain such an extreme violation of the Inverse Compton limit (10^{12} K). The variability of this source has been proven to be due to scattering in the Interstellar medium by a number of different experiments. Such an explanation requires a closeby scattering screen (few parsecs, Dennett-Thorpe & De Bruyn 2002 and Dennett-Thorpe & De Bruyn 2003) and it results in a brightness temperature of about 10^{14} K, still exceeding the inverse Compton limit and implying a very compact structure.

A number of VLBI campaigns have been carried out to image the innermost jet structures of J1819+3845. Three global VLBI experiments, including a Space VLBI epoch, and two observing sessions with the VLBA and the EVN arrays. However the main issue with imaging J1819+3845 has been its extreme variability. The data reduction is compromised by the difficulties in dealing with phase fluctuations and scintillation. Furthermore, the interstellar scintillation induced variations will cover the structure of the source due to the scattering broadening of the VLBI image.

2. The variability and the jet structure

The extreme variability of J1819+3548 has been observed since its first observation in 1999 and it has continued for more than 8 years. Due to its high brightness temperature and the compact nature, J1819+3845 was expected to expand quenching the scintillation or at least changing its variability characteristics following its internal structural variations. However, the variations have been observed continuously along the years showing also a clear annual modulation due to relative motion of the Earth and the scattering screen.

On one hand, the scintillation has prevented to image the milli-arcsecond structure of this puzzling object. On the other hand, it has been a powerful tool to study indirectly the source structure at μ-arcsecond scale. Macquart & de Bruyn (2007) have compared observations taken at identical epochs each year in order to disentangle the effects of source structural evolution and asymmetry in the scintillation pattern. Analysing the light curves and their power spectra, it was possible to indicate the evolution of the internal structure of J1829+3845. Comparing the scintillation characteristics in 2003, 2004 and 2006, Macquart & de Bruyn (2007) found an expanding double structure.

In 2008, EVN observations of J1819+3548 (Cimò *et al.* 2008) have shown no signs of any variability (see attached plots to compare the extreme scintillation to the quiescent phase). A possible explanation for the cessation of the fast variations in J1819+3548 is that the scattering screen has moved away from our line of sight. On the other hand, the disappearance of the extreme scintillations on J1819+3845 could be due to variations in the source structure at μ-arcsecond scale. A new component, which expands as it evolves would lead to episodic scintillations. The EVN data have shown however that the source is unresolved at the European baselines. We have analized the previous VLBI observations of J1819+3548 and we have not found a confirmation for any structural changes with time.

However our imaging process has been so far limited by the scintillation. A model of the source in its current quiescent phase is in progress. This will allow us to retroactively interpret the scintillation-limited data and can provide us with information on the structure of the innermost jet structures of the source and their evolution in the course of the previous 10 years.

References

Cimò, G. 2008, *PoS(IX EVN Symposium)*, 13
Dennett-Thorpe, J. & De Bruyn, A. G. 2000, *ApJ*, 529, L65
Dennett-Thorpe, J. & De Bruyn, A. G. 2001, *Nature*, 415, 57
Dennett-Thorpe, J., De Bruyn, A. G. 2003, *A&A*, 404, 113
Heeschen, D. S., Krichbaum, T., Schalinski, C. J., & Witzel, A. 1987, *AJ*, 94, 1493
Macquart, J.-P. & De Bruyn, A. G. 2007, *MNRAS*, 380, L20
Rickett, B. J., Witzel, A., Kraus, A., Krichbaum, T. P., & Qian, S. J. 2001, *ApJ*, 550, L11
Witzel, A., Heeschen, D. S., Schalinski, C., & Krichbaum, T. 1986 *Mitteilungen der Astronomischen Gesellschaft Hamburg*, 65, 239

Jets at all Scales
Proceedings IAU Symposium No. 275, 2011
G. E. Romero, R. A. Sunyaev & T. Belloni, eds.

© International Astronomical Union 2011
doi:10.1017/S1743921310016017

Evidence for shock-shock interaction in the jet of CTA 102

C. M. Fromm[1], M. Perucho[2], T. Savolainen[1], E. Ros[2,1], A. P. Lobanov[1] J. A. Zensus[1] and A. Lähteenmäki[3]

[1]Max-Planck-Institut für Radioastronomie,
Auf dem Hügel 69, D-53121 Bonn, Germany
email: `cfromm@mpifr.de`

[2]Departament d'Astronomia i Astrofísica, Universitat de València,
E-46100, Burjassot, València, Spain

[3]Aalto University, Metsähovi Radio Observatory,
FI-02540 Kylmälä, Finland

Abstract. We have found evidence for interaction between a standing and a traveling shock in the jet of the blazar CTA 102. Our result is based in the study of the spectral evolution of the turnover frequency-turnover flux density (ν_m, S_m) plane. The radio/mm light curves were taken during a major radio outburst in April 2006.

Keywords. galaxies: active, – galaxies: jets, – radio continuum: galaxies, – radiation mechanisms: non-thermal, – galaxies: quasars: individual: CTA 102

1. Introduction

The blazar CTA 102 (z=1.037) shows a curved jet which exhibits apparent velocities up to $15.4\,c$ (Lister *et al.* 2009). The analysis of single-dish light curves and $43\,\mathrm{GHz}$ VLBI observations of Hovatta *et al.* (2009) and Jorstad *et al.* (2005) yields bulk Lorentz factors, Γ between 15 and 17 and Doppler factors, δ between 15 and 22. These results lead to the picture that CTA 102 harbors a highly relativistic jet. The source underwent a major radio flare from millimetre to centimetre wavelength during spring 2006. We have analyzed the single-dish data available for this source and here we present results of our modeling of the spectral evolution during the radio flare.

2. Single-Dish Light Curves

For our analysis of the 2006 radio flare in CTA 102 we used radio/mm light curves spanning from $4.8\,\mathrm{GHz}$ to $340\,\mathrm{GHz}$. In order to obtain simultaneous spectra, we interpolated the observed light curves and performed a spectral analysis. Within this analysis we subtracted a quiescent spectrum from the obtained data and fitted a self-absorbed spectrum, defined by $S(\nu) = C(\nu/\nu_1)^{\alpha_t}\{1 - \exp[-(\nu/\nu_1)^{\alpha_0 - \alpha_t}]\}$, where $S(\nu)$ is the flux density, ν_1 is the frequency at which the opacity $\tau_s = 1$, and α_t and α_0 are the spectral indices for the optically thick and optically thin parts of the spectrum, respectively. The turnover frequency, ν_m, and the turnover flux density, S_m, can be calculated from the first and the second derivative of the synchrotron spectrum and they can be regarded as the characteristics of the spectrum. The uncertainties on the derived spectral values were obtained by performing Monte Carlo simulations. The derived spectral evolution showed a discrepancies from the standard evolution with in the shock-in-jet model (Marscher & Gear 1985), visible as a second hump in the $(\nu_m - S_m)$ plane (Fromm 2009).

3. Modeling

The shock-in-jet model assumes a power law relation between the turnover frequency, ν_m, and the turnover flux density, S_m, where the exponent depends on the dominant energy loss mechanism (Compton, Synchrotron and Adiabatic) and the evolution of the physical properties in the jet. The model assumes a power law distribution for the relativistic electrons, $N(E) \propto K E^{-s}$ and the evolution of the constant K follows the relation $K \propto R^{-k}$, where R is the radius of the jet. Furthermore the the evolution of the magnetic field can be expressed as $B \propto R^{-b}$, the evolution of the Doppler factor is $\delta \propto R^{-d}$ and of the jet radius, $R \propto L^r$, where L corresponds to the distance along the jet. Applying this model to the derived spectral behaviour leads to the following results for the evolution of the physical quantities.

time	stage	b	s	k	d	r
2005.6 - 2005.9	Comp/Adi	1.98	2.77	1.89	0.11	0.70
2005.9 - 2006.3	Comp	−3.58	1.20	2.13	1.58	−0.19
2006.3 - 2006.5	Adi	1.89	2.64	2.33	−0.22	1.00
2006.5 - 2006.8	Adi	1.89	2.64	3.05	−0.11	1.00

4. Discussion

The evolution of the parameter r (jet opening) could be an indication for a non-pressure matched jet and the behaviour of the power law index s could be interpreted as a re-acceleration of relativistic particles. Together with the variation in the magnetic field (b) and the Doppler factor (d) our result could be associated with a traveling-standing shock interaction. Standing shocks are common features in non-pressure matched jets. At the position of the standing shock, the traveling shock will encounter a region of increased particle density and magnetic field strength and will generate, due to shock acceleration at the shock front, a higher emissivity (Gómez *et al.* 1997). This increase in the emissivity could lead to the observed spectral behaviour. A more detailed discussion will be presented elsewhere. Future work should include the analysis of available multi-frequency VLBI observations in combination with relativistic magneto-hyrdodynamic simulations which could confirm our scenario of a shock-shock interaction in CTA 102.

References

Fromm, C. M. 2009, Diploma thesis

Gómez, J. L., Martí, J. M., Marscher, A. P., Ibañez, J. M., Alberdi, A. 1997 *Ap. Lett.*, 482, 33

Harris, D. E. & Roberts, J. A. 1960, *PASP*, 72, 237

Hovatta, T., Valtaoja, E., Tornikoski, M., & Lähteenmäki, A. 2009, *A&A*, 494, 527-537

Jorstad, S. G., Marscher, A. P., Lister, M. L., Stirling, A. M., Cawthorne, T. V., Gear, W. K., Gómez, J. L., Stevens, J. A., Smith, P. S., Forster, J. R., & Robson, E. I. 2005, *AJ*, 130, 1418–1465

Lister, M. L., Aller, H. D., Aller, M. F., Cohen, M. H., Homan, D. C., Kadler, M., Kellermann, K. I., Kovalev, Y. Y., Ros, E., Savolainen, T., Zensus, J. A., & Vermeulen, R. C. 2009 *ApJ*, 137, 3718–3729

Marscher, A. P. & Gear, W. K. 1985, *ApJ*, 298, 114–127

Jets at all Scales
Proceedings IAU Symposium No. 275, 2011
G. E. Romero, R. A. Sunyaev & T. Belloni, eds.

© International Astronomical Union 2011
doi:10.1017/S1743921310016029

Evolution of the parsec-scale jet in 3C 345

F. K. Schinzel[1]†, A. P. Lobanov[1], J. A. Zensus[1], G. B. Taylor[2], S. Jorstad[3] and A. Marscher[3]

[1]Max-Planck-Institut für Radioastronomie,
Auf dem Hügel 69, 53121 Bonn, Germany
email: `schinzel@mpifr.de`

[2]Dept. of Physics & Astronomy, University of New Mexico,
Albuquerque NM, 87131, USA

[3]Institute for Astrophysical Research, Boston University,
725 Commonwealth Avenue, Boston, MA 02215, USA

Abstract. The 16^{m} quasar 3C 345 is one of the best examples of an AGN showing structural and flux variability on parsec scales around a compact unresolved radio core. It has been observed from radio to γ-ray wavebands with a special focus on Very Long Baseline Interferometry (VLBI) observations in the range 1-100 GHz that cover a period of over 30 years. The complex pc-scale jet of 3C 345 exemplifies an archetypal "superluminal" jet with helical substructure. Existing VLBI observations of 3C 345 form an unprecedented database enabling a unique insight into the long-term evolution of the pc-scale radio emission. Here we present the latest results from our ongoing long-term VLBI monitoring of 3C 345, focusing on the morphological, kinematic, and spectral evolution of the pc-scale jet. Special attention will be given to the recent onset of a new period of high activity in the source that has been manifesting itself since 2008 from radio through γ-rays. Recent VLBI and high energy observations to study the relation between the radio emission and the production of high energy photons in 3C 345 are combined.

Keywords. galaxies: active, quasars: individual (3C 345), galaxies: jets

1. Introduction

The 16^{m} quasar 3C 345 has been observed at radio wavelengths for over 30 years, in particular with VLBI. The source still continues to be of special interest due to its complex, helical parsec-scale jet around a compact unresolved radio core and its pronounced multi-wavelength variability. A likely 8-10 year periodicity of the high activity phases in 3C 345 has been identified (Lobanov & Zensus 1999). Measurements of nuclear opacity and magnetic field strength (Lobanov 1998) yield a total mass for the central engine of $(4.0 \pm 2.4) \cdot 10^9 \ \mathrm{M_\odot}$.

3C 345 is known as a prominent variable source at high energies up to the X-ray band and only recently it was clearly identified as γ-ray source by the Large Area Telescope (LAT) aboard the *Fermi* satellite (Atwood *et al.* 2009, Schinzel *et al.* 2010a). This enabled continuous monitoring of γ-ray emission originating from the vicinity of 3C 345.

We have analyzed VLBI observations of the last three decades in order to understand the physics of the relativistic outflow and dynamics of central regions in 3C 345. After the launch of *Fermi*, a dense monthly VLBI monitoring was performed to study the new active period of 3C 345 and the combination of radio and γ-ray obserations of 3C 345 provide the opportunity to locate the sites of γ-ray emission and to study emission mechanisms.

† Member of the International Max Planck Research School (IMPRS) for Astronomy and Astrophysics at the Universities of Bonn and Cologne.

2. Results

The VLBI data collected on the pc-scale jet in 3C 345 revealed 19 bright features over a period of 30 years. Their brightness distribution was represented by Gaussian modelfits used for analysis of the kinematics and emission evolution of the features detected. At 15 GHz, the jet trajectories of such features were traceable up to a distance of 8 mas from the VLBI core. The pc-scale jet is initially directed westward, at a position angle of almost $90°$. The individual jet features trace a common channel of ~ 1 mas in width within a distance of ~ 5 mas from the core, beyond this distance the jet sharply turns north-ward. The component position angles measured at 0.5 mas radial separation from the VLBI core at 15 GHz offered an initial way to represent deviations from the $90°$ position angle. We saw no clear periodic trends like it was claimed in the past, but we saw long- and short-term variability of jet ejection angles on scales from 10 years to just a few years. A behavior like this might be expected for a helical, precessing jet.

Apparent speeds were determined for all jet features. After an accelaration phase, for a separation of > 0.7 mas, a constant characteristic speed of $\beta_{\mathrm{app}} \approx 12.2$ c is reached. With the exception of two features that showed an apparent speed of about 15.8 c.

A new moving emission region was detected on June 16, 2008, followed by detections of another new feature on January 24, 2009, and a third one on July 27, 2009. During 2009, the jet (average flux density: 3.6 Jy), at a distance of $\leqslant 0.3$ mas from the VLBI core, was brighter than the core (average flux density: 1.5 Jy) by a factor of 2.4. All newly ejected jet features showed similar apparent acceleration from $\sim 2 - 10$ c over a distance of 0.2 mas.

The γ-ray emission from 3C 345 was identified, based on correlations found between the optical and radio variability and major γ-ray events observed by *Fermi*/LAT in October 2009 (Schinzel *et al.* 2010a). A rising underlying trend in the γ-ray emission similar to the one in radio was observed, which suggests the γ-ray emission to originate over a distance of up to 10 pc. Out of six γ-ray events identified (2008–2010), two were associated with a new feature passing the 43 GHz VLBI core and one with a feature in the jet at a distance of 0.2 mas from the VLBI core, which showed a rapid multi-wavelength flare mid 2009. (Schinzel *et al.* 2010b).

3. Conclusions

Harvesting this unprecedented database on 3C 345 is going to give a unique insight into the long-term evolution of the pc radio emission and will further our understanding of the underlying physics of jet emission. We have observed that γ-rays are produced over large distances in the pc-scale radio jet, further constraining γ-ray emission scenarios of blazar jets.

References

Atwood, W. B., *et al.* 2009, *ApJ*, 697, 1071

Lobanov, A. P. 1998, *A&A*, 330, 79

Lobanov, A. P., & Zensus, J. A. 1999, *ApJ*, 521, 509

Marziani, P., *et al.* 1996, *ApJS*, 104, 37

Schinzel, F. K., *et al.* on behalf of the Fermi/LAT collaboration 2010a, *in prep.*

Schinzel, F. K., *et al.* 2010b, in: T. Savolainen, E. Ros, R. W. Porcas, & J. A. Zensus (eds.), *Radio flaring activity of 3C 345 and its connection to γ-ray emission*, Proc. Fermi meets Jansky (Bonn, Germany), pp. 175

Jets at all Scales
Proceedings IAU Symposium No. 275, 2011
G. E. Romero, R. A. Sunyaev & T. Belloni, eds.

© International Astronomical Union 2011
doi:10.1017/S1743921310016030

EVN monitoring observation of M 87 jet

Keiichi Asada[1] Masanori Nakamura[2] Akihiro Doi[3] Hiroshi Nagai[4] and Makoto Inoue[5]

[1]Institute of Astronomy and Astrophysics, Academia Sinica. P.O. Box 23-141, Taipei 10617, Taiwan, R.O.C.
email: asada@asiaa.sinica.edu.tw

[2]Allan C. Davis Fellow, Department of Physics and Astronomy, The Johns Hopkins University, 3400 North Charles Street, Baltimore, MD 21218; and Space Telescope Science Institute, 3700 San Martin Drive, Baltimore, MD 21218
email: nakamura@stsci.edu

[3]Institute of Space and Astronautial Science, Japan Aerospace Exploration Agency, 3-1-1 Yoshinodai, Sagamihara, Kanagawa, 229-8510, Japan
email: akihiro.doi@vsop.isas.jaxa.jp

[4]Institute of Space and Astronautial Science, Japan Aerospace Exploration Agency, 3-1-1 Yoshinodai, Sagamihara, Kanagawa, 229-8510, Japan
email: nagai@vsop.isas.jaxa.jp

[5]Institute of Astronomy and Astrophysics, Academia Sinica. P.O. Box 23-141, Taipei 10617, Taiwan, R.O.C.
email: inoue@asiaa.sinica.edu.tw

Abstract. We report results of our European VLBI Network observations towards M 87 jet at 1.6 GHz in order to study the velocity field. We revealed continuous jet up to 500 mas from the core and HST-1 component. We have not detected any proper motion for the components within first 160 mas from the core and significant superluminal motions from 2.5 to 3.5 c for the HST-1 components. Those are in good agreement with previous observations. We derived proper motions for the components about 160 to 500 mas from the core. Interestingly, the measured proper motions are faster than that of the inner components and slower than that of HST-1 components. It may suggest the possible acceleration region for superluminal features of M 87 jet.

Keywords. instrumentation: high angular resolution, techniques: interferometric, galaxies: active, galaxies: jets

1. Introduction

M 87 is one of the closest AGN jets (D = 16 Mpc) in the northern sky. At this distance, one mas corresponds to 0.08 pc, and the proper motion of 1 mas yr^{-1} corresponds to 0.25 c. There are many measurements of the proper motions using VLBI facilities, HST and VLA (Reid *et al.* (1989), Biretta *et al.* (1995), Biretta *et al.* (1999), Ly *et al.* (2004), Cheung *et al.* (2007), Kovalev *et al.* (2007)). With these measurements, we have known that the jet within the first 160 mas have had subluminal motion, while the jet components with the distance of more than 1 arcsec have shown superluminal motions up to 6 c. There is a missing link in our knowledge about a velocity field between inner part and outer part of the jet. In order to investigate the acceleration mechanism by revealing a velocity field, we have conducted three epoch monitoring observations towards intermediate scale of M 87 jet with European VLBI Network (EVN).

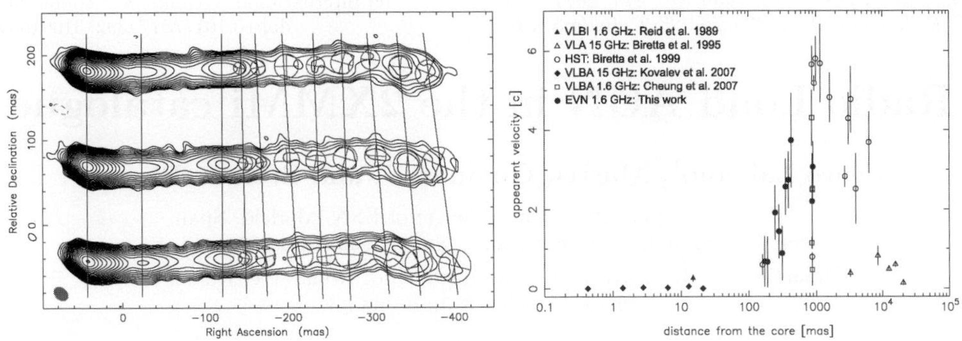

Figure 1. Images of M 87 jet with three epoch EVN observations (left). Velocity field of M 87 jet based on apparent proper motions. Apparent acceleration is clearly seen at the scale between 160 to 500 mas.

2. Observation

EVN observations were conducted on 12 March 2007, 2 March 2008 and 7 March 2009. All observations were made over 13 hrs at 1.6 GHz, with four IFs of 8-MHz bandwidth. An a priori amplitude calibration and Fringe fitting was performed using the AIPS. CLEAN and self-calibration was performed using Difmap.

3. Results & Discussion

We show the three epoch images of M 87 jet and the velocity field based on the measurements of the apparent proper motions in figure 1. We detected continuous jet up to 500 mas from the core and HST-1 component. Especially, the components about 160 to 500 mas from the core have not sufficiently been detected and investigated so far. We have not detected any proper motion for the components within first 160 mas from the core, while significant superluminal motions from 2.5 to 3.5 c for the HST-1 components. Those results are in good agreement with previous measurements (Reid *et al.* (1989), Cheung *et al.* (2007)). On the other hand, we succeeded to probe the proper motion at intermediate scale for the first time, and could reveal the proper motion of 1.0 to 3.8 c for the components at these scale. We succeeded to connect unconnected apparent velocities at the innermost and outermost region, and it probably suggests that M 87 jet is possible accelerated at this scale.

Acknowledgments

The European VLBI Network is a joint facility of European, Chinese, South African and other radio astronomy institutes funded by their national research councils.

References

Biretta, J. A., Zhou, F., & Owen, F. N. 1995 *ApJ*, 447, 582
Biretta, J. A., Sparks, W. B., & Macchetto, F. 1999 *ApJ*, 520, 621
Cheung, C. C., Harris, D. E., & Stawarz, Ł. 2007 *ApJ*, 663, L65
Kovalev, Y. Y., Lister, M. L., Homan, D. C., & Kellermann, K. I. 2007 *ApJ*, 668, L27
Ly, C., Walker, R. C., & Junor, W. 2004 *ApJ*, 660, 200
Reid, M. J., Biretta, J. A., Junor, W., Muxlow, T. W. B., & Spencer, R. E. 1989 *ApJ*, 336, 112

IAU Symposium 275: Jets at all Scales
Proceedings IAU Symposium No. 275, 2011
G. E. Romero, R. A. Sunyaev & T. Belloni, eds.

© International Astronomical Union 2011
doi:10.1017/S1743921310016042

Radio Loud AGN in the 2XMMi catalogue

Alvaro Labiano[1], Matteo Guainazzi[1], and Stefano Bianchi[2]

[1]European Space Astronomy Centre of ESA, Madrid, Spain
email: `Alvaro.Labiano@esa.int`, `Matteo.Guainazzi@esa.int`

[2]Dipartimento di Fisica, Università degli Studi Roma Tre, Rome, Italy
email: bianchi@fis.uniroma3.it

Abstract. We are carrying out a search for all radio loud Active Galactic Nuclei observed with *XMM-Newton*, including targeted and field sources to perform a multi-wavelength study of these objects. We have cross-correlated the Verón-Cetty & Verón (2010) catalogue with the *XMM-Newton* Serendipitous Source Catalogue (2XMMi) and found around 4000 sources. A literature search provided radio, optical, and X-ray data for 403 sources. This poster summarizes the first results of our study.

Keywords. Galaxies: active - Galaxies: Seyfert - quasars: general - X-rays: general

---

1. Introduction and sample

Bianchi *et al.* (2009a,b) presented the Catalogue of Active Galactic Nuclei (AGN) in the *XMM-Newton* Archive (CAIXA). They focused on the radio-quiet, X-ray unobscured ($NH < 2 \times 10^{22}$ cm^{-2}) AGN observed by XMM-Newton in targeted observations. We are carrying out a similar multiwavelength study, for both targeted and field radio-loud AGN observed by *XMM Newton*. We cross-correlated the Verón-Cetty & Verón (2010) catalogue (Quasars and Active Galactic Nuclei, 13th edition) with the *XMM-Newton* Serendipitous Source Catalogue (2XMMi, Watson *et al.* 2009) Third Data Release, and obtained a list of around 4000 sources. However, only 10% of the sources have published

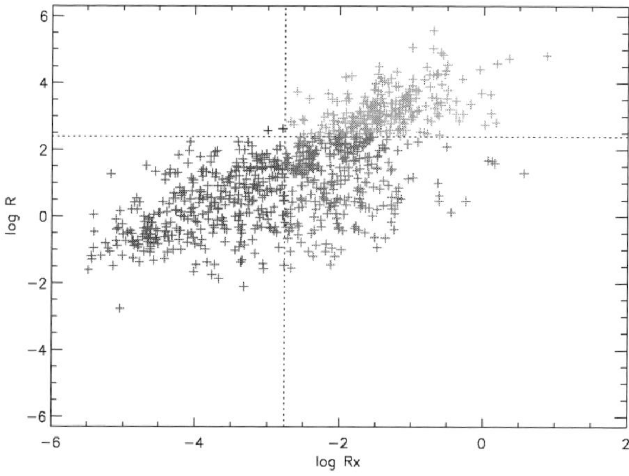

Figure 1. Radio loud (green) and radio quiet (blue and red) AGN. Rx is the ratio between 5 GHz and 2-10 keV emission. R is the ratio between 5 GHz and B-band emission. Boundares from Panessa *et al.* 2007. Blue points are sources classified as radio quiet by R and Rx parameters Red points are sources classified as radio quiet by the R parameter and radio loud by the Rx parameter

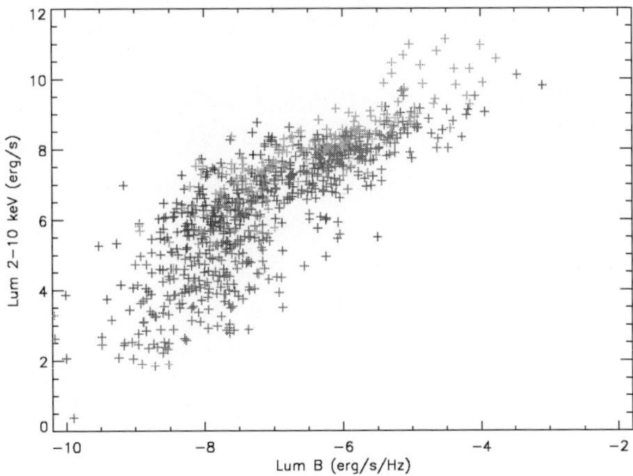

Figure 2. X-ray versus B-band luminosities. For optical luminosities higher than 10^{-6} erg/s/Hz radio loud (green) AGN are brighter in X-rays than radio quiet (red and blue). This effect is stronger at higher luminosities (10^{-5}-10^{-4} erg/s/Hz), where radio loud AGN deviate from the low luminosities correlation. X-ray emission in radio loud sources could have larger contributions from the jet.

optical and radio data. Our sample consists of all AGN (403 total, Figures 1 and 2) with available X-ray (2-10 keV), optical (B-band) and radio (5 GHz) data.

2. First results and ongoing work

Radio loud sources show jet contribution to optical and X-ray emission, and are brighter in X-rays than radio quiet. Optical and X-rays are AGN dominated with small contribution from host. For optical luminosities higher than 10^{-6} erg/s/Hz radio loud AGN are brighter in X-rays than radio quiet. This effect increases for higher luminosities ($10^{-5} - 10^{-4}$), where loud AGN deviate from the low luminosities correlation. X-rays in radio loud sources could have higher contributions from the jet. The sample seems to be missing faint, radio loud AGN, although at this point it is not clear if this is due to selection or astrophysical effects.

While X-rays in radio loud AGN seem to come mainly from jets, other mechanisms of X-ray emission are being studied (e.g. ADAF)? A complete spectral optical and X-ray analysis, including also the 0.2-2 kev band will bring light to the origin of the X-rays in these sources. We are currently studying the sample properties according to different classifications (Seyfert and QSO or FR I and FR II morphologies), and will include IR data when available, with the goal of carrying out a systematic analysis in as many wavelengths as possible.

References

Bianchi, S., Bonilla, N. F., Guainazzi, M., Matt, G., & Ponti, G. 2009a, *A&A*, 501, 915
Bianchi, S., Guainazzi, M., Matt, G., Fonseca Bonilla, N., & Ponti, G. 2009b, *A&A*, 495, 421
Panessa, F., Barcons, X., & Bassani L. *et al.*, 2009, *A&A*, 467, 519
Véron-Cetty, M. & Véron, P. 2010, *A&A*, 518, A10
Watson, M. G., Schröder, A. C., Fyfe, D., *et al.* 2009, *A&A*, 493, 339

Part 3. Microquasars

Jets at all Scales
Proceedings IAU Symposium No. 275, 2011
G. E. Romero, R. A. Sunyaev & T. Belloni, eds.

© International Astronomical Union 2011
doi:10.1017/S1743921310016054

Microquasars:
an observational review.

Stéphane Corbel[1]

[1]Université Paris 7 Denis Diderot and Service d'Astrophysique,
UMR AIM, CEA Saclay,
F-91191 Gif sur Yvette, France.
email: stephane.corbel@cea.fr

Abstract. In the past decade, several considerable achievements have been reached in the field of Galactic microquasars, especially in light of the extreme variability of their relativistic jets. These jets are now known to exist in at least three different flavours: the self absorbed compact jets in the hard state, the transient and discrete ejection events associated with the state transitions, and the emission associated with the interaction of the jets on the interstellar medium. Although their phenomenology is now starting to be rather well established, their emission and contribution to the total energy budget of microquasars is still the subject of active debate. One way to probe the origin of their emission at various wavelengths is to use the broadband correlations that may exist between different energy domains. Initiated in the radio and X-ray ranges, these broadband flux correlations now include optical and infrared observations of black hole candidates and also neutron star systems. In this review, I also outline the current observational status of the emission of relativistic jets at high energy.

Keywords. accretion, accretion disks – black hole physics – binaries: general – ISM: jets and outflows – radio continuum: stars – X-rays: stars.

1. Introduction

1.1. *Historical notes*

Eighteen years ago, microquasars were "discovered" based on the detection of large scale symmetric radio jets around the X-ray source 1E 1740.7−2942, located closed to the Galactic Center (Mirabel *et al.* 1992). The comparison with Active Galactic Nuclei (AGNs) was even more straightforward, after the detection of superluminal motions in the black hole accreting binary GRS 1915+105 (Mirabel & Rodriguez 1994), immediately followed by GRO J1655−40 (Tingay *et al.* 1995; Hjellming & Rupen 1995). These discoveries stimulated significant interest as microquasars represented ideal systems to study the inflow – outflow coupling in accreting systems on very short timescales (much faster than in AGNs), more compatible with human lifespans. As an additional comment, it is important to stress that microquasars are not only associated with black hole systems, as the most ultra-relativistic jets in our Galaxy are from the neutron star binary Cir X−1 (Fender *et al.* 2004a). Furthermore, jets from microquasars come in different forms or flavours (self absorbed compact jets, discrete ejection events and large scale lobes or hot spots), even if they are very rarely spatially resolved by the observations. This implies that almost all X-ray binaries (at least the ones with low magnetic fields) could potentially be considered as a microquasars if they were observed at the right time of their lifetime and with enough sensitivity to detect the emission of their jets. With the limited space available in this review, I will highlight only a few important results. More details could be found in the recent book review "The Jet Paradigm" (Belloni 2010 and references therein).

1.2. *Black holes spectral states*

Stellar mass black holes in accreting binary systems are known to undergo transitions between different spectral states, which are often defined by their X-ray spectral and timing behaviour (see McClintock & Remillard 2006 for a review), i.e. by the properties of the inner accretion flow. However, outflows in general (under their various forms), are now also crucial for the characterisation of these different spectral states. The two major spectral states can be defined as follows: 1) the hard state is generally characterised by a weak or absence (however see Tomsick *et al.* 2009) of thermal contribution from the accretion disk and the presence of powerful self-absorbed compact jets, and 2) the soft state is mostly dominated by thermal emission from the standard accretion disk and the absence of relativistic jets. In addition, the intermediate states are associated with many changes in the inflow - outflow coupling, including for example, the major and minor radio flares associated with relativistic jets (see Fender 2006) or the major changes in timing properties (e.g. Belloni 2010), usually ascribed to highest accretion rate in the inner part of the accretion disk. The Hardness-Intensity Diagram (HID) represents the evolution of the X-ray intensity versus the X-ray hardness, and is a convenient illustration for the evolution of the outburst of black hole transients (and also for neutron star or accreting white dwarf, see Körding *et al.* 2008) along the course of a typical outburst.

2. On the influence of compact jets in the hard state

2.1. *Powerful compact jets*

In addition to the more "spectacular superluminal" ejections from these systems (aka "à la GRS 1915+105"), new kind of jets have been introduced in these recent years that might have a significant impact on the spectral energy distribution of black holes (especially in the hard state). Indeed, the hard states are usually associated with weak (at the mJy level) unresolved radio sources, that are variable on daily to weekly timescales (Corbel *et al.* 2000; Fender 2001). The radio spectra are usually flat or slightly inverted (i.e. with a positive spectral index, α, and a definition of the radio flux density, S_ν, as $S_\nu \propto \nu^\alpha$). This is characteristic of the presence of conical self-absorbed compact jets, similar to that considered for flat spectrum AGNs (Blandford & Konigl 1979). These small-scale (a few tens of astronomical units) jets have been spatially resolved in only two cases: e.g. Cyg X$-$1 (Stirling *et al.* 2001) and GRS 1915+105 (Dhawan *et al.* 2000; Fuchs *et al.* 2003), but their presence is inferred in many other systems, including neutron star binaries (Migliari & Fender 2006).

The spectra of these compact jets are believed to be almost continuous from radio up to the near infrared/optical frequencies, in which a cut-off has been detected in black holes (e.g. Corbel & Fender 2002) and neutron stars (e.g. Migliari *et al.* 2010). With emission up to infrared, the compact jets are extremely powerful, implying that a significant fraction (at least 10%) of the accretion power could be released in these jets (Fender 2001; Corbel & Fender 2002). As a final note, the compact jets, believed to be ubiquitous in the hard state (however, see discussions on outliers in section 3.2), are quenched once the X-ray sources reach the soft X-ray state (Fender *et al.* 1999; Corbel *et al.* 2000).

2.2. *On the nature of the hard X-ray emission in the hard state*

The hard state of weakly accreting black holes has been the focus of many recent studies due to the presence of several emission processes that are very difficult to disentangle in this regime. Indeed, in this accretion state, the geometry is likely dominated by three main and different sites: the standard optically thick and geometrically thin accretion disk (Shakura & Sunyaev 1973), the self-absorbed compact jets (Fender 2001), and an optically thin corona of hot electrons (Sunyaev & Titarchuk 1980) that may or may not be the basis of the compact jets (e.g. Markoff *et al.* 2005). Emission from the companion star is usually negligible in low mass X-ray binaries. With the development of large campaigns of multi-wavelength observations, several attempts have been conducted to model the spectral energy distribution (SED) of black holes in their regimes of faint X-ray emission (e.g. McClintock *et al.* 2001; Markoff *et al.* 2003; Yuan *et al.* 2005). It turns out that optically thick synchrotron emission from the compact jets dominates the radio to near-infrared range (Corbel & Fender 2002; Buxton & Bailyn 2004; Homan *et al.* 2005; Russell *et al.* 2006; Coriat *et al.* 2009). When the accretion disk is detectable in the hard state, it is usually in optical and soft X-ray with the notable exception of XTE J1118+480 that was observed even in ultraviolet (McClintock *et al.* 2001). In hard X-rays, inverse Compton emission on the ambient photon field is the main emission process, with an active debate on whether it is thermal or non-thermal comptonisation, or synchrotron self-Compton (see for a review McClintock & Remillard 2006). Optically thin synchrotron emission from the compact jet may eventually play a role at high energy (Markoff *et al.* 2001; Homan *et al.* 2005; Russell *et al.* 2010). An alternative to the direct synchrotron case is the possibility of inverse comptonization (external or synchrotron self-Compton) in the base of the jets (e.g. Markoff *et al.* 2005; Migliari *et al.* 2007).

3. Broadband correlations in the hard state

3.1. *Radio / X-ray and Infrared / X-ray flux correlations*

Another way to probe the emission properties of these black hole systems is to search for peculiar correlations between various wavelengths. The tight non-linear correlation (with an index of ~ 0.7) between radio and X-ray emission in GX 339$-$4 (Corbel *et al.* 2000, 2003) demonstrated a strong coupling between the compact jets and the X-ray emitting media. This correlation could possibly indicate that X-rays in GX 339$-$4 originate from optically thin synchrotron emission from the compact jet (Corbel *et al.* 2003; Markoff *et al.* 2003) or from a radiatively inefficient accretion flow (Merloni *et al.* 2003; Heinz 2004, but see also Körding *et al.* 2006). Gallo *et al.* (2003) extended this correlation study to a larger sample of black holes and proposed a universal correlation (still with an index of ~ 0.7) between the radio and X-ray luminosity. The correlation observed in the hard state seems to be maintained down to quiescence (Corbel *et al.* 2003; Gallo *et al.* 2003, 2006). By taking into account the mass of the black holes, this correlation was extended to active galactic nuclei and led to the definition of the fundamental plan of black hole activity and possibly to universal scaling laws relating physics of accreting black holes on very different mass scales (Merloni *et al.* 2003; Falcke *et al.* 2004; Körding *et al.* 2006).

It is important to note that the universal radio/X-ray correlation presented by Gallo *et al.* (2003) is dominated by two sources (GX 339$-$4 and V404 Cyg) with few measurements from additional sources that are often close to transition to softer states. The fundamental plane relies on the correlation for the Galactic black holes, so it is important to review its validity. Corbel *et al.* (2008) revisited the correlation for V404 Cyg and confirmed the tight correlation from outburst down to quiescence with an index of ~ 0.6 that is more consistent with X-rays emanating from synchrotron self-compton emission at the base of the compact jets or from an inefficient accretion disk. New data from GX 339$-$4 also confirm the stability of the correlation (also with an index of ~ 0.6) over several different outbursts, though several peculiarities were also observed (Corbel *et al.* 2010, to be submitted).

Following these results, a similarly tight correlation was also found between X-ray and optical/near-infrared in a sample of black hole candidates and also in GX 339$-$4 over different outbursts (Homan *et al.* 2005; Russell *et al.* 2006; Coriat *et al.* 2009). In GX 339$-$4, Coriat *et al.* (2009) found tight OIR/X-ray correlations over four decades with

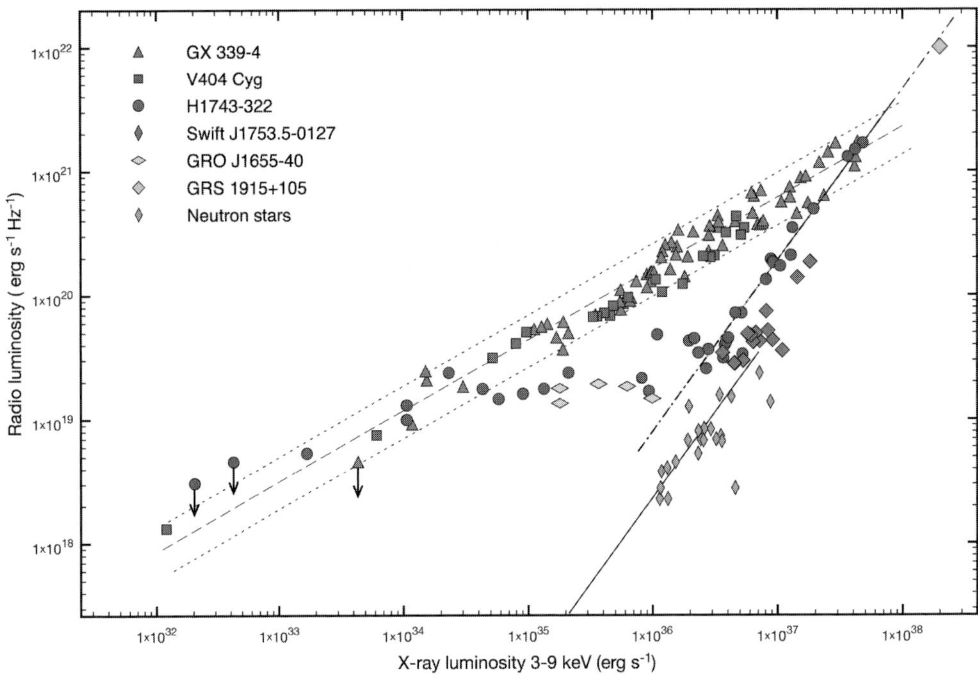

Figure 1. Radio 8.5 GHz luminosity against X-ray 3-9 keV luminosity from the hard state of H 1743−322 with a few other black hole candidates, as well as the hard state neutron stars sample of Migliari & Fender (2006). The dashed line is the fit to the high luminosity data of H 1743−322 and the neutron stars. Figure extracted from Coriat *et al.* submitted.

the presence of a break in the IR/X-ray correlation in the hard state. This correlation is the same for all four studied outbursts and could be interpreted in a consistent way by considering a synchrotron self-Compton origin of the X-rays in which the break frequency varies between the optically thick and thin regime of the jet spectrum. If correct, this interpretation would be consistent with the results based on the characterization of the SEDs (see section 2.2).

3.2. *On the nature of outliers*

However, in recent years, significant efforts have been achieved to observe most new black holes transients in outburst at radio frequencies. This led to the discovery of several sources lying significantly below and outside the scatter of the original radio/X-ray correlation (e.g. XTE J1650−500 Corbel *et al.* (2004); IGR J17497−2821 Rodriguez *et al.* (2007), Swift J1753.5−0127 Cadolle Bel *et al.* (2007); Soleri *et al.* (2010)). Until recently, It remained unclear how these outliers were related to typical black holes such as GX 339−4 or V404 Cyg. With a global study of H 1743−322 and other sources, Coriat *et al.* (2010, submitted) also found a very tight correlation for H 1743−322 in the bright hard state, but with a different slope of 1.38 ± 0.03 (compared to 0.6 for the standard correlation). At lower luminosity (below 5 × 10^{-3} L_{Edd}), H 1743−322 returned to the standard correlation (as defined by GX 339−4 and V404 Cyg). Coriat *et al.* (2010, submitted) proposed that the outliers could imply the presence of a radiatively efficient flow in the hard state at high accretion rate. In addition, below a critical accretion rate, the flow would have to become radiatively inefficient in order to account for the transition at lower luminosity. Alternatively, this could also imply a non linear coupling between

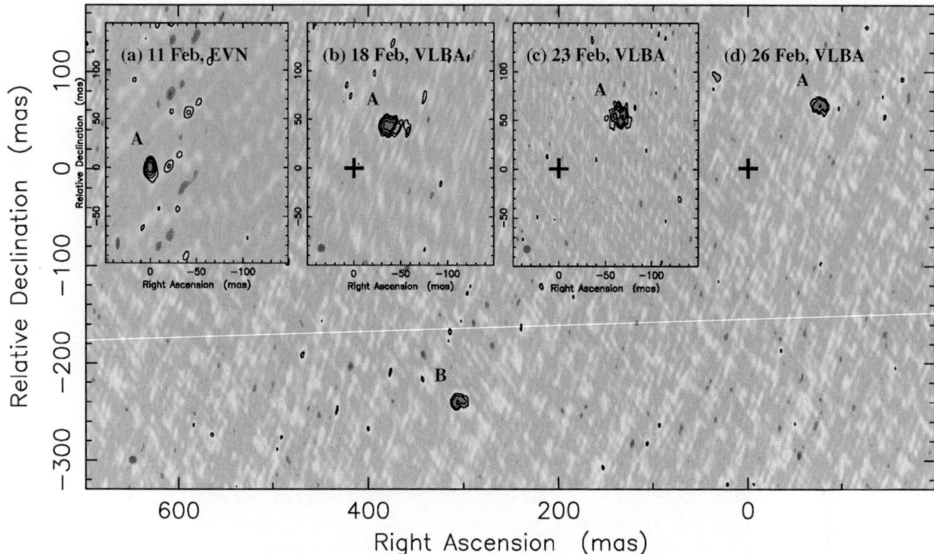

Figure 2. The decelerating jet of the X-ray transient XTE J1752−223. All the VLBI images are centered at the location of component A on 2010 February 11, indicated by a cross. Component B was not detected until 2010 February 26. See Yang *et al.* (2010) for more details (Figure extracted from this paper).

accretion power and jet power. Whatever the physical origin, the reason for finding these two groups of sources in the hard state at high accretion rate remains to be understood.

4. A unified model of disk jet coupling

The unified picture, which emerges from our current understanding, indicates that during the transition from the hard to soft states, a major radio flare may eventually be detected (if observed!) (Fender *et al.* 2004b; Corbel *et al.* 2004; Fender *et al.* 2009). This flare is usually interpreted as synchrotron emission from relativistic electrons ejected from the system with large bulk velocities. In a few cases, such jets have been directly imaged into one-sided (or two-sided) components moving away from the stationary core (e.g. Mirabel & Rodriguez 1994).

Broadly speaking, the jets on different scales (compact or discrete ejections) are rarely spatially resolved because of the lack of observations with sufficient resolution. The recent improvement of very long baseline interferometry should allow much better sampling, with faster reaction time, of the time evolution of relativistic jets. Recently, *RXTE* discovered the new transient XTE J1752−223, whose properties are very typical of black hole candidates. Yang *et al.* (2010) conducted in 2010 a series of very long baseline interferometry observations using the European VLBI Network (EVN) and the Very Long Baseline Array (VLBA). A moving jet component with significant deceleration is detected, as well as a likely receding jet in the last observation. The overall picture is consistent with an initially mildly relativistic jet, interacting with the interstellar medium or with swept-up material along the jet. The brightening of the receding ejecta at the final epoch can be well explained by initial Doppler deboosting of the emission in the decelerating jet.

The origin of the major radio flare can eventually be understood as a result of a variation of the Lorentz factor of the ejected material as the accretion disk reaches its innermost stable orbit, leading to the internal shocks within the flow. Alternatively, the

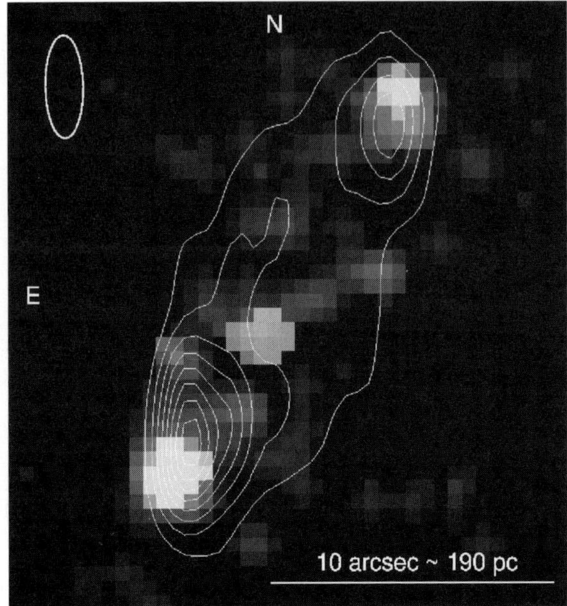

Figure 3. *Chandra*/ACIS color map of S26 (smoothed with a $1''$ Gaussian core), with ATCA 5.48-GHz intensity contours. The colour coding is: red: 0.2–1 keV; green: 1–2 keV; blue: 2–8 keV. See Soria *et al.* (2010) for more details (Figure extracted from this paper).

flare itself may originate from an ejection of the coronal material (Rodriguez *et al.* 2003, 2008). For that purpose, it is of interest to note the competition in GRS 1915+105, mediating through photo-ionization and Comptonization processes, between the jets in the hard state and the accretion disk winds in the soft state (Neilsen & Lee 2009).

5. The environment of jets: bubbles and hot spots

After discrete ejection events during the hard to soft state transition (Fender *et al.* 2004b; Corbel *et al.* 2004; Fender *et al.* 2009), the moving plasma condensations are observed in the radio range for a few weeks until their emission fades below detection levels due to adiabatic expansion losses. However, beside these sub-arcsecond scale transient jets seen during the initial part of the outburst of some black holes, large scale (from few tens of arcseconds to few arc-minutes) stationary radio jets have been observed later in a few systems, probably indicating the long term action of past ejection events on the surrounding interstellar medium (Mirabel *et al.* 1992; Martí *et al.* 2002). Of significant importance, *XMM-Newton* detected 7 arc-minutes (3 to 8 pc) slowly decaying X-ray jets centered on the position of 4U 1755−33, a black hole candidate (Angelini & White 2003; Park *et al.* 2005; Kaaret *et al.* 2006). The bubble nebula W50 around SS 433 also represents the archetypical example of the action of jets on the ISM, leading to an effective way to estimate the jets total mechanical power (above 10^{39} erg s^{-1}).

Similarly, this approach is now starting to be used for estimating the total luminosity of a few ultra-luminous X-ray sources (e.g. Soria *et al.* 2006; Lang *et al.* 2007), and in the extragalactic microquasar embedded in the nebula S26 in the nearby galaxy NGC 7793 (Pakull *et al.* 2010; Soria *et al.* 2010). S26 has a morphology reminiscent of Fanaroff-Riley type II (see Fig. 3) with a mechanical power around a few 10^{40} erg s^{-1} (i.e. 4 orders of magnitude larger than the X-ray luminosity of the core and above the Eddington limit for a stellar mass black hole). This points to the possibility of an accretion mode, which

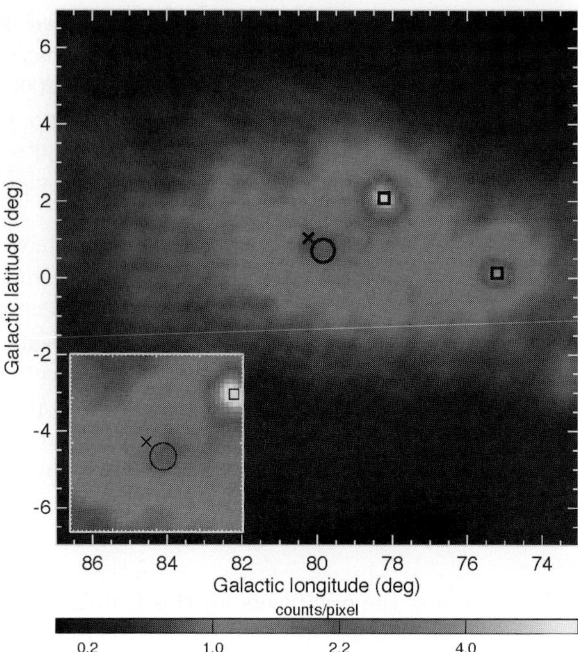

Figure 4. 200 MeV – 100 GeV *Fermi/LAT* gamma-ray counts map of a $14° \times 14°$ region centered on the position of Cyg X−3. It shows the region around Cyg X−3 encompassing emission from three bright gamma-ray pulsars, as well as the strong Galactic diffuse emission of the Cygnus region. Inset: Similar image for the central portion ($4° \times 4°$) corresponding to the off-pulse dataset from PSR J2032+4127 (which is not detected in the off-pulse data). See Fermi LAT Collaboration *et al.* (2009) for more details (Figure extracted from this paper).

channels most of the available power into jets, even at high mass accretion rates (Pakull *et al.* 2010; Soria *et al.* 2010).

Beside inflating bubble nebulae, the action of the jets on the ISM can also be revealed by the detection of (moving or not) hot-spots, the working sites of the jet interactions. In Sco X−1, an unseen underlying, highly relativistic, flow is energizing the hot spots (Fomalont *et al.* 2001b,a). The interaction may eventually lead to X-ray emission, as indicated by the *Chandra* detections of X-ray jets in XTE J1550−564, H 1743−322 or Cir X−1 (Corbel *et al.* 2002, 2005; Heinz *et al.* 2007; Soleri *et al.* 2009; Sell *et al.* 2010). The broadband spectra of these jets in XTE J1550−564 and H 1743−322 are usually consistent with synchrotron emission from high-energy (up to 10 TeV) particles accelerated in the shocks formed by the interaction of the jets with the ISM (Corbel *et al.* 2002; Kaaret *et al.* 2003; Tomsick *et al.* 2003; Hao & Zhang 2009).

6. Jets emission at high and very high energy

The detection of high (or very high) energy emission from a Galactic accreting binary has been a long awaited observation. Indeed, in the early 1970s and 1980s, detections of Cyg X−3 were reported in the MeV to PeV energies, which generated considerable excitements. However, the detections remained doubtful mainly because a more sensitive generation of telescopes and satellites failed to confirm these detections (Ong 1998). In 2009, the *Fermi* Large Area Telescope (*LAT*) and the *AGILE* satellite reported the detection of a variable high energy source (Fig. 4) coinciding with the position of the microquasar Cyg X−3 (Fermi LAT Collaboration *et al.* 2009; Tavani *et al.* 2009). The

212 S. Corbel

identification with Cyg X−3 is secured by the detection by *Fermi* of its orbital period in gamma rays, as well as the correlation of the *LAT* flux with radio emission from the relativistic jets of Cyg X−3 (Fermi LAT Collaboration *et al.* 2009). The gamma-ray emission can be explained by inverse Compton scattering of UV photons from the Wolf-Rayet (WR) star off of high energy electrons from the relativistic jets of Cyg X−3 (Fermi LAT Collaboration *et al.* 2009) or in a recollimation shocks associated with the strong wind from the WR star (Dubus *et al.* 2010).

In addition to Cyg X−3, low significance gamma-ray flares have been reported by MAGIC (Albert *et al.* 2007) and associated with the black hole binary Cyg X−1. According to the binary parameters of the system, this very high energy flare is believed to occur in a configuration with the black hole behind the high-mass star companion. The huge gamma-ray absorption should normally prevent any detection above few tens of GeV in this configuration, although Romero *et al.* (2010) propose a possible interpretation based on the interaction of the Cyg X−1 jets with the clumpy wind of the high-mass star. We note that AGILE has recently reported high energy gamma-ray flares from Cyg X−1 (e.g. Sabatini *et al.* 2010), that have, however, not been confirmed by Fermi †.

Acknowledgement

The author would like to thank the organizers for this fruitful conference and Richard Owen for a careful reading of this manuscript. Part of the research discussed in this review has received funding from the European Communitys Seventh FrameworkProgramme (FP7/2007-2013) under grant agreement number ITN 215212 Black Hole Universe.

References

Albert, J., Aliu, E., Anderhub, H., *et al.* 2007, *ApJL* , 665, L51
Angelini, L. & White, N. E. 2003, *ApJL* , 586, L71
Belloni, T., ed. 2010, Lecture Notes in Physics, Berlin Springer Verlag, Vol. 794, The Jet Paradigm
Blandford, R. D. & Konigl, A. 1979, *ApJ* , 232, 34
Buxton, M. M. & Bailyn, C. D. 2004, *ApJ* , 615, 880
Cadolle Bel, M., Ribó, M., Rodriguez, J., *et al.* 2007, *ApJ* , 659, 549
Corbel, S. & Fender, R. P. 2002, *ApJL* , 573, L35
Corbel, S., Fender, R. P., Tomsick, J. A., Tzioumis, A. K., & Tingay, S. 2004, *ApJ* , 617, 1272
Corbel, S., Fender, R. P., Tzioumis, A. K., *et al.* 2000, *A&A* , 359, 251
Corbel, S., Fender, R. P., Tzioumis, A. K., *et al.* 2002, Science, 298, 196
Corbel, S., Kaaret, P., Fender, R. P., *et al.* 2005, *ApJ* , 632, 504
Corbel, S., Koerding, E., & Kaaret, P. 2008, *MNRAS*, 0
Corbel, S., Nowak, M. A., Fender, R. P., Tzioumis, A. K., & Markoff, S. 2003, *A&A* , 400, 1007
Coriat, M., Corbel, S., Buxton, M. M., *et al.* 2009, *MNRAS*, 400, 123
Dhawan, V., Mirabel, I. F., & Rodríguez, L. F. 2000, *ApJ* , 543, 373
Dubus, G., Cerutti, B., & Henri, G. 2010, *MNRAS*, 404, L55
Falcke, H., Körding, E., & Markoff, S. 2004, *A&A* , 414, 895
Fender, R. 2006, Jets from X-ray binaries (Compact stellar X-ray sources), 381–419
Fender, R., Corbel, S., Tzioumis, T., *et al.* 1999, *ApJL* , 519, L165
Fender, R., Wu, K., Johnston, H., *et al.* 2004a, *Nature* , 427, 222
Fender, R. P. 2001, *MNRAS*, 322, 31
Fender, R. P., Belloni, T. M., & Gallo, E. 2004b, *MNRAS*, 355, 1105
Fender, R. P., Homan, J., & Belloni, T. M. 2009, *MNRAS*, 396, 1370
Fermi LAT Collaboration, Abdo, A. A., Ackermann, M., *et al.* 2009, Science, 326, 1512

† e.g. http://fermisky.blogspot.com/2010/03/lat-limit-on-cyg-x-1-during-reported.html

Fomalont, E. B., Geldzahler, B. J., & Bradshaw, C. F. 2001a, *ApJL* , 553, L27

Fomalont, E. B., Geldzahler, B. J., & Bradshaw, C. F. 2001b, *ApJ* , 558, 283

Fuchs, Y., Rodriguez, J., Mirabel, I. F., *et al.* 2003, *A&A* , 409, L35

Gallo, E., Fender, R. P., Miller-Jones, J. C. A., *et al.* 2006, *MNRAS*, 370, 1351

Gallo, E., Fender, R. P., & Pooley, G. G. 2003, *MNRAS*, 344, 60

Hao, J. F. & Zhang, S. N. 2009, *ApJ* , 702, 1648

Heinz, S. 2004, *MNRAS*, 355, 835

Heinz, S., Schulz, N. S., Brandt, W. N., & Galloway, D. K. 2007, *ApJL* , 663, L93

Hjellming, R. M. & Rupen, M. P. 1995, *Nature* , 375, 464

Homan, J., Buxton, M., Markoff, S., *et al.* 2005, *ApJ* , 624, 295

Kaaret, P., Corbel, S., Tomsick, J. A., *et al.* 2003, *ApJ* , 582, 945

Kaaret, P., Corbel, S., Tomsick, J. A., *et al.* 2006, *ApJ* , 641, 410

Körding, E., Falcke, H., & Corbel, S. 2006, *A&A* , 456, 439

Körding, E., Rupen, M., Knigge, C., *et al.* 2008, Science, 320, 1318

Lang, C. C., Kaaret, P., Corbel, S., & Mercer, A. 2007, *ApJ* , 666, 79

Markoff, S., Falcke, H., & Fender, R. 2001, *A&A* , 372, L25

Markoff, S., Nowak, M., Corbel, S., Fender, R., & Falcke, H. 2003, *A&A* , 397, 645

Markoff, S., Nowak, M. A., & Wilms, J. 2005, *ApJ* , 635, 1203

Martí, J., Mirabel, I. F., Rodríguez, L. F., & Smith, I. A. 2002, *A&A* , 386, 571

McClintock, J. E., Haswell, C. A., Garcia, M. R., *et al.* 2001, *ApJ* , 555, 477

McClintock, J. E. & Remillard, R. A. 2006, Black hole binaries (Compact stellar X-ray sources), 157–213

Merloni, A., Heinz, S., & di Matteo, T. 2003, *MNRAS*, 345, 1057

Migliari, S. & Fender, R. P. 2006, *MNRAS*, 366, 79

Migliari, S., Tomsick, J. A., Markoff, S., *et al.* 2007, *ApJ* , 670, 610

Migliari, S., Tomsick, J. A., Miller-Jones, J. C. A., *et al.* 2010, *ApJ* , 710, 117

Mirabel, I. F. & Rodriguez, L. F. 1994, *Nature* , 371, 46

Mirabel, I. F., Rodriguez, L. F., Cordier, B., Paul, J., & Lebrun, F. 1992, *Nature* , 358, 215

Neilsen, J. & Lee, J. C. 2009, *Nature* , 458, 481

Ong, R. A. 1998, *Physics Reports*, 305, 93

Pakull, M. W., Soria, R., & Motch, C. 2010, *Nature* , 466, 209

Park, S. Q., Miller, J. M., McClintock, J. E., & Murray, S. S. 2005, *ApJL* , 618, L45

Rodriguez, J., Bel, M. C., Tomsick, J. A., *et al.* 2007, *ApJL* , 655, L97

Rodriguez, J., Corbel, S., & Tomsick, J. A. 2003, *ApJ* , 595, 1032

Rodriguez, J., Shaw, S. E., Hannikainen, D. C., *et al.* 2008, *ApJ* , 675, 1449

Romero, G. E., Del Valle, M. V., & Orellana, M. 2010, *A&A* , 518, A12+

Russell, D. M., Fender, R. P., Hynes, R. I., *et al.* 2006, *MNRAS*, 371, 1334

Russell, D. M., Maitra, D., Dunn, R. J. H., & Markoff, S. 2010, *MNRAS*, 405, 1759

Sabatini, S., Tavani, M., Striani, E., *et al.* 2010, *ApJL* , 712, L10

Sell, P. H., Heinz, S., Calvelo, D. E., *et al.* 2010, *ApJL* , 719, L194

Shakura, N. I. & Sunyaev, R. A. 1973, *A&A* , 24, 337

Soleri, P., Fender, R., Tudose, V., *et al.* 2010, *MNRAS*, 865

Soleri, P., Heinz, S., Fender, R., *et al.* 2009, *MNRAS*, 397, L1

Soria, R., Fender, R. P., Hannikainen, D. C., Read, A. M., & Stevens, I. R. 2006, *MNRAS*, 368, 1527

Soria, R., Pakull, M. W., Broderick, J. W., Corbel, S., & Motch, C. 2010, *MNRAS*, 1324

Stirling, A. M., Spencer, R. E., de la Force, C. J., *et al.* 2001, *MNRAS*, 327, 1273

Sunyaev, R. A. & Titarchuk, L. G. 1980, *A&A* , 86, 121

Tavani, M., Bulgarelli, A., Piano, G., *et al.* 2009, *Nature* , 462, 620

Tingay, S. J., Jauncey, D. L., Preston, R. A., *et al.* 1995, *Nature* , 374, 141

Tomsick, J. A., Corbel, S., Fender, R., *et al.* 2003, *ApJ* , 582, 933

Tomsick, J. A., Yamaoka, K., Corbel, S., *et al.* 2009, *ApJL* , 707, L87

Yang, J., Brocksopp, C., Corbel, S., *et al.* 2010, *MNRAS*, L148+

Yuan, F., Cui, W., & Narayan, R. 2005, *ApJ* , 620, 905

Discussion

FENDT: What do we know about the collimation degree of compact jets?

CORBEL: The constraints are very pooor as only two sources have imaged compact jets. It is less than $2°$ in the case of Cyg X-1. This is also consistent with the compact jets in the plateau state for GRS 1915+105.

DE GOUVEIA DAL PINO: Regarding the point you reported about X-ray bubbles carried by jets at different scale (from galactoc compact sources to AGN FR II radiogalaxies) I just would like to comment that there is a poster outside by Falceta-Goncalves *et al.*, where we have performed 3D MHD simulations of the radio jet that seats in the center of the Perseus cluster of galaxies and inflate bubbles in the X-ray distribution of gas in the surrroundings, and distribute energy all over ~ 100 kpc scales - and stopping the cooling flow un the core of this galaxy cluster.

CORBEL: Thanks for the comment.

KUNDT: Pakull *et al.*'s energy estimate of S26 in the Sculptor galaxy may be overestimated by a factor of order 10^3 by the assumption that their lobes emit homogeneously, rather than by small-filling-factor thermal inclusions, of filling factor $f \approx 10^{-3}$.

CORBEL: Yes, I agree that it is a possibility.

YUAN: Most of the estimation of the power of the ejected blob are time-averaged values. Is there any estimation to a single ejection event?

CORBEL: Estimates do exist such as the ones done for the impulsive jets of GRS 1915+105 (see Fender *et al.* 1999) or even in the case of the decelerating jets in XTEJ1550-564 (Tomsick *et al.* 2003). With some cautions, this can be at least more than 10^{38}erg s^{-1}, i.e., a significant fraction of the accretion power.

Jets at all Scales
Proceedings IAU Symposium No. 275, 2011
G. E. Romero, R. A. Sunyaev & T. Belloni, eds.

© International Astronomical Union 2011
doi:10.1017/S1743921310016066

Nonthermal processes in microquasars

Valentí Bosch-Ramon

Dublin Institute for Advanced Studies, 31 Fitzwilliam Place, Dublin 2, Ireland
email: valenti@cp.dias.ie

Abstract. Microquasars are X-ray binaries that show extended radio jets. These jets can accelerate particles up to relativistic energies that produce non-thermal emission from radio to TeV, and could also make a non-negligible contribution to the galactic CRs in some energy ranges. The orbital motion and compactness of these sources allow the study of high-energy astrophysical phenomena in extreme conditions that change in accessible timescales. In this work, I briefly discuss the production of broadband non-thermal emission in microquasars, putting special emphasis on the high- and the very high-energy bands.

Keywords. Microquasars, non-thermal processes, gamma-rays

---

1. Introduction

Microquasars are X-ray binaries with non-thermal radio jets (see, e.g. Mirabel & Rodríguez 1999, Ribó 2005). They are called high-mass microquasars (HMMQ) when hosting a massive star, and low-mass microquasars (LMMQ) otherwise. The supply of energy in microquasars can be either accretion or an accreting rotating black-hole, and the power is channelled to a jet that is launched from the inner regions of the accretion disk (e.g. Blandford & Znajek 1977, Blandford & Payne 1982).

Historically, microquasars have been considered strong candidates to gamma-ray sources (e.g. Chadwick *et al.* 1985; see also Chardin & Gerbier 1989, Levinson & Blandford 1996, Paredes *et al.* 2000). However, it has not been until the most recent generation of ground-based Cherenkov telescopes arrived, like HESS, MAGIC and VERITAS, and satellite-borne instrumens, like *Fermi* and *AGILE*, that microquasars have become fully recognized as powerful gamma-ray emitters. The most relevant cases are the microquasars Cygnus X-1 and Cygnus X-3 (Albert *et al.* 2007, Sabatini *et al.* 2010, Tavani *et al.* 2009a, Abdo *et al.* 2009c; for upper-limits in some energies or source states, see Del Monte *et al.* (2010), Saito *et al.* 2009, Aleksic *et al.* 2010). Although the detected radio emission is evidence of particle acceleration in microquasar jets, the finding of microquasars being gamma-ray emitters is the proof that these sources can very efficient channelling accretion or black-hole rotational energy into radiation. In this paper, we review different aspects related to non-thermal emission in microquasars, focusing mainly in the production of gamma-rays.

2. Non-thermal emission in microquasars

Microquasar jets can produce non-thermal populations of relativistic particles via diffusive shock acceleration or other mechanisms at different spatial scales (for a recent review, see Bosch-Ramon 2010a; see also, e.g., Romero *et al.* 2003, Paredes *et al.* 2006, Rieger *et al.* 2007), generating photons of energies from radio to gamma-rays through synchrotron and inverse Compton (IC). Gamma-rays can be also produced through hadronic interactions, like proton-proton (pp) interactions, photomeson production, or photodisintegration. We refer the reader to the work by Bosch-Ramon & Khangulyan (2009)

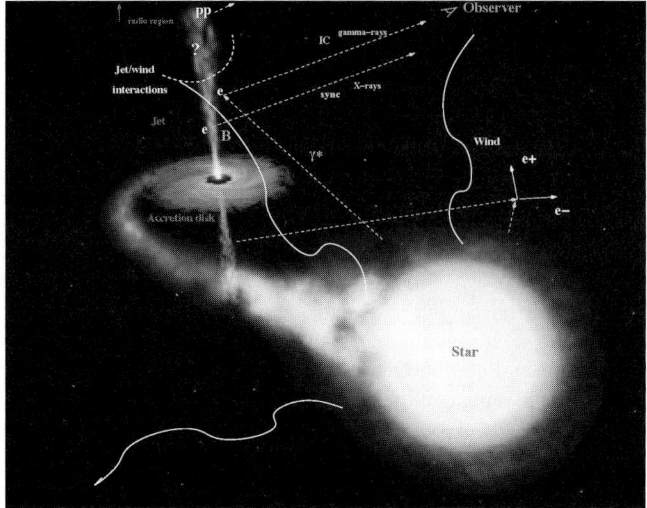

Figure 1. Illustrative picture of the microquasar scenario, in which relevant elements and processes are shown (background image from ESA, NASA, and Félix Mirabel).

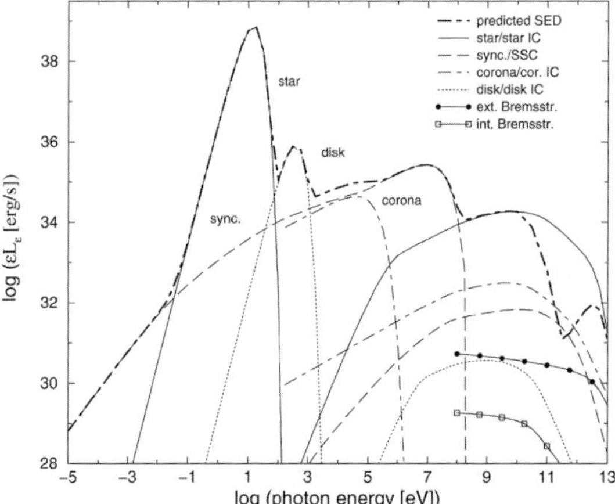

Figure 2. Computed spectral energy distribution for a typical HMMQ (see Bosch-Ramon *et al.* 2006a).

(and references therein) for details on non-thermal processes in the present context. In microquasars, the corona (or the base of the jet) could be also a non-thermal emitting region (e.g. Gierlinski *et al.* 1999; see also Romero *et al.* 2010), as well as the termination region of the jet in the ISM (e.g. Bordas *et al.* 2009). The relativistic protons that may be accelerated in microquasars can escape these systems enriching the CR content of our galaxy (e.g. Heinz & Sunyaev 2002). A sketch of the microquasar scenario, with the main elements and processes relevant to microquasars, is shown in Fig. 1.

2.1. *High-mass microquasars*

In the context of HMMQs, likely the most efficient gamma-ray mechanism would be IC scattering of photons from the stellar companion (e.g. Paredes *et al.* 2000, Dermer & Böttcher 2006, Bosch-Ramon *et al.* 2006a). As discussed in Sect. 3, the presence of the stellar photons may also imply strong attenuation of the gamma-ray emission. Hadronic

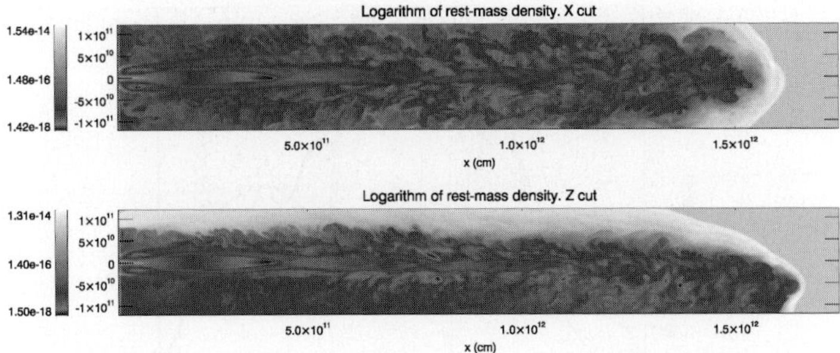

Figure 3. Two-dimensional map of the density of a jet interacting with the stellar wind, which is coming from the top of the image (see Perucho *et al.* 2010).

processes have been also discussed in the past (e.g. Romero *et al.* 2003; Aharonian *et al.* 2006b), although their efficiency is expected to be smaller than that of stellar photon IC. In Fig. 2 we show the computed broadband spectral energy distribution of HMMQs. In this particle instance, the dominant non-thermal emission processes are synchrotron and stellar photon IC.

Beside the jet itself, stellar wind-jet dynamical interactions at the binary spatial scales could also lead to non-thermal emission in HMMQs (Perucho & Bosch-Ramon 2008), and the likely clumpy nature of the wind (e.g. Owocki & Cohen 2006) could lead to HE and VHE flares (see, e.g., Araudo *et al.* 2009). An example of the importance of the wind-jet interaction is shown in Fig. 3, in which it is shown a 2-dimensional map of the density of a jet of power 10^{35} erg s^{-1} interacting with the stellar wind, which is coming from the top. Note that the jet is suffering significant bending and strong disruption not far beyond an asymmetric recollimation shock. Jets significantly more powerful, of up to 10^{37} erg s^{-1}, may be still affected by jet disruption due to the growth of non-linear instabilities triggered in the recollimation shock (see Perucho *et al.* 2010).

2.2. *Low-mass microquasars*

Low-mass microquasars have been also proposed to be HE and VHE emitters. In these objects, the most efficient gamma-ray mechanisms could be external IC with accretion photons, synchrotron self-Compton or hadronic interactions in the inner-most regions of the jet base (e.g. Atoyan & Aharonian 1999, Bosch-Ramon *et al.* 2006a, Romero & Vila 2009). In Fig. 4, we show the computed spectral energy distribution for a powerful jet of a low-mass microquasar. The dominant IC component is either IC scattering off corona photons or synchrotron self-Compton; note the soft spectrum above GeV energies due to the Klein Nishina effect. Gamma-ray absorption in the accretion disk and corona fields can be significant in this context (for details, see Bosch-Ramon *et al.* 2006b).

3. Impact of gamma-ray absorption

In high-mass microquasars, if gamma-ray absorption takes place in the photon field of the star (e.g. Ford 1984 , Protheroe & Stanev 1987, Moskalenko & Karacula 1994, Böttcher & Dermer 2005, Dubus 2006a, Khangulyan *et al.* 2008, Reynoso *et al.* 2008), the whole binary system could be an efficient broadband non-thermal emitter. In such a case, the created pairs can radiate through synchrotron and IC emission in the whole spectral band, interacting with the ambient magnetic field and stellar photons, respectively (Bosch-Ramon *et al.* 2008a). If the energy density of the magnetic field were much

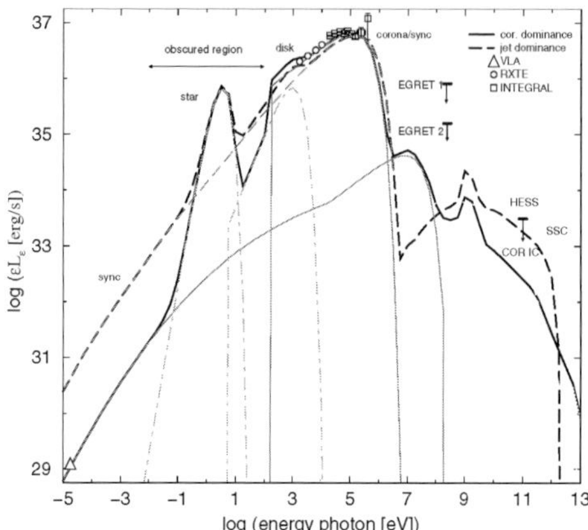

Figure 4. Computed spectral energy distribution of the non-thermal emission from 1E 1740.7−2942 for two situations. In one case, the hard X-rays come from a corona (long dashed line), whereas in the other, they are of synchrotron origin and come from the jet. Absorption in the accretion disk and corona photon fields has been taken into account (see the two dips below and above $\sim$ GeV energies). For details, see Bosch-Ramon *et al.* (2006b).

smaller than that of the stellar photon field, IC scatterings would be the dominant cooling channel of these pairs. Then these pairs would produce more gamma-rays that to their turn may be also absorbed, triggering what is called an electromagnetic cascade. In this way, the effective optical depth to gamma-rays would be significantly reduced. The deflection of the created pairs in the ambient magnetic field determines whether the cascade is linear or three dimensional (e.g. Bednarek 1997, Aharonian *et al.* 2006b, Orellana *et al.* 2007, Sierpowska-Bartosik & Torres 2007, Zdiziarski *et al.* 2009, Cerutti *et al.* 2010).

Under strong enough magnetic fields, electromagnetic cascades get suppressed (Khangulyan *et al.* 2008) and the X-ray (synchrotron) and GeV (single-scattering IC) emission from secondary pairs dominates the secondary energy output (Bosch-Ramon *et al.* 2008a). In Figs. 5 and 6 we show, as an example of this, the spectral energy distribution of the secondary radiation for a binary with system properties similar to those of LS 5039 (Aragona *et al.* 2009), computed for the phases when the compact object is in front (inferior conjunction) and behind the star (superior conjunction). The magnetic field in the stellar surface has been fixed to 20 G.

For stronger magnetic fields, the secondary radio emission may be also significant (Bosch-Ramon 2009, Bosch-Ramon 2010b). In Fig. 7, we show 5 GHz radio images for a similar case but adopting a star magnetic field of 200 G. The images for four different orbital phases are presented: 0.0 (superior conjunction), 0.25, 0.5 (inferior conjunction), and 0.75.

4. Microquasars in context: jet sources and gamma-ray binaries

The phenomenology at high energies of microquasars can be semi-quantitatively explained accounting for a set of ingredients or physical processes and effects: strong shocks; the angular dependence of gamma-ray absorption and IC scattering and the role of internal and external absorption (Bosch-Ramon *et al.* 2006a; Khangulyan *et al.* 2008), in some cases locating the emitter at some height above the orbital plane (for acceleration and modeling arguments, see Khangulyan *et al.* 2008; for absorption arguments, see Bosch-Ramon *et al.* 2008b); adiabatic losses (e.g. Bosch-Ramon *et al.* 2006c). The

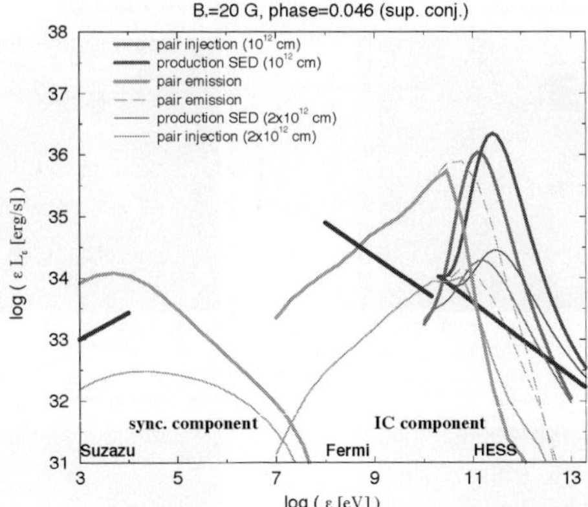

Figure 5. Computed spectral energy distribution of the radiation produced by secondary pairs in a high-mass binary (green lines) at the orbital phase associated to the superior conjunction of the compact object. Two different configurations have been considered, one in which the emitter is at a height of 10^{12} cm (production gamma-ray spectrum: thick solid blue line; corresponding pair injection: thick solid red line), and at 2×10^{12} cm (production gamma-ray spectrum: thin solid blue line; corresponding pair injection: thin solid red line) above the compact object. The production spectral energy distribution (thin long-dashed) of the secondary IC emission is also shown.

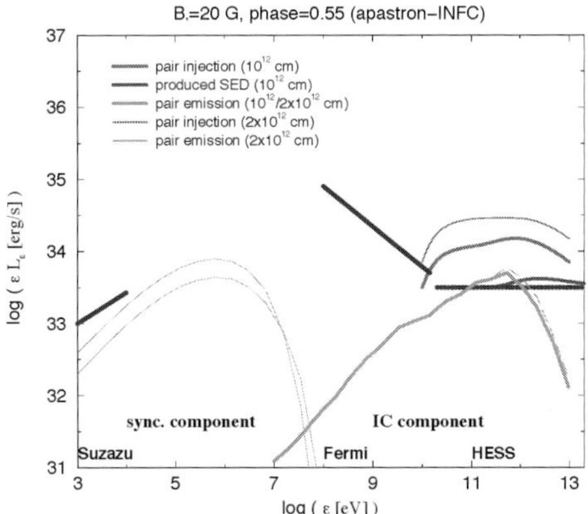

Figure 6. The same as in Fig. 5 but for the inferior conjunction of the compact object.

role of the magnetic field, e.g. through the impact of synchrotron cooling, is crucial for the VHE spectrum and luminosity. As noted above, one should not neglect either the impact of the stellar wind on the non-thermal processes (Perucho *et al.* 2010). Finally, the emitting region, generally the jet, can move fast enough to require the inclusion of Doppler boosting effects.

Beside microquasars, there are other jet sources in the Universe, like young stellar objects, active galactic nuclei and gamma-ray bursts. Together with microquasars, all these jet sources share the same basic elements: an accreting object; particle acceleration, possibly through strong shocks, Doppler boosting; and, with some exceptions, the same main suspects to be the dominant radiation processes (synchrotron, IC, and perhaps hadronic interactions). Nevertheless, despite being so similar, there are still important differences between the mentioned sources, like the accretor, the distance to us, the timescales, the accretion rate, and the properties of the jet itself, like its velocity and beaming relevance, density and temperature (e.g. efficient/unefficient thermal emission). There are as

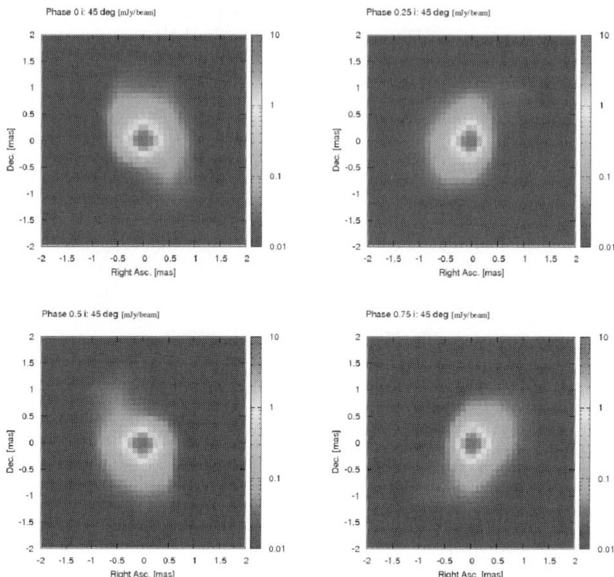

Figure 7. Computed image of the 5 GHz radio emission, in the direction to the observer, for different orbital phases. Units are given in mJy per beam, being the beam size ~1 milliarcsecond.

well the environmental factors, like external sources of photons and the external medium density and velocity. For instance, in the case of AGNs, different powers and medium densities give rise to different jet termination structures, with instabilities having a very important role (e.g. Fanaroff & Riley 1974; Perucho *et al.* 2010b). Also, GRBs can show very different phenomena during their afterglow phase, when the jet terminates, depending on the characteristics of the environment, since it may be determined by a massive star wind for a collapsar long GRB, or could be a much less dense medium in the case of coalescence of two compact objects (Chevalier & Li 1999). Finally, in young stellar objects, the jet termination region may be the best candidate to be a gamma-ray emitter, given the strong shocks occuring there (e.g. Araudo *et al.* 2007, Bosch-Ramon *et al.* 2010). The magnetization level of the jet also determines the acceleration processes and the emitting mechanism, and may be different in each source or emitting region.

Microquasars are a subclass of gamma-ray emitters that pertain to the more general class of gamma-ray binaries. These latter sources are binary systems in which gamma-ray production takes place, and generally consist of a normal (non-degenerated) star and a compact object or another normal star. In addition to microquasars (Cygnus X-1 and Cygnus X-3), there is one confirmed high-mass binary hosting a non-accreting pulsar, PSR B1259−63 (Aharonian *et al.* 2005b), possibly a massive star binary, η-Carina (Tavani *et al.* 2009b), two high-mass binaries harboring a compact object of unknown nature, LS 5039 and LS I +61 303, (Kniffen *et al.* 1997, Paredes *et al.* 2000, Aharonian *et al.* 2005a, Aharonian *et al.* 2006a, Albert *et al.* 2006, Abdo *et al.* 2009a, Abdo *et al.* 2009b; for discussions on the nature of these sources, see, e.g., Dubus 2006b, Chernyakova *et al.* 2006, Romero *et al.* 2007, Bosch-Ramon & Khangulyan 2009), and a high-mass binary candidate, HESS J0632+057 (Aharonian *et al.* 2007). Very recently, HE emission concurrent with a nova in the Symbiotic Binary V407 Cygni has been detected by *Fermi* Abdo *et al.* 2010, incorporating symbiotic binaries to the list of gamma-ray binary classes. All gamma-ray binaries share the main ingredients for gamma-ray production, i.e. powerful supersonic outflows, which may be collimated or quasi-spherical; strong shocks leading to particle acceleration; dense ambient photon fields for IC (although matter targets for hadronic processes may be important in some cases); and gamma-ray absorption in dense ambient photon fields.

As shown in the previous paragraph, physical effects and processes occurring in microquasars are also present to different extents in the known types of gamma-ray binaries and jet sources. In the context of gamma-ray binaries, these similarities make in some cases the identification of the specific nature of the object a difficult task, as it is shown by the debate on the nature of LS 5039 and LS I +61 303 (i.e. HMMQs versus non-accreting pulsar high-mass binaries, if not a really new class of object). In case of jet sources, similarities can help to understand sources whose (e.g.) timescales are not accessible to us. However, the similarities should not mask the differences, which put limits to analogy arguments that sometimes can be misleading. In order to avoid to carry analogies too far, and to shed light on complicated cases of source nature confusion, it is of primary importance to apply comprehensive approaches to study theoretically these objects, including semi-analytical modeling, numerical calculations (secondary emission) and (magneto)hydrodynamical simulations, all this encompassed of course by simultaneous multiwavelength observations with high angular, spectral and time resolution.

Acknowledgements

The author acknowledges support of the Spanish MICINN under grant AYA2007-68034-C03-1 and FEDER funds. The author also acknowledges support of the European Community under a Marie Curie Intra-European fellowship.

References

Abdo, A. A., *et al.* 2009b, *ApJ*, 706, L56

Abdo, A. A., *et al.* 2009b, *ApJ*, 701, L123

Abdo, A. A., *et al.* 2009c, *Science*, 326, 1512

Abdo, A. A., *et al.* 2010, *Science*, 2010, *Science*, 329, 817

Acciari V. A., *et al.*, 2008, *ApJ*, 679, 1427

Aharonian, F., *et al.* 2005a, *Science*, 309, 746

Aharonian, F., *et al.* 2005b, *A&A*, 442, 1

Aharonian *et al.* 2006a, *J. Phys. Conf. Ser.*, 39, 408 [astro-ph/0508658]

Aharonian, F., *et al.* 2006b, *A&A*, 460, 743

Aharonian, F., *et al.* 2007, *A&A*, 469, L1

Albert, J., *et al.* 2006, *Science*, 312, 1771

Albert, J., *et al.* 2007, *ApJ*, 665, L51

Aleksic, J. 2010, *ApJ*, 721, 843

Aragona, C., *et al.* 2009, *ApJ*, 698, 514

Araudo, A. T., *et al.* 2007, *A&A*, 476, 1289

Araudo, A. T., Bosch-Ramon, V., & Romero, G. E. 2009, *A&A*, 503, 673

Atoyan, A. M. & Aharonian, F. A. 1999, *MNRAS*, 302, 253

Bednarek, W. 1997, *A&A*, 322, 523

Blandford, R. D. & Znajek, R. L. 1977, *MNRAS*, 179, 433

Blandford, R. D. & Payne, D. G. 1982, *MNRAS*, 199, 883

Bordas *et al.* 2009, *A&A*, 497, 325

Bosch-Ramon, V., Romero, G E., & Paredes, J. M. 2006a, *A&A*, 447, 263

Bosch-Ramon, V. *et al.* 2006b, *A&A*, 457, 1011

Bosch-Ramon, V., Paredes, J. M., Romero, G. E., & Ribó, M. 2006, *A&A*, 459, L25

Bosch-Ramon, V., Khangulyan D., & Aharonian F. A. 2008a, *A&A*, 482, 397

Bosch-Ramon, V., Khangulyan D., & Aharonian F. A. 2008b, *A&A*, 489, L21

Bosch-Ramon, V. 2009, *A&A*, 493, 829

Bosch-Ramon, V., & Khangulyan, D. 2009, *Int. Jour, of Mod. Phys.* D, 18, 347

Bosch-Ramon, V. 2010a, *ASPC*, 422, 13

Bosch-Ramon, V. 2010b, *IJMPD*, 19, 741

Bosch-Ramon, V. 2010c, *A&A*, 511, 8

Böttcher, M. & Dermer, C. D. 2005, *ApJ*, 634, L81

Chadwick, P. M., *et al.* 1985, *Nature*, 318, 642

Chardin, G. & Gerbier, G. 1989, *A&A*, 210, 52

Chernyakova, M., Neronov, A., & Walter, R. 2006, *MNRAS*, 372, 1585

Chevalier, R. A. & Li, Z. 1999, *ApJ*, 520, 29

Dermer, C. & Böttcher, M., *ApJ* 643, 1081 (2006)

Dubus, G. 2006a, *A&A*, 451, 9

Dubus, G. 2006b, *A&A*, 456, 801

Fanaroff, B. L. & Riley, J. M. 1974, *MNRAS*, 167, 31

Ford, L. H. 1984, *MNRAS*, 211, 559

Gierlinski *et al.* 1999, *MNRAS*, 309, 496

Heinz & Sunyaev 2002, *A&A*, 390, 751

Khangulyan, D. *et al.* 2007, *MNRAS*, 380, 320

Khangulyan, D., Aharonian, F., & Bosch-Ramon, V. 2008, *MNRAS*, 383, 467

Kniffen, D. A. *et al.* 1997, *ApJ*, 486, 126

Levinson, A. & Blandford, R. D. 1996, *ApJ*, 456, L29

Mirabel, I. F. & Rodríguez, L. F. 1999, *ARA&A*, 37, 409

Del Monte, E. 2010, *A&A*, in press [astro-ph/1004.0849]

Moskalenko I. V. & Karakula S., 1994, *ApJ*, 92, 567

Orellana, M. *et al.* 2007, *A&A*, 476, 9

Owocki, S., Cohen, P., & David, H. 2006, *ApJ*, 648, 565

Paredes, J. M. *et al.* 2000, *Science*, 288, 2340

Paredes, J. M., Bosch-Ramon, V., & Romero, G. E. 2006, *A&A*, 451, 259

Perucho, M. & Bosch-Ramon, V. 2008, *A&A*, 482, 917

Perucho, M., Bosch-Ramon, V., & Khangulyan, D. 2010a, *A&A*, 512, L4

Perucho, M., *et al.* 2010b, *A&A*, in press [astro-ph/1005.4332]

Protheroe, R. J. & Stanev, T. 1987, *ApJ*, 322, 838

Ribó, M. 2005, in Future Directions in High Resolution Astronomy: *The 10th Anniversary of the VLBA*, (ASPC, 2005) 340, 421 [astro-ph/0402134]

Rieger, F. M., Bosch-Ramon, V., & Duffy, P. 2007, *Ap&SS*, 309, 119

Romero, G. E. *et al.* 2003, *A&A*, 410, L1

Romero, G. E. *et al.* 2007, *A&A*, 474, 15

Romero, G. E. & Vila G. S. 2009, *A&A*, 494, L33

Romero, G. E., Vieyro, F. L., & Vila G. S. 2010, *A&A*, in press [astro-ph/1006.5005]

Sabatini *et al.* 2010, *ApJ*, 712, L10

Saito *et al.* 2009, *Proc. 31st ICRC*

Sierpowska-Bartosik, A. & Torres, D. F. 2007, *ApJ*, 671, L145

Takahashi, T. *et al.* 2009, *ApJ*, 697, 592

Tavani, M., *et al.* 2009a, *Nature*, 462, 620

Tavani, M., *et al.* 2009b, *ApJ*, 698, L142

Zdziarski, A. A., Malzac, J., & Bednarek, W. 2009, *MNRAS*, 394, L41

Discussion

CASTRO-TIRADO: The compact sources detected at VHE are all high-mass X-ray binaries? Is this mainly due to the strong stellar winds from the companion stars?

BOSCH-RAMON: The strong stellar wind should be relevant but the difference brought by the star, I think, is the stellar radiation field in the context of VHE production.

KUNDT: You've been rightly stressing the impressive hardness of (some of) the micro-quasar spectra, with electron energies reaching PeV. An often mentioned "in-situ acceleration" conflict with the Second Law, beyond the test-particle limit, increasingly so for increasing Lorentz factors $\gamma \approx 10^9$. I encourage your consideration of an obliquely spinning magnetized neutron star as the exclusive booster, analogous to the Crab nebula.

BOSCH-RAMON: I agree with you that mechanisms of particle acceleration different from the canonical Fermi-type ones should be explored, since the conditions in these sources may indeed require something exotic to explain the very high energies of the observed photons.

Jets at all Scales
Proceedings IAU Symposium No. 275, 2011
G. E. Romero, R. A. Sunyaev & T. Belloni, eds.

© International Astronomical Union 2011
doi:10.1017/S1743921310016078

Investigating accretion disk – radio jet coupling across the stellar mass scale

James C. A. Miller-Jones[1], Gregory R. Sivakoff[2], Diego Altamirano[3], Elmar G. Körding[4], Hans A. Krimm[5], Dipankar Maitra[6], Ron A. Remillard[7], David M. Russell[3], Valeriu Tudose[8], Vivek Dhawan[9], Rob P. Fender[10], Sebastian Heinz[11], Sera Markoff[3], Simone Migliari[12], Michael P. Rupen[9] and Craig L. Sarazin[2]

[1]ICRAR - Curtin University of Technology, GPO Box U1987, Perth, WA 6845, Australia
email: james.miller-jones@curtin.edu.au

[2]Department of Astronomy, University of Virginia, P.O. Box 400325, Charlottesville, VA 22904, USA
email: grs8g@virginia.edu, sarazin@virginia.edu

[3]Astronomical Institute 'Anton Pannekoek', University of Amsterdam, P.O. Box 94249, 1090 GE Amsterdam, the Netherlands
email: d.altamirano@uva.nl, s.b.markoff@uva.nl, d.m.russell@uva.nl

[4]Université Paris Diderot and Service d'Astrophysique, UMR AIM, CEA Saclay, F-91191 Gif-sur-Yvette, France
email: elmar@koerding.eu

[5]NASA/Goddard Space Flight Center, Greenbelt, MD 20771, USA; and USRA, 10211 Wincopin Circle, Suite 500, Columbia, MD 21044, USA
email: hans.krimm@nasa.gov

[6]Department of Astronomy, University of Michigan, Ann Arbor, MI 48109, USA
email: dmaitra@umich.edu

[7]MIT Kavli Institute for Astrophysics and Space Research, Building 37, 70 Vassar Street, Cambridge, MA 02139, USA
email: rr@space.mit.edu

[8]Netherlands Institute for Radio Astronomy, Oude Hoogeveensedijk 4, 7991 PD Dwingeloo, the Netherlands
email: tudose@astron.nl

[9]NRAO Domenici Science Operations Center, 1003 Lopezville Road, Socorro, NM 87801, USA
email: vdhawan@nrao.edu, mrupen@nrao.edu

[10]School of Physics and Astronomy, University of Southampton, Southampton SO17 1BJ, UK
email: r.fender@soton.ac.uk

[11]Astronomy Department, University of Wisconsin-Madison, 475. N. Charter St., Madison, WI 53706, USA
email: heinzs@astro.wisc.edu

[12]European Space Astronomy Centre, Apartado/P.O. Box 78, Villanueva de la Canada, E-28691 Madrid, Spain
email: simone.migliari@sciops.esa.int

Abstract. Relationships between the X-ray and radio behavior of black hole X-ray binaries during outbursts have established a fundamental coupling between the accretion disks and radio jets in these systems. I begin by reviewing the prevailing paradigm for this disk-jet coupling, also highlighting what we know about similarities and differences with neutron star and white dwarf binaries. Until recently, this paradigm had not been directly tested with dedicated high-angular resolution radio imaging over entire outbursts. Moreover, such high-resolution monitoring campaigns had not previously targetted outbursts in which the compact object was either a neutron star or a white dwarf. To address this issue, we have embarked on the Jet Acceleration and

Collimation Probe Of Transient X-Ray Binaries (JACPOT XRB) project, which aims to use high angular resolution observations to compare disk-jet coupling across the stellar mass scale, with the goal of probing the importance of the depth of the gravitational potential well, the stellar surface and the stellar magnetic field, on jet formation. Our team has recently concluded its first monitoring series, including (E)VLA, VLBA, X-ray, optical, and near-infrared observations of entire outbursts of the black hole candidate H 1743-322, the neutron star system Aquila X-1, and the white dwarf system SS Cyg. Here I present preliminary results from this work, largely confirming the current paradigm, but highlighting some intriguing new behavior, and suggesting a possible difference in the jet formation process between neutron star and black hole systems.

Keywords. accretion, accretion disks, black hole physics, stars: white dwarfs, stars: neutron, ISM: jets and outflows, radio continuum: stars, X-rays: binaries

---

1. Introduction

Jets are found in accreting systems throughout the visible Universe. For stellar-mass accretors, the evolution of such jets occurs on human timescales, and can be probed by resolved monitoring observations. X-ray binaries are one such class of objects, in which two distinct types of jets are observed, with a clear connection between the X-ray state of the source and the observed radio emission. From the flat radio spectra seen in the hard X-ray state, the presence of steady, compact, partially self-absorbed outflows is inferred, which have been directly resolved in two black hole (BH) systems (Dhawan *et al.* 2000; Stirling *et al.* 2001). Brighter, optically-thin, relativistically-moving jets are associated with high-luminosity, soft X-ray states during outbursts (Mirabel & Rodríguez 1994).

1.1. *Black holes*

Our current understanding of the duty cycles and disc-jet coupling in black hole X-ray binaries (BH XRBs) derives from a compilation of X-ray spectral and timing information, together with radio flux density monitoring and a limited set of high-resolution radio imaging. The current paradigm, or 'unified model' (Fender *et al.* 2004) suggests that the jet morphology and power correlate well with position in an X-ray hardness-intensity diagram (HID). Steady, self-absorbed jets are inferred to exist in the very low luminosity quiescent state and the higher-luminosity hard state. As the X-ray intensity increases in the hard state, so too does the jet power, with the radio and X-ray luminosities following a non-linear correlation, $L_{\mathrm{Radio}} \propto L_{\mathrm{X}}^{0.7}$. At about $L_{\mathrm{X}} \sim 0.1$–$0.3 L_{\mathrm{Edd}}$, the X-ray spectrum begins to soften, and the jet speed increases as the inner disc radius moves inwards. Below a certain X-ray hardness (the 'jet line'), the core jet switches off and internal shocks develop in the flow, which are observed as bright, relativistically-moving radio ejecta. The source may remain at high luminosity for several weeks, making repeated transitions back and forth across the jet line, before the X-ray luminosity eventually decreases to $\sim 0.02 L_{\mathrm{Edd}}$ where the spectrum hardens (note the hysteresis effect compared to the higher-luminosity hard-to-soft transition), the core jet is re-established, and the source fades back into quiescence.

1.2. *Neutron stars*

Of the many different classes of neutron star (NS) X-ray binaries, only the low-magnetic field systems have shown evidence for radio emission. These systems are divided by mass accretion rate into two main classes; the Z-sources and the atoll sources, each with distinct X-ray spectral and timing characteristics. The Z-sources are consistently accreting at or close to the Eddington luminosity, whereas the atolls are accreting at a somewhat lower level. The X-ray spectral and timing properties of atolls show many similarities to black hole systems, with distinct soft (so-called 'banana') and hard ('extreme island') X-ray states, making them the best sources to compare with black hole X-ray binary outbursts.

To date, only a handful of atoll sources have been detected in the radio band during simultaneous radio/X-ray observations, showing them to be systematically fainter than the black hole sources at the same Eddington-scaled X-ray luminosity (Migliari & Fender 2006). However, it appears that a similar correlation between radio and X-ray luminosities holds in the hard-state atoll sources, but with a lower normalization and a steeper power-law index; $L_{\rm R} \propto L_{\rm X}^{1.4}$. At high X-ray luminosities, close to the Eddington limit where the sources show Z-type behavior, bright transient ejecta are thought to exist, just as in black hole systems, although there appears to be only mild suppression of radio emission in the atoll sources when they reach a soft X-ray state.

1.3. *White dwarfs*

A third class of interacting binaries where mass is transferred from a donor to a degenerate compact object has a white dwarf as the accretor. One class of such systems, the dwarf novae (a type of Cataclysmic Variable; CV), also have accretion discs, which periodically develop disc instabilities, leading to a sudden increase in the accretion rate and causing short-lived outbursts. During such outbursts, which recur at intervals of several weeks, these systems brighten by several magnitudes in the optical band. Despite similarities to XRBs, with accretion onto a compact object and outbursts triggered by disc instabilities, no jets have thus far been directly resolved in CVs.

Generalizing the HID to a 'disc-fraction luminosity diagram' (DFLD) by plotting the optical flux of the system against the fraction of emission arising from the power-law spectral component (as opposed to disc emission) shows that outbursts of dwarf novae follow a very similar track to BH and NS systems. Extending this analogy with the Unified Model then suggests that they should show flat-spectrum radio emission in the rise phase of an outburst, and resolved ejecta during the subsequent spectral softening. The former prediction was spectacularly confirmed during an outburst of the dwarf nova SS Cyg (Körding *et al.* 2008). The radio emission was highly variable, peaking at 1.1 mJy, and coincident with the optical outburst. During the decay, the radio spectrum was slightly inverted, suggestive of a compact jet, as resolved in BH XRBs. However, the existence of such a jet can only be directly verified with high-resolution imaging.

2. The Jet Acceleration and Collimation Probe of Transient X-ray Binaries (JACPOT XRB)

2.1. *Aims and strategy*

The Jet Acceleration and Collimation Probe of Transient X-ray Binaries (JACPOT XRB) project† aims to probe the similarities and differences in the jet launching process between these three different source classes by conducting intensive monitoring of an outburst of each of these different classes of accreting compact object. Time-resolved, high angular resolution monitoring in the radio band using the VLBA and (E)VLA, plus WSRT and EVN when available, allows us to directly image the evolution of the jets over the course of an outburst. Simultaneous multi-wavelength coverage using *Swift* BAT, *RXTE* PCA, and *MAXI* in the X-ray band, plus optical and infrared monitoring using the two Faulkes Telescopes, FanCam and CTIO (via the SMARTS consortium), then enables us to tie the jet behavior to the corresponding changes in the accretion flow.

† http://www.astro.virginia.edu/xrb_jets/

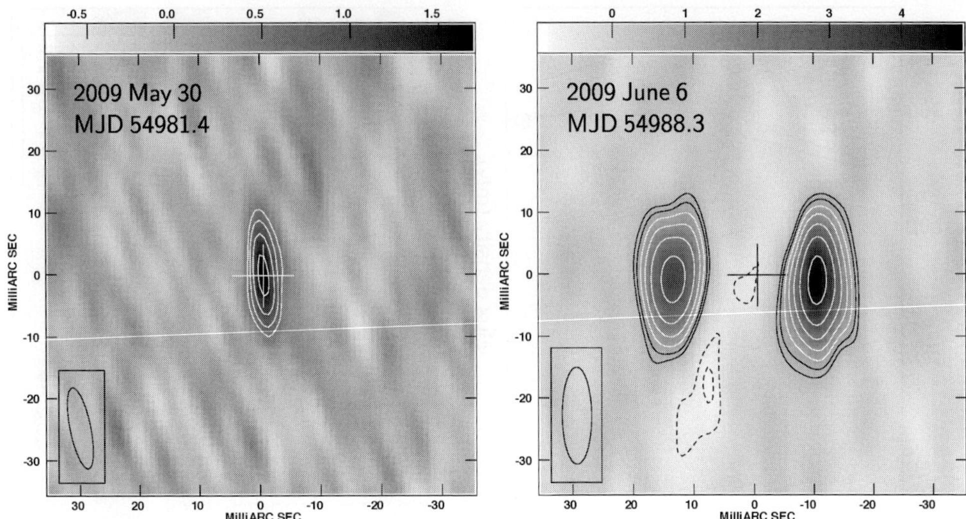

Figure 1. VLBA images of H1743-322 at 8.4 GHz. An unresolved source consistent with a compact, steady jet is seen in the early epochs, with the core quenching and giving rise to bright, optically-thin ejecta after the transition to the soft intermediate state. The grey scales denote flux density in mJy beam^{-1}, and the crosses mark the position of the central binary.

With such detailed, multi-wavelength monitoring, we aim to directly test the prevailing paradigm for black hole X-ray binary outbursts (Fender *et al.* 2004) and, by comparing the jet behavior across the three different source classes, to ascertain the role played in jet formation by the depth of the gravitational potential well and the presence or absence of a stellar surface and stellar magnetic field.

2.2. *Status*

One outburst of each class of accreting compact source has now been observed as part of the JACPOT XRB project, and data reduction and analysis are underway. Monitoring campaigns on one further black hole and one neutron star system have been approved by NRAO, and are awaiting triggering events.

2.3. *H 1743-322*

We triggered our first black hole observing campaign on the 2009 outburst of the black hole candidate source H 1743-322. The X-ray evolution of the outburst has already been analysed in detail by Motta *et al.* (2010) and Chen *et al.* (2010). We monitored the outburst with the VLA, ATCA and VLBA, tracking the evolution of the radio emission through the entire outburst. While the VLBA observations were hampered by the strong scattering and lack of good calibrators in the direction of the Galactic Center, we detected both the compact core during the initial rising hard state, and also the launching of ejecta immediately following the transition to the soft intermediate state (Fig. 1).

The disc-jet coupling in this system follows the standard pattern outlined by Fender *et al.* (2004), as shown in the hardness-intensity diagram (HID) in Fig. 2. We see flat-spectrum radio emission from a compact, unresolved core jet in the hard intermediate state (HIMS) at the beginning of the outburst. This core radio emission then quenches, giving rise to very faint, steep-spectrum emission as the source moves into the soft intermediate state (SIMS), following which we detect bright, optically-thin ejecta. The radio emission does not completely vanish during the high soft state (HSS), but shows a modest increase in flux density once the source returns to the HIMS, before fading with

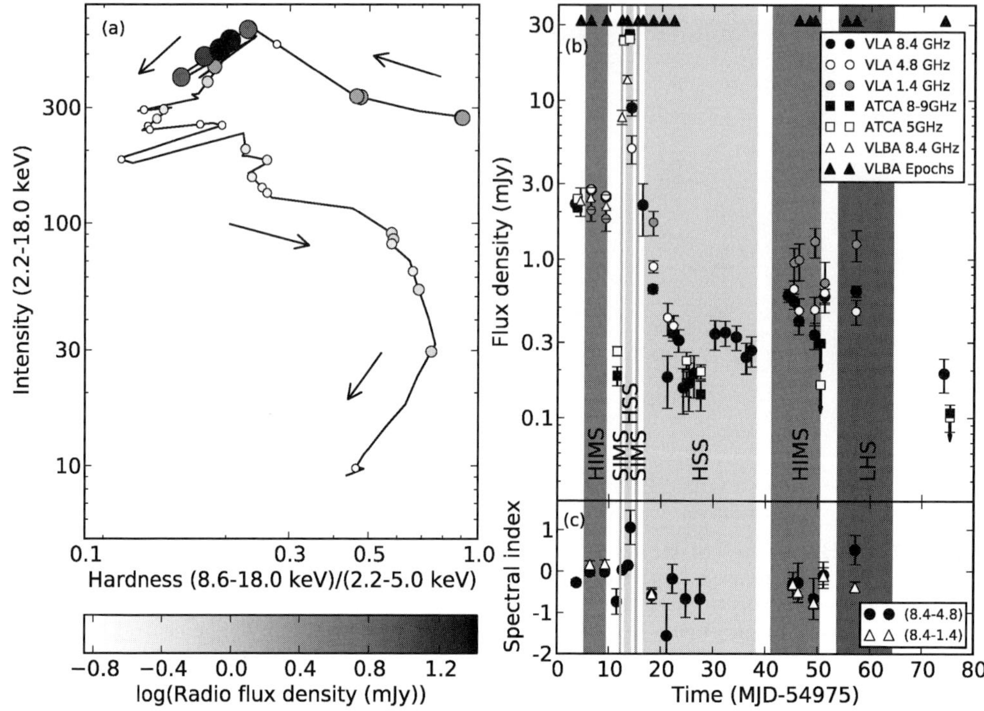

Figure 2. Disc-jet coupling during the 2009 outburst of H1743-322. (a) HID for the outburst, with circles indicating radio detections, whose size and color indicate the measured radio flux density. The X-ray state at the time of the radio observations has been interpolated from X-ray observations within 2 d. Arrows show how the source evolves with time. (b) Radio light curve, showing the state classifications of Motta *et al.* (2010): HIMS = Hard Intermediate State; SIMS = Soft Intermediate State; HSS = High Soft State; LHS = Low Hard State. (c) Radio spectral index as a function of time.

time during the decay phase in the low hard state (LHS). We draw particular attention to the pre-flare 'quench' phase seen during the hard to soft transition. Originally noted by Fender *et al.* (2004), this short-lived phase is often missed owing to limited temporal sampling in the radio band during outbursts of black hole X-ray binaries, but is surely key to understanding the nature of the transition from steady compact jets to bright, relativistically-moving ejecta. Over the reverse transition, the reactivation of the compact jet occurs at a hardness ratio of $0.26 < H < 0.58$, and appears to support the assertion of Fender *et al.* (2009) that the 'jet-line' is not vertical.

2.4. *Aql X-1*

The 2009 November outburst of Aql X-1 was observed using the VLA, VLBA, e-EVN and *RXTE* PCA, resulting in the first milliarcsecond-scale VLBI detection of the source. We obtained full spectral coverage of the outburst from the radio through infrared (FanCam, SMARTS), optical (Faulkes), ultraviolet (*Swift* UVOT) and X-ray bands. Modeling of the full spectral energy distribution and its evolution is underway.

The radio and X-ray lightcurves, plus the hardness-intensity diagram (HID) and color-color diagram (CCD) of the outburst are shown in Fig. 3. While full details of the observations and analysis may be found in Miller-Jones *et al.* (2010), we summarize the most salient points here. While hysteresis in the HID of Aql X-1 is well known (Maitra

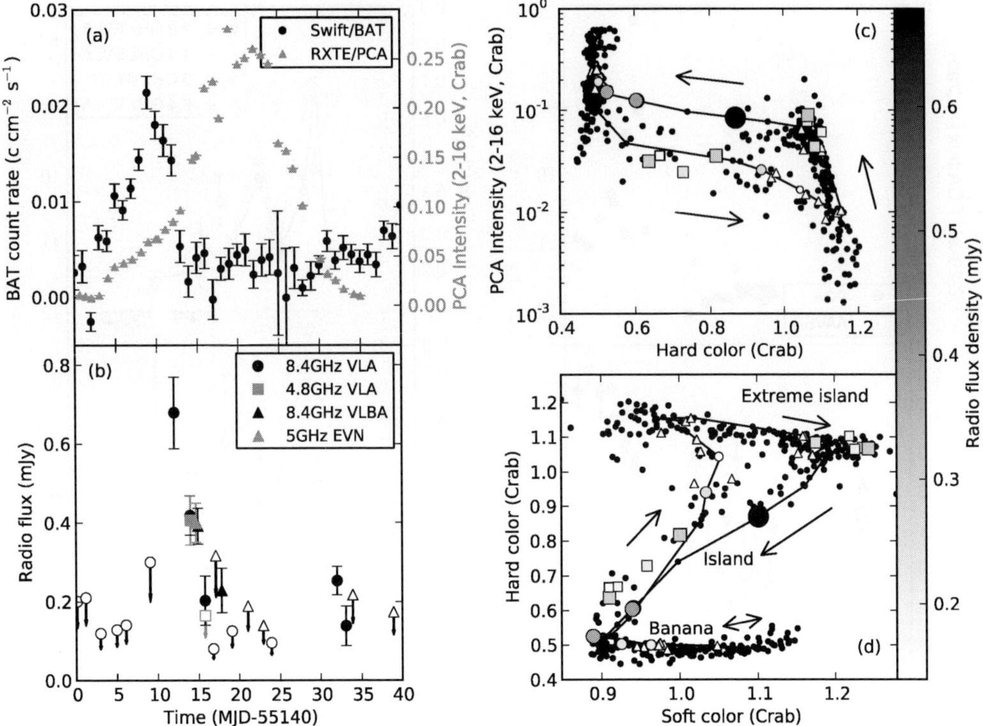

Figure 3. The 2009 November outburst of Aql X-1. (a) 15–50 keV *Swift* BAT and 2–16 keV *RXTE* PCA lightcurves. (b) Radio lightcurves from the VLA, VLBA and EVN. Open markers denote upper limits at the corresponding array and frequency. (c) HID showing the track of the 2009 November outburst, with dots representing all measured *RXTE* PCA points from previous outbursts. Filled circles represent the radio measurements during the 2009 outburst and filled squares denote archival radio measurements, with the size and color of these markers corresponding to the measured radio flux density. Hollow triangles denote non-detections. Arrows show how the source evolves with time. (d) CCD of the 2009 outburst, with the same symbols as in (c). Hard and soft colors are the count rate ratios (9.7–16.0 keV/6.0–9.7 keV) and (3.5–6.0 keV/2.0–3.5 keV) respectively, normalized by the Crab Nebula on a per-PCU basis.

& Bailyn 2004), there have been few previously-reported radio detections (Tudose *et al.* 2009). Our monitoring campaign provided the most complete radio coverage to date of an entire outburst of Aql X-1, finding the radio emission to be consistent with being activated by transitions from a hard spectral state to a soft state, and also by the reverse transition, just as seen in BH systems. A further similarity with black hole systems was the quenching of the radio emission above a certain X-ray luminosity ($\sim 10\%$ of the Eddington luminosity) while in a soft spectral state. However, in contrast to the BH systems, the VLBI observations combined with radio spectral information showed no evidence for steep-spectrum, optically-thin ejecta after the hard-to-soft state transition. The radio emission was at all times consistent with a compact, partially self-absorbed jet as seen in the hard states of BH systems. This may suggest a fundamental difference in the jet formation mechanism between the two classes of source.

2.5. *SS Cyg*

SS Cyg is the brightest dwarf nova, at a distance of only 166 pc (Harrison *et al.* 1999). It undergoes several outbursts each year, and in 2010 April, the AAVSO notified us that

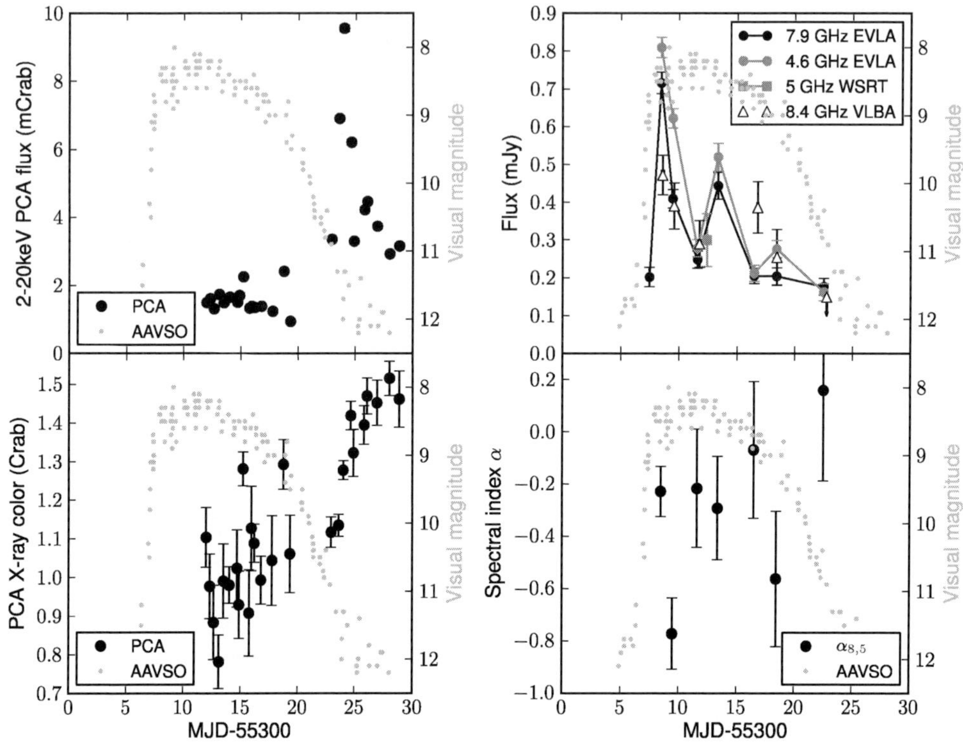

Figure 4. Light curves of SS Cyg during the 2010 April outburst. PCA X-ray color is the count rate ratio $(6{-}16\,\mathrm{keV})/(2{-}6\,\mathrm{keV})$, normalized by the Crab nebula on a per-PCU basis. Visual magnitudes from AAVSO are shown in light grey on each panel to mark the progress of the outburst. The outburst evolution follows the standard pattern in the optical and X-ray bands. EVLA radio detections were obtained at all epochs, and VLBA detections at the first 5 epochs.

a new outburst of the source was beginning. We were able to get on source with EVLA within 24 h, and followed up the ensuing detection with a full monitoring campaign using EVLA, WSRT, VLBA, *RXTE, Swift* and AAVSO. Unfortunately the source is too bright for *Swift* UVOT to provide coverage in the ultraviolet. The X-ray and radio light curves of the outburst are shown in Fig. 4. The outburst proceeded in a very similar fashion to that monitored by Wheatley *et al.* (2003), showing a sharp drop in the hard X-ray emission at the peak of the optical outburst, and a subsequent recovery as the optical emission faded. As seen by Körding *et al.* (2008), the radio emission rose rapidly at the beginning of the outburst, showing a steep radio spectrum ($\alpha = -0.77 \pm 0.14$, with $S_\nu \propto \nu^\alpha$) immediately after the peak, suggestive of optically-thin ejecta. During the decay phase later in the outburst, the observed flatter radio spectrum is more consistent with a partially self-absorbed compact jet.

The source was detected by the VLBA during 5 of the 6 epochs. These detections imply a brightness temperature in excess of 5×10^6 K. As argued by Körding *et al.* (2008), together with the radio spectral constraints this suggests the presence of an unresolved jet, as seen in accreting neutron star and black hole systems.

As an aside, the astrometric accuracy of the VLBA observations allowed us to fit for the source proper motion over the 10 days of observation. Assuming a distance of 166 pc (Harrison *et al.* 1999), we find a proper motion of $\mu_\alpha \cos\delta = 110 \pm 4\,\mathrm{mas\,y^{-1}}$ in R.A., and $\mu_\delta = 48 \pm 6\,\mathrm{mas\,y^{-1}}$ in Dec., fairly consistent with that given in UCAC3 (Zacharias *et al.* 2010), but significantly different from the proper motion derived by Harrison *et al.* (2000), with implications for the accuracy of the HST astrometric measurements. A future revision of the distance could remove the discrepancy between predictions of the disc instability model and the observed outburst luminosity (Schreiber & Lasota 2007).

3. Summary

We have embarked on an ambitious project to test the disc-jet coupling in three different classes of accreting stellar-mass compact objects, using high-resolution VLBI radio observations coupled with multiwavelength monitoring data. Using these data, we aim to (a) test the 'unified model' of Fender *et al.* (2004) for the disc-jet coupling in black hole X-ray binaries, and (b) to determine the role played in jet formation by the depth of the gravitational potential well, the stellar surface and the stellar magnetic field. We have already observed one outburst from each of a black hole, a neutron star and a white dwarf binary system, and while preliminary analysis of the data largely confirms the existing paradigm, some interesting similarities and differences between the evolution of black hole and neutron star outbursts have been found.

4. Acknowledgements

We are very grateful to the NRAO, *RXTE* and *Swift* schedulers for their flexibility and prompt responses which have made these observing campaigns feasible. We would also like to thank Mickael Coriat and Stephane Corbel for sharing their ATCA data on H1743–322, and Matthew Templeton and the worldwide network of AAVSO observers who triggered and supported our observing campaign on SS Cyg.

References

Chen, Y., Zhang, S., Torres, D., Wang, J., Li,. J., Li, T., & Qu, J. 2010, *A&A*, accepted (arXiv:1008.3059)

Dhawan, V., Mirabel, I. F., & Rodríguez, L. F. 2000, *ApJ*, 543, 373

Fender, R. P., Belloni, T. M., & Gallo, E. 2004, *MNRAS*, 355, 1105

Fender, R. P., Homan, J., & Belloni. T. M. 2009, *MNRAS*, 396, 1370

Harrison, T. E., McNamara, B. J., Szkody, P., McArthur, B. E., Benedict, G. F., Klemola, A. R., & Gilliland, R. L. 1999, *ApJ*, 515, L93

Harrison, T. E., McNamara, B. J., Szkody, P., & Gilliland, R. L. 2000, *AJ*, 120, 2649

Körding, E., Rupen, M. P., Knigge, C., Fender, R. P., Dhawan, V., Templeton, M., & Muxlow, T. W. B. 2008, *Science*, 320, 1318

Maitra, D. & Bailyn, C. D. 2004, *ApJ*, 608, 444

Migliari, S. & Fender, R. P. 2006, *MNRAS*, 366, 79

Miller-Jones, J. C. A., Sivakoff, G. R., Altamirano, D., Tudose, V., Migliari S., Dhawan, V., Fender, R. P., Garret, M. A., Heinz, S., Körding, E., G., Krimm, H. A., Linares, M., Maitra, D., Markoff, S., Paragi, Z., Remillard, R. A., Rupen, M. P., Rushton, A., Russell, D. M., Sarazin, C. L., & Spencer, R. E. 2010, *ApJ*, 716, L109

Mirabel, I. F. & Rodríguez, L. F. 1994, *Nature*, 371, 46

Motta, S., Muñoz-Darias, T., & Belloni, T. 2010, *MNRAS*, accepted (arXiv:1006.4773)

Schreiber, M. R. & Lasota, J.-P. 2007, *A&A*, 473, 897

Stirling, A. M., Spencer, R. E., de la Force, C. J., Garrett, M. A., Fender, R. P., & Ogley, R. N. 2001, *MNRAS*, 327, 1273

Tudose, V., Fender, R. P., Linares, M., Maitra, D., & van der Klis, M. 2009, *MNRAS*, 400, 2111

Wheatley, P. J., Mauche, C. W., & Mattei, J. A. 2003, *MNRAS*, 345, 49

Zacharias, N., *et al.* 2010, *AJ*, 139, 2184

Discussion

MIRABEL: Could you comment on your programme to determine proper motions and prove models on the formation of compact objects?

MILLER-JONES: We can only perfor accurate astrometry in the hard and quiscent spectral states. We can measure proper motions within a single outburst for the highest velocity systems. For other, slower jet systems, we need a larger time baseline, which requieres multiple outbursts. This can be done for all 3 systems presented here (H1743-322, AQL X-1, and SS CYG), since they all have repeated outbursts, separate by a few months to years. We also have an approved project to perform triggered astrometric observations of a larger sample of black hole systems to build up a larger sample of proper motions.

KAWAI: How fast is the response time of the radio observations triggered by detections of optical or X-ray transients?

MILLER-JONES: In some cases (such as campaign on SS CYG), we were able to get on source within 24 hours of triggering. In other cases, we have had to wait up to one week before the radio observations. It depends on how full the schedule is at our desired LST. NRAO is moving to a dynamical scheduling system which gives the more flexibility in scheduling and helps reduce response times following transient alerts.

Jets at all Scales
Proceedings IAU Symposium No. 275, 2011
G. E. Romero, R. A. Sunyaev & T. Belloni, eds.

© International Astronomical Union 2011
doi:10.1017/S174392131001608X

Jets from neutron star X-ray binaries: towards a unified scheme

Simone Migliari

European Space Agency/European Space Astronomy Center, Apartado/P.O. Box 78,
Villanueva de la Cañada, E-28691 Madrid, Spain, email: `smigliari@sciops.esa.int`

Abstract. Systematic multi-wavelength studies of neutron stars (NSs) have shown a jet and disk-jet coupling phenomenology which resembles, although with some important differences, that observed in black holes; ultra-relativistic transient ejection, steady compact jets, accretion-ejection cycles are indeed observed in NSs. I will review our observational knowledge of jet in NS X-ray binaries, focusing on the role of the parameters of the system which might be involved in the production of jets. First, I will discuss the role of the accretion rate, presenting a *unified scheme* for accretion-jet production throughout the different sub-classes of low-magnetic field NSs. Then, I will attempt to (make the first steps to) quantify the role of spin and magnetic field in powering the jet.

Keywords. stars: neutron, X-rays: binaries, radio continuum: general, ISM: jets and outflows.

1. Introduction

Jets are spectacular and powerful phenomena observed in a number of astrophysical objects, from non-relativistic to relativistic systems, such as Young Stellar Objects (YSOs), cataclismic variables (CV), X-ray binaries (XRBs) and active galactic nuclei (AGN; and are supposed to be at the origin of the powerful gamma-ray bursts). In the search for the elements and parameters contributing to the formation of jets, key is the cross comparison of the different jet systems. The accretion disk is a common denominator for all these systems and seems to be essential to the formation of their jets. However, differences in the jet behaviour among these systems, indicate that also parameters proper of the accreting star might play a role in the jet production.

1.1. *The sub-calsses of neutron star X-ray binaries*

Low-magnetic field NS XRBs are commonly divided into two broad classes, based on their X-ray spectral and timing properties, and named after the shape of their color-color diagram (CD) and hardness-intensity diagram (HID): atolls and Z (Fig. 1, left). Atoll sources have luminosities between 0.01 and 0.5 Eddington, while Z sources are persistently above 0.5 Eddington and sometimes even at super-Eddington luminosities. In their CD track, atolls show 'branches' (X-ray states) where the soft disk emission dominates more or less the spectrum over a non-thermal high-energy emission: a hard branch called island state, and a soft branch called banana state. The Z track typically show three branches, from softer to harder: Flaring Branch (FB), Normal Branch (NB) and Horizontal Branch (HB). Based on the orientation and presence of these branches, both atolls and Z sources can be further divided into sub-classes. In the following, I will divide the NSs into five sub-classes: Atoll-Hard (mostly island), Atoll-Soft (mostly banana), Atoll-GX (persistently bright, mostly banana), Z-Sco (or Sco-like) and Z-Cyg (or Cyg-like). It is still a matter of debate if variations of the mass accretion rate alone can explain the variety of NS sub-classes and the different branches within a sub-class.

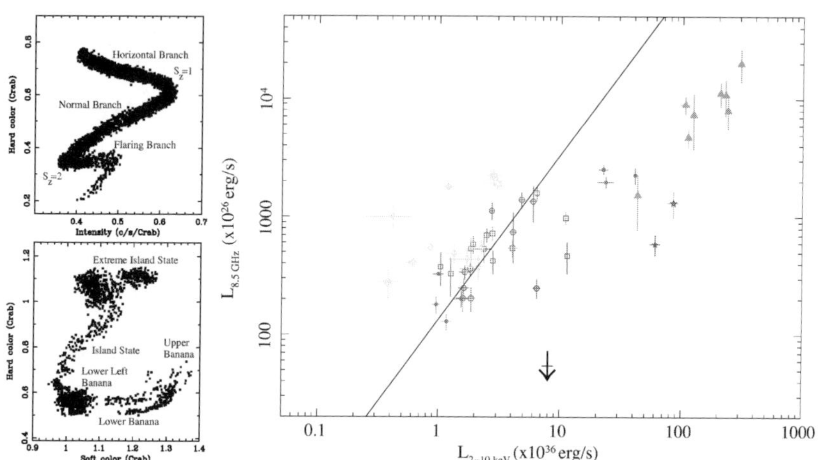

Figure 1. Left: typical CD of an Atoll and HID of a Z-type NS (van der Klis 2006). Right: Updated X-ray/radio luminosity plots of NSs: in red atolls in hard state and state transition; in blue atolls in soft state; in green Z sources; in yellow AMXPs. The solid line fits 4U 1728-34 with a power law with $\Gamma = 1.4$; The arrow shows the radio 3σ upper limit of GX 9+9, showing the quenching at $\sim 10\%$ Eddington.

1.2. *XTE J1701-462: the Rosetta Stone*

XTE J1701-462 is the first source showing the transition between Z-types and atoll-type, during an X-ray outburst, going through all the sub-classes and branches withing a few months (Homan *et al.* 2010 and references therein). Homan *et al.* (2010) showed that the average mass accretion rate (using the 2-2.9 keV count rate as a proxy) seems to be responsable for the different sub-classes, i.e. in order of increasing accretion rate: atoll-hard, atoll-soft, atoll-GX, Z-Sco and Z-Cyg. Nevertheless, it seems that variations in the accretion rate can trace the evolution of the CD thoughout all the NS sub-classes, the average mass accretion rate alone is not able to account for variations between the different branches *within* a sub-class (especially in Z sources), and at least an additional parameter seems to be needed.

2. The sample

The sample presented here consists of all the NS XRBs with i) a detected radio counterpart, ii) a quasi-simultaneous X-ray/radio coverage and iii) an estimated distance (e.g. Nelemans & Jonker 2004 for uncertainties). I report here only the radio detections (with the exception of GX9+9; see below). For the radio characteristics of most of the sources in the sample we refer to Migliari & Fender (2006) and references therein. In the following sections, I report new sources, updates and notes on key observations.

2.1. *Radio emission from Atolls*

4U 0614+091: The four radio detections of the source are reported in Migliari*et al.*(2010). The source was steady in its hard state and emission is consistent with being optically-thick synchrotron coming from a (slightly variable) compact jet.

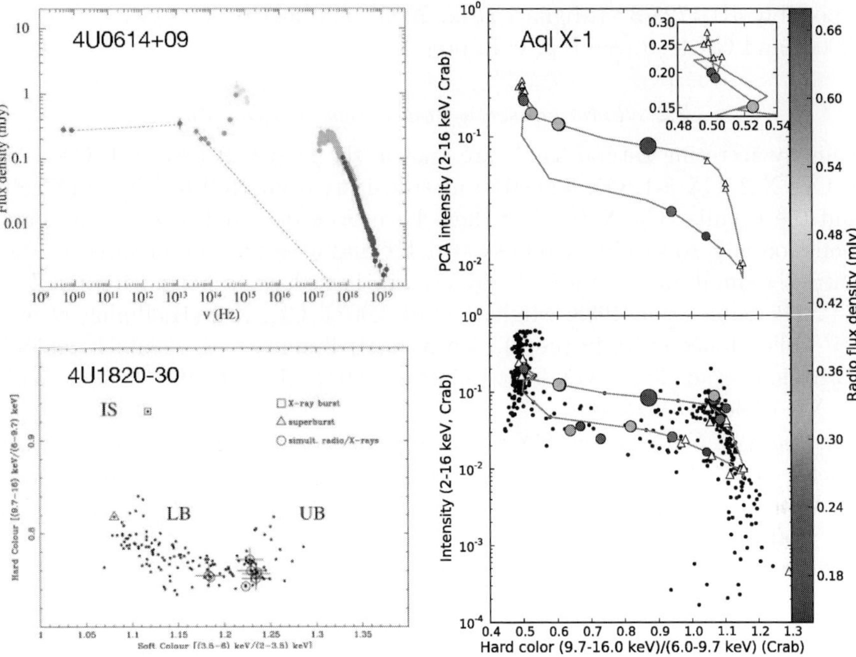

Figure 2. Radio detections of atolls. Upper left: broadband spectrum of 4U 0614+101 when steadily in its hard state (Migliari *et al.* 2010). Lower left: CD of 4U 1820-30 showing with circles the radio detection when steadily in its soft state (Migliari *et al.* 2004). Right: HID of AqlX-1 showing the evolution of the radio emission during X-ray outbursts. The top panel shows the 2009 outburst, while the lower panel all the available observations (Miller-Jones *et al.* 2010).

Aql X-1: Miller-Jones *et al.* (2010) report on daily VLA and VLBA observations of Aql X-1 during the 2009 outburst, the best radio coverage of a NS XRB outburst to date†. The source has been detected 14 times in the radio band (see also Tudose *et al.* 2009). The strongest radio flux density is observed during the X-ray state transition. Also a radio quenching during the soft state and after the state transition is observed, similar to that found in BHs. The radio emission is always optically-thick along the whole X-ray outburst, indicating that it arises from a compact jet.

4U 1820-30 and Ser X-1: Migliari *et al.* (2004) reported radio emission when they were steady in their soft state (banana). The radio luminosity is comparable to that observed in the hard state from other sources. This is different from what has been observed so far in BHs, whose radio emission is at least an order of magnitude lower than that in hard state. The radio spectrum was consistent with being optically thick, indicating a compact jet.

MXB 1730-335 (the Rapid Burster): The source has been detected in its soft state, near the peak of two X-ray outbursts (Rutledge *et al.* 1998; Moore *et al.* 2000). Contrary to what is expected from BHs, the radio spectra of the two dual-band observations are consistent with a compact jet.

Quenching: in atolls, radio emission is observed to increase with X-ray luminosity and then quench at a few to ten per cent the Eddington limit: Aql X-1 (Miller-Jones *et al.*

† JACPOT collaboration: http:www.astro.virginia.edu/xrb_jets

2010), possibly 4U 1728-34 (Migliari *et al.* 2003), and atoll-GX as GX 3+1 (Berendsen *et al.* 2000) and GX 9+9 (see Fig. 1; in prep.).

2.2. *Radio emission from Z-type neutron stars*

Eight† highly-accreting Z-type XRBs are known: six persistent (Sco X-1, GX 17+2, GX 349+2, Cyg X-2, GX 5-1, GX 340+0), one transiting from atoll to Z type (XTE J1701-462) and the peculiar Cir X-1: all of them have been detected in the radio band. The radio emission in Z souces in extremely variable and it seems to be related to the X-ray state. Strictly simultaneous radio/X-ray studies have been reported for six Z sources: GX 17+2 (Penninx *et al.* 1988; Migliari *et al.* 2007), Cyg X-2 (Hjellming *et al.* 1990a; Migliari, Miller-Jones *et al.* in prep.), Sco X-1 (Hjellming *et al.* 1990b; Fomalont *et al.* 2001; Bradshaw *et al.* 2003), GX 5-1 (Tan *et al.* 1992), 4U 1701-462 (Fender *et al.* 2007) and Cir X-1 (Soleri *et al.* 2009).

Z-Sco: Cyg X-2, Sco X-1 and GX 17+2: Simultaneous X-ray radio observations show a link between the X-ray branch and the intensity of the radio emission for the three sources. Within each source, the radio flux density increases from the HB to the NB and it is weakest in the FB (see a sketch inset in Fig. 3). Radio brightenings have been observed during state transitions between HB and NB (Sco X-1) and between FB to NB (GX 17+2). Both optically-thick and optically-thin radio spectra have been observed. Therefore, Z sources, contrary to atolls, can launch transient jets.

Z-Cyg: Cyg X-2 and GX 5-1: The radio coverage for these two sources is sparse and only part of the HID track is observed simultaneously with X-rays. GX 5-1 shows radio emission in the HB and an enhanced flux density when the source transits to the NB (Tan *et al.* 1992). The increase in radio emission is possibly associated to the HB-NB transition. Cyg X-2 shows stronger radio emission at the FB-NB apex, while the jet seems to be quenched when it is steadily in the NB (in prep.). No dual-band observations are available to investigate the nature of the jet in Cyg X-2. The radio spectra in GX 5-1 show both optically thick and optically thin synchrotron emission.

3. A Unified Scheme: jet power vs accretion rate

In Fig. 3, I present with a unified scheme, the puzzled radio behaviour of low-magnetic field NSs observed up to now throughout the different sub-classes, i.e. as a function of the mass accretion rate. The results can be summarized as follows.

- Radio emission has been detected from both atolls and Z sources.
- On average, the radio flux density detected increases with accretion rate, from atolls to Z sources.
- Radio emission has been detected in *all branches* of atolls and Z sources.
- Atolls show only optically-thick synchrotron radio emission, consistent with a compact jet, even during X-ray state transitions and outbursts.
- At a few per cent of the Eddington limit, the compact jet radio emission seems to be quenched. (The lowest radio upper limit comes from the 'atoll-GX' GX 9+9 at approximately 10% Eddington.)
- Z sources show optically-thick and optically-thin synchrotron radio emission, evidence for the presence of a powerful compact jet and transient jets.
- In both atolls and Z sources, there is evidence for an association between X-ray state (branch)transitions and enhanced jet activity.

† Nine if we also count GX 13+1, which is an hybrid atoll-Z source

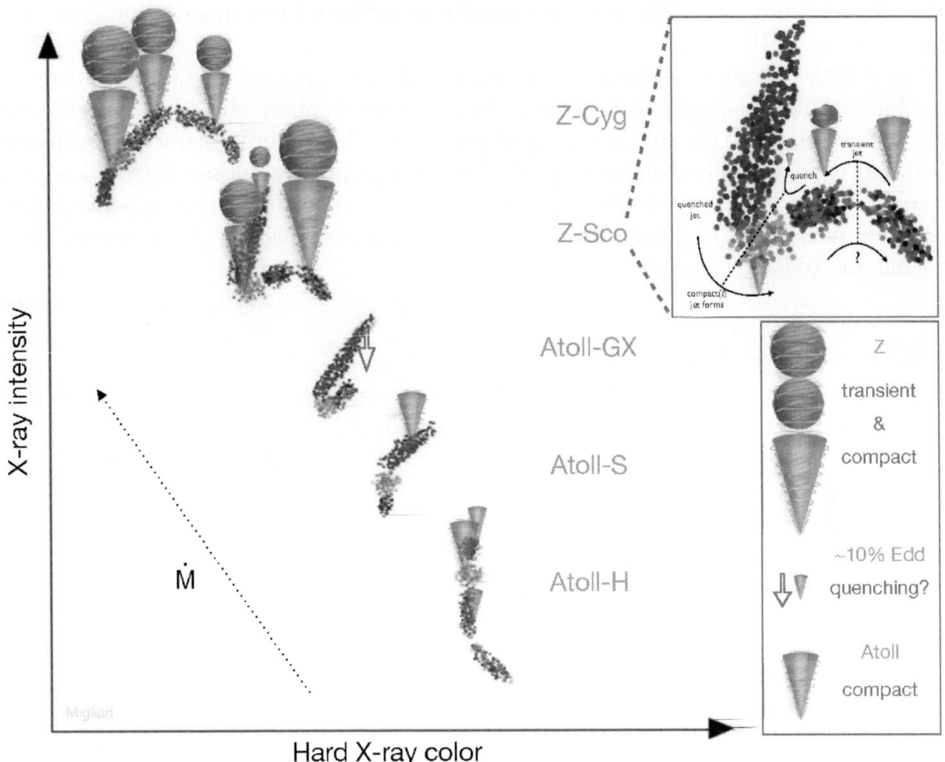

Figure 3. A unified scheme for low-magnetic field NSs: sub-classes and radio emission as a function of the accretion rate. We show the typical HID track of five main sub-classes. In blue we represent the FB and the banana states and in orange the island-banana apex which evolves into the FB-NB apex. The red cones represent the observed compact jet emission: the brighter the jet, the bigger the cone. Transient jets are shown as detached blobs.

4. Neutron star magnetic field vs jet power

Contrary to BHs, NSs in XRB systems show the presence of strong surface magnetic fields. The magnetic field can be directly measured only by observing cyclotron emission lines in the X-ray spectrum. Cyclotron lines have been observed in 15 X-ray binary pulsars, giving a magnetic field near the poles in the range $10^{11} - 10^{13}$ G.

4.1. *Radio upper limits from high-magnetic field neutrons*

We made a systematic analysis of radio observations of high-magnetic field neutron star X-ray binaries with estimated stellar magnetic field and known distance (Fig. 4, right; Migliari, Miller-Jones *et al.*, in prep.). The figure includes new radio observations of Her X-1, 4U 0115+634, 4U 1907+09 and A 0535+26, which give the most stringent radio upper limits of the class to date. Only one high-magnetic field NS have been detected in the radio band (4U 1223-624; Pestalozzi *et al.* 2009). However, the radio flux density is consistent with emission from the wind of the companion star which should dominate the radio spectrum. Furthermore, the reported variability between orbital phases which Pestalozzi *et al.* (2009) speculate it is coming from a transient jet, show large uncertainties, i.e. less than 3σ significance, and therefore must be confirmed. Tudose *et al.* (2010) reported the only constraining X-ray radio upper limits to date for A 0535+26 during the X-ray outburst of July-August 2010: our WSRT observations show upper limits an order of magnitude below the radio luminosity of atolls in hard state at the same X-ray luminosity (at $\sim 10^{36}$ erg/s).

Does a high-magnetic field quench the jet production? The picture is still blurred. No detections of high-magnetic field X-ray pulsars have been observed so far. However, if we take into account also the soft X-ray luminosity, the radio upper limits are still consistent with what expected in low-magnetic field NSs (X-ray/radio luminosity correlation). Only one observation is off, i.e. A 0535+26. Therefore, although hints that a high magnetic field may quench the jet production are growing, further investigation is needed before drawing any reliable conclusion.

5. Spin vs jet power

In NSs, the spin might affect the jet production through the interaction of the accretion disk and the magnetic field anchored to the NS poles.

5.1. *Measuring the spin of the neutron star*

Unlike in BHs, the spin in NS XRBs can be directly observed by non-thermal emitting particles accelerated by the polar magnetic field of the rotating NS. This is the case for high-magnetic field XRB pulsars and accretins millisec X-ray pulsars (AMXPs), where a coherent pulsation is directly detected. In low-magnetic field NS XRBs, the spin period can be observed indirectly as 'burst oscillation', i.e. quasi-coherent oscillations during type-I X-ray bursts. In the three sources where both direct pulsation and burst-oscillation have been observed, the burst oscillation was consistent with the spin frequency ν_s (Chakrabarty *et al.* 2003; Strohmayer *et al.* 2003; Altamirano *et al.* 2008). Twin kHz quasi periodic oscillations (QPOs) in the X-ray power spectra have been observed for a number of NS XRBs. The difference between the peak frequencies ν_{kHz} of the two QPOs seems to be in some relation with the spin frequency (either consistent with or half ν_s deending on the source; e.g. van der Klis 2006). Below, for those sources where a direct measurement of the pulsation of the NS is not observed, we choose to equal the spin to the average burst oscillation frequency or the highest $\Delta\nu_{kHz}$ measured.

5.2. *Estimating the jet power*

Following Fender *et al.* (2010), I estimated the (ranking) jet power of the sample, using as a proxy the normalization of the power law fitting the sources in the X-ray/radio luminosity plane (Fig. 1, right). Only two atoll sources, 4U 1728-34 and Aql X-1, span an order of magnitude in X-ray luminosity and show a correlation in the radio/X-ray luminosity plane. The slope is different for the two sources: $\Gamma \sim 1.4$ and $\Gamma \sim 0.6$, respectively for 4U 1728-34 and Aql X-1 (Migliari *et al.* 2003; Tudose *et al.* 2009; Miller-Jones *et al.* 2010). I chose to treat the average NS radio emission as dominated by a compact jet. I fitted each source in the X-ray/radio luminosity plane with a power law, fixing the slope first to 1.4 and then to 0.6, and plot both the normalizations against the spin frequencies.

5.3. *Spin-jet power correlation?*

The rank in jet power L_{Jet} among the NSs, estimated as the rank in the normalization of the law $L_R \propto L_X^{1.4}$ (i.e. fit to 4U 1728-34 and expected theoretically for radiatively efficient accretion when $\dot{M} \propto L_{Jet}$; see also Körding *et al.* 2007), show an apparent correlation to the spin: the non-parametric Spearman rank statistics gives a probability of 3e-2 against the null hypothesis of an uncorrelated data set (Fig. 4). However, given the outlier (4U 0929-314 with $\nu_s = 185$ Hz) and the uncertainties on the measurements

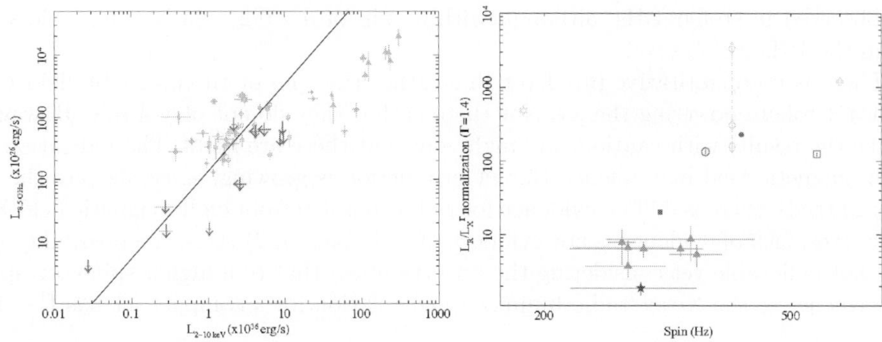

Figure 4. Left: Upper limits of high magnetic field X-ray pulsars (purple) superimposed to the other NSs (grey). Right: Spin vs X-ray/radio luminosity Normalization of a fitting power law with $\Gamma = 1.4$.

discussed above, a much better statistics is necessary to confirm the relation. If we plot the spin frequency against the normalization fixing the law $L_R \propto L_X^{0.6}$ (fit to Aql X-1), there is no obvious correlation.

A lack of evidence of a relation between spin and jet power, suggests that the main driver of wind/jet is not strictly linked to the accreting object. This would point towards the extended-disk model as the most likely mechanism (Blandford & Payne 1982). The same conclusion for YSOs has been derived by Anderson *et al.* (2003) and Ferreira *et al.* (2006), measuring, at least for the outer portion of the jet, a large launching disk radius for a sample of T-Tauri stars. They also discuss the existence of a coaxial, faster flow inside the outer jet, possibly powered by one of the other two mechanisms (X-wind or stellar-wind). Note that also in NS XRBs, a faster, relativistic jet superimposed to a slower radio ejection has been observed in two NS systems (Sco X-1 and Cir X-1; Fomalont *et al.* 2001; Fender *et al.* 2004) and possibly also in SS 433 (where the nature of compact object is still unknown; Migliari *et al.* 2005). In these cases, the jet may tap power from the angular momentum of compact object itself, and the spin and stellar magnetic field may in fact be crucial elements.

6. Models for jet production in NS XRBs, what are we looking for?

Based on the results described in this work, I herewith list some results that I think a model for jet formation in NSs should address:

- when liberated, the average jet power correlates with the average accretion rate.
- within a sub-class, the observed jet power depends on (at least) a second parameter, which regulate the channels through which the power is liberated (jet, wind,...) and is related to X-ray state transitions (e.g. instantaneous accretion rate at a certain radius? disk instabilities?).
- transient, optically-thin synchrotron radio emission is observed only in Z sources, i.e. at a high accretion rate, above a few tens per cent of the Eddington luminosity limit.
- faster, underlying jet emission, superimposed to a slower jet, is observed in at least two highly-accreting NS sources. Is there a two-modes jet production? are two processes for jet formation coexisting as suggested for YSOs?
- *if* the process of transient/faster jet is powerded by the spin in both NSs and BHs, then the NS must find a way to produce a jet which is comparable or even faster than

that observed in stellar BHs, although with a spin of $a <0.2$, which is likely less than that in the BHs we observe.

- There is no quantitative proof yet that either the spin or the magnetic field play a significant role in powering the jet. Nor there is that they do not play a role. However, if we take the results with caution, we might say that there are hints. The indication that a high magnetic field may quench the jet production is growing, since, despite the (still ongoing) trials, there is still no evidence for radio emission from high-magnetic field X-ray pulsars (yes, lack of evidence is not evidence of lack, but...); 2) there is a possibility (very weak, not believable yet considering the uncertainties) that to a higher spin corresponds a stronger jet, if the X-ray/radio luminosity correlation for a compact jet has $\Gamma \sim 1.4$.

Acknowledgements

I am thankful to James Miller-Jones for the many discussions and for comments on a draft of this work.

References

Anderson, J. M., Li, Z.-Y., Krasnopolsky, R., & Blandford, R. D. 2003, *ApJL*, 590, L107

Blandford, R. D., & Payne, D. G. 1982, *MNRAS*, 199, 883

Bradshaw, C. F., Geldzahler, B. J., & Fomalont, E. B. 2003, *ApJ*, 592, 486

Chakrabarty, D.*et al.* 2003, *Nature*, 424, 42

Fender, R. *et al.* 2004, *Nature*, 427, 222

Fender, R. *et al.* 2007, *MNRAS*, 380, L25

Fender, R. P., Gallo, E., & Russell, D. 2010, *MNRAS*, 406, 1425

Fender, R. 2010, Lecture Notes in Physics, Berlin Springer Verlag, 794, 115

Ferreira, J., Dougados, C., & Cabrit, S. 2006, *A&A*, 453, 785

Fomalont, E. B., Geldzahler, B. J., & Bradshaw, C. F. 2001, *ApJ*, 558, 283

Hjellming, R. M., Han, X. H., Cordova, F. A., & Hasinger, G. 1990, *A&A*, 235, 147

Hjellming, R. M., *et al.* 1990, *ApJ*, 365, 681

Homan, J., *et al.* 2010, *ApJ*, 719, 201

Körding, E. G. *et al.* 2006, *MNRAS*, 369, 1451

Migliari, S., & Fender, R. P. 2006, *MNRAS*, 366, 79

Migliari, S. *et al.* 2003, *MNRAS*, 342, L67

Migliari, S. *et al.* 2004, *MNRAS*, 351, 186

Migliari, S., *et al.* 2007, *ApJ*, 671, 706

Migliari, S., *et al.* 2010, *ApJ*, 710, 117

Miller-Jones, J. C. A., *et al.* 2010, *ApJL*, 716, L109

Penninx, W. *et al.* 1988, *Nature*, 336, 146

Pestalozzi, M. *et al.* 2009, *A&A*, 506, L21

Psaltis, D., & Chakrabarty, D. 1999, *ApJ*, 521, 332

Rutledge, R., Moore, C., Fox, D., & Lewin, W. 1998, ATel, 8, 1

Shu, F., Najita, J., Ostriker, E., Wilkin, F., Ruden, S., & Lizano, S. 1994, *ApJ*, 429, 781

Soleri, P., Tudose, V., Fender, R., van der Klis, M., & Jonker, P. G. 2009, *MNRAS*, 399, 453

Tan, J. *et al.* 1992, *ApJ*, 385, 314

Tudose, V., Migliari, S., Miller-Jones, J. C. A. *et al.* 2010, ATel, 2798, 1

Tudose, V., Fender, R. P., Linares, M., Maitra, D., & van der Klis, M. 2009, *MNRAS*, 400, 2111

van der Klis, M. 2006, Compact stellar X-ray sources, 39

Discussion

DAVID MEIER: Marina Romanova has simulated the magnetic propeller mechanism which produces jets from the magnetic neutron star's rotation. This too gives a jet power $P_J \propto \dot{M}^\beta$, where $\beta \sim 1$. So, just because $P_J \propto \dot{M}$ does not mean that the jet cannot be pulsar driven.

MIGLIARI: Absolutely correct. As a note, I must add that the slope of the X-ray/radio luminosity correlation from which a $P_J \propto \dot{M}$ can be inferred assuming a radiatively efficient disk emission (i.e. $L_X \propto \dot{M}$), is not certain yet. One source, 4U1728-34 shows a relation $L_R \propto L_X^\Gamma$ with $\Gamma \sim 1.4$, while Aql X-1 shows $\Gamma \sim 0.6$.

ELISABETE M. DE GOUVEIA DAL PINO: My question is also related to Romanova and collaborators' propeller models for magneto-rotational launching. You said that you looked for very high (B$> 10^{12}$ G) NSs in order to see a possible evidence for a jet quenching, but actually normal NSs can have even weaker magnetic fields and B can actually help launching. Why is that so? were you looking for magnetar-like NSs?

MIGLIARI: When we started searching systematically for a possible radio synchrotron emission from high-magnetic field X-ray pulsars, we were looking for any possible effect of the magnetic field on jet emission, either quenching or active (as in the case of the propeller effect). We did not find any convincing quantitative proof for either the effect, yet. The *indication* for a quenching effect comes from the fact that you just mentioned: jets have been detected in weak magnetic field NSs, but not in high magnetic field NSs. However, again, there is no strong quantitative proof for a quenching effect, if we take into account the X-ray luminosity of the observed sources. We did not look for magnetars yet, which are even more extreme in magnetic field strength.

FENG YUAN: Is the transient jet detected in AqlX-1 associated with X-ray state transition?

MIGLIARI: yes, with transitions from hard to soft and from soft to hard states. In both directions.

CHRISTIAN FENDT: What is the role of the stellar magnetic field inclination? if the inclination is too high, jet formation will be disturbed/suppressed. Jet formation requires a high degree of axisymmetry.

MIGLIARI: Yes, I agree. Unfortunately, in our sample of high-magnetic field NSs we do not have information on the magnetic axis inclination, except for some possible constrain/selection bias coming from the fact that we selected the sources where we observe pulsations and cyclotron lines (the only way to measure directly the magnetic field), both coming from regions near the magnetic poles.

Jets at all Scales
Proceedings IAU Symposium No. 275, 2011
G. E. Romero, R. A. Sunyaev & T. Belloni, eds.

© International Astronomical Union 2011
doi:10.1017/S1743921310016091

Suzaku studies of microquasars

Yoshihiro Ueda and Megumi Shidatsu

Department of Astronomy, Kyoto University,
Kyoto 606-8502, Japan
email: `ueda@kusastro.kyoto-u.ac.jp`

Abstract. We report our recent results from multiwavelengths studies of microquasars, focusing on X-ray data of GX 339–4 and GRS 1915+105 obtained with *Suzaku* and other observatories. The broad band coverage and high energy resolution achieved with *Suzaku* (or a combination of *Chandra*/HETGS and *RXTE*) enable us to perform the most reliable spectral analysis both on iron-K features and continuum, and thus to best constrain the accretion disk structure of microquasars and its relation to the jet formation at various mass accretion rates.

Keywords. accretion, accretion disks — black hole physics — X-rays: individual (GX 339–4, GRS 1915+105)

1. Suzaku results on GX 339–4 in the low/hard State

1.1. *Introduction*

GX 339–4 is a transient Galactic black hole binary (BHB) discovered in the early 1970s. Though being one of the best studied canonical BHBs, the accretion flow geometry as a function of mass accretion rate is not firmly established, and is still a subject of intensive investigation. In the very high state, Miller *et al.* (2008) reported the detection of an extremely broad iron-K emission line from *Suzaku* spectra implying a rapid spin of the black hole. This was questioned by Yamada *et al.* (2009), however, who showed that the same data are consistent with a non-rotating black hole, rather supporting the earlier result by Makishima *et al.* (1986) in the high/soft state. Miller *et al.* (2006) suggested that the standard disk of GX 339–4 extends to the innermost stable circular orbit (ISCO) in bright phases of the low/hard state, based on the detection of a relativistically broadened iron-K line, whereas the Tomsick *et al.* (2009) result using *Suzaku* data at a fainter flux level indicates that the disk is truncated at a much larger radius than the ISCO.

To reveal the inner disk structure and the origin of the continuum emission in the low/hard state, we observed GX 339–4 using the X-ray satellite *Suzaku* in 2009 March, when it became active and stayed in the low/hard state. Quasi-simultaneous near-infrared observations were performed with the *Infrared Survey Facility* (*IRSF*) 1.4 m telescope to reveal the jet activity. We refer the readers to Shidatsu *et al.* (2010) for details.

1.2. *X-Ray Spectra*

With *Suzaku*, we have obtained the best quality simultaneous, broad band X-ray spectra ever obtained from GX 339–4 in the low/hard state, covering the 0.5–310 keV band. The overall continuum in the 2–100 keV band can be roughly approximated by a power law of a photon index of ≈ 1.5, above which a sharp decline consistent with an exponential cutoff is present (see Figure 1, middle panel). This strongly suggests a thermal origin for the continuum, although small contribution from non-thermal components is not completely ruled out. Following the previous studies of BHBs in the low/hard state with *Suzaku*, we adopt models consisting of emission from a standard accretion disk and its thermal Comptonization by a hot corona. The disk emission is represented by the multi-color disk (MCD) model (Mitsuda *et al.* 1984). For the Comptonization component, we employ the `compPS` model in XSPEC (Poutanen & Svensson 1996), which is able to take into account a reflection component from the disk.

We find that the broad band X-ray spectra is dominated by the thermal Comptonization component with an electron temperature of $\approx$170 keV. The contribution from the direct disk component is very small, indicating that the inner disk is almost fully covered by the hot corona. The Comptonizing corona has at least two optical depths, $\tau = 0.91$ and 0.25. This implies an inhomogeneous structure in the clouds. Similar results have been obtained from other canonical BHBs in the low/hard state with *Suzaku* (e.g., Makishima *et al.* 2008). The best-fit model and unfolded spectrum is plotted in the top panel of Figure 1.

The analysis of the iron-K line profile yields an inner disk radius of $(15^{+10}_{-7})R_{\rm g}$ ($R_{\rm g}$ represents the gravitational radius GM/c^2, where G, M, and c is the gravitational constant, black hole mass, and light velocity, respectively), with an inclination angle of $i = 52^{+8}_{-5}$ degrees. This radius is consistent with that estimated from the continuum fit by assuming the conservation of the photon number in Comptonization. Our results indicate that the standard disk of GX 339–4 is truncated in the low/hard state. Comparing our $R_{\rm in}$ value with that derived from the 2008 *Suzaku* observation (Tomsick *et al.* 2009), when the 1–100 keV flux is 14 times fainter than in our epoch, we can study how the inner edge of the accretion disk evolves during the low/hard state as a function of luminosity. The XIS spectra, best-fit model, and ratio between the data and continuum model are plotted in Figure 2; it is seen that the iron-K line in the 2008 data is significantly narrower than that observed in 2009. The resulting inner disk radius is found to be $R_{\rm in} = 680^{+740}_{-310}R_{\rm g}$ when we assume an inclination angle of $i = 52°$ and an emissivity index of -2.3 as obtained from our observations. These results suggest that the inner edge significantly moved inward as the luminosity increased from 2008 to 2009 even if the source was commonly observed in the low/hard state.

1.3. *Spectral Energy Distribution (SED)*

We examine the origin of the multiwavelength SED of GX 339–4 in the low/hard state with different emission components, utilizing our quasi-simultaneous near infrared and X-ray spectra combined with previous data (Figure 3). There we plot the estimated contribution of the "intrinsic" MCD component including photons that are Compton-scattered in the corona, through the photon number conservation. The upper limit of the contribution from the companion star is also plotted.

Our *IRSF* spectrum shows a slightly steeper slope (an energy index of $\alpha \approx -0.1$) than the radio one; the corresponding $J - K_{\rm s}$ color is similar to that of the 1981 observations compiled by Corbel & Fender (2002). As shown by them, this color is consistent with a superposition of two different components with $\alpha = -0.6$ and $\alpha = 2.1$, which can be explained by the optically *thin* synchrotron radiation from the jets and the reprocessed, thermal emission from irradiated outer parts of the disk, respectively (see also Markoff *et al.* 2003 and Gandhi *et al.* 2010). It is thus indicated that the transition point from optically thick to thin regime of the synchrotron emission must appear at frequencies just below or around the near-infrared band, $\sim 10^{14}$ Hz. Based on a simple analysis, we estimate the magnetic field and size of the jet base to be $\sim 10^5$ G and $\sim 10^9$ cm, respectively. The synchrotron self Compton component is estimated to be $< 1\%$ of the total X-ray flux.

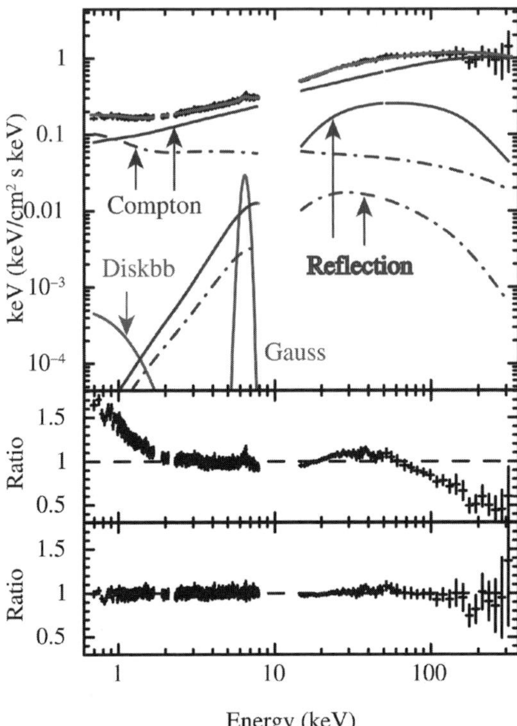

Figure 1. (Top) Unfolded spectrum of GX 339–4 in the low/hard state observed with *Suzaku* in 2009 March. The continuum is described by the double `compPS` + MCD model. The Galactic absorption is corrected. The reflection component for each Comptonized continuum and the iron-K emission line (Gaussian) are separately plotted. (Middle) Spectral ratio between the data and a single power law model fitted to the 2–100 keV range. (Bottom) Spectral ratio between the data and the best-fit model.

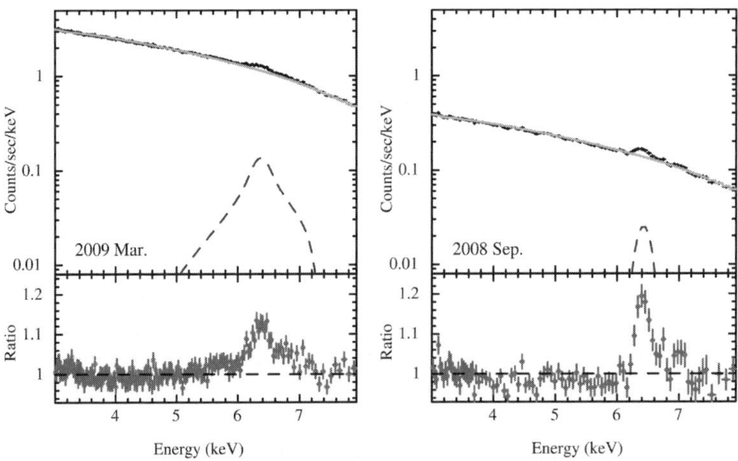

Figure 2. (Upper) *Suzaku* XIS spectra of GX 339–4 in the 3–8 keV band, fitted with the `diskline` model for the iron-K emission line. (Lower) The spectral ratio between the data and the continuum model from which the diskline component is excluded. (Left) The 2009 March data. (Right) The 2008 September data (Tomsick *et al.* 2009).

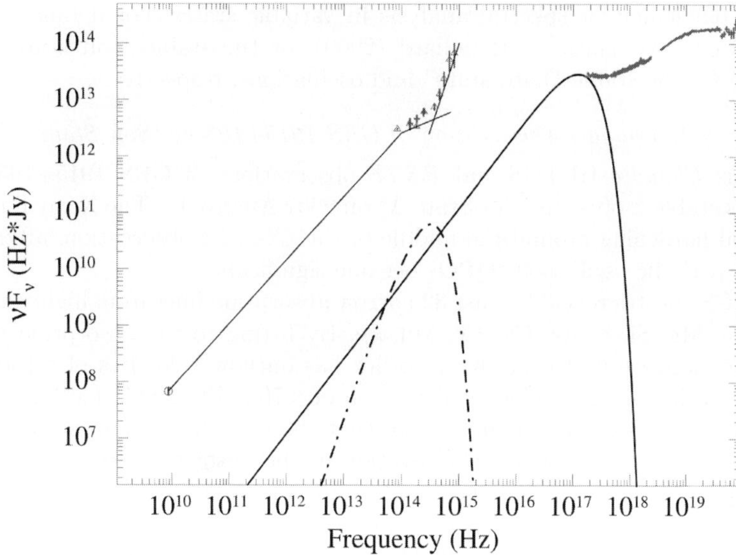

Figure 3. Multiwavelength SED of GX 339–4 in 2009 March, corrected for the interstellar absorption/extinction by assuming $A_V = 3.7$. The radio flux at 8640 MHz is estimated from the empirical relation with the 20–100 keV flux obtained by Corbel *et al.* (2003). In the infrared to optical band, the data from our *IRSF* observations (crosses) are plotted together with those obtained in 1981 compiled by Corbel & Fender (2002) whose flux level is adjusted to fit our K_s-band flux (triangles). The thick solid curve shows an estimated contribution from the intrinsic MCD emission that includes Compton-scattered photons. The dot-dashed curve represents the upper limit of the contribution from the companion star. The broken power law (thin solid lines) corresponds to an estimated contribution from the compact jet, while the steep power law component represents the thermal emission from the irradiated outer disk.

2. Recent Progress in X-ray Studies of GRS 1915+105

2.1. *Introduction*

GRS 1915+105 is the brightest microquasar in our Galaxy, providing us with a unique opportunity to study the accretion flow onto a black hole at high fractions of Eddington ratio (for a review, see Fender & Belloni 2004). Belloni *et al.* (2000) identified 12 "Classes" of the X-ray variability pattern, and found that the instantaneous spectral state of GRS 1915+105 can be divided into three basic states, States A, B, and C, where the source undergoes frequent transition (oscillations) or stays stable. The correspondence between the three states of GRS 1915+105 and canonical states observed in normal black holes (i.e., the low/hard, intermediate or very high, high/soft states) is not completely understood, however.

Here we report our recent results on GRS 1915+105 observed in the stable "soft state" (Class ϕ) with *Chandra*/HETGS and *RXTE*, and in the stable "hard state" (Class χ) and limit-cycle oscillations (Class θ) with *Suzaku*. Instead of standard models usually applied to canonical black hole binaries (e.g., MCD plus a power law), we consider more physically motivated models for the continuum to discuss the nature of the accretion flow. The good energy resolution achieved with *Chandra*/HETGS and *Suzaku*/XIS enable us to examine the disk geometry from the iron-K emission line, as well as the property of the disk wind from the absorption lines. We stress that for correct spectral modelling of GRS 1915+105, accurate interstellar absorptions with non Solar abundances should be taken into account; otherwise, an artificial broad iron-K line feature would easily appear, for instance. Also, one needs to take the effects by the dust-scattering component into account, in particular for spectral analysis in variable states. For details, we refer the readers to Ueda, Yamaoka, & Remillard (2009) for the stable "soft state" and Ueda *et al.* (2010) for the stable "hard state" and oscillations, respectively.

2.2. *Chandra Observation of GRS 1915+105 in "Soft State"*

Simultaneous *Chandra* HETGS and *RXTE* observations of GRS 1915+105 were performed in its stable "soft state" (or State A) on 2007 August 14. The X-ray flux increased with spectral hardening around the middle of the *Chandra* observation, after which the 67 Hz quasi periodic oscillation (QPO) became significant.

The HETGS spectra reveal at least 32 narrow absorption lines from highly ionized ions, including Ne, Mg, Si, S, Ar, Ca, Cr, Mn, Fe. By fitting to the absorption line profiles by Voigt function, we find that the absorber has outflow velocities of ≈ 150 and ≈ 500 km s^{-1} with a line-of-sight velocity dispersion of ≈ 70 and ≈ 200 km s^{-1} for the Si XVI and Fe XXVI ions, respectively, indicating that the wind has a non-uniform dynamical structure along the line-of-sight. The location of the absorber is estimated at $\sim 10^5 R_{\mathrm{g}}$ from the source, consistent with thermally and/or radiation driven winds. The mass outflow rate carried by the wind is estimated to be $\sim 10^{19}$ g s^{-1} for an assumed solid angle of $\Omega/4\pi = 0.2$. Neilsen & Lee (2009) discuss a possibility that such wind regulates the jet formation.

The continuum spectra obtained with *RXTE* in the 3–25 keV band can be well described with a thermal Comptonization with an electron temperature of ≈ 4 keV and an optical depth of ≈ 5 from seed photons from the standard disk extending down to $(4-7)R_{\mathrm{g}}$. In this interpretation, most of the radiation energy is produced in the Comptonization corona, which completely covers the inner part of the disk. This is consistent with the absence of a very broad component in the iron-K emission line.

2.3. *Suzaku Observations of GRS 1915+105 in "Hard State" and Oscillations*

Suzaku observed GRS 1915+105 from 2005 October 16 to 18 for a net exposure of ≈ 80 ks. The light curves are shown in Figure 4. Around the epoch, a large multiwavelength campaign was conducted involving other space and ground observatories (for the preliminary results, see Ueda *et al.* 2006). We made spectral analysis of the *Suzaku* data for four distinct states, "stable" state (State C in Class χ), "oscillation high" state (State C in Class θ), and "oscillation med" and "oscillation low" states (State A in Class θ), sorted by the hard X-ray flux (see Figure 4). The XIS spectra in the 5–9 keV band in these states are plotted in Figure 5. The main results from the *Suzaku* observation are summarized as follows:

(a) The *Suzaku* broad band spectra of GRS 1915+105 band both in the stable and oscillation states are commonly represented with a model consisting of a MCD component, its strong Comptonization (modelled by *eqpair*, Coppi 1999), and a reflection component, over which the opacity of the highly ionized disk wind (including iron-K absorption lines and edge) is applied. The Comptonized component dominates the flux above ~ 3 keV.

(b) The Comptonization is made by optically-thick ($\tau \approx 7$–10), non-thermal / thermal ($T \approx 2$–3 keV) hybrid plasmas. The non-thermal electrons are produced with 10–60% of the total energy input to the plasma with a power law index of ≈ 3–4, which account for the hard X-ray tail above ~ 50 keV.

(c) During the limit-cycle oscillation, the reflection component as estimated by the iron-K emission line is the largest during the dip phase but disappears in the flare phase. We interpret this as evidence for self-shielding effects that Comptonized photons are obscured by the surrounding cool region of the expanded disk when viewed at a very high inclination angle from the outer disk. This is illustrated in Figure 6. The evolution of the inner disk geometry is in accordance with theoretical predictions. The idea also explains the anti-correlation between the soft X-ray and near-IR fluxes reported by Arai *et al.* (2009) on a long time scale ($\sim$days), and by Ueda *et al.* (2006) on a short time scale ($\sim$minutes) during the oscillations.

(*d*) The disk wind, traced by iron-K absorption lines of Fe XXV and Fe XXVI, always exists during the oscillation and its ionization well correlates with the X-ray flux. This supports the photoionization origin. In the stable state, the iron-K absorption lines are not detected probably because it is too highly ionized and/or the scale height is small.

(*e*) The disk parameters suggest that the inner disk structure of GRS 1915+105 is similar to that in the very high state of canonical black hole binaries. The spectral variability in the oscillation state is explained by the change of the disk geometry and of physical parameters of Comptonizing corona, particularly the fractional power supplied to the acceleration of non-thermal particles.

2.4. *Summary*

The unprecedentedly high quality spectra of GRS 1915+105 obtained with *Chandra* and *Suzaku* have revealed the complex accretion disk structure and its evolution during the limit-cycle oscillations for the first time. We suggest that optically thick and low temperature Comptonization process, sometimes dominated that by non-thermal electrons, is a key phenomenon to understand black hole accretion flows at high fractions of Eddington ratio, and hence must be considered in theoretical models. In the near future, *Astro-H* will greatly contribute to this field with the excellent energy resolution and large effective area in the iron-K band.

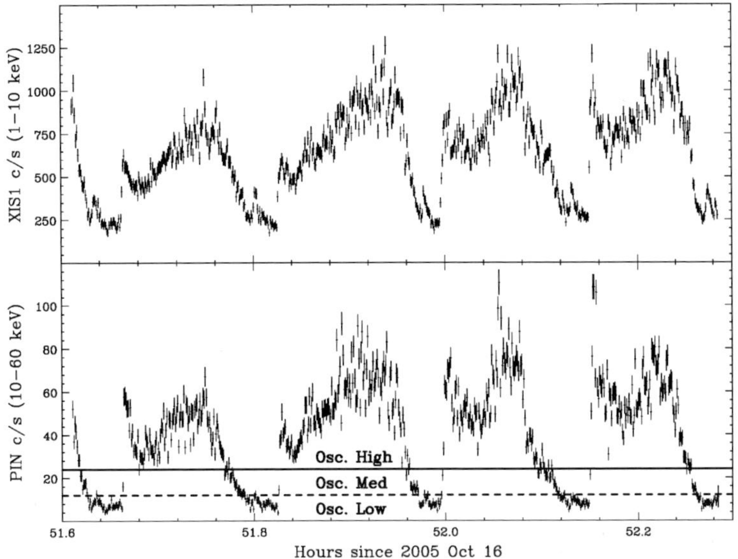

Figure 4. A representative example of *Suzaku* light curves in the 1–10 keV (XIS, upper) and 10–60 keV (PIN, lower) bands during oscillations (Class θ). The bin size is 16 seconds. The horizontal lines in the lower panel correspond to the thresholds between oscillation high/med and med/low states.

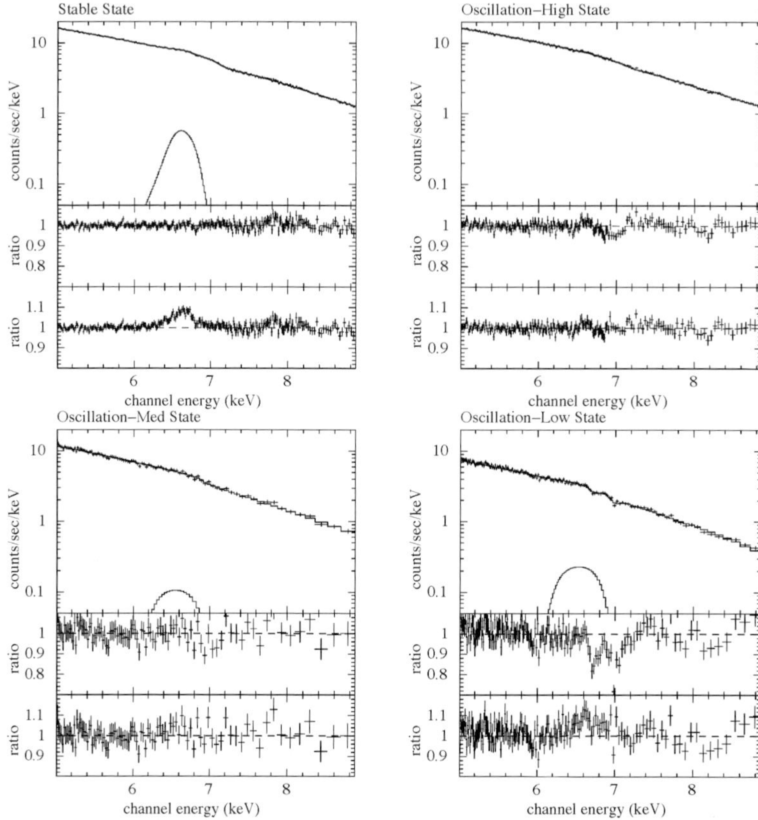

Figure 5. The XIS spectra in the 5–9 keV band. The solid curve represents the best-fit model and the contribution from the iron-K emission is separately plotted. The middle (bottom) panel plots the spectral ratio between the data and the model when the iron-K absorption (emission) lines are excluded. From left to right and top to bottom, stable, oscillation-high, oscillation-med, and oscillation-low state.

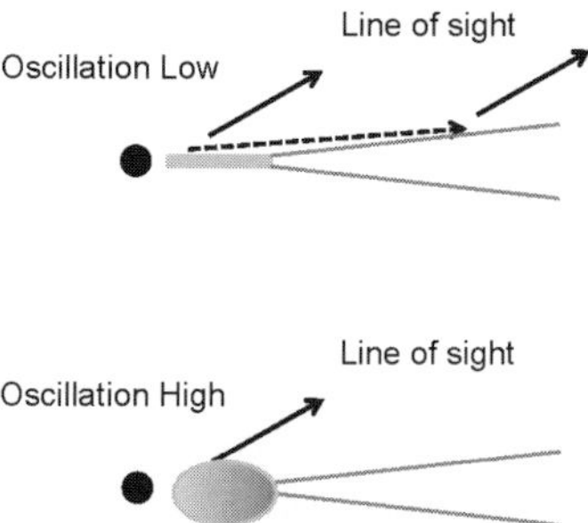

Figure 6. The schematic view of the accretion disk structure of GRS 1915+105 in the oscillation low state (dip, State A) and oscillation high state (flare, State C). The filled circle and gray area shows the black hole and inner accretion disk, respectively. During the flare phase, expansion of the inner disk causes self-shielding as viewed from the outer disk, which eliminates the iron-K emission feature, the reprocessed IR emission, and possibly the launch of the disk wind.

References

Arai, A., *et al.* 2009, *PASJ*, 61, L1

Belloni, T., *et al.* 2000, *A&A*, 355, 271

Coppi, P. S., 1999, in "High Energy Processes in Accreting Black Holes", ASP Conference Series 161, ed. Poutanen, J., & Svensson, R. (San Francisco: Astron. Soc. Pac.), p. 375

Corbel, S., & Fender, R. P. 2002, *ApJ*, 573, L35

Corbel, S., *et al.* 2003, *A&A*, 400, 1007

Fender, R., & Belloni, T. 2004, *ARAA*, 42, 317

Gandhi, P., *et al.* 2010, *MNRAS*, 407, 2166

Makishima, K., *et al.* 1986, *ApJ*, 308, 635

Makishima, K., *et al.* 2008, *PASJ*, 60, 585

Markoff, S., *et al.* 2003, *A&A*, 397, 645

Miller, J. M., *et al.* 2006, *ApJ*, 653, 525

Miller, J. M., *et al.* 2008, *ApJ*, 679, L113

Mitsuda, K., *et al.* 1984, *PASJ*, 36, 741

Neilsen, J., & Lee, J. C. 2009, *Nature*, 458, 481

Poutanen, J., & Svensson, R. 1996, *ApJ*, 470, 249

Shidatsu, M., *et al.* 2010, *ApJ*, submitted

Tomsick, J. A., *et al.* 2009, *ApJ*, 707, L87

Ueda, Y., *et al.* 2006, in "VI Microquasar Workshop: Microquasars and Beyond", ed. T. Belloni (Como: Proc. Science), p. 23.1

Ueda, Y., Yamaoka, K., & Remillard, R. 2009, *ApJ*, 695, 888

Ueda, Y., *et al.* 2010, *ApJ*, 713, 257

Yamada, S., *et al.* 2009, *ApJ*, 707, L109

Discussion

YUAN: How precisely we know about the stable luminosity of GRS 1915? This is potentially the only super-Eddington BH. This is importan since almost all quasars are sub-Eddington.

UEDA: Typically, 30% of $L_{\rm Edd}$, but not always constant.

FALCKE: You ruled out the jet/non-thermal model from Markoff *et al.* for your Suzaku data in favor of thermal Comptonization. What is the difference in chi-squared if you compare both models?

UEDA: I din not perform fitting to our SED yet, but see a recent paper by Gandhi *et al.* (2010).

KYLAFIS: I disagree with your statement that the spectrum of 339-4 can be fitted by thermal Comptonization but not by non-thermal. On the contrary, if thermal Comptonization can fit a spectrum, non-thermal (as in a jet) can do it more easily.

UEDA: I don't agree, although fitting with a "pure" thermal Comptonization would be an extreme case and we may need to consider a non-thermal component as well.

Jets at all scales
Proceedings IAU Symposium No. 275, 2011
G. E. Romero, R. A. Sunyaev & T. Belloni, eds.

© International Astronomical Union 2011
doi:10.1017/S1743921310016108

Fitting along the Fundamental Plane: New comparisons of jet physics across the black hole mass scale

Sera Markoff[1], Michael A. Nowak[2], Dipankar Maitra[3], Jörn Wilms[4], Elena Gallo[3], Robert Hynes[5] and Richard Plotkin[1]

[1] Astronomical Institute "Anton Pannekoek", University of Amsterdam
Science Park 904, 1098XH, Amsterdam, the Netherlands
email: `s.b.markoff,r.m.plotkin@uva.nl`

[2] Kavli Institute for Astrophysics and Space Research
Cambridge, MA 02139, USA
email: `mnowak@space.mit.edu`

[3] Department of Astronomy, University of Michigan
Ann Arbor, MI 48109, USA
email: `dmaitra,egallo@umich.edu`

[4] Dr. Karl Remeis-Observatory & ECAP, University of Erlangen-Nuremberg, Bamberg,
Germany
email: `joern.wilms@sternwarte.uni-erlangen.de`

[5] Department of Physics and Astronomy, Louisiana State University, Baton Rouge, LA 70803
email: `rih@theory.phys.lsu.edu`

Abstract. Correlations between the radio and X-ray bands in the hard state of black hole X-ray binaries (BHBs) have led to the discovery of the Fundamental Plane of black hole accretion, linking accretion-driven radiative attributes to black hole mass. Although this discovery has led to new constraints on radiative efficiencies, there is still significant degeneracy in terms of understanding the governing physics. I present several new results exploring the processes driving the Fundamental Plane over the black hole mass range. These include the first ever homogeneous fits of sources at approximately the same Eddington luminosity but millions of times different in mass, which I focus on for this proceeding article.

Keywords. black hole physics; X-rays: binaries; galaxies: jets; accretion, accretion disks; radiation mechanisms: nonthermal

1. Introduction

By now the existence of a Fundamental Plane of black hole (BH) accretion has been established with high statistical significance (e.g., Merloni *et al.* 2003; Falcke *et al.* 2004; Körding *et al.* 2006). In observational terms, the Fundamental Plane (FP) can be described as a log-linear correlation between the radio and X-ray luminosities of both stellar mass and supermassive BHs, with a mass dependent normalization term between them. This behavior, which holds over orders of magnitude in power, indicates that accreting BHs experience at least one mode in which they regulate their radiative and mechanical luminosity similarly at a given Eddington accretion rate $\dot{m} = \dot{M}/\dot{M}_{\mathrm{Edd}}$, regardless of their mass.

The regression methods used to derive the coefficients of the FP generally result in significant scatter around the correlation, anywhere from $\sim 0.38 - 0.65$dex, depending on the sample of active galactic nuclei (AGN) used (Körding *et al.* 2006). While new results on the intrinsic behavior of BH binaries (BHBs) (where the radio/X-ray correlation was originally discovered, see, e.g., Corbel *et al.* 2003) have revealed deviations from a perfect power-law (see Coriat *et al.*, this volume), the scatter could also be due to the sample selection. One major difference between the samples used in the two FP "discovery" papers is that Merloni *et al.* (2003) included Seyferts while Falcke *et al.* (2004) did not. If one believes in a direct accretion state mapping, then the radio/X-ray correlations observed in the so-called "hard state" of BHBs would not be expected to apply to Seyferts, which are in a radio-quiet state more similar to the "soft state" of BHBs. It is thus rather strange that Seyferts land so close to the correlation at all. When they are excluded, the scatter is significantly reduced (Körding *et al.* 2006; Gültekin *et al.* 2009), while samples of BL Lacs adhere to the correlation very well (see Plotkin *et al.*, this volume). Newer work including lower luminosity Seyferts (King, Miller *et al.*, in prep.) may indicate a separate correlation for Seyferts corresponding to their expected higher accretion efficiency, but this is clearly a point that still needs to be resolved.

The FP is generally understood to be driven by ratios of radiative efficiencies (see, e.g., Markoff *et al.* 2003; Heinz & Sunyaev 2003), but there is still significant degeneracy in understanding the exact physical drivers. For instance, different X-ray producing mechanisms ranging from direct synchrotron, synchrotron self-Compton and inverse Compton from radiatively inefficient flows can all account reasonably for the correlations observed. Thus to move beyond simple scalings, it is desirable to test physical models of the broad-band spectra that can scale with mass.

2. Results

We first tried this test a few years ago for the nearby low-luminosity AGN M81, using an outflow dominated model originally fit to many different observations of hard state BHBs (Markoff *et al.* 2005; Gallo *et al.* 2007; Migliari *et al.* 2007; Maitra *et al.* 2009). In the BHB fits, the main free parameters were the input power N_j in $L_{\rm Edd}$ units, the size of the jet base/corona R_0 (in mass-scaling units of r_g), the temperature of the assumed thermal distribution of electrons injected at the base, T_e, the location of the start of particle acceleration in the jets $z_{\rm acc}$, the power law index of the accelerated particle tail p and the ratio of magnetic to internal energy density k. Another parameter included in the model is the inner edge of the accretion disk $R_{\rm in}$. The accretion disk cannot often be directly constrained by the fits, and contrary to what occurs with thermal Comptonization models, it is not the only contributor of the upscattered photon field. However it does play a role, and the inner radius can be somewhat constrained by the X-ray fit, though it is not part of the core model. While N_j varied appropriate to the luminosity of the observations, most parameters were found to fall in rather small ranges for a variety of sources.

When we applied the model used for the BHBs to our simultaneous radio through X-ray data from a coordinated campaign on M81 (Markoff *et al.* 2008), we found that the best fit yielded parameter values in exactly the same range as that found for the stellar-mass sources. The only exception was the jet base/corona temperature T_e, which was $2 - 5$ times hotter in M81. We are currently investigating the reality of assuming a thermal distribution in the first place (Pe'er, Markoff *et al.*, in prep.) but one explanation is that the (in our case, non-bound) plasma cools much faster in the denser environment near a stellar-mass black hole, with the same relative accretion rate $\dot{m}$ fed into the same scaled size in r_g units.

In this article we present our newest attempt to test the underlying physics of the FP, by performing the homogeneous modeling of two sources separated by over six orders of magnitude in mass. If the FP is truly driven by shared physics as a function of $\dot{m}$, then

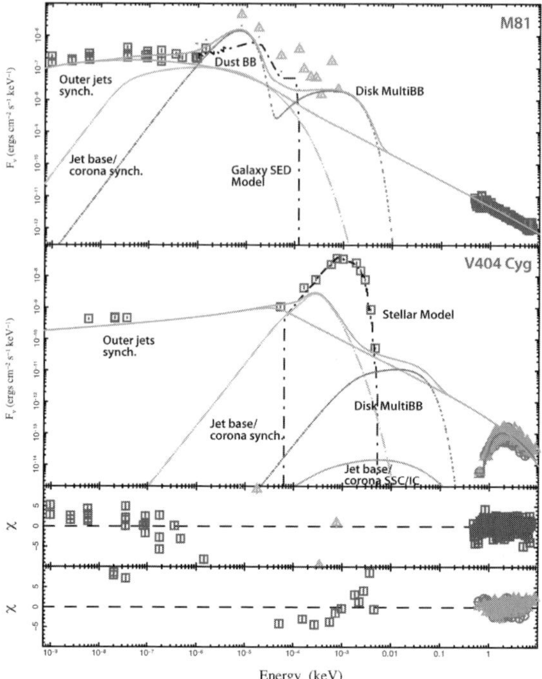

Figure 1. First attempted joint modeling of an AGN and BHB, in this case M81 and V404 Cyg (see Markoff *et al.* 2008; Hynes *et al.* 2009 for details about the data sets). The same physical model, based on Markoff *et al.* (2005), was applied to both sources, with individual physical parameters such as mass, distance and inclination fixed at values determined from observations. The free parameters R_0, p, $z_{\rm acc}$, k and $R_{\rm in}$ (see text for explanation) were fixed to vary together for the two sources. Additional components for star/stars, and dust components dominate the opt/IR range. This new approach, although preliminary, already results in promising fits to both data sets.

two accreting BHs at roughly the same scaled power should be well explained by the same physical model, if all size and power scales are expressed in mass-scaling units (r_g, and $\dot{m}$ or $L_{\rm Edd}$).

The difficulty is in identifying two sources with the same $\dot{m}$, since part of the degeneracy in the models involves uncertainty about the relationship between $L_{\rm bol}$ or $L_{\rm X}$ and $\dot{M}$, as well as uncertainties in the actual BH mass. M81 is a good source to start with on the supermassive side, because its mass and distance are very well constrained, as well as the availability of a high-quality simultaneous broadband spectrum (Markoff *et al.* 2008). In our campaign we found M81 to be in a relatively low state compared to its prior average, at $L_{\rm X} \sim 2 \times 10^{-6}$, about an order of magnitude lower than some earlier epochs. For the BHB we chose V404 Cyg, specifically a simultaneous multiwavelength data set during quiescence presented in Hynes *et al.* (2009). The source was about 10 times fainter in the X-rays, in Eddington units, compared to the observation of M81. On the other hand, the FP correlation slope tells us that $L_{\rm X} \propto \dot{M}^{2-2.4}$ (Markoff *et al.* 2003; Heinz & Sunyaev 2003), so the difference in $\dot{m}$ is only a factor of a few. We felt that this was close enough to assess whether the same mass-scaled model could apply to two sources at vastly different physical scales.

In this proceedings we present a very preliminary attempt, with all the caveats that entails because we have not yet had time to explore parameter space, but it is already quite interesting. The core model used is the outflow dominated model developed in Markoff *et al.* (2005) and Maitra *et al.* (2009), supplemented for M81 with a plasma model for the X-ray line emission as described in Young *et al.* (2007), and with a stellar model for V404 Cyg presented in Hynes *et al.* (2009). Further, the IR/optical data for M81 are non-simultaneous and include some contamination from the galaxy stellar population and dust, so we are exploring models which combine nuclear emission with additional galactic components in this region. For the key parts of the spectrum as far as the FP is concerned, the radio and the X-ray bands, clearly a reasonable solution has been found that can explain both the SEDs as well as the broadband correlations. In the model presented here, both sources are dominated by synchrotron emission in the X-ray band, which is compatible with recent results by Russell *et al.* (2010), but other solutions may yet be found.

3. Discussion

We are currently exploring the parameter space for this new class of joint models, to see which combinations of tied parameters result in the most sensible fits across the mass scale (Markoff, Nowak *et al.*, in prep.). This approach offers interesting, new possibilities for exploring the interplay of accretion physics and the environment. For instance, one could use joint models with appropriate microquasars to study the effect of the environment in the FRI/FRII differentiation. The technique could also be exploited to help "fill" the gaps in multiwavelength campaigns where coverage was limited, to help make predictions for further observations, or to constrain models. At the very lowest accretion rates, where weak luminosities and small count rates lead to poor statistics, a combined fit with a closer and thus brighter source of comparable accretion rate could be an invaluable new strategy.

References

Corbel, S., Nowak, M., Fender, R. P., Tzioumis, A. K., & Markoff, S. 2003, *A&A*, 400, 1007

Falcke, H., Körding, E., & Markoff, S. 2004, *A&A*, 414, 895

Gallo, E., Migliari, S., Markoff, S., Tomsick, J. A., Bailyn, C. D., Berta, S., Fender, R., & Miller-Jones, J. C. A. 2007, *ApJ*, 670, 600

Gültekin, K., Cackett, E. M., Miller, J. M., Di Matteo, T., Markoff, S., & Richstone, D. O. 2009, *ApJ*, 706, 404

Heinz, S. & Sunyaev, R. A. 2003, *MNRAS*, 343, L59

Hynes, R. I., Bradley, C. K., Rupen, M., Gallo, E., Fender, R. P., Casares, J., & Zurita, C. 2009, *MNRAS*, 399, 2239

Körding, E., Falcke, H., & Corbel, S. 2006, *A&A*, 456, 439

Maitra, D., Markoff, S., Brocksopp, C., Noble, M., Nowak, M., & Wilms, J. 2009, *MNRAS*, 398, 1638

Markoff, S., Nowak, M., Corbel, S., Fender, R., & Falcke, H. 2003, *A&A*, 397, 645

Markoff, S., Nowak, M., Young, A., Marshall, H. L., Canizares, C. R., Peck, A., Krips, M., Petitpas, G., Schödel, R., Bower, G. C., Chandra, P., Ray, A., Muno, M., Gallagher, S., Hornstein, S., & Cheung, C. C. 2008, *ApJ*, 681, 905

Markoff, S., Nowak, M. A., & Wilms, J. 2005, *ApJ*, 635, 1203

Merloni, A., Heinz, S., & di Matteo, T. 2003, *MNRAS*, 345, 1057

Migliari, S., Tomsick, J. A., Markoff, S., Kalemci, E., Bailyn, C. D., Buxton, M., Corbel, S., Fender, R. P., & Kaaret, P. 2007, *ApJ*, 670, 610

Russell, D. M., Maitra, D., Dunn, R. J. H., & Markoff, S. 2010, *MNRAS*, 405, 1759

Young, A. J., Nowak, M. A., Markoff, S., Marshall, H. L., & Canizares, C. R. 2007, *ApJ*, 669, 830

Discussion

PROGA: What is the value of T_e/T_i in your spectrum fitting method?

MARKOFF: At the moment it's set $T_e = T_i$, but in the future simulations we can play with that parameter.

PROGA: Can you determine this from the spectral fits from your simulations?

MARKOFF: We just started doing these, so it's something we will look into in the near future.

BOSCH-RAMON: You may include thermal free-free emission in your models to get a better estimate of the proton temperature ratio.

MARKOFF: For Sgr A*, free-free emission from the inner part we simulate is negligible, but for the scaling models applied to XRB/AGN, we plan to include protons and hadronic interactions, but it is not yet in the model.

Jets at all Scales
Proceedings IAU Symposium No. 275, 2011
G. E. Romero, R. A. Sunyaev & T. Belloni, eds.

© International Astronomical Union 2011
doi:10.1017/S174392131001611X

Accretion-outflow connection in the outliers of the "universal" radio/X-ray correlation

M. Coriat,[1]† S. Corbel,[1] L. Prat,[1] J. C. A. Miller-Jones,[2] D. Cseh,[1] A. K. Tzioumis,[3] C. Brocksopp,[4] J. Rodriguez,[1] R. P. Fender[5] and G. R. Sivakoff[6]

[1]Laboratoire AIM, CEA-IRFU/CNRS/Université Paris Diderot,
CEA Saclay, F-91191 Gif-sur-Yvette, France
email: mickael.coriat@cea.fr, stephane.corbel@cea.fr, lionel.prat@cea.fr,
david.cseh@cea.fr, jerome.rodriguez@cea.fr

[2]NRAO Headquarters, 520 Edgemont Road, Charlottesville, VA 22903, USA
email: james.miller-jones@curtin.edu.au

[3]Australia Telescope National Facility, CSIRO, P.O. Box 76, Epping, NSW 1710, Australia
email: tasso.tzioumis@csiro.au

[4]Mullard Space Science Laboratory, University College London,
Holmbury St. Mary, Dorking, Surrey RH5 6NT, UK
email: cb4@mssl.ucl.ac.uk

[5]School of Physics and Astronomy, University of Southampton, Southampton SO17 1BJ, UK
email: r.fender@soton.ac.uk

[6]Department of Astronomy, University of Virgina, P.O. Box 400325,
Charlottesville, VA 22904-4325, USA
email: grs8g@virginia.edu

Abstract. In recent years, numerous efforts have been devoted to unravel the connection between accretion flow and jets in accreting compact objects. Here we report new constraints on these issues, through the long term study of the radio and X-ray behaviour of the black hole candidate H 1743−322. This source is known to be one of the "outliers" of the universal radio/X-ray correlation, i.e. a group of stellar mass accreting black holes displaying fainter radio emission for a given X-ray luminosity, than expected from the correlation. In this work we find, at high X-ray luminosity in the hard state, a tight radio/X-ray correlation with an unusual steep slope of $b = 1.38 \pm 0.03$. This correlation then breaks below $\sim 5 \times 10^{-3} L_{\rm Edd} (M/10 M_\odot)^{-1}$ in X-rays and becomes shallower. When compared with radio/X-ray data from other black hole X-ray binaries, we see that the deviant points of H 1743−322 join the universal correlation and seem to follow it at low luminosity. Based on these results, we investigate several hypotheses that could explain both the $b \sim 1.4$ slope and the transition toward the universal correlation.

Keywords. black hole physics; X-rays: binaries; galaxies: jets; accretion, accretion disks; radiation mechanisms: nonthermal

1. Introduction

Observations of several black hole X-ray binaries (BHXBs) have brought evidence that a strong correlation exists, during the hard X-ray state, between radio and X-ray emissions (e.g. Corbel *et al.* 2003; Gallo *et al.* 2003). This connection takes the form of a non-linear flux correlation, $F_{\rm Rad} \propto F_{\rm X}^{b}$, where $F_{\rm Rad}$ is the radio flux density, $F_{\rm X}$ is the X-ray flux and $b \sim 0.5 - 0.7$. This correlation highlight the strong link between the

† Present address: School of Physics and Astronomy, University of Southampton, Southampton SO17 1BJ, UK. email: m.coriat@soton.ac.uk

compact jets and the accretion flow (disk and/or corona). This radio/X-ray correlation, initially established for the source GX 339-4, has been extended to other galactic black holes (mainly V404 Cyg, Gallo *et al.* 2003; Corbel *et al.* 2008) and even active galactic nuclei (Merloni *et al.* 2003; Falcke *et al.* 2004; Körding *et al.* 2006).

However in the following years, a few galactic black hole candidates (BHCs) have been found to lie outside the scatter of the original radio/X-ray correlation, (e.g. XTE J1720-318, Brocksopp *et al.* 2005; XTE J1650-500, Corbel *et al.* 2004; IGR J17497−2821, Rodriguez *et al.* 2007; Swift J1753.5-0127, Cadolle Bel *et al.* 2007, Soleri *et al.* 2010) thus either increasing its scatter, or challenging the universality of the correlation itself. For a given X-ray luminosity, these outliers show a radio luminosity fainter than excepted from the correlation (and thus are sometimes dubbed "radio-quiet" black holes). However, for most of these outliers, there are no radio measurements available at low X-ray luminosities. Therefore we do not know whether they remain "under luminous" in radio at low accretion rates. The current lack of data also prevents to precisely measure the slope of the correlation (if any) for the outliers. It is thus unclear if the outliers follow a correlation similar to the "standard" BHXBs but with a lower normalisation or if their inflow/outflow connection is intrinsically different. Moreover, we do not know if their behaviour is recurrent over several outbursts.

These are the issues that motivated the work presented here on the outlier H 1743−322. This source was discovered during a bright outburst in 1977 (Kaluzienski & Holt 1977). Since then, it underwent six outbursts between 2003 and 2010. All were followed by various X-ray satellites and ground-based radio observatories, providing a comprehensive set of data to study the accretion/ejection coupling in this source. The purpose of this work is to investigate this connection in detail through the study of the radio/X-ray correlation over the 6 outbursts. Consequently, we analyse almost all available radio data covering these outbursts, together with all the X-ray data from the Rossi X-ray Timing Explorer (*RXTE*) instruments. The sequence of observations and data reduction processes are detailed in Coriat *et al.* (submitted). In the following we very briefly present the main results of this study.

2. Results and discussion

Figure 1 shows the radio 8.5 GHz luminosity against the X-ray 3-9 keV luminosity of H 1743−322 in the hard states of the 6 outbursts. Above $\sim 10^{36}$ erg s^{-1} in the 3–9 keV band, we note a clear correlation over one orders of magnitude. We fit the data with a powerlaw of the form $L_{\rm Rad} \propto L_{\rm X}^{b}$, where $L_{\rm Rad}$ is the radio luminosity and $L_{\rm X}$ is the X-ray luminosity. We obtain $b = 1.38 \pm 0.03$. This correlation index clearly differs from the one usually observed for BHXBs (i.e. 0.5–0.7). We also note that some data points clearly depart from the correlation below $\sim 2 \times 10^{36}$ erg s^{-1} in X-ray. This could indicate a significant evolution of the inflow - outflow connection when the source reaches low luminosities. To locate H 1743−322 with regards to the universal radio/X-ray correlation of black holes and neutron stars X-ray binaries, we also plot on Fig. 1 the data from the black holes GX 339-4 (Corbel *et al.*, in prep), V404 Cyg (Gallo *et al.* 2003; Corbel *et al.* 2008), Swift J1753.5-0127 (Soleri *et al.* 2010) and GRO J1655-40 and from the neutron stars Aql X-1 (Tudose *et al.* 2009; Miller-Jones *et al.* 2010) and 4U 1728-34 (Migliari & Fender 2006). Fig. 1 shows that H 1743−322 lies significantly below the correlation for BHXBs but is still more "radio-loud" than neutron stars. As noted by Soleri *et al.* (2010), the data from Swift J1753.5-0127 are consistent with a correlation index of 1.4. In addition, we note the deviant points at low luminosity that join the standard correlation driven by GX 339−4 and V404 Cyg, and seem to follow it (cf.

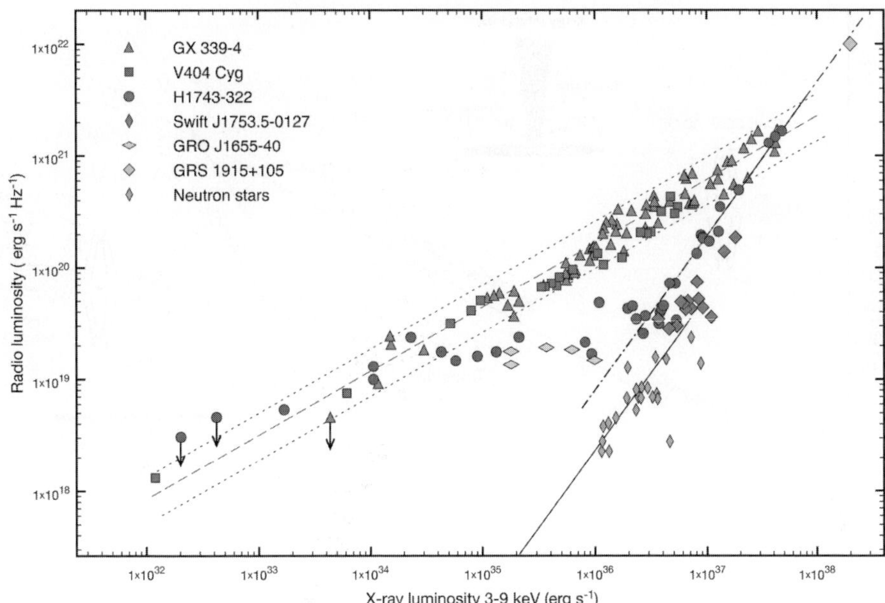

Figure 1. Radio 8.5 GHz luminosity against X-ray 3-9 keV luminosity from the hard state BHCs H 1743−322, GX 339-4, V404 Cyg, Swift J1753.5-0127, GRO J1655-40 and the hard state neutron stars Aql X-1 and 4U1728-34. A representative data point of GRS 1915+105 in the plateau state is also plotted. The dashed-dotted line is the fit to the high luminosity data of H 1743−322 with a function of the form $L_{\rm Rad} \propto L_{\rm X}^b$, with $b \sim 1.4$. The dashed line is the fit to the GX 339-4 and V404 Cyg data with $b \sim 0.6$. The dotted lines define the scattering around the fit. The solid line is the fit to the neutron stars data with $b \sim 1.4$

points below 2×10^{34} erg s^{-1}). We also note that GRO J1655-40 seems to follow the same deviation. This supports the idea of a significant transition in the coupling between the jets and the X-ray emitting component. To complete this figure, we add a "representative" point of GRS 1915+105 in the plateau state (Fender *et al.* 2010). We see that it lies on the extrapolation of the correlation of the outliers. This could be purely fortuitous but given the peculiar behaviour of this source, we thought it was worth highlighting its possible association with the outliers.

As mentioned in the introduction, some of the numerous questions associated with the outliers are whether they follow the same correlation as the other BHXBs but with a lower normalisation, and whether they remain under luminous in radio at low accretion rates. In the case of H 1743−322, we obtain a correlation coefficient of $b = 1.38 \pm 0.03$. This is the first precise measurement of the radio/X-ray correlation of an outlier. If H 1743−322 is representative of these radio-quiet Galactic black holes, our results suggest that their radio sub-luminosity is due to a different correlation coefficient rather than a different normalisation constant. Concerning the behaviour at low luminosity, the data suggest that the outliers do not remain sub-luminous in radio but rather seem to dovetail with the "standard" correlation.

As we show in Coriat *et al.* (submitted), one of the possible interpretation of these results is to consider that the difference between the standards BHXBs and the outliers arise from the radiative efficiency of their accretion flow. It is usually accepted that accreting black holes in the hard state display a radiatively inefficient inner accretion flow with $L_X \propto \dot{M}^{2-3}$. If we consider that the outlier rather display a radiatively efficient accretion flow (with $L_X \propto \dot{M}$) we can reproduce the observed correlation. However, below

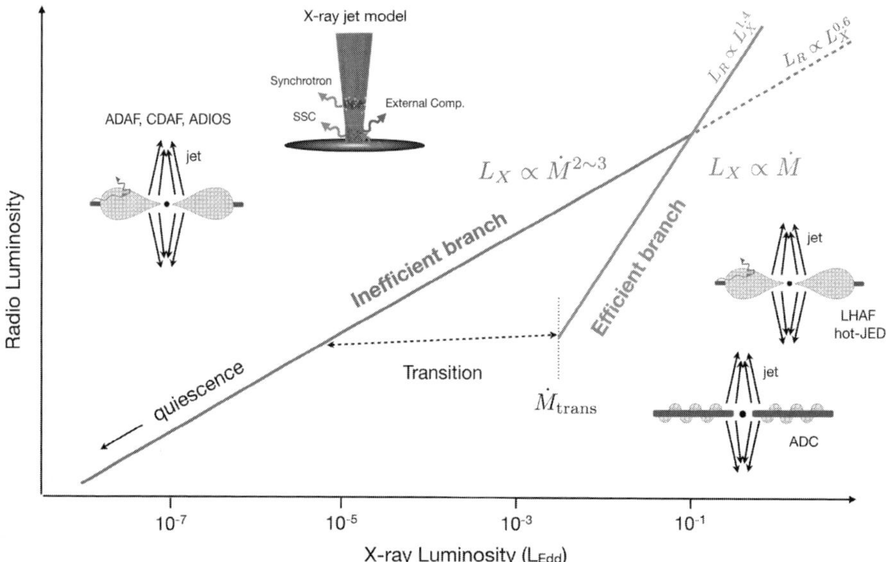

Figure 2. A schematic drawing of the global radio/X-ray correlation for galactic black holes. The X-ray luminosity is expressed in terms of the Eddington luminosity for a 10 $M_\odot$ black hole. On this figure, we illustrate the case where the steep correlation of the outliers is a consequence of the coupling between a radiatively efficient accretion flow and a steady compact jet. We can then distinguish two branches in the radio/X-ray diagram of BHXBs, according to the efficiency of the accretion flow and the consequent scaling of the X-ray luminosity with the mass accretion rate. We also illustrate the possibility of a transition between branches below a critical accretion rate $\dot{M}_{\mathrm{trans}}$. Some examples of models (figures adapted from Markoff *et al.* 2005 and M. Nowak) that could explain the accretion/ejection properties of each branch are also represented.

a critical accretion rate $\dot{M}_{\mathrm{trans}}$, the flow would have to become radiatively inefficient to account for the transition at low luminosity. This idea is summarised by the sketch of Fig. 2. Due to the limited number of pages, we refer the interested reader to Coriat *et al.* (submitted) for a more comprehensive and detailed discussion and alternative interpretations.

References

Brocksopp, C., Corbel, S., Fender, R. P., *et al.* 2005, *MNRAS*, 356, 125

Cadolle Bel, M., Ribó, M., Rodriguez, J., *et al.* 2007, *ApJ*, 659, 549

Corbel, S., Fender, R. P., Tomsick, J. A., Tzioumis, A. K., & Tingay, S. 2004, *ApJ*, 617, 1272

Corbel, S., Körding, E., & Kaaret, P. 2008, *MNRAS*, 389, 1697

Corbel, S., Nowak, M. A., Fender, R. P., Tzioumis, A. K., & Markoff, S. 2003, *A&A*, 400, 1007

Falcke, H., Körding, E., & Markoff, S. 2004, *A&A*, 414, 895

Fender, R., Gallo, E., & Russell, D. 2010, ArXiv e-prints

Gallo, E., Fender, R. P., & Pooley, G. G. 2003, *MNRAS*, 344, 60

Kaluzienski, L. J. & Holt, S. S. 1977, IAU Circ., 3099, 3

Körding, E., Falcke, H., & Corbel, S. 2006, *A&A*, 456, 439

Markoff, S., Nowak, M. A., & Wilms, J. 2005, *ApJ*, 635, 1203

Merloni, A., Heinz, S., & di Matteo, T. 2003, *MNRAS*, 345, 1057

Migliari, S. & Fender, R. P. 2006, *MNRAS*, 366, 79

Miller-Jones, J. C. A., Sivakoff, G. R., Altamirano, D., *et al.* 2010, *ApJL*, 716, L109

Rodriguez, J., Cadolle Bel, M., Tomsick, J. A., *et al.* 2007, *ApJL*, 655, L97

Soleri, P., Fender, R. P., Tudose, V., *et al.* 2010, MNRAS, in press (arXiv:1004.1066)

Tudose, V., Fender, R. P., Linares, M., Maitra, D., & van der Klis, M. 2009, *MNRAS*, 400, 2111

Discussion

MEIER: Do you have a good idea of the type of power spectra in your RQ BHs? GRO J1655-40 and GRS 1915+105 are on your transition line and alternate correlation line, respectively, and have peculiar power spectra, flat with a cutoff at ~ 10 Hz.

CORIAT: Indeed, we see similar power spectra in H1743, but maily when the source is in the hard intermediate state. We see a flat part with a cutoff at around ~ 5 Hz (bur it evolves between 2 and 10 Hz). Around the cutoff frequency, we usually see a strong type C QPO with several harmonics. But this kind of PDS seem to be typical of the hard intermediate state.

YUAN: A comment on disk-corona model for hard X-ray emission. Observationally, the high/soft state is described by the disk, but hard the X-ray emission is very weak. Theoretically, simulations show that the corona is very weak to produce hard X-rays.

CORIAT: From your observational argument, I don't know if it really excludes the possibility that hard X-ray emission in the hard state originates from a magnetic corona. Indeed, the accretion conditions could be different. I am not an expert in simulations of corona, but it seems that several simulations show strong hard X-ray emission from corona. So, as it is not clear whether a corona could be responsible or not or the hard X-ray emission, we just don't want to exclude this possibility.

Jets at all Scales
Proceedings IAU Symposium No. 275, 2011
G. E. Romero, R. A. Sunyaev & T. Belloni, eds.

© International Astronomical Union 2011
doi:10.1017/S1743921310016121

Jet launching and field advection in quasi-Keplerian discs

Jonathan Ferreira and Pierre Olivier Petrucci

Laboratoire d'Astrophysique de Grenoble, CNRS, Université Joseph Fourier, B.P. 53, F-38041 Grenoble, France

Abstract. The fact that self-confined jets are observed around black holes, neutron stars and young forming stars points to a jet launching mechanism independent of the nature of the central object, namely the surrounding accretion disc. The properties of Jet Emitting Discs (JEDs) are briefly reviewed. It is argued that, within an alpha prescription for the turbulence (anomalous viscosity and diffusivity), the steady-state problem has been solved. Conditions for launching jets are very stringent and require a large scale magnetic field B_z close to equipartition with the total (gas and radiation) pressure. The total power feeding the jets decreases with the disc thickness: fat ADAF-like structures with $h \sim r$ cannot drive super-Alfvénic jets. However, there exist also hot, optically thin JED solutions that would be observationally very similar to ADAFs.

Finally, it is argued that variations in the large scale magnetic B_z field is the second parameter required to explain hysteresis cycles seen in LMXBs (the first one would be $\dot{M}_a$).

Keywords. accretion, accretion disks; (magnetohydrodynamics:) MHD; (stars:) binaries: general; ISM: jets and outflows; galaxies: jets

1. Jet Emitting Discs

Self-confined jets are observed around a wide variety of astrophysical objects, namely from young forming stars (YSOs), neutron stars or stellar black holes (X-ray binaries) and supermassive black holes (AGN). Although the underlying emission mechanisms are quite different (molecular and atomic lines in YSOs, synchrotron radiation from supra-thermal particles in LMXBs and AGN), they share common properties: a high degree of collimation (opening angle of a few degrees), a systematic link with the accretion disc and carry away a sizable fraction of the released accretion power. It is thus natural to seek for a model where accretion and ejection processes are interdependent *regardless of the central object* (be it a star or a compact object).

The most promising and so far developed model is ejection from a quasi-Keplerian accretion disc, commonly referred to Blandford & Payne (1982) jets (hereafter BP). In this model, a large scale magnetic field of bipolar topology is assumed to thread a near-Keplerian accretion disc. This field extracts angular momentum and energy from the disc and transfers them back to the ejected material. This material is then accelerated to infinity and maintained self-confined by the hoop-stress due to the strong toroidal field. This self-confinement effect has been shown by Blandford & Payne (1982) and discussed by many other subsequent authors. However, what remained to be done was to compute the connection with the underlying disc, since the BP model was using ad-hoc boundary conditions at the disc surface.

This work has been done more than 10 years ago by Ferreira & Pelletier (1993, 1995); Ferreira (1997). Using a self-similar ansatz, it is indeed possible to solve exactly the *full set* of non-relativistic MHD equations describing a resistive, viscous accretion disc thread by a large scale field. Both viscosity ν_v and magnetic diffusivity ν_m are assumed to arise from a MHD turbulence and were modeled with an alpha-prescription, namely $\nu_m = \alpha_m V_A h$ where V_A is the Alfvén velocity on the disc midplane, h the disc vertical scale height and $\nu_v = \mathcal{P}_m \nu_m$ where $\mathcal{P}_m$ is the effective magnetic Prandtl number. The set of MHD equations have been solved from the disc midplane to the jet termination point, subjected to the constraint of smoothly crossing the critical points of the flow (mainly slow-magnetosonic and Alfvén). These two regularity conditions are unbiased by the self-similar ansatz and severely constrain the parameter space of these Jet Emitting Discs (hereafter JEDs). Indeed, defining the disc magnetization $\mu = B_z^2 / P$ where $P = P_{gas} + P_{rad}$, it has been shown that powerful jets can only be obtained with $0.1 < \mu < 1$, namely a large scale field close to equipartition. This result was in severe contrast with previous works (e.g. Ogilvie & Livio (1998, 2001)) and explains why the latter could *not* obtain super-Alfvenic, steady-state jets from thin discs. Solutions linked to the disc and crossing the three MHD critical points were also obtained, but it was shown that self-similarity introduces a strong bias on the fast-magnetosonic point that is probably unphysical (Ferreira & Casse 2004).

Because of the mass loss, the disc accretion rate varies as $\dot{M}_a \propto r^\xi$, where ξ is the disc ejection efficiency: the higher ξ the more mass is being loaded onto the jets. Models with isothermal or adiabatic magnetic surfaces achieve typical values of $\xi \sim 0.01$ (Casse & Ferreira 2000a), whereas more heavily loaded jets with $\xi \sim 0.1$ can be obtained when a surface heating is present (Casse & Ferreira 2000b). The asymptotic velocity reached along a given field line is a direct function of ξ. In a non-relativistic jet, it writes $v_j = v_K \sqrt{2\lambda - 3}$, where $\lambda \simeq 1 + \frac{1}{2\xi}$ is the magnetic lever arm and v_K the Keplerian speed at the line footpoint. Thus, contrary to the usual belief, jets from near-Keplerian accretion discs can have velocities much larger than v_K. However, because the mass loss is not that tiny, highly relativistic jets are also difficult. Indeed, the jet magnetization, namely the ratio of the MHD Poynting flux to the kinetic energy flux at the disc surface is $\sigma \simeq \xi^{-1}$ and never reaches the high values (10^4 to 10^6) sometimes invoked. As a consequence, only mildly relativistic flows are expected, but not flows with bulk Lorentz factors around 10 or 20. For those, another mechanism must be found, either extracting energy from a rotating black hole (Blandford & Znajek 1977) or a two-component jet (see e.g. Boutelier *et al.* (2008) and references therein).

Although exact and solving the long standing problem of the jet launching from accretion discs, these solutions did not catch much attention in the AGN and LMXB communities. This is probably due to the fact that the ADAF solution (e.g. Narayan & Yi 1994) drew all the attention because of its success in explaining low luminosity hard spectra (Narayan *et al.* 1995), while hoping to explain also jet formation (Narayan & Yi 1995). Although no clear proof of this latter statement was ever produced, this idea remained deeply embedded in the two aforementioned communities. Self-similar JED models are actually able to address this question, since all terms are included in the calculations. Indeed, Casse & Ferreira (2000a) remarked that they could not find any steady-state JED solution with a disc aspect ratio $\varepsilon = h/r > 0.3$. This is because, as the disc gets thicker, the effect of the radial magnetic tension increases drastically so as to cancel rotation at the disc surface. As a consequence, the toroidal field vanishes and so does the MHD Poynting flux: there is no more power to feed the jets. Figure (1) shows the total power P_{jets} (mechanical, thermal and electromagnetic) leaving the disc normalized to the accretion power P_{acc} as a function of the disc aspect ratio h/r. Geometrically thin discs give rise to jets that carry away up to 99% of the released accretion power whereas slim discs with $\varepsilon = 0.1$ can only provide 50% to their jets. This power goes rapidly to zero so that no super-Alfvénic isothermal jet can be found from thick discs with $\varepsilon = h/r > 0.3$. Hence, ADAFs are unable to drive powerful jets, although thermally driven winds would

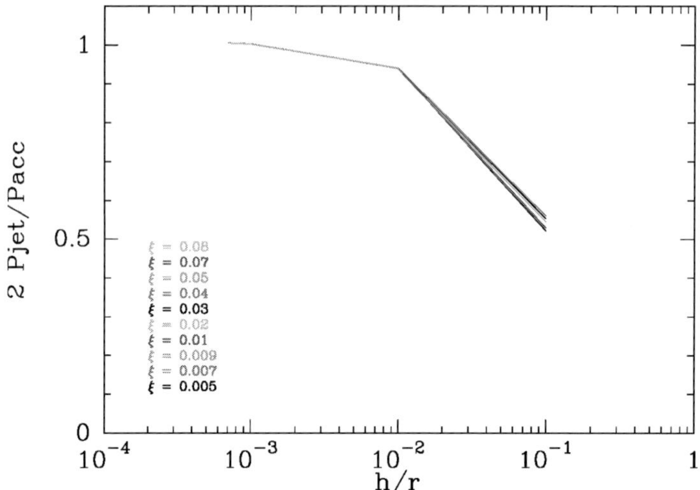

Figure 1. Ratio of the total jet power to the total accretion power as a function of the disc thickness, for all isothermal models of Ferreira (1997). Note that the power decreases rapidly, regardless of the value of the disc ejection efficiency ξ.

be a natural outcome (ADIOS, Blandford & Begelman 1999). Global MHD simulations do show ejection from thick disc structures, but they are probably of this kind, carrying a small fraction only of the released power. On the other hand, 2.5D MHD simulations of alpha discs recover all of the results given by self-similar solutions (see Tzeferacos $et\ al.$ 2009 and references therein).

Finally, Petrucci $et\ al.$ (2010) solved the JED energy equation and showed that JEDs also harbor the three characteristic thermal solutions: two stable (a cold, geometrically thin and optically thick solution and a hot, thicker and optically thin one) and one intermediate unstable solution. It seems therefore that JEDs could account for most of the successes of the ADAFs, while explaining jet formation.

2. JEDs in LMXBs

LMXBs are known to exhibit hysteresis cycles which cannot be explained by the sole variation of the disc accretion rate $\dot{M}_a(t)$. Hysteresis is a well known phenomenon that requires the variation of at least two independent parameters. Ferreira $et\ al.$ (2006) argue that it could be the large scale magnetic field B_z, or magnetic flux Φ available in the disc. Indeed, LMXBs show cycles where they switch from low luminosity hard states with a jet signature (radio emission) to high soft states with a sudden radio quenching. This can be interpreted as cycles between a JED (in its optically thin, hot branch) and a Standard Accretion Disc (hereafter SAD). Now, what determines the change from one accretion solution to another is mainly the disc magnetization $\mu \propto \Phi^2/\dot{M}_a$. While JEDs exist only for a narrow interval in μ, the disc accretion rate is seen to vary for at least two orders of magnitude. Within our framework, this readily means that the disc magnetic flux Φ must be readjusted in the disc, implying a local variation of the field strength and/or of the JED radial extent (Petrucci $et\ al.$ 2008).

There are various processes that could lead to a variation of the vertical $B_z(r,t)$ field distribution: (1) dynamo action, (2) variation of the field polarity provided by the donor star (Tagger $et\ al.$ 2004) and (3) the interplay between turbulent diffusion and advection. The first two possibilities are of course reasonable but would be somewhat too parametric yet. The third one can be computed in the framework of alpha discs. Lubow *et al.* (1994) showed that a SAD would get rid off its magnetic field unless the turbulent viscosity is much larger than the diffusivity (precisely, $\mathcal{P}_m \sim r/h$). This is kind of problematic as our current view of turbulence would suggest instead $\mathcal{P}_m \sim o(1)$ Lesur & Longaretti (2009). However, they did not take into account the disc vertical equilibrium and, as pointed out by Rothstein & Lovelace (2008), advection could still be occurring. In fact, what is requested in order to switch from an outer SAD to an inner JED is only to reach a large enough magnetization μ. Above a threshold, jets can be triggered and the field no longer diffuses away.

Murphy, Ferreira & Zanni (in prep) have made long term 2.5D MHD numerical simulations of an alpha disc threaded by a large scale magnetic field (for details see Murphy *et al.* 2010). It turns out that the steady-state field distribution in the outer SAD is very steep, leading to an increase of the magnetization towards the center. However, this process takes a long time, of the order of the accretion time scale. This is potentially of large importance as the hysteresis cycles observed in LMXBs are very long and, in any case, on much longer time scales than the dynamical time scales inferred. *This implies that the disc magnetic field distribution $B_z(r)$ is never in a steady-state.* As a consequence, it is affected by both the initial and boundary (donor) conditions. This offers a possibility to explain why the same object can display different cycles while in outburst.

To conclude, it seems that the presence of a large scale vertical magnetic field is an unavoidable ingredient for jet launching, regardless of the nature of the central object, and a key parameter to explain LMXB long term variations.

References

Blandford, R. D. & Begelman, M. C. 1999, *MNRAS*, 303, L1
Blandford, R. D. & Payne, D. G. 1982, *MNRAS*, 199, 883
Blandford, R. D. & Znajek, R. L. 1977, *MNRAS*, 179, 433
Boutelier, T., Henri, G., & Petrucci, P. 2008, *MNRAS*, 390, L73
Casse, F. & Ferreira, J. 2000a, *A&A*, 353, 1115
Casse, F. & Ferreira, J. 2000b, *A&A*, 361, 1178
Ferreira, J. 1997, *A&A*, 319, 340
Ferreira, J. & Casse, F. 2004, *ApJL*, 601, L139
Ferreira, J. & Pelletier, G. 1993, *A&A*, 276, 625
Ferreira, J. & Pelletier, G. 1995, *A&A*, 295, 807
Ferreira, J., Petrucci, P.-O., Henri, G., Saugé, L., & Pelletier, G. 2006, *A&A*, 447, 813
Lesur, G. & Longaretti, P.-Y. 2009, *A&A*, 504, 309
Lubow, S. H., Papaloizou, J. C. B., & Pringle, J. E. 1994, *MNRAS*, 267, 235
Murphy, G. C., Ferreira, J., & Zanni, C. 2010, *A&A*, 512, A82+
Narayan, R. & Yi, I. 1994, *ApJL*, 428, L13
Narayan, R. & Yi, I. 1995, *ApJ*, 444, 231
Narayan, R., Yi, I., & Mahadevan, R. 1995, *Nature*, 374, 623
Ogilvie, G. I. & Livio, M. 1998, *ApJ*, 499, 329
Ogilvie, G. I. & Livio, M. 2001, *ApJ*, 553, 158
Petrucci, P. O., Ferreira, J., Henri, G., Malzac, J., & Foellmi, C. 2010, ArXiv 1007.1478
Petrucci, P.-O., Ferreira, J., Henri, G., & Pelletier, G. 2008, *MNRAS*, 385, L88
Rothstein, D. M. & Lovelace, R. V. E. 2008, *ApJ*, 677, 1221
Tagger, M., Varnière, P., Rodriguez, J., & Pellat, R. 2004, *ApJ*, 607, 410
Tzeferacos, P., Ferrari, A., Mignone, A., *et al.* 2009, *MNRAS*, 400, 820

Discussion

DE GOUVEIA DAL PINO: Have you tested the advection of B_z component including turbulence? I believe that this transport may behave completely differently in presence of 3D MHD turbulence.

FERREIRA: The modeling is done in 2.5D only but within an alpha prescription for the magnetic diffusivity. Whether or not 3D MHD turbulence in accretion discs can indeed be described by a local prescription is an open question. However, recent work on 3D MRI in shearing box has shown that a fully turbulent medium displays both a viscosity and a magnetic diffusivity, such that the effective magnetic Prandtl number lies between 2 and 5 (Lesur & Longaretti 2009). Moreover, the functional dependency of the transport coefficients on the magnetic field is the same as the one we used ($\nu_v \propto \nu_m \propto V_A h$). This is a good hint that, for our purposes, it might not be unreasonable to model 3D turbulence with an alpha.

FENDT: What is the role of the alpha in your simulations? You don't need any viscosity to transport angular momentum as the magnetic field does that.

FERREIRA: Both viscosity and magnetic diffusivity have been included in the self-similar calculations, allowing therefore in principle to model a SAD (where viscosity is in the main agent of angular momentum transport) or a JED (where it is the jets). What comes out of the calculation is that the torque due to the jets must be the dominant one. This is not assumed but comes out as a result for cold flows to become super-slow magnetosonic. Magnetic diffusivity is absolutely required to build up steady-state models.

Jets at all scales
Proceedings IAU Symposium No. 275, 2011
G. E. Romero, R. A. Sunyaev & T. Belloni, eds.

© International Astronomical Union 2011
doi:10.1017/S1743921310016133

On the nature of the "radio quiet" black hole binaries

Paolo Soleri[1] and Rob Fender[2]

[1] Kapteyn Astronomical Institute, University of Groningen,
PO BOX 800, 9700 AV Groningen, the Netherlands
email: `soleri@astro.rug.nl`

[2] School of Physics and Astronomy, University of Southampton,
Hampshire SO17 1BJ Southampton, United Kingdom
email: `R.Fender@soton.ac.uk`

Abstract. The accretion/ejection coupling in accreting black hole binaries has been described by empirical relations between the X-ray/radio and X-ray/optical-infrared luminosities. These correlations were initially supposed to be universal. However, recently many sources have been found to produce jets that, given certain accretion-powered luminosities, are fainter than expected from the correlations. This shows that black holes with similar accretion flows can produce a broad range of outflows in power Here we discuss whether typical parameters of the binary system, as well as the properties of the outburst, produce any effect on the energy output in the jet. We also define a jet-toy model in which the bulk Lorentz factor becomes larger than ~ 1 above $\sim 0.1\%$ of the Eddington luminosity. We finally compare the "radio quiet" black holes with the neutron stars.

Keywords. X-rays: binaries, ISM: jets and outflows, accretion, accretion disks

1. Introduction

Relativistic ejections (jets) are a common consequence of accretion processes onto stellar-mass black holes. In the low/hard state (LHS) and in the quiescent state of black hole candidates (BHCs) a compact, steady jet is on. The jet is highly quenched in the high/soft state (HSS) of BHCs (see Fender 2010 for a review).

Corbel *et al.* (2003) and Gallo *et al.* (2003) found that the radio luminosity of many BHCs in the LHS correlates over several orders of magnitude with the X-ray luminosity. They proposed that a correlation of the form $L_X \propto L_R^{0.58\pm0.16}$ (where L_X and L_R are the X-ray and radio luminosities) could be universal and also valid for sources in quiescence (Gallo *et al.* 2006). This relation describes a coupling between the accretion processes and the ejection mechanisms. A similar correlation also hold between the the X-ray and the optical/infrared (IR) luminosities (Russell *et al.* 2006).

However, in the past few years, several "radio quiet" outliers have been found (Xue & Cui 2007; Gallo 2007). These sources seem to feature similar X-ray luminosities to other BHCs but are characterized by a radio emission that, given a certain X-ray luminosity, is fainter than expected from the radio/X-ray correlation. It is possible that a correlation with similar slope but lower normalization than the other BHCs could describe this discrepancy, at least in a few sources (e.g. Soleri *et al.* 2010). If confirmed, this would suggest that some other parameters might be tuning the accretion-ejection coupling, allowing accretion flows with similar radiative efficiency to produce a broad range of outflows.

Casella & Pe'er (2009) suggested that different values of the jet magnetic field can cause a quenching of the radio emission, without influencing the energy output in the X

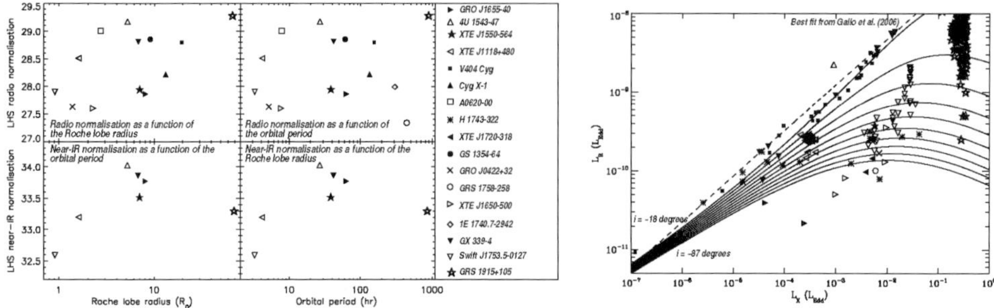

Figure 1. *Left-hand panel:* Radio and near-IR normalizations as a function of the orbital period and the size of the Roche lobe of the BHCs. Our BHC sample is listed in the inset. The inset also shows a key to the symbols. *Right-hand panel:* Values of the radio luminosity expected from our toy model for 10 viewing angles, in Eddington units. See the left-hand panel for a key to the symbols.

rays. Fender *et al.* (2010) showed that, if our measures of the spin and the estimates of the jet power are correct, the spin does not play any role in powering jets from BHCs.

In this conference Proceedings we investigate whether there is a connection between the values of some binary parameters and properties of the outburst of 17 BHCs (listed in Figure 1, left-hand panel) and the compact steady-jet power. We will follow the approach presented in Fender *et al.* (2010) to use the normalizations of the radio/X-ray and IR/X-ray correlations as a proxy for the jet power. The data used to calculate the normalizations are from Gallo *et al.* (2003), Gallo *et al.* (2006), Gallo (2007), Russell *et al.* (2006), Russell *et al.* (2007) and Soleri *et al.* (2010). We develop a jet-toy model to study whether de-boosting effects can explain the scatter around the radio/X-ray correlation. We also compare the "radio quiet" BHCs to the accreting neutron star (NS) X-ray binaries.

2. BHC properties and jet power

Since the accretion disc occupies $\sim 70\%$ of the Roche lobe of the black hole, we calculated the size of the Roche lobe of the accretor as a measure of the disc size. Figure 1 (left-hand panel) shows the radio and near-IR normalizations as a function of the size of the Roche lobe of the black hole and the orbital period of the binary. To test whether there is any correlation between the jet power and these two orbital parameters, we calculated the Spearman rank correlation coefficients. The values of the correlation coefficients ρ, as well as the null hypothesis probabilities (the probability that the data are not correlated), are reported in Table 1. Clearly no correlation is present.

We also investigate the dependence of the radio and near-IR normalizations on the inclination between the jet axis and the line of sight. We will refer to this angle i as either inclination or viewing angle. Casella *et al.* (2010) recently showed that compact-steady jets from BHCs can have rather high bulk Lorentz factor $\Gamma > 2$. This result suggests that de-boosting effects can become important, not only at high viewing angles. In our analysis we are assuming that the X-ray emission is un-beamed (but see Fender 2010 and references therein). To test if a correlation exists, we calculated the Spearman coefficients ρ. We show them in Table 1. In the case of the near-IR normalizations, we obtained $\rho \sim -0.9$, with a probability for the null hypothesis of $\sim 2\%$. This suggests that there is an anticorrelation between the inclination angle and the near-IR normalization. However, the lack of data points (only 7) might have biased this result.

During an outburst, BHCs usually show a transition to the HSS. However, some sources spend the whole outburst in the LHS (or in the LHS and in the intermediate states),

Table 1. Spearman rank correlation test for our data points. The number of data points, as well as the probabilities for the null hypothesis are reported.

Normalization	Number of points	Spearman coefficient ρ	Probability for the null hypothesis (%)
		Size of the Roche lobe	
radio	13	0.5	11.0
near IR	7	0.3	43.0
		Orbital period	
radio	15	0.2	54.2
near IR	7	0.4	33.2
		Inclination angle	
radio	13	-0.2	44.7
near IR	7	-0.9	2.4

without transiting to the soft states (see Belloni 2009 and references therein). We investigated whether the type of outburst (LHS only or with a transition to the soft states) affects the jet power, but we could not find any obvious dependence.

3. A jet-toy model

Here we define a jet-toy model. The aim is to test whether a dependence of the bulk Lorentz factor of the jet Γ on the accretion powered X-ray luminosity might qualitatively describe the scatter around the radio/X-ray correlation and the "radio quiet" BHCs population. We will consider a Γ Lorentz factor that becomes larger than ~ 1.4 above $\sim 0.1\%$ of the Eddington luminosity L_{Edd}. This assumption is based on the fact that compact steady jets are usually thought to be mildly relativistic ($\Gamma \leqslant 2$ but see Casella *et al.* 2010) in the LHS while major relativistic ejections ($\Gamma \geqslant 2$) are tentatively associated with the transition from the hard to the soft states. We considered a uniform distribution of 10 values of $\cos i$ between 0 and 1. Figure 1 (right-hand panel) illustrates the results from our toy model: it clearly results in a distribution in the (L_X, L_R) plane which broadens at higher luminosities. The toy model can qualitatively reproduce the scatter around the radio/X-ray correlation.

4. Comparison with neutron stars

NSs are known to be fainter in radio than BHCs, given a certain X-ray luminosity, by a factor $\gtrsim 30$ (see e.g. Migliari & Fender 2006). This difference in radio power can be reduced to a factor $\gtrsim 7$ if a mass correction from the fundamental plane of black hole activity (Merloni *et al.* 2003) is applied. We will now compare the "radio quiet" BHCs to the population of NSs that have been detected in radio. Our sample of NSs includes the same data points as in Migliari & Fender (2006) with the addition of points from recent observations of Aql X-1, 4U 0614-091 and IGR J00291+5934 (Tudose *et al.* 2009, Migliari *et al.* 2010 and Lewis *et al.* 2010, respectively). To test whether the "radio quiet" BHCs and the NSs are statistically distinguishable in the (L_X, L_R) plane, we performed a two-dimensional Kolmogorov-Smirnov (K-S) test. The K-S test shows that the probability that the "radio quiet" BHCs and the NSs are statistically indistinguishable (i.e. the probability of the null hypothesis) is different from 0, despite being small ($P \sim 0.13\%$), if a mass correction is applied. If we do not apply a mass correction, the probability of the null hypothesis is consistent with 0: the two groups constitute two different populations.

5. Conclusions

We examined three characteristic parameters of BHCs and the properties of their outbursts to test whether they regulate the energy output in the jet. If our estimates of the jet power are correct, both the orbital period and the size of the accretion disc are not related to the radio and near-IR jet power. We could also not find any association between the jet power and the type of outburst (with or without a transition to the HSS). We did not find any association between the viewing angles and the jet power inferred from radio observations. The jet power obtained from near-IR measurements decreases when the inclination angle increases. This result could favour a scenario in which the jet decelerates moving from the IR-emitting to the radio-emitting part.

We defined a jet-toy model in which the jet Lorentz factor becomes larger than $\sim$ 1 above $0.1\% L_{Edd}$. The model results in a distribution in the (L_X, L_R) plane which broadens at high luminosities. However, the model has several limitations, for instance it can not reproduce the measured inclination angles of the BHCs in our sample.

We finally compared the "radio quiet" BHCs to the NSs. A two-dimensional K-S test can not completely rule out the possibility that the two families are statistically indistinguishable in the (L_X, L_R) plane, if a mass correction is applied. This result suggests that some "radio quiet" BHCs could actually be NSs; alternatively it suggests that some BHCs feature a disc-jet coupling similar to NSs.

References

Belloni, T. M. 2010, in: T. Belloni (ed.), *The Jet Paradigm, Lecture Notes in Physics* (Berlin Springer Verlag), p. 53

Casella, P. & Pe'er, A. 2009, *ApJ* (Letters), 703, L63

Casella, P., Maccarone, T. J., O'Brien, K., Fender, R. P., Russell, D. M., van der Klis, M., Pe'er, A., Maitra, D., Altamirano, D., Belloni, T., Kanbach, G., Klein-Wolt, M., Mason, E., Soleri, P., Stefanescu, A., Wiersema, K., & Wijnands, R. 2010, *MNRAS*, 404, L21

Corbel, S., Nowak, M. A., Fender, R. P., Tzioumis, A. K., & Markoff, S. 2003, *A&A*, 400, 1007

Fender, R. P 2010, in: T. Belloni (ed.), *The Jet Paradigm, Lecture Notes in Physics* (Berlin Springer Verlag), p. 115

Fender, R. P., Gallo, E., & Russell, D. M. 2010, *MNRAS*, 406, 1425

Gallo, E., Fender, R. P., & Pooley, G. G. 2003, *MNRAS*, 344, 60

Gallo, E., Fender, R. P., Miller-Jones, J. C. A., Merloni, A., Jonker, P. G., Heinz, S., Maccarone, T. J., & van der Klis, M. 2006, *MNRAS*, 370, 1351

Gallo, E. 2007, in T. di Salvo *et al.* (eds.), *The Multicolored Landscape of Compact Objects and Their Explosive Origins* (AIPC series), p. 715

Lewis, F., Russell, D. M., Jonker, P. G., Linares, M., Tudose, V., Roche, P., Clark, J. S., Torres, M. A. P., Maitra, D., Bassa, C. G., Steeghs, D., Patruno, A., Migliari, S., Wijnands, R., Nelemans, G., Kewley, L. J., Stroud, V. E., Modjaz, M., Bloom, J. S., Blake, C. H., & Starr, D. 2010, *A&A*, 517, 72

Merloni, A., Heinz, S., & di Matteo, T. 2003, *MNRAS*, 345, 1057

Migliari, S. & Fender, R. P. 2006, *MNRAS*, 366, 79

Migliari, S., Tomsick, J. A., Miller-Jones, J. C. A., Heinz, S., Hynes, R. I., Fender, R. P., Gallo, E., Jonker, P. G., & Maccarone, T. J. 2010, *ApJ*, 710, 117

Russell, D. M., Fender, R. P., Hynes, R. I., Brocksopp, C., Homan, J., Jonker, P. G., & Buxton, M. M. 2006, *MNRAS*, 371, 1334

Russell, D. M., Maccarone, T. J., Körding, E. G., & Homan, J. 2007, *MNRAS*, 379, 1401

Soleri, P., Fender, R. P., Tudose, V., Maitra, D., Bell, M., Linares, M., Altamirano, D., Wijnands, R., Belloni, T., Casella, P., Miller-Jones, J. C. A., Muxlow, T., Klein-Wolt, M., Garrett, M., & van der Klis, M. 2010, *MNRAS*, 406, 1471

Tudose, V., Fender, R. P., Linares, M., Maitra, D., & van der Klis, M. 2009, *MNRAS*, 400, 211

Xue, Y. Q. & Cui, W. 2007, *A&A*, 466, 1053

Discussion

SAMBRUNA: Do you have enough data to place these rq sources on the Fender hard/soft diagram?

SOLERI: No, we did not do that. They are supposed to be in a hard state.

YUAN: This is a comment on X-ray/radio correlation. We predict Yuan & Cui (2005) that the correlation should steepen when $L_X < L_{\mathrm{crit}}$. This prediction was confirmed in Yuan, Yu, & Ho (2009).

SOLERI: For what concerns black hole X-ray binaries there are also detections of systems in quiescence at very low X -ray luminosities ($10^{-8} L_{\mathrm{Edd}}$, Gallo *et al.* 2006). These systems follow the "normal" correlation with a slope of ~ 0.6.

Jets at all scales
Proceedings IAU Symposium No. 275, 2011   © International Astronomical Union 2011
G. E. Romero, R. A. Sunyaev & T. Belloni, eds.   doi:10.1017/S1743921310016145

GRS 1915+105 "celebrates its majority" (1992–2010)

Alberto J. Castro-Tirado

Instituto de Astrofísica de Andalucía (IAA-CSIC), Glorieta de las Astronomía s/n, E-18080
Granada, Spain
email: ajct@iaa.es

Abstract. Over the 18 years since its discovery, GRS 1915+105 has continuously brightened in the X/γ-ray sky. It is considered the prototypical microquasar. Most of these are LMXBs that show sporadic ejection of matter at apparently superluminal velocities. In these the three basic ingredients of quasars are found: a black hole, an accretion disc and collimated jets of high energy particles, but in microquasars the black hole is only a few $M_\odot$ instead of several $\times\ 10^6$ $M_\odot$; the accretion disc had mean thermal temperature of several $\times\ 10^6$ K instead of several $\times$ 10^3 K, and the particles ejected at relativistic speeds travel distances of a few ly only, compared to few $\times\ 10^6$ ly as in radio galaxies. However many open issues remain to be addressed.

Keywords. X-rays: individual (GRS 1915+105), accretion disks, acceleration of particles, black hole physics

1. The discovery

The *Granat* satellite was launched in Dec 1989, carrying the SIGMA gamma-ray telescope, the ART-P X-ray telescopes and the WATCH all-sky X-ray monitor. WATCH was scrutinizing the sky with the four units and on 15 Aug 1992, after returning from Crimean Astrophysical Observatory (where we managed to discover during the night the optical counterpart to X-ray Nova Per 1992), a very unusual, hard X-ray source outshone the rest of sources in the WATCH sky: the GRanat Source located at RA=19h15m, Dec=+10°.5 (Castro-Tirado, Brandt & Lund 1992, Castro-Tirado *et al.* 1994). See Fig. 1.

At our request, SIGMA observed the field, improving the 1° WATCH error box, reporting a much more accurate position leading to the discovery of a variable radiosource months after (Mirabel *et al.* 1993a), whose position could be accurately placed on the sky with respect to field stars in the optical. At this precise location, an IR counterpart was reported (Mirabel *et al.* 1993b, Castro-Tirado *et al.* 1993), which was later confirmed by means of near-IR spectroscopy, showing variable, non Doppler-shifted emission lines, with the system suggested to be a low-mass X-ray binary (LMXB) (Castro-Tirado *et al.* 1996), which was finally confirmed in 2001. See also Fig. 2.

Although activity was already seen by BATSE/*CGRO* since Mar 1992 (Paciesas *et al.* 1995), GRS 1915+105 remained undetected in an archival *Einstein*/HRI pointed observation in the 1970s (Castro-Tirado 1994). The first two years of WATCH observations indeed revealed a highly variable object (Sazonov *et al.* 1994).

2. The superluminal source

GRS 1915+105 is the first superluminal source discovered in the Galaxy (Mirabel and Rodriguez 1994): in 1994, clouds of plasma were ejected in opposite directions from the central source at speeds close to (but less than) that of light (0.92c), and the relativistic effects led to the apparent superluminal motion. The energetics and bulk Lorentz factor were consistent with radiation pressure acceleration of pair plasmoids (Liang and Li 1995) at a distance D = 12.5 ± 1.5 kpc (Chaty *et al.* 1996). The jet power was estimated to be $\sim10^{38}$ erg/s (Fender *et al.* 1999).

VLBI imaging resolved the nucleus as a compact jet of length $\sim$10 AU. Its properties were better fitted by a conically expanding synchrotron jet model rather than a thermal jet. The nuclear jet varied in $\sim$30 min during minor X-ray/radio outbursts and reestablished within $\sim$18 hr of a major outburst, indicating the robustness of the X-ray/radio (or disk/jet) system to disruption. A limit of < 100 km/s was placed on its proper motion with respect to its neighborhood (Dhawan *et al.* 1999).

3. The jet-disk connection

Multi-wavelength observations revealed the probable synchrotron origin of the IR emission (Fender & Pooley 1998, MNRAS 300, 576), and this was consistent with a scenario wherein the IR flux originates in a relativistic plasma that has been ejected from the inner accretion disk (Eikenberry *et al.* 1998, Mirabel *et al.* 1998). Analysis of 165 *RXTE* observations allowed the observations to be classified into 12 classes, which could be ascribed to 3 basic states: a hard state corresponding to the non-observability of the innermost parts of the accretion disk, and two softer states with a fully observable disk (Belloni *et al.* 2000).

4. A massive black hole

Evidence for GRS 1915+105 containing a Kerr rotating black hole (a* = 0.98) leading to a frame dragging effect was proposed already in 1998 (Cui *et al.* 1998). The mass function was determined in 2001, thanks to ESO´s VLT mid-res near-IR spectroscopy (Greiner *et al.* 2001). The compact object´s mass (M_x = 14.0 ± 4.4 M$_\odot$) was derived assuming an inclination of i = 66° ± 2° from the orientation of the radio jets and a more accurate distance. The mass for the early K-type giant star is M_d = 0.81 ± 0.53 M$_\odot$, consistent with a more evolved stripped-giant donor star in GRS 1915+105 (Harlaftis and Greiner 2004). 7 years of K-band monitoring of the low-mass X-ray binary GRS 1915+105 allowed to detect orbital and superhump signatures. Positive correlations between the

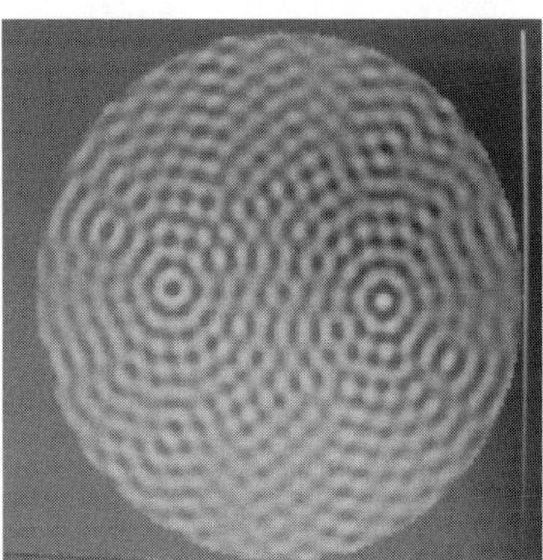

Figure 1. The WATCH/*Granat* cross-correlation map on 15 Aug 1992, showing GRS1915+105 disputing with Sco X-1 the supremacy in the X-ray sky.

infrared flux and the X-ray flux and X-ray hardness were demonstrated. Analysis of the frequency spectrum showed that the orbital period of the system is $P_{orb} = 30.8 \pm 0.2$ d (Neil *et al.* 2007).

5. Selected recent results

For an extensive review on the data gathered until 2003 (and the corresponding interpretation), we refer the reader to the review work published in 2004 (Fender and Belloni 2004). In this section some selected results obtained since then are briefly mentioned.

Based on radio and X-ray observations, the semi-quantitative model for the state transition proposed by Fender & Belloni (2004) could be extended to other microquasars (Fender *et al.* 2004). *RXTE* and *CGRO*/OSSE spectra were well fitted by Comptonization of disc blackbody photons, with very strong evidence for the presence of a non-thermal electron component in the Comptonizing plasma (Zdiarski *et al.* 2005). Based on a spectral analysis of the X-ray continuum it was confirmed that the compact primary in GRS 1915+105 is a rapidly rotating Kerr black hole (McClintock *et al.* 2006). X-ray spectral and temporal analysis of four observations showed strong radio to X-ray correlations (Rodriguez *et al.* 2008). However, GRS 1915+105 is considered an outlier on the L_X-L_R relation like H 1743-322 (Coriat *et al.* 2010).

In addition, phase-resolved *CXO* X-ray spectroscopy gives support to a highly variable disk wind (Neilsen *et al.* 2010) and *Suzaku* observations give evidence for "self-shielding"; i.e. the Comptonizing corona becomes geometrically thick in the flare phase (Ueda *et al.* 2010). The plateau states can be described by the same broadband model with a steady outflow and an accretion inflow (van Oers *et al.* 2010).

Accretion disk winds as the jet suppression mechanism in GRS 1915+105 have been proposed (Neilsen and Lee 2009) and HESS upper limits on fluxes > 410 GeV imply that the radiative efficiency of the compact jet at VHE is $< 0.01\%$ (Acero *et al.* 2009).

GRS 1915+105 has been in a state of constant outburst since its discovery in 1992, an eruption which has persisted $\sim$100 times longer than those of more typical LMXBs. It has been proposed that the outburst could last $\sim$100 yr (Deegan *et al.* 2009).

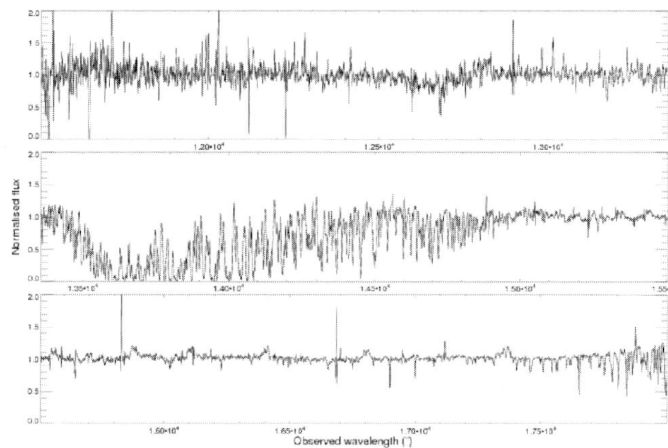

Figure 2. VLT/X-Shooter JH spectrum of GRS1915+105 obtained in June 2009 by A. de Ugarte Postigo. H I emission lines in the range 1.59-1.74 μm are noticeable. Based on data obtained from the ESO Science Archive Facility.

6. Summary and open questions

GRS1915+105 is the first highly relativistic ($\Gamma > 2$) jet source known in the Galaxy and can be considered as the "holy grail" for studying the jet-disk coupling in accreting black-hole (BH) systems (and also testing General Relativity effects) as it is located at an extreme range of parameter space for an outflow dominated model, requiring near- or super- Eddington accretion rates and maximal jet power (($\Gamma < 30$), and a high level of magnetic domination (at least in core $r < 100r_g$). Amongst the many open questions that remain to be addressed, here are few of them:

1. Magnetic fields and disk winds are ingredients finally being considered in the models. What about the MHD simulations?

2. Why are the near-IR emission lines not Doppler-shifted as in SS 433?

3. Is the slow and steady jet always present in hard state? If so, why we do not see its long-term action on the local ISM?

4. Is this the first hypernova we are aware of in the Milky Way? A WR-progenitor?

5. The outburst may last 100 yr. What is the duty cycle? In any case, have we just seen in 1992 the onset of a persistent LMXB?

References

Acero, F. *et al.* 2009, *A&A*, 508, 1135
Belloni T. *et al.* 2000, *A&A*, 355, 271
Castro-Tirado, A. J., Brandt, S., & Lund, N. 1992, *IAU Circ.*, 5590
Castro-Tirado, A. J. *et al.* 1993, *IAU Circ.*, 5830
Castro-Tirado, A. J. 1994, *Ph.D. Thesis*, Copenhagen Univ.
Castro-Tirado, A. J. *et al.* 1994, *ApJ* (Supplement Series), 92, 469
Castro-Tirado, A. J. *et al.* 1996, *ApJ* (Letters), 491, L99
Coriat, M. *et al.* 2010, *These Proceedings*
Chaty, S. *et al.* 1996 *A&A*, 318, 825
Cui, W. *et al.* 1998 *ApJ* (Letters), 492, L53
Deegan, P. *et al.* 2009 *MNRAS*, 400, 1337
Dhawan, V. *et al.* 2000 *ApJ*, 543, 373
Eikenberry, S. *et al.* 1998 *ApJ* (Letters), 494, L61
Fender, R. *et al.* 1999 *MNRAS*, 304, 865
Fender, R. & Belloni, T. 2004 *ARA&A*, 42, 317
Fender, R., Belloni, T., & Gallo, E. 2004 *A&A*, 355, 1105
Greiner, J., Cuby & McCaughrean 2001 *Nature*, 414, 522
Harlaftis, E. & Greiner, J. 2004 *A&A*, 414, L13
Liang, E. P. & Li, H. 1995, *A&A*, 248, L45
McClintock, J. *et al.* 2006, *ApJ*, 652, 518
Mirabel, I. F. *et al.* 1993a, *IAU Circ.*, 5773
Mirabel, I. F. *et al.* 1993b, *IAU Circ.*, 5830
Mirabel, I. F. & Rodríguez, L. F. 1994, *Nature*, 371, 46
Mirabel, I. F. *et al.* 1998, *A&A*, 330, L9
Neil, E. T. *et al.* 2007, *ApJ*, 657, 409
Neilsen, J. & Lee, J. C. 2009, *Nature*, 458, 481
Neilsen, J. *et al.* 2010, *These Proceedings*,
Paciesas, W. *et al.* 1995, *NYASA*, 759, 308
Rodriguez, J. *et al.* 2008, *ApJ*, 675, 1449
Sazonov, S. *et al.* 1994, *PAZh*, 20, 901
van Oers, P. *et al.* 2010, *MNRAS*, in press
Ueda, Y. *et al.* 2010, *ApJ*, 713, 257
Zdziarski, A. *et al.* 2005, *MNRAS*, 360, 825

Discussion

NEILSEN: Regarding the infrared emission lines, I believe thath the interpretation is that when discrete ejection events occur, you have blobs of plasma above the disk that illuminate the outer disk and pump the infrared emission lines, so I think it's a good explanation.

CASTRO-TIRADO: Yes, this is a plausible interpretation. When the near-infrared emission lines were first reported (Castro-Tirado *et al.* 1996), it was surprising not to find them Doppler-shifted and we also proposed at that time that the IR flux cloud arise from free-free emission in a wind flowing out o the accretion disk.

DE GOUVEIA DAL PINO: Regarding your remark 7 on a potential connection between GRS 1915+105 and hypernova/WR late evolution, do you know any tentative modelling that tries to construct this link?

CASTRO-TIRADO: Not in the particular case of GRS 1915+105, although for Wolf-Rayet progenitors to GRBs, has been suggested in several cases (e.g. Castro-Tirado *et al.* 2010 and references therein).

Jets at All Scales
Proceedings IAU Symposium No. 275, 2011
G. E. Romero, R. A. Sunyaev & T. Belloni, eds.

© International Astronomical Union 2011
doi:10.1017/S1743921310016157

Jet-disk connection in OJ287

Mauri J. Valtonen[1,2], Tuomas Savolainen[3] and Kaj Wiik[2]

[1] Helsinki Institute of Physics, FIN-00014 University of Helsinki, Finland
email: `mvaltonen2001@yahoo.com`

[2] Dept. of Physics & Astronomy, Tuorla Observatory, University of Turku, 21500 Piikkiö,
Finland
email: `kaj.wiik@utu.fi`

[3] Max-Planck-Institut für Radioastronomie, Auf dem Hügel 69, D-53121 Bonn, Germany
email: `tsavolainen@mpifr-bonn.mpg.de`

Abstract. A model for OJ 287 consisting of two orbiting black holes has been constructed using optical light curve data. The model has successfully predicted the occurrence of sharp optical outbursts of OJ 287 for the past 15 years. Here we test if also the variations in the radio jet position angle can be explained within the framework of this same model, which has most of its parameters fixed by the timing of the optical flares. The model applied here has only three free parameters left, the (trivial) zero point of the jet position angle, the time lag between changes in the disk and jet orientations, and the zero point of the viewing angle. Despite its simplicity and the small number of free parameters, the model appears to be able to reproduce the main properties of the observed position angle variations during the past 30 years. The best fits are obtained when the time lag is either ~ 4 or ~ 14 years. However, the jet orientation seems to be unrelated to the direction of the spin of the primary black hole. This implies, assuming that the basic model is correct, that the mean orientation of the jet is determined by the orientation of the inner accretion disk, not by the spin axis of the black hole.

Keywords. quasars, jets, binary black holes

---

1. Introduction

Many quasars and other extragalactic radio sources possess parsec scale jets where the jet direction varies with time – often wobblingly. There are several possible explanations for this behaviour, including rotation of the flow, propagation of current-driven or Kelvin-Helmholtz instabilities, and precession (and/or nodding motion) of the black hole's accretion disk due to a periodic perturbation by a possible binary companion.

BL Lac object OJ 287 is one of the best known candidates for harbouring a supermassive binary black hole (BBH). In the case of OJ 287 the primary evidence for a BBH comes from the optical light curve which shows periods of outbursts at regular 12 year intervals (Sillanpää *et al.* 1988). The binary model for OJ 287 has predicted the optical variations in OJ 287 from 1995 until 2030 (Lehto & Valtonen 1996, Sundelius *et al.* 1997). For the first fifteen years the brightness variations have followed the model with great accuracy (Valtonen *et al.* 2011). The flares related to the two sharp impacts in this period were predicted correctly within a few days. The first sharp flare was expected to begin on Nov 3, 1995 (as reported in a meeting in Oxford six weeks earlier) and the prediction was correct to the day (Sundelius *et al.* 1996), while the second one was expected to begin on Sept 10, 2007 and the prediction was an equal success (Valtonen & Lehto 1997, Valtonen *et al.* 2008). The ability to predict the optical light curve so accurately presents a great challenge to any alternative model.

Valtonen *et al.* (2006) showed that at least three different mechanisms have to be included in a satisfactory BBH model for OJ 287: impacts of the secondary BH on the primary disk, tidal influence on the accretion flow, and the jet wobble. The first mechanism is responsible for the double peak structure of the sharp outbursts, the second one causes flares of longer duration, while the third one is responsible for the 60 yr cycle, with a 12 yr modulation. It is the latter mechanism which is of interest here since it also changes the jet orientation angle in the sky.

2. VLBI data

We have collected published VLBI data from the literature with observations dating back to the early 1980s (Roberts *et al.* 1987, Gabuzda *et al.* 1989, Gabuzda & Cawthorne 1996, Vicente *et al.* 1996, Fey *et al.* 1996, Tateyama *et al.* 1996, 1999, Fey & Charlot 1997, Rantakyrö *et al.* 1998, Gabuzda & Gomez 2001, Tateyama & Kingham 2004, Ojha *et al.* 2004, Jorstad *et al.* 2001, 2005, Piner *et al.* 2007, D'Arcangelo *et al.* 2009). In addition, we have analysed 79 epochs of 15 GHz VLBA observations from the MOJAVE database (Lister *et al.* 2009a), resulting in altogether 196 epochs of VLBI observations. These data were used to build a record of variations in the position angle (PA) of the parsec scale jet over a 30-year period.

The data were divided in two categories: 2-6 cm observations and 0.3-1.3 cm observations. The division has practical significance since the angular resolution and also the core position depend on the observing frequency (Blandford & Königl 1979). At higher frequencies the knots emerging in the jet are seen earlier than at low frequencies. Because of the longer time coverage of the low frequency observations, we use them in the following discussion. There is also little deviation in the PA between different wavelengths in the 2-6 cm data, which justifies stacking these observations together.

We have not tried to back-extrapolate the ejection epochs of the knots since it easily leads to ambiguous results for inhomogeneous data sets with significant time gaps. Instead, at each epoch, we measured the mean position angle of the jet within the first milliarcsecond from the core and then took a yearly average of these values. Lister *et al.* (2009b) report angular speeds of 0.5-0.8 mas yr^{-1} for the components near the core in OJ 287, which means that we measure changes in the PA with a delay of at least $\sim 1-2$ yr. We note here that our approach does not take into account any asymmetry in the brightness distribution across the jet, which may introduce additional uncertainty about the mean jet direction.

3. Disk model

The accretion disk of the primary BH is modelled by discrete non-interacting particles, 37200 in number. They are placed in circular coplanar orbits between 8 and 20 Schwarzschild radii of the primary black hole. This is the region which is mostly influenced by the gravitational effects of the secondary black hole. The dynamical model is described in detail in Valtonen *et al.* (2010). Note that all the parameters of the binary orbit in this model are fixed by the timing of the optical outbursts. We calculate the orbital elements of the disk particles at each time step, and take their average values for every calendar year. Then a sliding average over 10 years is calculated for every annual point. This smoothing process is necessary since the sound crossing time of the ring is of the order of ten years, and the smoothing does not happen automatically in our simulation, unlike in real gaseous disks. We note that this model is clearly oversimplified as it neglects viscosity and pressure in the disk, but we use it here to give a first approximation of the disk bending. A more physical model including viscosity is currently being worked on.

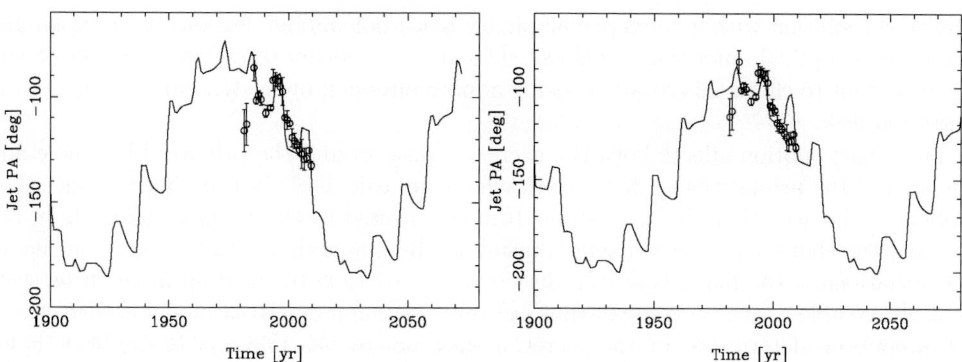

Figure 1. The line shows the variation of the jet angle as a function of time, based on the binary model of OJ 287. The viewing angle of the jet varies between 1 and 2 degrees. The circles represent the yearly average PA observed in the 2-6 cm wavelength range. The error bars represent the standard deviation of the PA when more than two observations per year were available. *Left:* A 4-year lag is introduced between changes in the accretion disk orientation and jet reorientation. *Right:* The same as the left-hand side, but the lag is 14 years.

The basic cycle of the disk precession due to perturbations of the secondary BH appears to be 120 yr. We may now assume that the jet lies along the rotation axis of the disk and when the disk orientation changes, the jet axis also changes, with some time lag. Figure 1 shows the predicted PA changes (solid line). Note that we have only three free parameters since everything else has been fixed by the optical model. The first parameter is the zero point of the radio position angle. It can be changed by sliding the theoretical curve up and down vertically in Figure 1. The second parameter is related to the time lag, and it is changed by sliding the theoretical curve horizontally along the time axis. The third parameter is the zero point of viewing angle, varying of which stretches or compresses the vertical scale. This last parameter can be constrained by requiring that the model is consistent with the viewing angles derived from long-term flux density variations.

Figure 1 shows the two best fits. Both have the same viewing angle range of 1–2 degrees, which is consistent with radio variability studies (e.g., Savolainen *et al.* 2010), but different values of time lags. The model with a 4-yr delay between the changes in the accretion disk orientation and the re-orientation of the pc scale jet fits the data surprisingly well, except for the two earliest data points which are from four-station observations with the US VLBI Network. The poor (u, v)-coverage of these observations leaves the jet orientation more uncertain than in the rest of the data.

The model predicts that the jet PA turns clockwise by ~ 50 degrees during the next 10–20 years, with the exact timing depending on the chosen time lag. Interestingly, Agudo *et al.* (2010) reported that a new jet component recently appeared at PA = 200 degrees in the 43 GHz VLBA data. However, during the last few years there have been multiple ejections with widely differing position angles in the 43 GHz jet (D'Arcangelo *et al.* 2009, Agudo *et al.* 2010), which may indicate that these components are moving kinks in the jet and therefore their apparent path may deviate from the mean jet direction. It will be important to follow the jet PA with 15 GHz VLBI observations during the next few years in order to see if the mean direction of the jet continues to turn as predicted.

4. Discussion

As shown in Figure 1, the direction of the parsec scale radio jet of OJ 287 shows significant evolution on the plane of the sky. We have tested if it is possible to fit this observed evolution with a very specific binary black hole model developed for explaining the periodic optical outbursts of OJ 287. The answer appears to be yes. However, it will be important to check this result by using a more physical, hydrodynamical model of the accretion disk.

The binary motion affects both the accretion disk around the primary black hole and the spin of the primary black hole. Both effects are calculable in the binary model. The wobble of the accretion disk axis seems to be connected to the jet position angle in the sky and the changes are most likely transmitted by magnetic fields. On the other hand, the variations in the black hole spin direction are too slow to show up in the time scale that we are able to study. We may thus say that assuming our BBH model is correct, the jet direction is determined by the accretion disk, not by the spin axis of the black hole.

It has been previously noticed that the optical outbursts lag behind tidal influences in the disk by about three months (Valtonen *et al.* 2006). Considering the distance of the perturbed ring from the central black hole, the speed of transmission of particles from the disk to the jet would be of the order of the orbital speed of the disk. Thus it appears that in this model particles travel freely along magnetic field lines from the perturbed disk to the jet. The tilting of the accretion disk in the perturbed region (8 – 20 Schwarzschild radii from the central black hole) causes also the tilting of the radio jet.

References

Agudo, I., *et al.* 2010, in *"Fermi meets Jansky - AGN in radio and gamma-rays"*, Savolainen, T., Ros, E., Porcas, R. W., & Zensus, J. A., eds., 143

Blandford & Königl 1979, *ApJ*, 232, 34

D'arcangelo, F. D., *et al.* 2009, *ApJ*, 697, 985

Fey, A. L., Clegg, A. W., & Fomalont, E. B. 1996, *ApJS*, 105, 299

Fey, A. L. & Charlot, P. 1997, *ApJS* 111, 95

Gabuzda, D. C., Wardle, J. F. C., & Roberts, D. H. 1989, *ApJL*, 336, L59

Gabuzda, D. C. & Cawthorne, T. V. 1996, *MNRAS*, 283, 759

Gabuzda, D. C. & Gomez, J. L. 2001, *MNRAS*, 320, 49

Jorstad, S. G., *et al.* 2001, *ApJS*, 134, 181

Jorstad, S. G., *et al.* 2005, *AJ*, 130, 1418

Lehto, H. J., & Valtonen, M. J. 1996, *ApJ*, 460, 207

Lister, M. L., *et al.* 2009a, *AJ*, 137, 3718

Lister, M. L., *et al.* 2009b, *AJ*, 138, 1874

Ojha, R. *et al.* 2004, *ApJS*, 150, 187

Piner, G., *et al.* 2007, *AJ*, 133, 2357

Rantakyrö, F., *et al.* 1998, *A&AS*, 131, 451

Roberts, D. H., Gabuzda, D. C., & Wardle, J. F. C. 1987, *ApJ*, 323, 536

Savolainen, T., *et al.* 2010, *A&A*, 512, A24

Sillanpää, A., *et al.* 1988, *ApJ*, 325, 628

Sundelius, B., *et al.* 1996, *ASP Conf.Ser.* 110, 99

Sundelius, B., *et al.* 1997, *ApJ*, 484, 180

Tateyama, C. E., *et al.* 1996, *PASJ*, 48, 37

Tateyama, C. E. & Kingham, K. A. 2004, *ApJ*, 608, 149

Tateyama, C. E., *et al.* 1999, *ApJ*, 520, 627

Valtonen, M. J. & Lehto, H. J. 1997, *ApJL*, 481, L5

Valtonen, M. J., *et al.* 2006, *ApJ*, 646, 36

Valtonen, M. J., *et al.* 2008, *Nature*, 452, 851

Valtonen, M. J., *et al.* 2010, *ApJ*, 709, 725

Valtonen, M. J., *et al.* 2011, *ApJ*, in press

Vicente, L., Charlot, P., & Sol, H. 1996, *A&A*, 312, 727

Discussion

EMMANOULOPOULOS: Is the model that you suggest stable from a MHD point of view?

VALTONEN: We have not carried out MHD simulations. However, the particle disk that we use is stable over 10000 years.

DE GOUVEIA DAL PINO: Is there any further evidence of SMBH binary systems?

VALTONEN: OJ287 is the only one that came up in the Tuorla monitoring program of about 50 sources. But there are reasons to expect, based on cosmological models, that more will be discovered if the monitoring is extended to larger samples, say 1000. In addition, there are many rotating radio jets which may indicate a binary system.

FENDT: The binary BH is responsible for precession. May it also affect the path of the jet by gravitational interactions (binding)?

VALTONEN: In the binary model, the secondary goes close to the jet once in every period. It has been noted that the optical flux drops dramatically at those times for a few weeks. One explanation is, as you mentioned, the gravitational jet bending at those times.

Jets at all Scales
Proceedings IAU Symposium No. 275, 2011
G. E. Romero, R. A. Sunyaev & T. Belloni, eds.

© International Astronomical Union 2011
doi:10.1017/S1743921310016169

X-ray radiation of the jets and the supercritical accretion disk in SS 433

Sergei Fabrika[1] and Alexei Medvedev[2]

[1]Special Astrophysical Observatory,
369167, Nizhij Arkhyz, Russia
email: fabrika@sao.ru

[2]Moscow State University,
119992, Moscow, Russia
email: a.s.medvedev@gmail.com

Abstract. The observed X-ray luminosity of SS 433 is $\sim 10^{36}$ erg/s, it is known that all the radiation is formed in the famous SS 433 jets. The bolometric luminosity of SS 433 is $\sim 10^{40}$ erg/s, and originally the luminosity must be realized in X-rays. The original radiation is probably thermalized in the supercritical accretion disk wind, however the missing more than four orders of magnitude is surprising. We have analysed the XMM-Newton spectra of SS 433 using a model of adiabatically and radiatively cooling X-ray jets. The multi-temperature thermal jet model reproduces very well the strongest observed emission lines, but it can not reproduce the continuum radiation and some spectral features. We have found a notable contribution of ionized reflection to the spectrum in the energy range from ~ 3 to 12 keV. The reflected spectrum is an evidence of the supercritical disk funnel, where the illuminating radiation comes from deeper funnel regions, to be further reflected in the outer visible funnel walls ($r \geqslant 2 \cdot 10^{11}$ cm). The illuminating spectrum is similar to that observed in ULXs, its luminosity has to be no less than $\sim 10^{39}$ erg/s. A soft excess has been detected, that does not depend on the thermal jet model details. It may be represented as a BB with a temperature of $T_{bb} \approx 0.1$ keV and luminosity of $L_{bb} \sim 3 \cdot 10^{37}$ erg/s. The soft spectral component has about the same parameters as those found in ULXs.

Keywords. X-rays: individual (SS 433), accretion, accretion disks, black hole physics

---

SS 433 is the only known persistent superaccretor in the Galaxy – a source of relativistic jets (Fabrika (2004) for review). This is a massive close binary, where the compact star is most probably a black hole. Its intrinsic luminosity is estimated to be $\sim 10^{40}$ erg/s, with its maximum located in non-observed UV region. Almost all the observed radiation is formed in the supercritical accretion disk, and the donor star contributes less than 20 % of the optical radiation. The extreme luminosity of the object is supported by a very well measured kinetic luminosity of the jets, $\sim 10^{39}$ erg/s, both in direct X-ray and optical studies of the jets and in the studies of the jet-powered nebula W 50. At the same time we know that practically all the energy at the accretion onto a relativistic star is released in X-rays. This means that the observed radiation of SS 433 is a result of thermalization of the original radiation in the strong wind forming in the supercritical disk.

The observed X-ray luminosity of SS 433 is $\sim 10^{36}$ erg/s and it is believed that all the X-rays come from the cooling X-ray jets (Kotani *et al.* 1996; Marshall, Canizares, & Schulz 2002; Brinkmann, Kotani, & Kawai 2005). The observed X-ray radiation is at least four orders of magnitude less than the bolometric luminosity, which originally must be released in X-rays. Both the orientation of SS 433 and the visibility conditions do not allow us to see any deep areas of the supercritical accretion disk funnel. However, the strong mass loss of the accretion disk gives us a hope for detecting some indications of

280

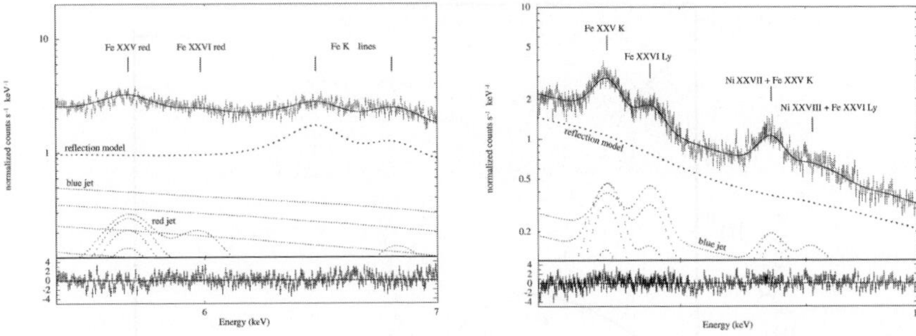

Figure 1. Observed spectrum of SS 433 (orbit 707, $\psi \sim 0$) with the thermal jet model components and the additional reflection model in two spectral regions 5.0–7.0 (left) and 7.0–10.0 keV (right). The total model spectrum is shown with solid line. The reflection model explains well the fluorescence iron line at 6.4 keV, the recombination iron Kα line of Fe XXV at 6.7 keV and the iron absorption edge (Kubota *et al.* 2007) at $\sim$8–9 keV.

the funnel radiation. The goal of our study is to find these indications of the funnel in the SS 433 X-ray radiation.

We have analysed XXM spectra of SS 433 with a well-known standard model of the adiabatically cooling X-ray jets, taking into account cooling by radiation (Medvedev & Fabrika 2010). We have selected only those observations (Brinkmann, Kotani, & Kawai 2005) with highest S/N, where disk was not eclipsed by the donor and was the most open to the observer (precessional phases $\psi = 0.04$ and 0.84). We confirm that the jet model reproduces the iron emission line fluxes quite well. However, the thermal jet model alone can not reproduce the continuum radiation in the XMM spectral range. We use then the multi-temperature thermal jet model together with the REFLION ionized reflection model (Ross & Fabian 2005). We divide the whole spectrum into four parts ($0.8 - 2.0$, $2.0 - 4.0$, $4.0 - 7.0$ and $7.0 - 12.0$ keV) and find the reflection model parameters independently in each part, but with the same jet parameters and the same IS absorption N_H.

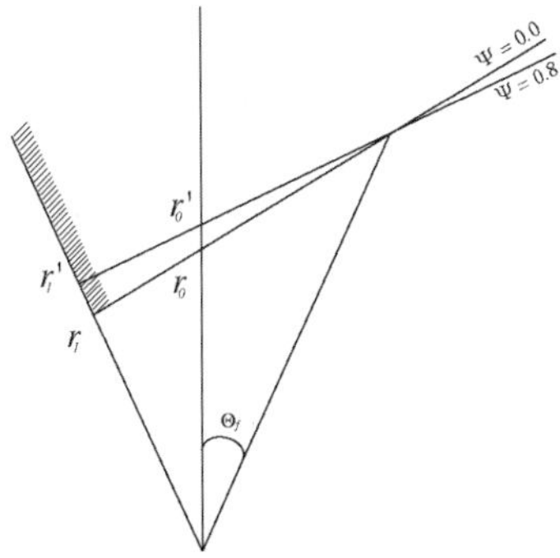

Figure 2. A sketch of the funnel and the blue jet visibility at two precessional phases considered. r_0 (r_0') is the distance between the base of the visible jet and the top of the cone (the bottom of the funnel). r_1 (r_1') is the distance between the base of the visible funnel wall and the top of the cone.

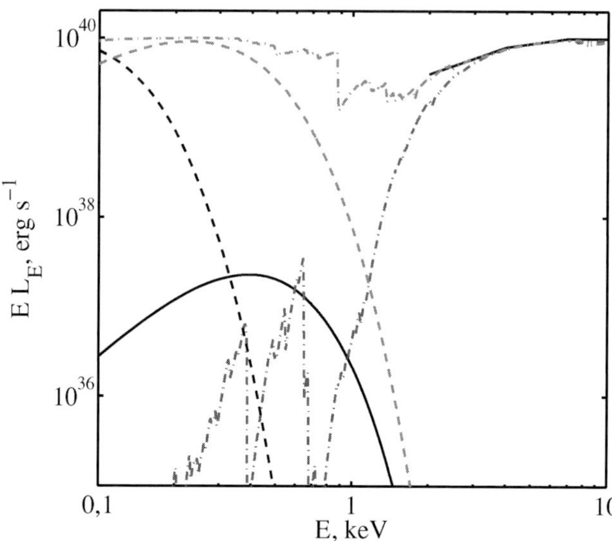

Figure 3. Additional components in X-ray spectrum of SS 433 (the thermal jet component is not shown). The illuminating incident radiation found in reflection model is shown by solid lines (arbitrary scaled to $ELE = 10^{40}$ erg/s) in three energy ranges: $2.0-4.0$ keV ($\Gamma = 1.0$), $4.0-7.0$ keV ($\Gamma = 1.6$) and $7.0-12.0$ keV ($\Gamma = 2.0$). The soft BB component ($T_{bb} = 0.1$ keV, $L_x \sim 3{\cdot}10^{37}$ erg/s) detected in the $0.8-2.0$ keV energy range is shown by solid line. There are two originally the same flat spectra after absorption in ABSORI model with $N_H = 10.0 \times 10^{22}$ cm^{-2}, $\xi = 150$ (grey dash-dotted line) and $N_H = 6.0 \times 10^{22}$ cm^{-2}, $\xi = 3$ (bold dash-dotted line). There are two MCF (multi-colour funnel model) non-absorbed spectra with $r_1 = 1.7 \cdot 10^{11}$ cm (bold dotted line) and $r_1 = 5 \cdot 10^{10}$ cm (grey dotted line).

Fig. 1 shows two spectral regions with the "thermal jet + reflection" model, we find quite a good representation of the spectra. The reflection model explains well the spectral features unexplained before. We find that the ionization parameter is about the same in the different X-ray ranges, $\xi \sim 300$, which indicates a highly ionized reflection surface. The illuminating radiation photon index changes from flat, $\Gamma \approx 2$, in the $7-12$ keV range to $\Gamma \approx 1.6$ in the range $4-7$ keV and to $\Gamma \leqslant 1$ in the range $2-4$ keV.

Fig. 2 shows the jet and the probable funnel wall visibility at the two precessional phases studied. We found the visible blue jet base is $r_0 \approx 2 \cdot 10^{11}$ cm, the gas temperature at r_0 is $T_0 \approx 17$ keV. The IS gas column density is $N_H \sim 1.5 \cdot 10^{22}$ cm^{-2}, which is in good agreement with the IS extinction found in optical and UV observations. We confirm the previous finding (Kotani *et al.* 1996; Brinkmann, Kotani, & Kawai 2005) that the red jet is probably not seen (blocked) in the soft energy range of $0.8-2.0$ keV. However, both the model with blocked the red jet portions, and the model with the whole red jet visible (but with some higher N_H) give the same result, that the thermal jet model alone can not explain the soft continuum. We also confirm that Nickel is highly overabundant, at least 10 times, in the jets (Kotani *et al.* 1996; Brinkmann, Kotani, & Kawai 2005).

We have not found any evidences of the reflection in the soft $0.8-2.0$ keV energy range, instead the soft excess is detected in spectra. The soft excess does not depend on the thermal jet model details, this model alone can not explain the soft continuum. We suppose that in the soft X-rays we observe direct radiation of the visible funnel wall. We represented this component (Fig. 3) as a black body radiation with a temperature of

$T_{bb} \approx 0.1$ keV and a total luminosity (being observed face-on at $r > r_1 = 1.7 \cdot 10^{11}$ cm) of $L_{BB} \sim 3 \cdot 10^{37}$ erg/s. The soft excess may be also fitted with a multicolour funnel (MCF) model (Fabrika *et al.* 2006). The soft excess is observed in the spectra of ULXs with the soft component temperature of $T \sim 0.1$ keV (Stobbart, Roberts, & Wilms 2006), which is similar to that found in SS 433. If the ULXs or some of them are nearly face-on versions of SS 433, one may adjust their soft X-ray components to the outer funnel walls radiation (Poutanen *et al.* 2007).

We conclude that the additional reflected spectrum is an indication of the funnel radiation. The illuminating radiation spectrum is flat in the range of $7-12$ keV, as it is expected in supercritical accretion disks (Poutanen *et al.* 2007). Comptonization may extend and flatten the spectrum to higher energies. With multiple scatterings in the funnel the hard radiation may survive absorption. The observed reflected luminosity in this $7-12$ keV spectral range is $L_{refl} \sim 10^{36}$ erg/s. We find that the illuminating luminosity has to be no less than $\sim 10^{39}$ erg/s (Medvedev & Fabrika 2010).

In the range $2.0-7.0$ keV the additional reflected spectrum is curved (Fig. 3). Therefore we expect an existence of an absorbing (reflecting) medium inside the funnel, which produces such a curved spectrum of radiation, illuminating the outer reflected surface. Presumably this medium is the deep funnel regions. The recent data show (Stobbart, Roberts, & Wilms 2006; Berghea *et al.* 2008) that ULXs posses curvature and rather flat X-ray spectra, which are difficult to interpret with a single-component or any other simple model. We note that the main properties of the ULXs X-ray spectra are similar to those restored additional spectral components in SS 433, the ULXs may be "face-on" versions of SS 433.

The softer $(2-7$ keV) part of the illuminating spectrum (Fig. 3) carries a trace of absorption. This is assumed to be due to the multiple scatterings in the funnel. It is important to note that an existence of He- and H-like absorption edges at about zero velocity is a mandatory property of the funnel spectrum to be able to produce the observed jet velocity due to the line-locking mechanism. The jet velocity value, $v_j \approx 0.26c$, and its unique stability, where the velocity does not depend on the activity state, indicate that the jet acceleration must be controlled by the line-locking mechanism (Shapiro, Milgrom, & Rees 1986; Fabrika 2004).

References

Berghea C. T., Weaver K. A., Colbert E. J. M., & Roberts T. P., 2008, *ApJ*, 687, 471
Brinkmann W., Kotani T., & Kawai N., 2005, *A&A*, 431, 575
Fabrika S., 2004, *ASPRv*, 12, 1
Fabrika S., Karpov S., Abolmasov P., & Sholukhova O., 2006, *IAUS*, 230, 278
Kotani T., Kawai N., Matsuoka M., & Brinkmann W., 1996, *PASJ*, 48, 619
Kubota K., Kawai N., Kotani T., Ueda Y., & Brinkmann W., 2007, *ASPC*, 362, 121
Marshall H. L., Canizares C. R., & Schulz N. S., 2002, *ApJ*, 564, 941
Medvedev A. & Fabrika S., 2010, *MNRAS*, 402, 479
Poutanen J., Lipunova G., Fabrika S., Butkevich A. G., & Abolmasov P., 2007, *MNRAS*, 377, 1187
Ross R. R. & Fabian A. C., 2005, *MNRAS*, 358, 211
Shapiro P. R., Milgrom M., & Rees M. J., 1986, *ApJS*, 60, 393
Stobbart A.-M., Roberts T. P., & Wilms J., 2006, *MNRAS*, 368, 397

Discussion

KUNDT: In my 2005 Springer book "Astrophysics, a new approach", I have listed seven independent distance estimates for SS 433, all of order 3kpc; which leave the system in the ballpark of galactic neutron-star energetics. What distance to SS 433 did you adopt? It is important for luminosity estimates.

FABRIKA: We adopted a distance 5.1 kpc according to results from ratio imaging and the kinetic model (5.0 − 5.5 kpc).

MILLER-JONES: A comment: Lockman, Blundell, & Goss (2007) took higher resolution HI spectra of SS 433 and found a distance consistent with that derivd from the radio kinetic model, i.e., 5.5 kpc.

Jets at all Scales
Proceedings IAU Symposium No. 275, 2011
G. E. Romero, R. A. Sunyaev & T. Belloni, eds.

© International Astronomical Union 2011
doi:10.1017/S1743921310016170

The disk/jet connection in the enigmatic microquasar Cygnus X-3

Karri I. I. Koljonen[1], Diana C. Hannikainen[1,2], Michael L. McCollough[3], Guy G. Pooley[4], Sergei A. Trushkin[5], Marco Tavani[6] and Robert Droulans[7]

[1]Aalto University Metsähovi Radio Observatory, Metsähovintie 114, 02540 Kylmälä, Finland
email: `karri@kurp.hut.fi`

[2]Finnish Centre for Astronomy with ESO (FINCA), University of Turku, Väisäläntie 20, FI-21500 Piikkiö, Finland

[3]Smithsonian Astrophysical Observatory, 60 Garden Street, Cambridge, MA 02138-1516, USA

[4]Astrophysics Group, Cavendish Laboratory, 19 J. J. Thomson Avenue, Cambridge CB3 0HE, UK

[5]Special Astrophysical Observarory RAS, Karachaevo-Cherkassian res, Nizhnij Arkhyz, 36916, Russia

[6]INAF-IASF, I-00133 Rome, Italy

[7]CESR/CNRS – Université de Toulouse, 9 Av. du Colonel Roche, 31028 Toulouse Cedex 04, France

Abstract. Simultaneous multi-wavelength observations are crucial for understanding the physics of microquasars, especially the accretion disk/jet connection. The enigmatic microquasar Cygnus X-3 exhibits strong, relativistic jet ejection events producing radio flares up to 20 Jy. These events are preceded by a very soft X-ray state with quenched emission in the radio and hard X-ray bands. Recently, GeV flux was observed by the *AGILE* and *Fermi* γ-ray observatories during the newly-identified hypersoft state. By using an extensive database of simultaneous multi-wavelength observations gathered from Cygnus X-3 we can form a more unified picture of the nature of the source and show how the recent γ-ray observations fit into it.

Keywords. Accretion, accretion disks – Binaries: close – Gamma rays: observations – Radio continuum: stars – Stars: winds, outflows – X-rays: binaries – X-rays: individual: Cygnus X-3

1. Introduction

Cygnus X-3 (Cyg X-3) is a well-known X-ray binary (XRB) discovered in 1966 (Giacconi *et al.* 1967) but whose true nature remains a mystery despite extensive multi-wavelength observations throughout the years. Its X-ray lightcurve (Parsignault *et al.* 1972) shows strong 4.8-hour orbital modulation and the infrared lightcurve (Mason *et al.* 1986) weaker modulation, typical of low-mass XRBs. However, infrared observations suggest that its mass-donating companion is a Wolf-Rayet (WR) star (van Kerkwijk *et al.* 1992), which would make it a high-mass XRB. Also, unlike most other XRBs, Cyg X-3 is relatively bright in the radio virtually all of the time and it undergoes giant radio outbursts with strong evidence of jet-like structures moving away at relativistic speeds (e.g. Mioduszewski *et al.* 2001). The nature of the compact object is not certain, but it is thought to be a black hole due to its spectral resemblance to other black hole XRB systems, such as GRS 1915+105 and XTE J1550−564. In addition, there is no evidence of a neutron star system producing such massive radio outbursts (up to $\sim$20 Jy, Waltman *et al.* 1995) as observed from Cyg X-3. Cyg X-3 is a unique system in our Galaxy, but

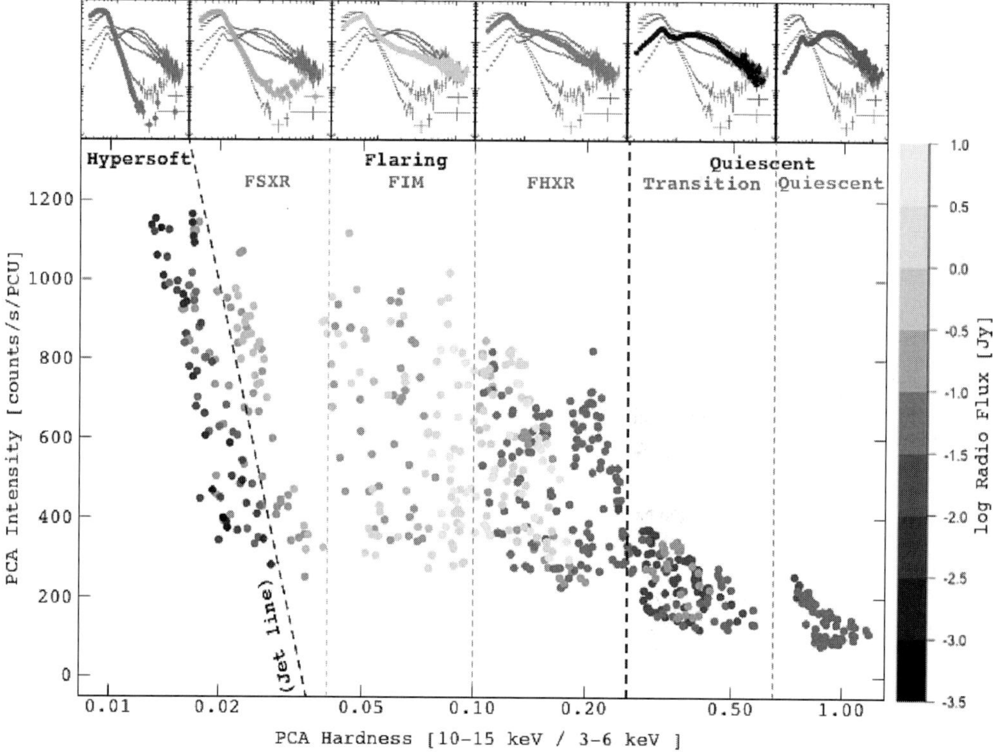

Figure 1. The hardness-intensity diagram showing the added information gained when including the radio dimension (Koljonen *et al.* 2010). The radio flux is represented by the hue of the points, with lighter colors referring to higher flux and darker to lower flux. The six different spectral states of Cyg X-3 are clearly demarcated in the plot, and the X-ray spectra are plotted above their respective areas. The jet line, between the hypersoft and the flaring soft X-ray state, is marked: a major flare occurs when the source transits across this line from the hypersoft state to the FSXR state.

recent observations of two XRBs in IC 10 (Prestwich *et al.* 2007) and NGC 300 (Carpano *et al.* 2007), both containing a WR companion, show strong evidence of a black hole primary. It is worth noting that these systems may represent a crucial link towards the evolution of a double black hole binary.

Here we shall review the radio/X-ray states of Cyg X-3 and show where the GeV emission fits into this picture.

2. Unified picture of Cyg X-3: the radio/X-ray states

It was clear early on that in order to make sense of the spectral evolution of Cyg X-3, multi-wavelength observations are necessary (e.g. McCollough *et al.* 1999; Szostek, Zdziarski & McCollough 2008; Koljonen *et al.* 2010). In Fig. 1 (reproduced from Koljonen *et al.* 2010), the HID for Cyg X-3 is plotted using additional data from (near) simultaneous radio coverage. In this way the radio/X-ray states can be divided according to the X-ray hardness and/or radio flux into three distinct areas (hypersoft, flaring, quiescent). In addition, the flaring state can be further subdivided according to X-ray hardness and the shape of the spectra: flaring/soft X-ray (FSXR), flaring/intermediate (FIM) and flaring/hard X-ray (FHXR). The quiescent state can also be further subdivided into two regions: quiescent and transition. These subdivisions are rather ad hoc since the flaring spectra and the quiescent spectra are more or less continuous, but we nevertheless make these distinctions so as to preserve the legacy of previous studies (e.g. Szostek, Zdziarski & McCollough 2008).

3. Major flares in Cyg X-3

In the following we describe the characteristics of a major flaring event in Cyg X-3:

(a) Once Cyg X-3 enters the radio/HXR quenched state, denoted as the hypersoft state in Fig. 1, it will remain in that state for a couple of days up to a month. It is a rare state, comprising $\sim$ 2–3 % of the overall monitoring live time. Upon emerging from the hypersoft state, Cyg X-3 will always exhibit major flaring (Szostek, Zdziarski & McCollough 2008; Koljonen *et al.* 2010).

(b) The HXR and radio switch from an anti-correlation to a correlation when entering the hypersoft state and the HXR flare together with the radio (McCollough *et al.* 1999). However, major flares differ depending on how strong the accompanying HXR flare is (Koljonen *et al.* 2010). Variations can also occur within the same flaring episode.

(c) During a major flaring episode, the X-ray spectra evolve from soft to hard (Szostek, Zdziarski & McCollough 2008; Koljonen *et al.* 2010). The episode begins in the hypersoft state that is then followed by a rising non-thermal tail in the X-ray spectra (i.e. FSXR–FIM states) until the tail exhibits a cutoff (FHXR state). However, the FHXR and FIM states can occur in the opposite order after the peak of the major flare.

(d) The SEDs after a major flare can be fitted with a microquasar model consisting of soft X-ray seed photons from the inner accretion disk, a Comptonizing population of hot electrons and a synchrotron emitting diluted jet. Recently, we have found that the radio emission shows a correlation with the K-band (McCollough *et al.*, in prep.), strengthening the notion of jet emission extending into the IR. Preliminary results from the fitting shows an optically thin radio spectra, strong absorption from IR to SXR and decreasing optical depth as the major flare decay (Koljonen & McCollough 2008; Koljonen *et al.*, in prep.).

(e) Interestingly, major flares show up as multiple-peaked in the radio (see possible explanation for double-peaked flares in Tammi & Hovatta 2010).

(f) QPOs are sometimes associated with a decaying major flare; see Fig. 2 where the QPOs detected in van der Klis & Jansen (1985) are shown in the context of GBI radio monitoring (Koljonen *et al.*, in prep.).

4. γ-rays in Cyg X-3

GeV emission has recently been detected from Cyg X-3 by *AGILE* (Tavani *et al.* 2009) and *Fermi* (Fermi LAT Collaboration *et al.* 2009; Corbel, these proceedings). The γ-ray emission arises during specific times characterized by low HXR (Fig. 3) and radio flux and high SXR flux, i.e. the hypersoft state. The γ-ray flares are observed during the declining phase as well as the rising phase to/from the hypersoft state (Tavani *et al.* 2009). It is still under debate whether the γ-rays arise from leptonic (Dubus, Cerutti & Henri 2010) or hadronic processes (e.g. Romero *et al.* 2003).

5. Summary

Cyg X-3 exhibits complicated spectral behavior and it is only through multi-wavelength observations that more comprehensive insight is gained from this system. Here we have shown that by taking into an account simultaneous radio, soft X-ray, hard X-ray and

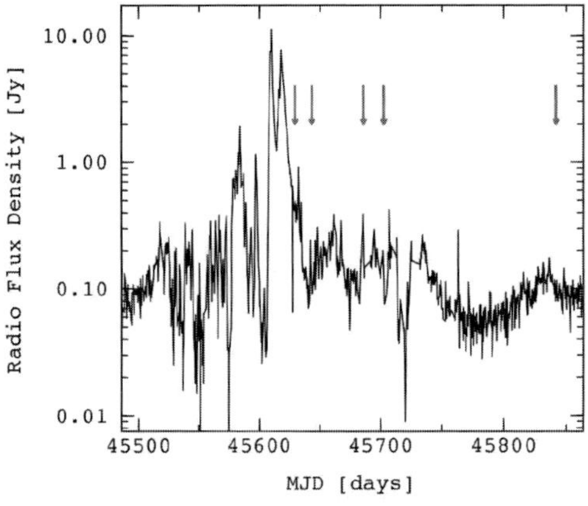

Figure 2. GBI 2.25 GHz lightcurve with red arrows indicating when *EXOSAT*/ME (van der Klis & Jansen 1985) observations exhibited QPOs (Koljonen *et al.*, in prep.).

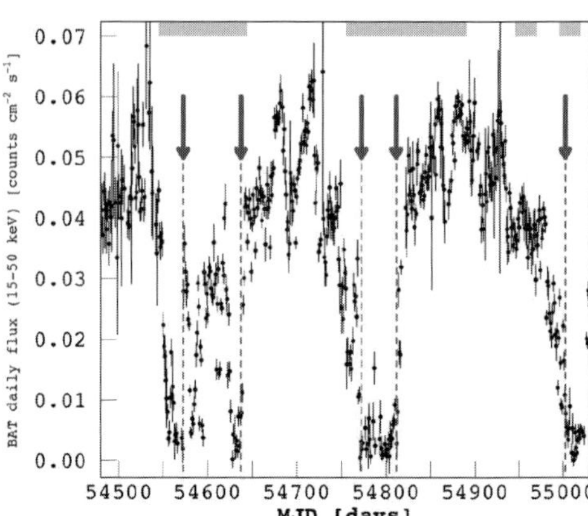

Figure 3. The BAT lightcurve with *AGILE* coverage (shaded gray bars) and the GeV detections (blue arrows).

γ-ray observations we can form an unified picture of the overall spectral evolution of the system. The most interesting spectral state probed by this approach is the hypersoft state, a prelude to major flare event in the system, characterized by quenched emission in the radio and hard X-rays, and peak emission in the soft X-rays and γ-rays. The study of the theoretical implications of these observations is ongoing and a large multi-wavelength collaboration is at the ready to catch this hypersoft state and the accompanying major flaring event over the whole electromagnetic spectrum.

References

Carpano, S., Pollock, A. M. T., Prestwich, A., Crowther, P., Wilms, J., Yungelson, L., & Ehle, M. 2007, *A&A*, 466, L17

Dubus, G., Cerutti, B., & Henri, G. 2010, *MNRAS*, 404, L55

Fermi LAT Collaboration *et al.* 2009, *Science*, 326, 1512

Giacconi, R., Gorenstein, P., Gursky, H., & Waters, J. R. 1967, *ApJ*, 148, L119

Koljonen, K. I. I., Hannikainen, D. C., McCollough, M. L., Pooley, G. G., & Trushkin, S. A. 2010, *MNRAS*, 406, 307

Koljonen, K. I. I. & McCollough, M. L. 2008, *in proceedings of 7th INTEGRAL Workshop*, PoS (Integral08)075

Mason, K. O., Cordova, F. A., & White, N. E. 1986, *ApJ*, 309, 700

McCollough, M. L., Robinson, C. R., Zhang, S. N. *et al.* 1999, *ApJ*, 517, 951

Mioduszewski, A. J., Rupen, M. P., Hjellming, R. M. *et al.* 2001, *ApJ*, 553, 766

Parsignault, D. R., Gursky, H., Kellogg, E. M. *et al.* 1972, *Nature*, 239, 123

Prestwich, A. H., *et al.* 2007, *ApJL*, 669, L21

Romero, G. E., Torres, D. F., Kaufman Bernadó, M. M., & Mirabel, I. F. 2003, *A&A*, 410, L1

Szostek, A., Zdziarski, A. A. & McCollough, M. 2008, *MNRAS*, 388, 100

Tammi, J. & Hovatta, T. 2010, *International Journal of Modern Physics D*, 19, 971

Tavani, M., Bulgarelli, A., Piano, G. *et al.* 2009, *Nature*, 462, 620

van der Klis, M. & Jansen F. A. 1985, *Nature*, 313, 768

van Kerkwijk, M. H., Charles, P. A., Geballe, T. R. *et al.* 1992, *Nature*, 355, 703

Waltman, E. B., Ghigo, F. D., Johnston, K. J., Foster, R. S., Fiedler, R. L., & Spencer, J. H. 1995, *AJ*, 110, 290

Discussion

KALEMCI: For all the other transients, the flares happen during the hard to soft state transition. Why does it happen the other way around for Cyg X-3?

KOLJONEN: This is true and Cyg X-3 indeed is a rather bizarre object. At this point, I would speculate the physical scenario behind the observations, but this might be due to the disappearence of the inner disk into the jet during the jet ejection.

RODRIGUEZ: Mirabel argued that HMXRBs would be stationary with respect to the surrounding medium, How much is known of the motions of Cyg X-3?

KOLJONEN: I don't know about the proper motion Cyg X-3, but maybe James Miller-Jones can say something more about it.

Jets at all Scales
Proceedings IAU Symposium No. 275, 2011
G. E. Romero, R. A. Sunyaev & T. Belloni, eds.

© International Astronomical Union 2011
doi:10.1017/S1743921310016182

The physics of disk winds, jets, and X-ray variability in GRS 1915+105

Joseph Neilsen[1,2], Julia C. Lee[1,2] and Ron Remillard[3]

[1]Harvard University Department of Astronomy, Cambridge, MA 02138, USA

[2]Harvard-Smithsonian Center for Astrophysics, Cambridge, MA 02138, USA

[3]MIT Kavli Institute for Astrophysics and Space Research, Cambridge, MA 02139, USA

Abstract. We present new insights about accretion and ejection physics based on joint *RXTE/Chandra* HETGS studies of rapid X-ray variability in GRS 1915+105. For the first time, with fast phase-resolved spectroscopy of the ρ state, we are able to show that changes in the broadband X-ray spectrum (*RXTE*) on timescales of seconds are associated with measurable changes in absorption lines (*Chandra* HETGS) from the accretion disk wind. Additionally, we make a direct detection of material evaporating from the radiation-pressure-dominated inner disk. Our X-ray data thus reveal the black hole as it ejects a portion of the inner accretion flow and then drives a wind from the outer disk, all in a bizarre cycle that lasts fewer than 60 seconds but can repeat for weeks. We find that the accretion disk wind may be sufficiently massive to play an active role in GRS 1915+105, not only in quenching the jet on long timescales, but also in possibly producing or facilitating transitions between classes of X-ray variability.

1. Introduction

Of all the known Galactic black holes, GRS 1915+105 is undoubtedly the most prolific source of state transitions. Discovered as a transient by GRANAT in 1992 (Castro-Tirado *et al.*), it has remained in outburst for the last 18 years and is typically one of the very brightest sources in the X-ray sky. It is also one of the most variable: its X-ray lightcurve consists of at least 14 different patterns of variability, most of which are high-amplitude and highly-structured (Belloni *et al.* 2000, Klein-Wolt et al. 2002, Hannikainen *et al.* 2005). It is believed that many of these variability classes, which are labeled with Greek letters (Belloni *et al.* 2000), are limit cycles of accretion and ejection in an unstable disk (Belloni et al. 1997, Mirabel *et al.* 1998, Tagger *et al.* 2004).

Of its many classes of X-ray variability, three of the best studied are the χ state, which produces steady optically thick jets (see, e.g. Dhawan *et al.* 2000, Klein-Wolt *et al.* 2002), the β state, a wild 30-minute cycle with discrete ejection events (Mirabel *et al.* 1998), and the ρ state, which is affectionately known as the heartbeat state for the similarity of its lightcurve to an electrocardiogram (see Figure 1, right panel) and which consists of a slow rise followed by a short bright pulse, repeating with a period of roughly 50 s (Taam, Chen, & Swank 1997, Belloni *et al.* 2000). And while these variability classes have clearly established strong connections between the accretion disk and the jet, the physical processes behind this disk-jet connection have yet to be completely revealed.

2. Results and Discussion

In an effort to quantify the physics of these unusual variability classes and the disk-jet connection in GRS 1915+105, we recently analyzed ~ 10 years of high-resolution *Chandra* HETGS observations of this black hole X-ray binary (Neilsen & Lee 2009) and performed follow-up with detailed variability analysis (Neilsen *et al.* 2010b). In our long-term study, we found that during states dominated by jet activity, the X-ray spectra revealed a

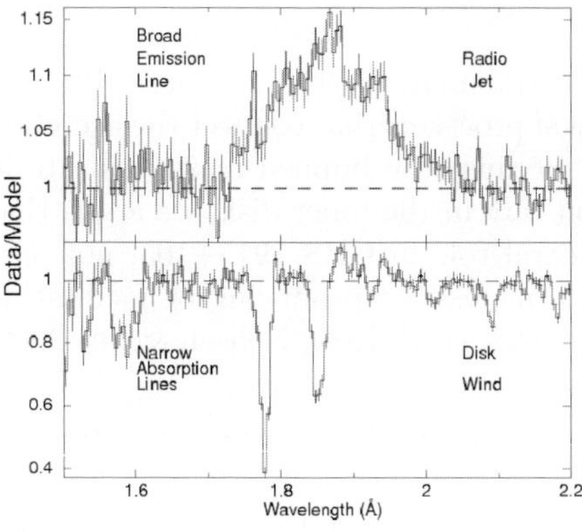

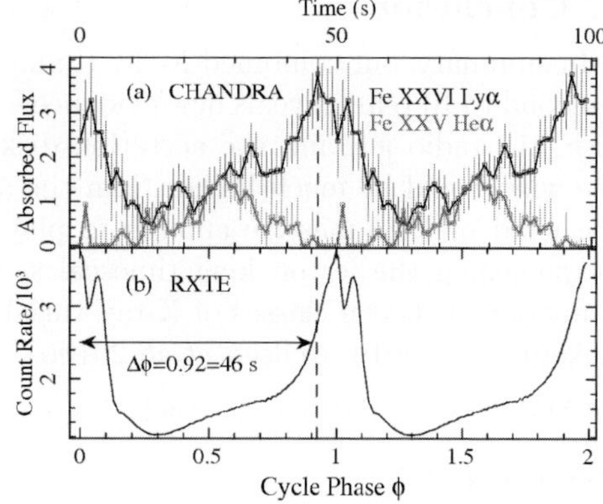

Figure 1. Two results from high-resolution X-ray spectroscopy of GRS 1915+105, clearly indicating links between the disk, wind, and jet on many different timescales. **Left:** The wind-jet connection on long timescales, based on Neilsen & Lee (2009), where we show that the disk wind may carry enough gas away from the black hole to halt the flow of matter into the jet. We detect the wind and the jet via their characteristic spectral signatures: (top) a broad iron emission line produced when the base of the jet illuminates the inner disk, and (bottom) narrow, blueshifted absorption lines originating in the highly-ionized disk wind. **Right:** Phase-resolved spectroscopy of the accretion disk wind in the ρ state from Neilsen *et al.* (2010b), showing strong flux and ionization changes in the disk wind on timescales of 5 seconds. The X-ray lightcurve, phase–folded over many cycles, is shown for comparison. Since the X-ray luminosity variations are insufficient to produce the observed ionization variability, our phase-resolved spectral analysis requires changes in the structure of the wind on timescales much less than one minute.

broad iron emission line, which we argued must originate when the jet illuminates the inner accretion disk. In contrast, we found that states where the jet is quenched display strong, narrow, blueshifted absorption lines from a highly-ionized accretion disk wind (see also Miller *et al.* 2008). We were able to demonstrate that the strength of this wind is anticorrelated with the *fractional* hard X-ray flux, which is therefore a useful diagnostic of both the accretion state and outflow physics. Furthermore, we discovered that the wind carries enough matter away from the black hole to suppress the jet, so that mass ejection is regulated in GRS 1915+105 (Figure 1, left panel).

To explore the implications of this result on short timescales and reveal fast (1 s) changes in the accretion flow, we have also performed the very first phase-resolved spectroscopy of the ρ variability class in GRS 1915+105 (Neilsen et al. 2010b) using a joint *RXTE/Chandra* observation. Through a combination of X-ray timing and both broadband and high-resolution X-ray spectroscopy, we show for the first time that changes in X-ray continuum on timescales of seconds (as probed by *RXTE*) are associated with measurable changes in absorption lines (*Chandra* HETGS) from the accretion disk wind (Fig. 1, right panel). Because the X-ray luminosity does not change enough to produce the observed ionization, the density of the disk wind must be modulated periodically on these timescales (see Lee et al. 2002 for additional evidence of wind structural changes).

From our broadband spectral analysis, we find spectroscopic evidence for a *local* Eddington limit (Fukue 2004, Lin *et al.* 2009) and the radiation pressure instability in the inner accretion disk (e.g. Lightman & Eardley 1974, Belloni *et al.* 1997). Our phase-resolved spectral analysis also allows us to detect bremsstrahlung emission from material evaporating from the inner accretion flow (see also Janiuk & Czerny 2005). Follow-up comprehensive analysis of all *RXTE* observations of the ρ state (Neilsen *et al.* 2010a) suggest that this periodic evaporation process is an essential component of this strange class of variability.

3. Conclusion

In summary, our combined X-ray timing and spectral analysis probing timescales from 1 second to 10 years reveals new evidence for physical processes that connect the accretion disk, the radio jet, and the accretion disk wind. Because the implied mass loss rate in the wind could be much higher than the accretion rate in the inner disk, we argue that the wind may be massive enough to play an integral role in GRS 1915+105, not only in quenching the jet on long timescales, but also in possibly producing or facilitating transitions between classes of X-ray variability (Shields *et al.* 1986, Neilsen & Lee 2009, Luketic *et al.* 2010, Neilsen *et al.* 2010b).

References

Belloni, T., Mendez, M., King, A. R., van der Klis, M., & van Paradijs, J. 1997, *ApJ Lett.*, 479, L185

Belloni, T., Klein-Wolt, M., Mendez, M., van der Klis, M., & van Paradijs, J. 2000, *A&A*, 355, 271

Castro-Tirado, A. J., Brandt, S., & Lund, S. 1992, *IAU Circ.* 5590

Fender, R. P. & Belloni, T. 2004, *ARAA*, 42, 317

Dhawan, V., Mirabel, I. F., & Rodriguez, L. F. 2000, *ApJ*, 543, 373

Fukue, J. 2004, *PASJ*, 56, 569

Hannikainen, D. *et al.* 2005, *A&A*, 435, 995

Janiuk, A. & Czerny, B. 2005, *MNRAS*, 356, 205

Lee, J. C., Reynolds, C. S., Remillard, R. A., Schulz, N. S., Blackman, E. G., & Fabian, A. C. 2002, *ApJ*, 567, 1102

Lin, D., Remillard, R. A., & Homan, J. 2009, *ApJ*, 696, 1257

Lightman, A. & Eardley, D. M. 1974, *ApJ Lett.*, 187, L1

Luketic, S., Proga, D., Kallman, T. R., Raymond, J. C., & Miller, J. M. 2010, *ApJ*, 719, 515

Klein-Wolt, M., *et al.* 2002, *MNRAS*, 331, 745

Miller, J. M., Raymond, J., Reynolds., C. S., Fabian, A. C., Kallman, T. R., & Homan, J. 2008, *ApJ*, 680, 1359

Mirabel, I. F. *et al.* 1998, *A&A*, 330, L9

Neilsen, J. & Lee, J. C. 2009, *Nature*, 458, 481

Neilsen, J., Lee, J. C., & Remillard, R. A. 2010a, *ApJ Lett.*, submitted

Neilsen, J., Remillard, R. A., & Lee, J. C. 2010b, *ApJ*, submitted

Shields, G. A., McKee, C. F., Lin, D. N. C., & Begelman, M. C. 1986, *ApJ*, 306, 90

Taam, R. E., Chen, X., & Swank, J. H. 1997, *ApJ*, 485, L83

Tagger, M., Varniere, P., Rodriguez, J., & Pellat, R. 2004, *ApJ*, 607, 410

Discussion

FALCKE: Did you ever try to phase-stack radio data? E.g., with the VLM you should be able to get second-scale integration times.

NEILSEN: We can't actually phase-fold the radio data because the time resolution is 32 seconds for a 50 second oscillation. But whe have an approved EVLA observation for the next summer where will have 3-second resolution and we will be able to do this.

KALEMCI: What can you say about the power law index and disk/power law flux ratio when the strong broad band noise is observed?

NEILSEN: I don't know the powe law/disk ratio off the top of my head. But I can tell you that this one interval that looks like the hard state (but lasts only 5 seconds) corresponds to the minimum of the cycle, and it is spectrally hard. However, the power law contributes less than about 50% of the luminosity.

MIRABEL: What are the speeds of these winds?

NEILSEN: The X-ray absorption is blueshifted by around 1000 km/s, but it varies from around 400 km/s to as much as 1600 km/s. In general, for our data, the higher the ionization, the faster the wind.

Jets at all Scales
Proceedings IAU Symposium No. 275, 2011
G. E. Romero, R. A. Sunyaev & T. Belloni, eds.

© International Astronomical Union 2011
doi:10.1017/S1743921310016194

GRS1915+105: a comparison of the plateau state to the canonical hard state

Pieter van Oers[1] and Sera Markoff[2]

[1]School of Physics & Astronomy, University of Southampton,
Highfield, Southampton SO17 1BJ, United Kingdom
email: pvo1g09@soton.ac.uk

[2]Astronomical institute "Anton Pannekoek", University of Amsterdam,
Science Park 904, 1098 XH, The Netherlands
email: S.B.Markoff@uva.nl

Abstract. GRS 1915+105 is a very peculiar black hole binary that exhibits accretion-related states that are not observed in any other stellar-mass black hole system. One of these states, however – referred to as the plateau state – may be related to the canonical hard state of black hole X-ray binaries. Both the plateau and hard state are associated with steady, relatively lower X-ray emission and flat/inverted radio emission, that is sometimes resolved into compact, self-absorbed jets. To investigate the relationship between the plateau and the hard state, we fit two multi-wavelength observations using a steady-state outflow-dominated model, developed for hard state black hole binaries. The data sets consist of quasi-simultaneous observations in radio, near-infrared and X-ray bands. Interestingly, we find both significant differences between the two plateau states, as well as between the best-fit model parameters and those representative of the hard state. We discuss our interpretation of these results, and the possible implications for GRS 1915+105's relationship to canonical black hole candidates.

Keywords. black hole physics, accretion, accretion disks, radiation mechanisms: general, X-rays: binaries, galaxies: active, galaxies: jets

1. Introduction

GRS 1915+105 is a hard X-ray transient located in the constellation of Aquila, at $l = 45.37°$, $b = -0.22°$. The obvious parallels to the jets in Active Galactic Nuclei (AGN) led to this source being classified as a "microquasar" (Mirabel & Rodríguez 1998). Observations with instruments onboard the *Rossi X-ray Timing Explorer* (*RXTE*) have revealed a richness in variability, distinguishing GRS 1915+105 from every other known BHB, over which astronomers are still puzzling to this day.

Belloni *et al.* (2000) were able to classify all variability patterns stretching over more than a year into only twelve classes, based on colour-colour diagrams and light curves. Most classes are understood as the interplay of two or three of three basic states (A, B and C). Class χ is reserved for state C exclusively and is usually called the *plateau* state. As both are associated with compact self-absorbed steady outflows (e.g. Homan & Belloni 2005, Remillard & McClintock 2006), it is tempting to compare the plateau state to the "canonical" hard state (HS). However, although the plateau state and the HS share many similarities, it does display some distinct properties that cannot be ignored. For instance, while the BHBs in the HS usually have a luminosity of $\lesssim 10\%$ $L_{\rm Edd}$, the average luminosity observed in the plateau state is $\sim L_{\rm Edd}$. Moreover, the plateau X-ray photon index is never as hard (Fender & Belloni 2004).

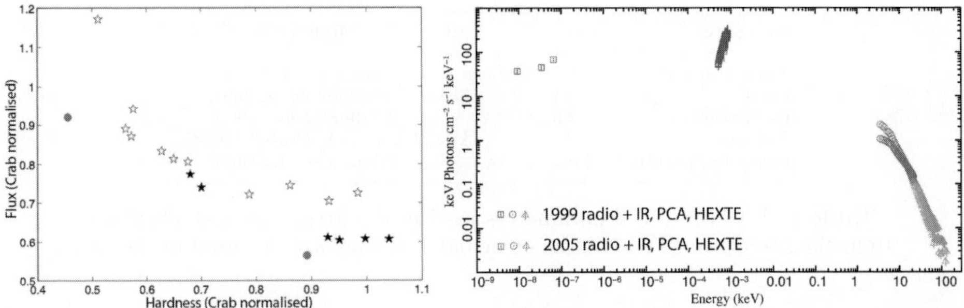

Figure 1. left: HID comparing our observations (filled circles) to the plateau states from Belloni *et al.* (2000) (data courtesy of T. Belloni). The 2005 observation is softer and of higher-luminosity. **right:** Comparison of the multi-wavelength SEDs for the 1999 and 2005 data set.

2. Observations

Based on the lightcurves and its position in the Hardness Intensity Diagram (HID) diagram (see Figure 1), GRS 1915+105 was in the plateau state on July 8th 1999 and April 13th 2005. To obtain multi-wavelength spectra for these episodes, quasi-simultaneous radio and infrared data were added to X-ray data, taken with *RXTE*: The 1999 data set comprises *RXTE* ObsID 40403-01-09-00, H and K bands from UKIRT and 2.25 & 8.3 GHz radio observations from GBI. The 2005 data set consists of *RXTE* ObsID 90105-07-02-00, K band from CTIO and 15 GHz radio from Ryle telescope.

3. Model and physical parameters

As we are dealing here with multi-wavelength spectra we will fit our datasets using the outflow dominated model from Markoff *et al.* (2005) (hereafter: MNW05), which was specifically designed for simultaneous radio through X-ray datasets. MNW05 has already been applied successfully to various Galactic sources in the HS: Cyg X-1, GX 339-4, XTE J1118+480, GRO J1655-40 and A0620-00 (MNW05, Maitra *et al.* (2009), Migliari *et al.* 2007, Gallo *et al.* 2007). In addition MNW05 has been successfully used to fit spectra from Supermassive Black Holes M81*, SGR A* and NGC4258 (Markoff *et al.* 2008, Markoff *et al.* 2007, Yuan *et al.* 2002). There exists, therefore, a solid framework for the modelling of a variety of BH systems, against which to test fits of data from the GRS 1915+105 plateau state.

In addition to the jet MNW05 includes a simple black body to account for the companion star and a "standard" geometrically thin accretion disc Shakura & Sunyaev (1973). The main physical parameters in our jet are: The jet luminosity (N_j) that represents the power going into the jet and scales with the accretion power at the inner edge of the accretion disc. The power is equally divided to supply the internal and kinetic pressures. We assume that the kinetic energy is carried by cold protons, while the leptons do the radiating. The energy involved in the internal pressure goes into the particle and magnetic energy densities, with a ratio determined by $k = U_{\mathrm{B}}/U_{\mathrm{e}}$). The radius of the jet-base is also a free parameter, r_0. The particles start of in a quasi-thermal (relativistic) Maxwellian distribution, the peak energy of which is determined by the electron temperature (T_e). In a segment of the jet located at some variable distance (z_{acc}) from the base, we assume 75 % of the leptons is accelerated into a power-law distribution with slope p.

All fits are done using the MNW05 model with an additive `Gaussian` line between 6 and 7 keV. These models are either convolved with `reflect` (Magdziarz & Zdziarski 1995) or multiplied with `smedge` (Ebisawa 1991), and consequently multiplied with a

parameter	value	units	Reference
column density	4.7	10^{22} cm^{-2}	Chaty *et al.* (1996)
mass	14	M$_\odot$	Greiner *et al.* 2001
inclination	66	$^\circ$	Fender *et al.* (1999)
distance	11	kpc	Chapuis & Corbel (2004)
donor temperature	4455	K	Alonso *et al.* (1999)

Table 1. Fixed physical parameters used in the fitting process, obtained from the literature. We omit the error bars, as they are not used in the fits.

		similar				distinct	
parameter	units	HS range	GRS 1915+105	parameter	units	HS range	GRS 1915+105
r_0	r$_{\rm g}$	3.5-20.2	20.4	$k(=U_{\rm B}/U_{\rm e})$	r$_{\rm g}$	1-7	692
T_e	10^9	20-52	9.2	$z_{\rm acc}$	10^3 r$_{\rm g}$	0.007-0.4	30
p		2.1-2.9	2.3	$N_{\rm j}$	$10^{-2} L_{\rm Edd}$	0.03-7	0.5

Table 2. 1999 parameters found to be similar (left) to and distinct (right) from canonical HS.

photo-electric absorption model (**phabs**) to account for the interstellar medium. Because including the **reflect** model was found to yield inferior statistics to including the **smedge** model and often yielded reflection component normalisation factors not deemed physical (MNW05) we did not pursue this line of investigation in great detail.

Table 1 shows the physical parameters we took from the literature.

4. Results

4.1. *1999*

From the 1999 data set fits (Figure 2) trends for the definitive plateau state are revealed. Qualitatively we see that, while in the HS we see a synchrotron domination below $\sim$10 keV, the plateau state synchrotron is no longer dominant at these energies, but rather has a contribution of about 1% that of the inverse Compton up to $\sim$50 keV. This result is qualitatively similar to the blazar sequence as found by e.g. Ghisellini *et al.* (1998) and Fossati *et al.* (1998), who find an anti-correlation between the frequency where the synchrotron power peaks and the power going in to the jet. A more quantitative comparison is made in Table 2. Indeed, while some parameters are quite similar to those found for the HS, other are quite distinct. The ratio k is more than 2 orders in magnitude larger in GRS 1915+105and $z_{\rm acc}$ increases an order in magnitude in our source. As may have been expected, the jet power approaches the Eddington luminosity in the plateau state. The relation between k and $z_{\rm acc}$ is further looked into in Polko *et al.* (in press). An correlation between k and the jet luminosity is also suggested by results from Markoff *et al.* (2008).

The bottom 2 plots in Figure 2 show the results of an attempt to find a statistically convincing fit with less extreme parameters. This fit is in fact a different local minimum with only slightly worse statistics ($\chi^2_{\rm red} \sim$1.44), representing a redistribution of the energy budget over the energy related parameters ($k \sim 29$, $N_{\rm j} \sim 1\ L_{\rm Edd}$, $p \sim 1.8$). Indeed, the magnetic domination has dropped by a factor 15, almost comparable to values found in the HS. The reduction in k is mainly compensated by an increase in jet luminosity. In addition we see a decrease in p, which can perhaps be understood considering cooling break arguments and the fact that z_{acc} is so much further out in our source. Because of this increase in distance to the first acceleration zone, the magnetic fields involved in the acceleration are almost expected to have decreased further then usual. Hence electron cooling may be negigible in this case and we could be looking directly at an uncooled

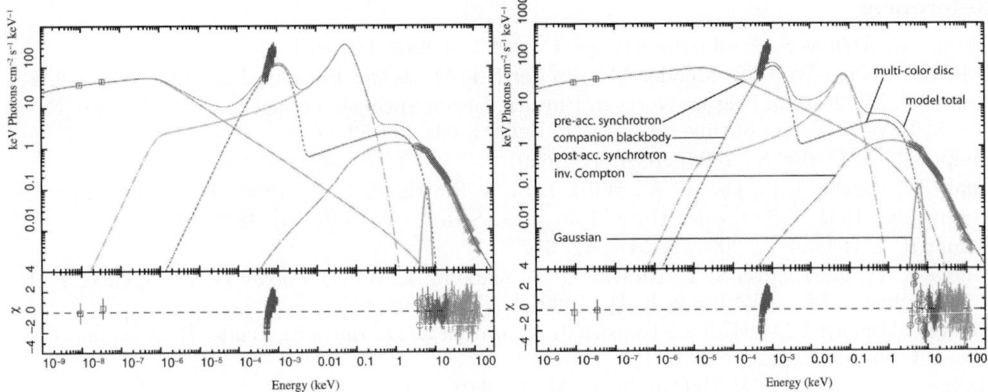

Figure 2. Multi-wavelength best fit (left; $\chi^2_{red} \sim 1.16$) and alternative (right: $\chi^2_{red} \sim 1.44$) fit model spectra for the 1999 data set. The alternative fit is still statistically convincing but reflects less extreme parameters (see text). Individual contributions from the MNW05 and Gaussian models are shown in the plot to the right. The "model total" curve is not forward-folded through the detector response matrices and does not include absorption due to the interstellar hydrogen column density or the `smedge` model.

particle distribution, which would decrease the spectral slope by ~ 0.5 (Heavens & Drury 1988).

4.2. *2005*

Because of the increased X-ray slope compared to the 1999 data set fitting this data set is increasingly difficult. Although two statistically convincing fits ($\chi^2_{red} \sim 1$) have been obtained for this data set, the results were found to have other physical uncertainties. One solution to getting an increased slope is to increase the decaying slope of the inverse Compton component. Reducing the electron temperature T_e is found to have exactly this effect. However the temperature has to be reduced to $\sim 4 \times 10^9$ K which will leave most particles at sub-relativistic speeds with Lorentz factor unity (consistent with thermal Comptonisation). As our model uses a relativistic Maxwellian this is not strictly justified.

Another solution to fitting the increased X-ray slope is to fit this regime with the exponential decay of the pre-acceleration synchrotron component. Although this theoretical shape is normally used, it may in reality be different. In addition this method of fitting requires an extremely hard synchrotron spectrum, normally only found in extremely relativistic sources.

Because of these issues we will not base any firm conclusions on the 2005 data set.

5. Conclusion

The MNW05 model seems to approximate rather well the 1999 data set, but clearly all the parameters need to be pushed into an extreme regime in order to fit to data well. The 2005 data set can be fit, but only marginally, suggesting already a difference between individual plateau states. Perhaps the plateau state would be better compared to the Hard Intermediate State (Belloni 2010), but in this case no reference framework is available yet. Recent evidence however suggests that in contrary to most canonical sources, GRS 1915+105 may be moving on a radiatively efficient track (Corbel *et al.*, in prep.), perhaps explaining some of the extreme differences found between this source and other stellar mass black hole binaries.

References

Alonso A., Arribas S., & Martínez-Roger C., 1999, *A&AS*, 140, 261

Belloni T., Klein-Wolt M., Méndez M., van der Klis M., & van Paradijs J., 2000, *A&A*, 355, 271

Belloni T. M., 2010, in Lecture Notes in Physics, Berlin Springer Verlag, Vol. 794, Lecture Notes in Physics, Berlin Springer Verlag, T. Belloni, ed., pp. 53–+

Chapuis C. & Corbel S., 2004, *A&A*, 414, 659

Chaty S., Mirabel I. F., Duc P. A., Wink J. E., & Rodriguez L. F., 1996, *A&A*, 310, 825

Ebisawa K., 1991, PhD thesis,, Univ. Tokyo; ISAS Research Note No. 483, (1991)

Fender R. & Belloni T., 2004, *ARAA*, 42, 317

Fender R. P., Garrington S. T., McKay D. J., Muxlow T. W. B., Pooley G. G., Spencer R. E., Stirling A. M., & Waltman E. B., 1999, *MNRAS*, 304, 865

Gallo E., Migliari S., Markoff S., Tomsick J. A., Bailyn C. D., Berta S., Fender R., & Miller-Jones J. C. A., 2007, *ApJ*, 670, 600

Greiner J., Cuby J. G., & McCaughrean M. J., 2001, *Nature*, 414, 522

Heavens A. F. & Drury L. O., 1988, *MNRAS*, 235, 997

Homan J. & Belloni T., 2005, *Ap&SS*, 300, 107

Magdziarz P. & Zdziarski A. A., 1995, *MNRAS*, 273, 837

Maitra D., Markoff S., Brocksopp C., Noble M., & Nowak M., Wilms J., 2009, *MNRAS*, 398, 1638

Markoff S., Bower G. C., & Falcke H., 2007, *MNRAS*, 379, 1519

Markoff S., Nowak M., Young A., Marshall H. L., Canizares C. R., Peck A., Krips M., Petitpas G., Schödel R., Bower G. C., Chandra P., Ray A., Muno M., Gallagher S., Hornstein S., & Cheung C. C., 2008, *ApJ*, 681, 905

Markoff S., Nowak M. A., & Wilms J., 2005, *ApJ*, 635, 1203

Migliari S., Tomsick J. A., Markoff S., Kalemci E., Bailyn C. D., Buxton M., Corbel S., Fender R. P., & Kaaret P., 2007, *ApJ*, 670, 610

Mirabel I. F. & Rodríguez L. F., 1998, *Nature*, 392, 673

Remillard R. A. & McClintock J. E., 2006, *ARAA*, 44, 49

Shakura N. I. & Sunyaev R. A., 1973, *A&A*, 24, 337

Yuan F., Markoff S., Falcke H., & Biermann P. L., 2002, *A&A*, 391, 139

Discussion

MEIER: GRS 1915+105 shows a power spectrum with barndwidth limited noise +QPO when in the plateau state. GROJ 1655-40 also has such power spectrum when it produces a steady jet. Have you investigated whether GROJ 1655-40 was also in a plateau state at that time and if it has extreme parameters?

VAN OERS: GROJ1655-40 has been fitted by Migliari *et al.* (2007) but parameters are less extreme.

KALEMCI: GROJ 1655-40 will be extensively discussed in my talk tomorrow.

Jets at all scales
Proceedings IAU Symposium No. 275, 2011
G. E. Romero, R. A. Sunyaev & T. Belloni, eds.

© International Astronomical Union 2011
doi:10.1017/S1743921310016200

Connections between jet formation and multiwavelength spectral evolution in black hole transients

Emrah Kalemci[1], Yoon-Young Chun[1], Tolga Dinçer[1], Michelle Buxton[2], John A. Tomsick[3], Stephane Corbel[4] and Philip Kaaret[5]

[1]Sabanci University, Tuzla
İstanbul, Turkey
email: `ekalemci@sabanciuniv.edu`

[2]Department of Astronomy, Yale University, P.O. Box 208101
New Haven, CT 06520-8101, USA

[3]Space Sciences Laboratory, University of California, Berkeley
7 Gauss Way, Berkeley, CA 94720-7450, USA

[4]Laboratoire Astrophysique des Interactions Multi-echelles (UMR 7158)
CEA/DSM-CNRS-Universite Paris Diderot
CEA Saclay, F-91191 Gif sur Yvette, France

[5]Department of Physics and Astronomy, University of Iowa
Van Allen Hall, Iowa City, IA 52242, USA

Abstract. Multiwavelength observations are the key to understand conditions of jet formation in Galactic black hole transient (GBHT) systems. By studying radio and optical-infrared evolution of such systems during outburst decays, the compact jet formation can be traced. Comparing this with X-ray spectral and timing evolution we can obtain physical and geometrical conditions for jet formation, and study the contribution of jets to X-ray emission.

In this work, first X-ray evolution - jet relation for XTE J1752-223 will be discussed. This source had very good coverage in X-rays, optical, infrared and radio. A long exposure with *INTEGRAL* also allowed us to study gamma-ray behavior after the jet turns on. We will also show results from the analysis of data from GX 339-4 in the hard state with *SUZAKU* at low flux levels. The fits to iron line fluorescence emission show that the inner disk radius increases by a factor of >27 with respect to radii in bright states. This result, along with other disk radius measurements in the hard state will be discussed within the context of conditions for launching and sustaining jets.

Keywords. accretion, black hole physics, line: profiles

---

1. Introduction

Hard state of GBHTs is of great interest because emission from all fundamental accretion components are present: the accretion disk, the corona and the jet. In this state, there is also rich timing information as evident from QPOs and broad band timing features in the power spectra. Our group is especially interested in the hard state during the decays of outbursts, as the decays not only involve state transitions in temporal and spectral properties, but the appearance of jet signatures in the radio and near infrared (NIR) also occurs during decays.

Thanks to intense monitoring campaigns in X-rays with *RXTE* accompanied by radio (especially with ATCA and VLA) and NIR (with SMARTS) of GBHTs, we were able to get a general picture of evolution during decays: A fast timing transition (sudden increase in rms amplitude of variability accompanied by an increase in power-law flux) is followed

299

300   Emrah Kalemci *et al.*

by a hardening a few days to weeks later. Close to the end of hardening, jet signatures are observed in NIR and radio. In some cases, a softening is observed at very low flux levels (see Kalemci, *et al.* 2006a for detailed description). Moreover, observations with HEXTE on *RXTE* indicate disappearance of breaks in the high energy spectra of GBHTs after jet turn on (Kalemci, *et al.* 2005 and 2006b). We continue to monitor GBHTs during decays to improve the overall picture, while extending our studies to the evolution of iron line features and timing at low flux levels.

2. Multiwavelength evolution of XTE J1752−223

GBHT XTE J1752−223 was discovered by *RXTE* in October 2009 (Markwardt, *et al.* 2009). The outburst evolution was monitored intensely by *RXTE* (Shaposhnikov, *et al.* 2010), *MAXI* (Nakahira, *et al.* 2010) and *SWIFT* (Curran, *et al.* 2010). Its X-ray properties are consistent with being a black hole (Munoz Darias, *et al.* 2010).

The source was also monitored in radio (Brocksopp, *et al.* 2009) and observed with EVN and VLBA in radio (Yang, *et al.* 2010). Finally, it is monitored with the Faulkes Observatory in optical (http://staff.science.uva.nl/~davidr/faulkes/xtej1752.html) and also with the SMARTS CTIO 1.3m Telescope in optical (I band) and in NIR (Buxton, *et al.* 2009).

We concentrated on the decay of outburst and characterized the X-ray spectral evolution using the PCA on *RXTE*. We fit the X-ray spectrum with absorption, disk blackbody,

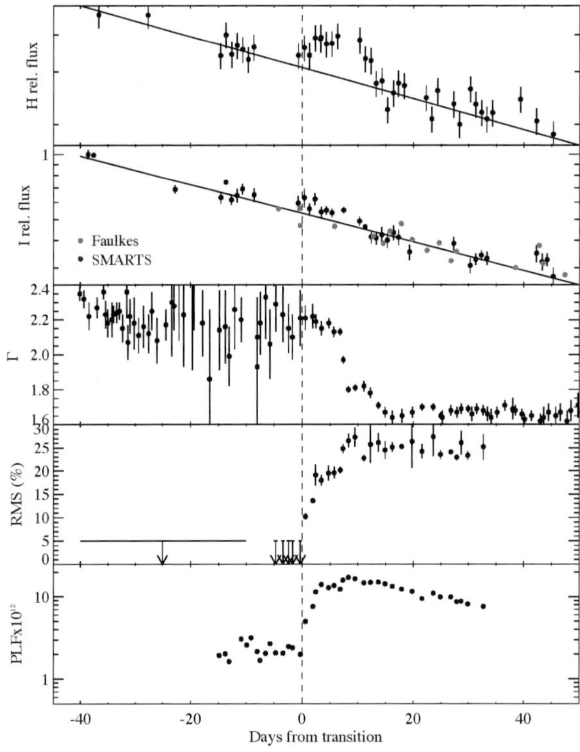

Figure 1. Evolution of NIR (H band) and optical (I band) relative flux with respect to X-ray spectral (spectral index, Γ and power-law flux, PLF) and timing (RMS) evolution. Time 0 is chosen at the beginning of fast timing transition.

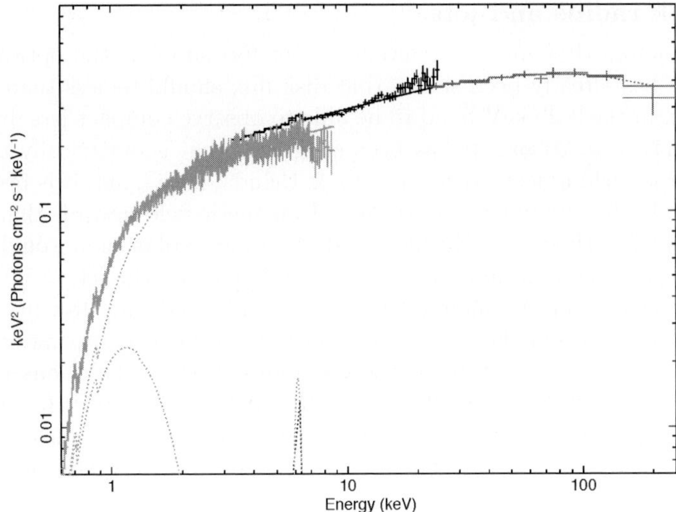

Figure 2. ISGRI, XRT and PCA combined spectrum.

power-law and a smeared edge. If necessary we added a Gauss to represent the iron line. We also created power spectra in 3-25 keV band. On MJD 55282.6 we observed timing features in the power density spectrum (PDS) with 10% rms variability. For observations before this date, the source was in the soft state with no significant timing features in the PDS. In time scale of 2 days, stronger broad band noise filled the PDS and the rms amplitude of variability increased to 20% (see Fig. 1).

In addition we extracted H and I band differential magnitudes using data from the SMARTS CTIO Telescope. The H band observations were fitted with an exponential decay to bring out the brightening in flux. This kind of brightening during outburst decays are common in GBHTs and are associated with optically thin synchrotron radiation from compact jets (Buxton, & Bailyn, 2004, Kalemci, *et al.* 2005, Russell, *et al.* 2006, Coriat, *et al.* 2009) but other possibilities exist (thermal emission from disk, irradiation of secondary star Homan, *et al.* 2005). This is the first time that we observe an increase in NIR flux coinciding with timing transition, well before the hardening of spectrum. It is interesting to note that the optical data do not show a strong (if any) increase in flux. We are investigating the possibility that this increase is due to irradiation, rather than synchrotron, however, lack of strong brightening in optical is in favor of jet interpretation (see Fig. 1, top two panels, which also includes data from Faulkes Observatory, acknowledgment Dave Russell).

Finally we also investigated whether breaks disappear in the hard X-ray spectrum after the jet turns as observed in some other sources (Kalemci, *et al.* 2005, Caballero-Garcia, *et al.* 2007), by modeling the *INTEGRAL* ISGRI spectrum together with *SWIFT* XRT and *RXTE* PCA. The ~300 ks *INTEGRAL* observation was taken around 20 days after the transition. Fig. 2 shows the spectrum and the fit model that consists of absorption, disk blackbody, a power-law with a cutoff and a gaussian. The spectrum clearly requires a break, with folding energy of ~270 keV. If these folding energies are typical for GBHTs after jet turn on, this may explain why HEXTE would miss such breaks for typical short observing observations with *RXTE* for other sources.

3. Inner disk radius and jets

One of the factors that may be relevant for jet formation is the optically thick disk inner radius. It has already been shown that disk flux should be less than a few percent of the total flux in the 3-25 keV band to be able to observe compact jets during outburst decays (Kalemci, *et al.* 2006a). It has been suggested that geometrically thick disks are prerequisite for launching jets (Meier, Koide, & Uchida, 2001), and it is been shown that a coronal mechanism is more likely to transport magnetic field inwards than the optically thick disk (Beckwith, Hawley, & Krolik, 2009). A simple explanation would be a receding accretion disk, and a corona filling the gap before launching the jet.

If this is the case, then we should see evidence of a recessing disk in the hard state before the jet turns on. On the other hand, several work concentrating on the iron line reflection features, and diskblackbody normalization provide result consistent with disk staying close to the innermost stable orbit (Reis, *et al.* 2010, and references therein, Tomsick, *et al.* 2008, but also see Done, & Diaz-Trigo, 2010). In most of these cases, there is evidence of a compact jet. Therefore, the jets can form when the disk is close to the last stable orbit.

We also investigate the relation between jets and inner disk radius further by the analysis of iron line profile at very low fluxes in the hard state. We detected a narrow line at 0.14% Eddington Luminosity for GX 339-4 with *SUZAKU*, and placed the inner disk radius $105R_g$ at $18°$ inclination. The disk recede, while the jet is still active. We conclude that the development of the corona plays an important role on formation of jets, and low disk fluxes are required for this, however the inner disk radius does not play a role on the formation of jets.

References

Beckwith, K., Hawley, J. F & Krolik, J. H. 2009, *ApJ* 707, 428

Brocksopp, C. *et al.* 2009, *ATEL* 2278.

Buxton, M. M. & Bailyn, C. D. *et al.* 2004, *ApJ* 615, 880

Buxton, M. M. *et al.* 2009, *ATEL* 2549.

Caballero Garcia, M. D. *et al.* 2007, *ApJ* 669, 534

Curran, P. A. *et al.* 2010, *MNRAS*, accepted for publication

Coriat, M. *et al.* 2009, *MNRAS*, 400, 123

Done, C. & Diaz-Trigo, M. 2010, *MNRAS*, 407, 2287

Homan, J. *et al.* 2005, *ApJ*, 624, 295

Kalemci, E. *et al.* 2005, *ApJ*, 622, 508.

Kalemci, E. *et al.* 2006a, *in Proceedings of the 6th Microquasar Workshop*, 12

Kalemci, E. *et al.* 2006b, *ApJ*, 639, 340.

Nakahira, N. *et al.* 2010, *PASJ*, accepted for publication

Meier, D. L. Koide, S., & Uchida, Y. 2001, *Science* 291, 84

Markwardt, C. B. *et al.* 2009, *ATEL* 2258.

Munoz-Darias, T. *et al.* 2010, *MNRAS*, 404, L94

Reis, R. C. *et al.* 2010, *MNRAS*, 402, 836

Russell, D. M. *et al.* 2006, *MNRAS*, 371, 1334

Russell, D. M. & Fender, R. *et al.* 2010a, Invited chapter for the edited book "Black Holes and Galaxy Formation", Nova Science Publishers, Inc., at press

Shaposhnikov, N. *et al.* 2010, *ApJ*, accepted for publication

Tomsick, J. A. *et al.* 2008, *ApJ*, 680, 593

Tomsick, J. A. *et al.* 2009, *ApJL*, 707, L87

Yang, J. *et al.* 2010, *MNRAS*, accepted for publication

Jets at all Scales
Proceedings IAU Symposium No. 275, 2011
G. E. Romero, R. A. Sunyaev & T. Belloni, eds.
© International Astronomical Union 2011
doi:10.1017/S1743921310016212

A new model of emission from microquasar jets, and possible explanation to the outliers of the fundamental plane

Asaf Pe'er[1] and Piergiorgio Casella[2]

[1]Harvard-Smithsonian Center for Astrophysics, 60 Garden Street, Cambridge, MA 02138, USA
email: apeer@cfa.harvard.edu

[2]School of Physics and Astronomy, University of Southampton,
Southampton, Hampshire, SO17 1BJ, UK

Abstract. We present a new model of emission from jets in microquasars, which implements elements from the study of jets in gamma-ray bursts to these objects. By assuming that electrons are accelerated once at the base of the jet to a power law distribution above a low energy Maxwellian, and are cooled by synchrotron emission and possible adiabatic energy losses along the jet, a wealth of spectra can be obtained. We show our theoretical results which can explain some of the key observations. In particular, we show that: (I) a flat radio spectrum, as is frequently seen, is a natural outcome of the model; (II) Strong magnetic field results in a flux decay in the optical/UV band as $F_\nu \sim \nu^{-1/2}$, irrespective of many of the uncertainties of the model. (III) An increase of the magnetic field above a critical value of $\sim 10^5$ G leads to a sharp decrease in the flux at the radio band, while the flux at higher frequencies saturates to a constant value. We conclude that scatter in the values of the magnetic field may provide a natural explanation to the observed scatter in the radio/X ray luminosity correlation seen in these objects.

Keywords. plasmas, radiation mechanisms: non-thermal, stars: winds, outflows, X-rays: binaries

1. Introduction

The common theoretical modeling of emission from microquasar jets is based on the seminal work by Blandford & Königl (1979). According to this model, the flat radio spectrum is explained as due to synchrotron emission from power law distributed electrons inside a conical jet, with a decaying magnetic field, $B(r) \propto r^{-1}$. In Pe'er & Casella (2009), we have extended this model in several ways. The two most notable extensions are: (1) We have included a low energy cutoff in the distribution of the accelerated electrons, i.e., a power law is assumed only at a limited range $\gamma_{\min}..\gamma_{\max}$. (2) We added cooling of the electrons, due to synchrotron emission and possible adiabatic energy losses, inherently into the calculations. These two additions result in numerous non-trivial spectra.

2. Electron energy distribution inside the jet

As the electrons propagate inside the jet they lose their energy. As was shown by Kaiser (2006) and Pe'er & Casella (2009), the decay of the magnetic field results in two qualitatively different regimes in electron distribution: † Close to the base of the jet, the magnetic field is strong, resulting in an initial rapid cooling. Further along the jet, the

† The analysis presented here is valid for wide jets, defined by $r(x) \propto x^{a_{jet}}$ with $a_{jet} > 1/2$.

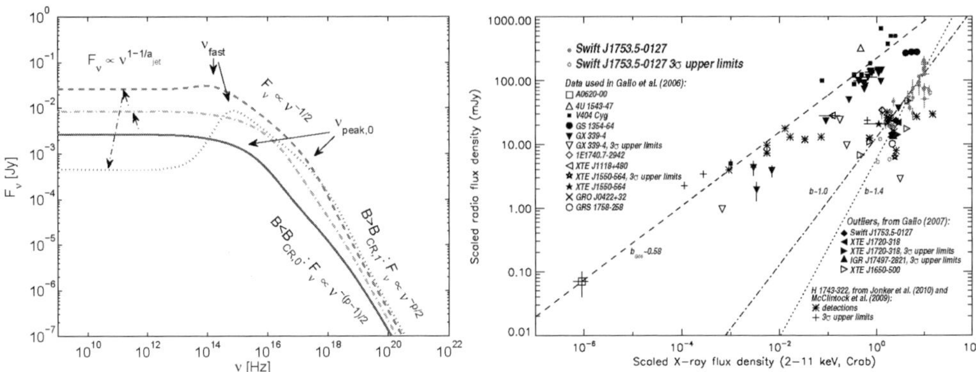

Figure 1. Left: spectrum produced for various values of the magnetic field. For high B-field, the flux at the radio frequencies is suppressed. This may provide a natural explanation to the outliers to the fundamental plane, which are shown in the right pannel (from Soleri *et. al.*,2010.)

magnetic field decays, and as the electrons propagate they maintain their energy, or lose it adiabatically. This fact enables to define a critical value to the magnetic field: for weak B field, $B < B_{cr,0} \approx 10^3$ G, the electrons do not lose their energy initially, while for stronger magnetic field the initial rapid cooling is significant.

3. The resulting spectrum

As a wealth of spectra can be obtained, we focus here on a few key results and refer the reader to Pe'er & Casella (2009) for further details. (1) The existence of two characteristic energies. The inclusion of a low energy cutoff in the electron distribution implies that there are two characteristic breaks in the spectrum: ν_{peak}, the characteristic energy of synchrotron emission from electrons at the peak of the distribution (at $\gamma_{\min}$); and ν_{break}, the transition between optically thin and optically thick emission. For weak magnetic field, $\nu_{break} < \nu_{peak}$. For conical jets, both frequencies decay in a similar way, resulting in a flat radio spectrum, due to the decay of ν_{peak} along the jet. (2) For high magnetic field, the electrons rapidly cool close to the jet base. In this region, the magnetic field is nearly constant, and the resulting spectrum, typically in the optic/X band is $F_\nu \propto \nu^{-1/2}$. Since the cooling occurs close to the jet base, this result is independent on the jet geometry, or on the exact nature of the particle distribution. (3) For very high magnetic field ($B > B_{cr,1} \approx 10^5$ G), following the rapid cooling, the peak of the emission (ν_{peak}) is below ν_{break}. This results in a suppression of the radio flux, which is **not** accompanied by a suppression of the flux at higher frequencies, since the energetic photons are emitted close to the jet base. Casella & Pe'er (2009) showed that this result may provide a natural explanation to the outliers often seen to the known X/radio flux correlation observed in the low/hard state in microquasars (the "fundamental plane").

References

Blandford, R. D. & Königl, A. 1979, *ApJ*, 232, 34
Casella, P. & Pe'er, A. 2009, *ApJ*, 703, L63
Kaiser, C. R. 2006, *MNRAS*, 367, 1083
Pe'er, A. & Casella, P. 2009, *ApJ*, 699, 1919
Soleri, P., *et. al.* 2010, *MNRAS*, 406, 1471

Title of your IAU Symposium
Proceedings IAU Symposium No. 275, 2011
G. E. Romero, R. A. Sunyaev & T. Belloni, eds.

© International Astronomical Union 2011
doi:10.1017/S1743921310016224

Gamma-rays and neutrinos from accreating neutron stars

Włodek Bednarek

Department of Astrophysics, University of Łódź,
ul. Pomorska 149/153, 90-236 Łódź, Poland
email: bednar@astro.phys.uni.lodz.pl

Abstract. Dense wind of a massive star can be partially captured by a companion neutron star (NS) creating a very turbulent and magnetized transition region at some distance from the NS surface. We consider the consequences of electron and hadron acceleration at such a transition region. Electrons lose energy on the synchrotron process and the inverse Compton (IC) scattering of thermal radiation from the NS surface and/or the massive star. We calculate the synchrotron spectra (from X-rays to soft γ-rays) and IC spectra in the case of sources accreting the matter under the accretor and propeller scenarios. It is argued that a population of accreting massive binaries, recently discovered by the INTEGRAL observatory, can be detectable by the *Fermi* LAT telescope. On the other hand, TeV γ-ray emission from other class of massive binaries can be interpreted in terms of a magnetar accreting matter in the propeller scenario. We also calculate the expected neutrino event rates in a km^2 detector produced by relativistic hadrons accelerated in such scenario.

Keywords. Neutron stars, radiation processes, gamma-rays, neutrinos

---

We consider a compact binary system containing rotating neutron star and a massive companion of the O,B type star. It is assumed that the mass from the companion is effectively captured by a strong gravitational potential of the NS. Depending on the rotational period and surface magnetic field of NS, the accretion process can occur in the phase of accretor, propeller, or ejector. A very turbulent and magnetized transition region is formed when the magnetic pressure balances the pressure of infalling matter. In the case of accretor, most of the matter falls onto the NS surface creating a small hot region on the NS surface (for details see Bednarek 2009a).

The matter penetrate below the light cylinder radius for the NS periods, $P_{\rm I} > 0.084 B_{12}^{4/7} M_{16}^{-2/7}$ s, where $B = 10^{12} B_{12}$ G is the surface magnetic field of NS, and $\dot{M} = 10^{16} M_{16}$ g s^{-1} is the accretion rate. The accretion onto NS surface (accretor phase) occurs for the NS periods $P_{\rm II} > 3.7 B_{12}^{6/7} M_{16}^{-3/7}$ s. The accretion occurs in the propeller phase for $P_{\rm I} < P < P_{\rm II}$.

Particles gain energy from the acceleration mechanism, $\dot{P}_{\rm acc} = \xi c E / r_{\rm L} \approx 2.6 \times 10^4 \xi_{-1} M_{16}^{6/7} B_{12}^{-5/7}$ erg s^{-1}, where $\xi = 10^{-1} \xi_{-1}$ is the acceleration parameter, $r_{\rm L}$ is the Larmor radius of particles, E is the particle energy. Electrons lose energy on the IC process in the Thomson (T) and the Klein-Nishina (KN) regimes. The photon energy density from the polar cap ($\rho_{\rm cap}$) at the acceleration region is, $\rho_{\rm cap} \approx 1.4 \times 10^8 M_{16}^{11/7} B_{12}^{-8/7}$ erg cm^{-3}. The radius of the polar cap region on the NS surface is $R_{\rm cap} = (R_{\rm NS}^3 / R_{\rm A})^{1/2} \approx 5 \times 10^4 B_{12}^{-2/7} M_{16}^{1/7}$ cm, and its surface temperature is, $T_{\rm cap} = (L_{\rm X} / \pi R_{\rm cap}^2 \sigma)^{1/4} \approx 4.7 \times 10^7 B_{12}^{1/7} M_{16}^{5/28}$ K, where $R_{\rm A} = 4 \times 10^8 B_{12}^{4/7} M_{16}^{-2/7}$ cm is the Alfven radius. The energy density of the magnetic field at $R_{\rm A}$ is, $\rho_{\rm B} = B_{\rm A}^2 / 8\pi \approx 10^7 M_{16}^{12/7} B_{12}^{-10/7}$ erg cm^{-3}. The energy losses of electrons for the synchrotron and IC in the T regime can be calculated from, $\dot{P}_{\rm loss} = (4/3)c\sigma_{\rm T}\rho\gamma^2 \approx 2.7 \times 10^{-14}\rho_{\rm (cap,B,\star)}\gamma^2$ erg s^{-1}. By balancing energy gains and losses we get the maximum energy of accelerated electrons, $\gamma_{\rm max} \approx 3 \times 10^5 \xi_{-1}^{1/2} B_{12}^{5/14} M_{16}^{-3/7}$. The maximum power transfered from the rotating NS to the matter can be estimated on $L_{\rm acc} = \dot{M}_{\rm acc} v_{\rm rot}^2/2 \approx 3 \times 10^{34} B_{12}^{8/7} M_{16}^{3/7} P_1^{-2}$ erg s^{-1}, where $v_{\rm rot}$ is the rotational velocity of the magnetic field of NS at $R_{\rm A}$. The high energy γ-rays produced in the ICS of thermal photons from the polar cap can be also absorbed in this same radiation field and the massive star radiation initiating an IC $e^\pm$ pair cascade.

We consider γ-ray production in the accretor and propeller scenarios. Due to the large optical depths, γ-rays are produced in the IC $e^\pm$ pair cascade initiated by relativistic electrons in the radiation of the polar cap. In order to follow the process of γ-ray production, we developed the numerical code which simulate the cooling process of electrons taking into account not only the γ-ray production in the IC process but also the synchrotron energy losses of primary electrons and secondary cascade $e^\pm$ pairs. Two models for injection of relativistic electrons are considered: (1) injection with the power law spectrum up to the maximum Lorentz factor $\gamma_{\rm max}$, and (2) monoenergetic injection with the Lorentz factors $\gamma_{\rm max}$. The example spectra are shown for the sources with different values of $\dot{M}$, ξ and B and for specific binary IGR J19140+0951 (see Bednarek 2009a). We also consider a neutron star with a superstrong magnetic field accreting the matter in the propeller scenario. The γ-ray spectrum is calculated from the interaction of electrons with the radiation of the massive star. As an example, we consider the TeV γ-ray binary system, LSI +61 303, which was suggested to contain magnetar. The calculations are compared with the X-ray to TeV γ-ray spectrum of this source (see Bednarek 2009b).

The maximum energies of the accelerated hadrons are determined by the balance between the acceleration time scale and the escape time scale of hadrons from the acceleration site. It can be defined as, $\tau_{\rm esc} = R_{\rm A}/v_{\rm f} \approx 0.4 B_{12}^{6/7} M_{16}^{-3/7}$ s, where $v_{\rm f}$ is the free fall velocity of the matter at $R_{\rm A}$. We estimate the Lorentz factor of hadrons on, $\gamma_{\rm p} \approx 2.7 \times 10^6 \xi_{-1} B_{12}^{1/7} M_{16}^{3/7}$. We conclude that hadrons are at first accelerated to the maximum energies and after that they interact with thermal photons during their fall onto the NS surface. We calculate the spectra of muon neutrinos coming from the decay of pions produced in collisions of hadrons with the thermal radiation form the NS surface. The expected neutrino event rates from the example sources are reported in Bednarek (2009c).

In summary, we calculated the X-ray to γ-ray, and neutrino emission from the accreating neutron stars inside the massive binary systems. As an example, we consider the high mass X-ray binary systems: IGR J19140+0951, the TeV γ-ray binary LSI +61 303, and two X-ray pulsars: GRO 1744-28 and A 0535+262. We conclude that massive binaries, recently discovered by INTEGRAL, may be detected in the γ-rays by the Fermi telescope. The accreting magnetar scenario can be responsible for the TeV γ-ray emission from LSI +61 303. The nearby X-ray pulsars within binary systems can have a chance to be detected by the future large scale neutrino telescopes.

References

Bednarek, W. 2009a, *A&A* 495, 919

Bednarek, W. 2009b, *MNRAS* 397, 1420

Bednarek, W. 2009c, *PRD* 79, 123010

Jets at all Scales
Proceedings IAU Symposium No. 275, 2011
G. E. Romero, R. A. Sunyaev & T. Belloni, eds.

© International Astronomical Union 2011
doi:10.1017/S1743921310016236

The pertinence of jet emitting discs in microquasars.
Theory and comparison to observations

Pierre-Olivier Petrucci[1] **Jonathan Ferreira**[1] **Gilles Henri**[1]
Julien Malzac[2] **and Cedric Foellmi**[1]

[1]Laboratoire d'Astrophysique de GrenOble,
Université Joseph Fourier - Grenoble 1 / CNRS
UMR 5571, BP 53, 38041 Grenoble Cedex 09, France
email: `pierre-olivier.petrucci@obs.ujf-grenoble.fr`

[2]Centre d'Etude Spatial des Rayonnement
UMR 5187
9, av du Colonel Roche BP 44346 31028 Toulouse Cedex 4 France

Abstract. Jet Emitting Disc (JED) has dynamical properties quite different from both the standard and advection dominated discs. It also exhibits three different thermal equilibrium branches at a given radius: two stable (cold and hot) and one intermediate unstable. The hot solution has all the characteristics of the so-called "hot corona" generally invoked in XrB sytems in the Low/Hard states. We detail the energetics and radiative expectations of our model and show their good agreement with those observed in Cygnus X-1 in terms of jet power, jet velocity and spectral emission.

Keywords. X-rays: binaries, accretion, accretion disks, magnetic fields: MHD

1. Introduction

Jet emitting discs (JED herafter) were originaly studied by Ferreira & Pelletier (1993) in their magnetized accretion ejection structures model (MAES herafter). This model was developed so as to treat both the accretion disc and the jet it generates consistently. The idea is the same as in earlier studies of magneto-centrifugally launched disc winds (Blandford & Payne 1982). However, in the MAES, the solution starts from the midplane of the resistive MHD disc and evolves outwards in the ideal MHD wind/jet. This differs drastically from other studies where the disc was only treated as a boundary condition, hence forbidding any precise quantication of the effect of the MHD wind on the disc.

The resolution of the thermal equilibrium of a JED gives three branches, a cold and hot ones, both thermally and viscously stable and an intermediate unstable one. The hot branch, that would be observationally very similar to ADAFs, corresponds to the JED solution discussed in this paper and applied to Cyg. X-1 (Petrucci *et al.* 2010). Thus JEDs could account for most of the successes of the ADAFs, while explaining jet formation (see Ferreira's talk, this proceeding).

2. Comparison to Cygnus X-1

The energetics of the jets and the X-ray corona of Cygnus X-1 have been investigated recently by Malzac *et al.* (2009 herafter M09). Observations constraint the ratio

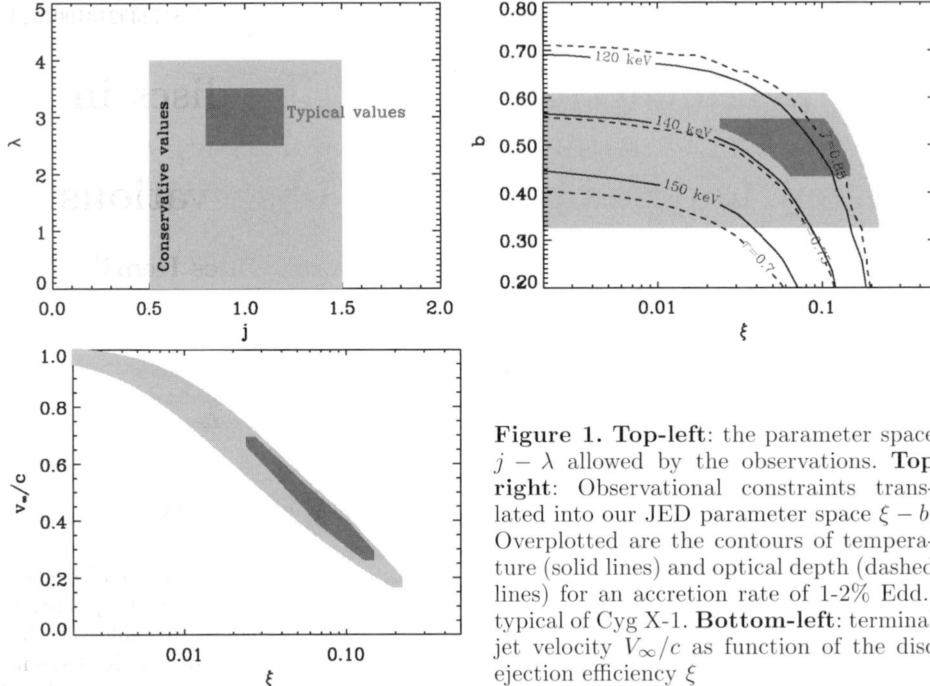

Figure 1. Top-left: the parameter space $j - \lambda$ allowed by the observations. **Top right**: Observational constraints translated into our JED parameter space $\xi - b$. Overplotted are the contours of temperature (solid lines) and optical depth (dashed lines) for an accretion rate of 1-2% Edd., typical of Cyg X-1. **Bottom-left**: terminal jet velocity V_∞/c as function of the disc ejection efficiency ξ

$j = P_{jets}/L_h$ of the total jet kinetic power to the typical X-ray luminosity in the hard state as well as the ratio λ of the soft to hard radiative efficiencies to be in the ranges (M09):

$$0.45 \leqslant j = \frac{P_{jets}}{L_h} \leqslant 1.5 \text{ and } \lambda = \frac{L_s}{\dot{M}_s}\frac{\dot{M}_h}{L_h} \leqslant 4 \tag{2.1}$$

These observational constraints on j and λ can be easily translated into constraints on our JED parameters b and ξ, where $P_{jets} \simeq bP_{acc}$ and $\dot{M}_a(R) \propto R^\xi$. The top-left panel of Fig. (2) displays the domain in the observed parameter space $j - \lambda$ allowed by the observations. The top right panel shows the same constraints but translated into our JED parameter space $\xi - b$. Overplotted are the contours of the JED temperature and optical depth for an accretion rate of 1-2% Edd., typical of Cyg X-1. The bottom-left panel of Fig. (2) shows the terminal jet velocity V_∞/c as function of the disc ejection efficiency ξ which are in good agreement with observations.

The accretion and ejection properties of JEDs agree with the observations of the prototypical black hole binary Cygnus X-1. The JED solutions are likely to be relevant to the whole class of microquasars.

References

Blandford, R. D. & Payne, D. G. 1982, *MNRAS*, 199, 883

Ferreira, J. & Pelletier, G. 1993, *A&A*, 276, 625+

Malzac, J., Belmont, R., & Fabian, A. C. 2009, *MNRAS*, 400, 1512

Petrucci, P.-O., Ferreira, J., Henri, G., Malzac, J., & Foellmi, C. 2010, *A&A*, in press (arXiv:1007.1478)

Jets at all scales
Proceedings IAU Symposium No. 275, 2011
G. E. Romero, R. A. Sunyaev & T. Belloni, eds.

© International Astronomical Union 2011
doi:10.1017/S1743921310016248

Conditions for jet formation in accreting neutron stars: the magnetic field decay

Federico García[1], Deborah N. Aguilera[2] and Gustavo E. Romero[1]

[1]Instituto Argentino de Radioastronomía, Argentina
email: fgarcia@iar-conicet.gov.ar

[2]Ohio University, USA & Laboratorio Tandar, CNEA, Argentina
email: aguilera@ohio.edu, aguilera@tandar.cnea.gov.ar

Abstract. Accreting neutron stars can produce jets only if they are weakly magnetized ($B \sim 10^8$ G). On the other hand, neutron stars are compact objects born with strong surface magnetic fields ($B \sim 10^{12}$ G). In this work we study the conditions for jet formation in a binary system formed by a neutron star and a massive donor star once the magnetic field has decayed due to accretion. We solve the induction equation for the magnetic field diffusion in a realistic neutron star crust and discuss the possibility of jet launching in systems like the recently detected Supergiant Fast X-ray Transients.

Keywords. stars: neutron, magnetic fields, accretion, X-rays: binaries

Magnetic field decay and jet production in SFXTs

Several Galactic bright X-ray transients presumable composed by a compact object (a neutron star (NS) or a black hole) and an OB super-giant companion have recently been discovered by *INTEGRAL*/IBIS and *Swift*. The so-called Supergiant Fast X-ray Transients (SFXTs) show short flares of typically few hours to days reaching $10^{36}-10^{37}$ erg.s^{-1} returning to quiescent levels of $\sim 10^{32}$ erg.s^{-1} emission (Sguera *et al.* 2009). It has been stated that such systems could also emit jets only if they are weakly magnetized ($B \sim 10^8$ G) (Massi & Kaufman Bernadó 2008, Migliari & Fender 2006). We investigate here if such condition can be met for a realistic NS model born with pulsar's strength field ($B \sim 10^{12}$ G) for which the magnetic field is diffused in the NS crust in about 10^{6-7} yrs (Geppert & Urpin 1994).

Magnetic field decay in the neutron star crust

We consider that the composition of the neutron star crust is modified by the accreted material and we construct two background models as in Aguilera *et al.* 2008: a low mass NS (LM) with $M = 1.4$ M$_\odot$ and $R_{\mathrm{NS}} = 11.72$ km and a high mass NS (HM) with $M = 1.8$ M$_\odot$ and $R_{\mathrm{NS}} = 11.35$ km. For both models the crust thickness is in the range $0.5 - 1$ km, which defines a characteristic length scale for the confinement of the crustal magnetic field. The thermal evolution of the NS crust is taken from previous work Aguilera *et al.* 2008 considering different states for the $^3P_2$ neutron superfluidity in the core: strong superfluidity (SF) or no pairing at all (no SF).

We assume a dipolar magnetic field confined to the NS crust (Urpin & Konenkov 2008) described by the Stokes function $s(r,t)$ in radial coordinates which is solution of the induction equation for the magnetic field. The electrical conductivity σ is a function of the temperature T and the density ρ and includes all the relevant electron scattering processes (Potekhin 1999). The surface magnetic field at any instant can be obtained by evaluating $B(t) = B_0 s(R_{\mathrm{NS}}, t)$, where $B_0 = 10^{12}$ G is the initialfield strength.

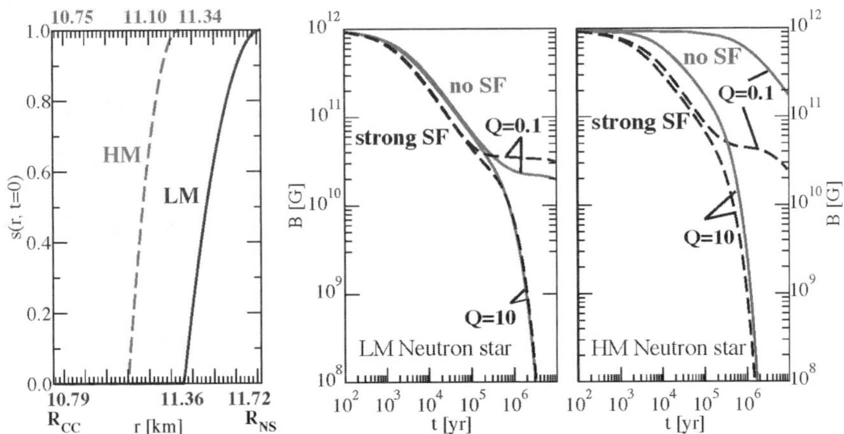

Figure 1. Initial Stokes function (left) and magnetic field evolution (middle and right).

The initial distribution we choose for the Stokes function in the NS crust is shown in Fig. 1 (left panel) and it matches with an external dipole at the NS surface imposing the condition on its derivative $ds/dr + 1/R_{\mathrm{NS}} = 0$.

Our results show that the decay of the initial field depends on the underlying NS properties like e.g. its mass and impurity content of the crust Q, as well as on the NS cooling history (SF or no SF in the core) in the relevant timescale of 10^{6-7}yr (Fig. 1, middle and right panels). As a NS cools down and $T \lesssim 10^7$ K, electron scattering on impurities becomes important and this results in a persisting magnetic field decay for the most impure crusts ($Q \sim 10$). The NS surface field could be as low as 10^8 G if the crust is very resistive (very impure) and the star is old enough ($> 10^6$ yrs). On the other hand, the magnetic field decays only about 1-2 orders of magnitude in a crust that is a nearly perfect lattice ($Q \sim 0.1$) in the same timescale. The results are less sensitive to the NS superfluidity in the core.

Conclusions

The magnetic field decay in accreting NSs depends on the NS underlying model through the cooling evolution and the crust composition. In particular, it depends strongly on the impurity content of the NS crust: NSs with high impurity content are much resistive and therefore the magnetic field can decay up to 3-4 orders of magnitude in about 10^{6-7} yrs. Thus, the magnetic field condition ($< 10^8$ G) for jet formation in SFXTs would be more likely achieved in high mass systems with high metallicity winds.

Acknowledgments DNA is supported by CONICET and PICT-2007-00818 (Argentina), and DE-FG02-93ER40756 (Ohio U., USA). FG is supported by Training Fellowship from CIC (Buenos Aires, Argentina) and GER by Conicet and PIP-2010-0078 (Argentina).

References

Aguilera, D. N., Pons, J. A., & Miralles, J. A. 2008, *A&A* 486, 255

Geppert, U. & Urpin, V. 1994, *MNRAS*, 271, 490

Massi, M. & Kaufman Bernadó, M. 2008, *A&A* 477, 1

Migliari, S. & Fender, R. P. 2006, *MNRAS*, 366, 79

Potekhin, A. Y. 1999, *A&A*, 351, 787; and correspoding routines for the electronic conductivity available at www.ioffe.rssi.ru/astro/conduct/condmag.html

Sguera, V. *et al.* 2009, *ApJ* 697, 1194

Urpin, V. & Konenkov, D. 2008, *A&A* 483, 223

Jets at all Scales
Proceedings IAU Symposium No. 275, 2011
G. E. Romero, R. A. Sunyaev & T. Belloni, eds.

© International Astronomical Union 2011
doi:10.1017/S174392131001625X

Do cataclysmic variables produce jets?

G. Tovmassian[1], B. Gänsicke[2], S. Zharikov[1], A. Ramirez[1] and M. Diaz

[1]Instituto de Astronomía, UNAM, México; [2]Dept of Physics, University of Warwick, UK;
[3]IAGCA, Universidade de São Paulo, Brazil

Abstract. It is believed that CVs do not produce jets unlike other, more massive interactive binaries. Here we present spectrophotometric observations of the super-outburst of V455 And. We show here that a strong wind perpendicular the the accretion disc at the maximum of the super-otburst, i.e a jet, can probably explain the observed spectroscopic behavior of this system.

Keywords. (stars:) binaries: spectroscopic, (stars:) novae, cataclysmic variables

1. Introduction

V455 And is a 81.08 min, eclipsing dwarf nova (DN). Its primary is a magnetic WD with 1.12 min spin period. It is also pulsating. The most intriguing feature of this object is its ≈ 3.5 h spectroscopic period detected in the wings of emission lines (Araujo-Betancor et al., 2005). In 2007, V455 underwent an 8^m superoutburst. We report here the spectral observations during the rise and possible interpretation of its unusual behavior.

The early rise to the super-outburst spectral observations were obtained on 4.2-m WHT, whereas the rest of the data comes from 2.1 m telescope of OAN SPM. The photometric data used here was obtained from the AAVSO data base.

2. Photometric and Spectral behavior

The object at quiescence is V≈ 16.3. It was slightly fainter than V=14 when the first spectrum was obtained. From that moment and through the rise to the max 8.5 mag 160 spectra covering more than 10 hours were taken. The light curve (*lc*) of the outburst is presented in Fig. 1a. For comparison, we over-plotted in blue the *lc* of a similar object

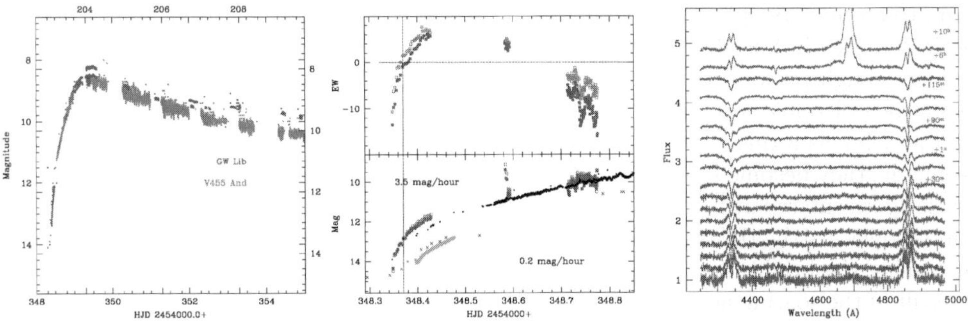

Figure 1. The rise to the superoutburst: a. the superposition of *lc*s of V455 (red) and GW Lib (blue); b. the zoom on early rise. On the lower panel, the black points are from photometry, while the blue and magenta points are measurements of the continuum level around H$_\gamma$ & H$_\beta$ lines correspondingly. The green points and the red crosses are measurements of GW Lib in V and visual correspondingly. On the upper panel the measurements of equivalent widths of the same lines are presented, showing the rise of absorption first, then turn around and return of emission; c. the progression of spectra from the start of observations. The time since the start is marked on the right side of spectra.

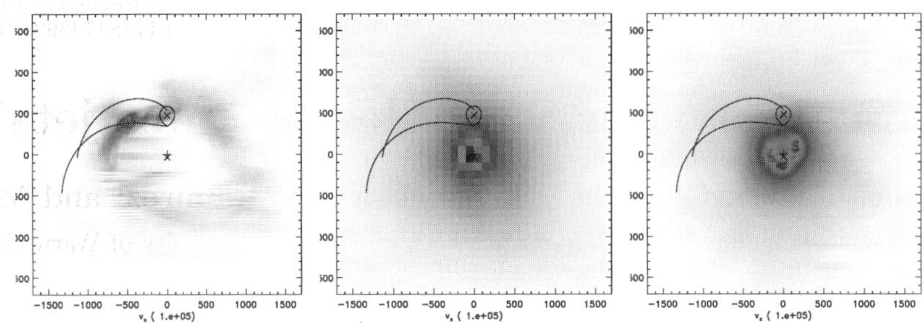

Figure 2. The Doppler tomograms of three orbits taken at the very beginning, several hours later after emission lines re-emerged and at the top of the super outburst.

GW Lib (axes are shifted to coincide c;osely with the V455). GW is only the second object of that class observed during the rise. One notable difference between two *lc*s is that the V455 was rising very steeply at first, which has changed into a slower rise later. This switch in growth rate was accompanied by the reemergence of the emission lines.

From the onset of observations the spectra of V455 evolved in a common to DN way. Emission lines gradually convert into absorptions. Usually the emission lines do not disappear completely, but display small peaks inside absorptions. Such condition of the disk usually prevails during most of the rise, maximum and plateau of the outburst. Only after about a dozen of days the lines start to return to emission, indicating that the disk clears out, becoming optically thin.

The 2007 superoutburst of V455 was different. After 6h from the start of observations, and well before the max of outburst, the emission lines reemerge and by the end of the night (10h later) they were extremely intensive with the addition of a very strong He II line (Fig. 1c). The line profiles evolved from markedly double-peaked to single peaked, and the corresponding tomograms changed from showing extended ring of accretion disk into a single concentrated spot close to the center of the mass of the binary. The system remained in that condition for ~ 10 days since the start of the outburst. Only 13 days later wide double-peaked lines and donut-shaped tomograms returned.

3. A possible scenario

Among possible explanations of such behavior we find a strong wind from the accretion disk, to be the most plausible. Ribeiro & Diaz (2008) demonstrated that a biconic wind, optically thick in lines from the accretion disk, will produce single peaked emission lines strong enough to dominate over the hot optically thick accretion disk. Fig. 2 shows the evolution of Doppler tomograms of the object at the rise and its maximum. The accretion disk appears as a ring on tomograms at the very beginning. But after the emission lines reemerge, they become single peaked and their corresponding tomograms exhibit a rather dense spot centered at zero velocity. We interpret the observed behavior in context of the wind model. As the the outburst starts, the disk turns hotter and optically thick, rapidly increasing the brightness. At some point however, the wind commensurates, possibly as a result of the soaring accretion rate and being reinforced by a wobbling magnetic field of the white dwarf and a portion of matter is blown perpendicular to the plane of the disk, giving rise to the emission lines. We again see the disk in the lines after the outburst plateau, when the wind lines turn back optically thin.

References

Ribeiro, F. M. A. & Diaz, M. P. 2008, *PASJ*, 60, 327
Araujo-Betancor *et al.*, 2005, *A&A*, 430, 629

Jets at all Scales
Proceedings IAU Symposium No. 275, 2011
G. E. Romero, R. A. Sunyaev & T. Belloni, eds.

© International Astronomical Union 2011
doi:10.1017/S1743921310016261

Transient high-energy flares from accreting black holes

Florencia L. Vieyro[1,2] and Gustavo E. Romero[1,2]

[1]Instituto Argentino de Radioastronomía (CCT La Plata, CONICET),
C.C.5, 1894 Villa Elisa, Buenos Aires, Argentina
email: fvieyro@iar-conicet.gov.ar

[2]Facultad de Ciencias Astronómicas y Geofísicas, Universidad Nacional de La Plata,
Paseo del Bosque, 1900 La Plata, Argentina

Abstract. We present a model for high-energy flares in accreting black holes based on the injection in a magnetized corona of a non-thermal population of relativistic particles. Coupled transport equations are solved for all species of particles and the electromagnetic and neutrino output is predicted for the case of Galactic black holes.

1. Introduction

Several Galactic gamma-ray sources have been recently found to be rapidlyvariable. One of this sources is the well-known microquasar Cygnus X-1, which is a binary system composed by a massive star and an accreting black hole (Poutanen *et al.* 1997). The flaring nature of Cyg X-1 in gamma rays has been confirmed with the AGILE satellite (Sabatini *et al.* 2010).Different models have been proposed to explain the origin of these episodes (e.g., Romero *et al.* 2010; Zdziarski *et al.* 2009; Bosch-Ramon *et al.* 2008). These works are mainly focused on the analysis of gamma-ray emission that might be produced in the jet. The approach of this work is different: we consider the effect of the injection of non-thermal particles in the magnetized corona around the black hole. These particles can be locally accelerated by reconnection events and subsequent diffusive processes taking place on the plasma. Our main goal is to study coronal flares and their output in both electromagnetic radiation and neutrinos. In what follows, we outline the basic model and present the results of our calculations.

2. Static corona model

The model considered here is a static corona where the relativistic particles can be removed by diffusion. We consider a two-temperature corona in steady state, with a thermal emission characterized by a power-law with an exponential cutoff at high energies, as observed in several X-ray binaries in the low-hard state. For more details, an extensively description of this model can be found in Romero *et al.* (2010).

We solved the transport equation in steady state obtaining particle distributions. Then the spectral energy distributions of all radiative processes were estimated. We also took into account the effect of secondary pairs created by photon-photon pair production. Figure (1) shows the total SED together with the spectrum of Cygnus X-1, detected by COMPTEL McConnell *et al.* (2000) and the radio detection of the jet by Stirling *et al.* (2001).

It can be seen that emission in the range 100 MeV to 1 TeV is suppressed by absorption in the soft photon fields. All emission detected in this range should be produced in the jet at some distance from the blak hole (e.g., Bosch-Ramon *et al.* 2008; Romero *et al.* 2010).

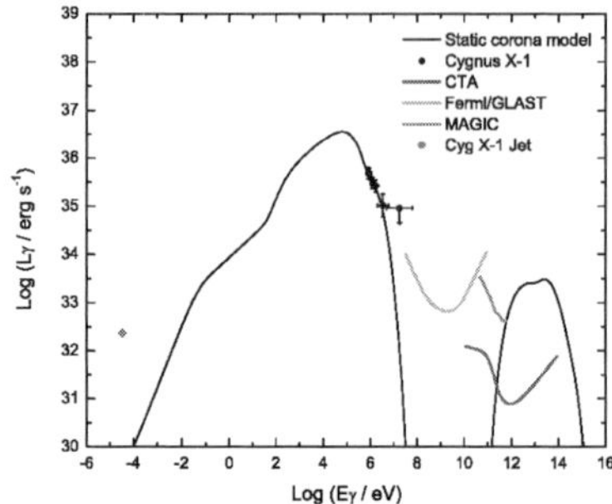

Figure 1. Spectral energy distribution obtained with the steady state model of a static corona. The prediction fits the observation made by COMPTEL of Cygnus X-1 (McConnell *et al.* 2000) and the detection of the jet by Stirling *et al.* (2001).

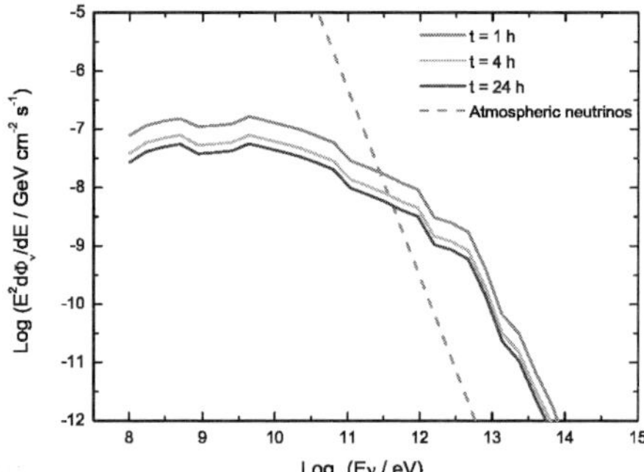

Figure 2. Evolution of the neutrino flux and atmospheric neutrinos.

However, if a sudden injection of relativistic protons occurs, for instance as a consequence of major reconnection events, a neutrino burst may be produced. We explore this scenario in this work.

3. Flare model

The temporal dependence of the particle injection is characterized by a FRED (Fast Rise and Exponential Decay) behavior, whereas the energy dependence is a power-law.

We consider neutrino injection by charged pion and muon decay. The differential flux of neutrinos arriving at the Earth can be seen in Fig. (2).

According to our results, a neutrino burst might be detectable in this kind of flares in accreting black holes. The electromagnetic part of the flare can be also detectable at some energies, depending of the optical depth for photon annihilation. A complete discussion will presented elsewhere.

References

Bosch-Ramon, V., Khangulyan, D., & Aharonian, F. A., 2008, *A&A*, 489, L21
McConnell, M. L., *et al.* 2000, *ApJ*, 543, 928
Poutanen, J., Krolyk, J. H., & Ryde, F., 1997, *MNRAS*, 292, L21
Romero, G. E., del Valle, M. V., & Orellana, M., 2010, *A&A*, 518, A12
Romero, G. E., Vieyro, F. L., & Vila, G. S., 2010, *A&A*, 519, A109
Stirling, A. M., *et al.*, 2001, *MNRAS*, 327, 1273
Sabatini, S., *et al.* 2010, *ApJ*, 712, L10
Zdziarski, A. A., Malzac, J., & Bednarek, W., 2009, *MNRAS*, 394, 1, L41

Jets at all Scales
Proceedings IAU Symposium No. 275, 2011
G. E. Romero, R. A. Sunyaev & T. Belloni, eds.

© International Astronomical Union 2011
doi:10.1017/S1743921310016273

A leptonic/hadronic jet model for the low-mass microquasar XTE J1118+480

Gabriela S. Vila[1] and Gustavo E. Romero[1,2]

[1]Instituto Argentino de Radioastronomía, (CCT La Plata - CONICET), C.C. N°5 (1894),
Villa Elisa, Buenos Aires, Argentina
email: gvila@iar-conicet.gov.ar

[2]Facultad de Ciencias Astronómicas y Geofísicas (FCAG, UNLP), Paseo del Bosque s/n
(1900), La Plata, Buenos Aires, Argentina
email: romero@iar-conicet.gov.ar

Abstract. We present a one–zone jet model that fits the data from simultaneous broadband radio-to-X-rays observations of XTE J1118+480. We calculate the radiative contribution to the non-thermal spectrum of both relativistic electrons and protons, as well as that from secondary muons, charged pions and electron-positron pairs produced at high-energy hadronic interactions. The distributions in energy of all the particle species are obtained taking into account the energy losses, injection, decay and escape from the emission region. We also include absorption effects on the emission spectrum due to photon-photon annihilation. Finally, we discuss the detectability of XTE J1118+480 at high energies with the present instruments according to the predictions of our model for the gamma-ray band.

Keywords. Radiation mechanisms: nonthermal, X-rays: individual (XTE J118+480).

1. Introduction

The source XTE J1118+480 is an X-ray binary located at a distance of ~ 1.7 kpc (Gelino *et al.* 2006) at high galactic latitudes ($b = +62.3°$). The mass of compact object is larger than $6M_\odot$ (McClintock *et al.* 2000) being, therefore, a firm black hole candidate in the galactic halo. Although no relativistic jets have been directly imaged yet, the characteristics of the radio emission (e.g. Fender *et al.* 2001) strongly suggest that the source is a low-mass microquasar (LMMQ).

Several leptonic models have been proposed to explain the broadband radio-to-X-rays spectrum of XTE J1118+480 during the low-hard state (LHS), see for example Esin *et al.* (2001), Markoff *et al.* (2001), Yuan *et al.* (2005) and Maitra *et al.* (2009). In this work we present the application of a lepto-hadronic jet model to this source. As well as fitting the existing data, we intend to asses the possibility of gamma-ray emission from the jet due to different interactions processes involving relativistic particles.

2. Model

The jet model applied here is presented in detail in Vila & Romero (2010) and references therein. We assume that the jet is injected at a distance $z_0 = 50R_g$ from the black hole and expands as a cone of semi aperture $\sim 6°$. The outflow carries a 10% of the accretion power. The magnetic field at z_0 is $\sim 10^6 - 10^7$ G, and decays as $B \propto z^{-1.5}$. At a certain distance $\sim 10z_0$ a fraction $q = 0.1$ of the jet power is transferred to relativistic protons and leptons through diffusive shock acceleration. These particles are injected with a hard power-law spectrum $\propto E^{-1.5}$, and then suffer radiative and non-radiative losses until their energy distribution reaches the steady state. We calculate the electromagnetic emission from relativistic primary (p and e^-) and secondary ($\pi^\pm$, $\mu^\pm$ and $e^\pm$ pairs) particles due to several processes. Also added is the thermal emission spectrum from an standard accretion disc.

3. Results and conclusions

Figure 1 shows two least-squares fits to the LHS spectrum of XTE J1118+480 obtained with our model, corresponding to observations carried out during the outbursts of 2000 and 2005. Accretion rates of $0.1 - 0.2\dot{M}_{\rm Edd}$ were necessary to achieve the observed X-ray emission levels. The power injected in relativistic protons and leptons in the jet is approximately the same, with a ratio $L_p/L_e = 1 - 3$.

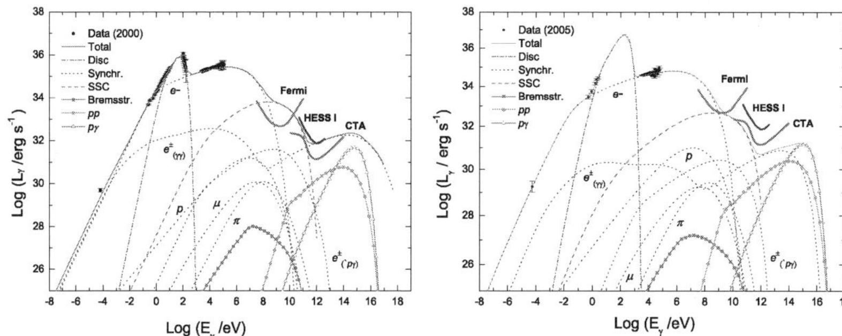

Figure 1. Model fits to the observed LHS spectrum of XTE J1118+480. The sensitivity curves of the Fermi satellite and the Cherenkov arrays HESS I and the future CTA are indicated.

The radio and X-ray data are well fitted with synchrotron emission from primary electrons. The direct hadronic contribution and electromagnetic radiation of secondary pions and muons are not significant. Synchrotron emission from $e^\pm$ pairs injected through $\gamma\gamma$ annihilation is relevant at energies above $\sim 0.1 - 1$ TeV.

According to our results, the source would be detectable by Fermi in the GeV range with a soft spectrum. At TeV energies the synchrotron radiation of secondary pairs falls just below the sensitivity of HESS I (after 50 hours of exposure), but might be observed by the future CTA array. Our model also predicts that electron-positron pairs are produced at a significant rate. This is interesting in connection with the results of Weidenspointner *et al.* (2008), who found that the 0.511 MeV annihilation line emission detected outside the galactic plane traces the asymmetric spatial distribution of low-mass X-ray binaries in the Galaxy (see, however, Bouchet *et al.* 2010).

References

Bouchet L. *et al.* 2010, *ApJ*, 720, 1772

Esin A. *et al.* 2001, *ApJ*, 555, 483

Fender R. *et al.* 2001, *MNRAS*, 322, L23-L27

Frontera F. *et al.* 2001, *ApJ*, 561, 1006

Gelino D. *et al.* 2006, *ApJ*, 642, 438

Maitra *et al.* 2009, *MNRAS*, 389, 1638

Vila G. S. & Romero G. E. 2010, *MNRAS*, 403, 1457

Weidenspointner *et al.* 2008, *Nature*, 451, 159

Yuan F. *et al.* 2005, *ApJ*, 620, 905

Jets at all Scales
Proceedings IAU Symposium No. 275, 2011
G. E. Romero, R. A. Sunyaev & T. Belloni, eds.

© International Astronomical Union 2011
doi:10.1017/S1743921310016285

Isolating the jet in broadband spectra of XBs

David M. Russell[1], Fraser Lewis[2,3,4], Dipankar Maitra[5], Robert J. H. Dunn[6], Sera Markoff[1], Peter G. Jonker[7,8,9], Manuel Linares[10] and Valeriu Tudose[11,12,13]

[1]Astronomical Institute 'Anton Pannekoek', University of Amsterdam, PO Box 94249, 1090 GE Amsterdam, the Netherlands;   email: d.m.russell@uva.nl
[2]Faulkes Telescope Project, School of Physics and Astronomy, Cardiff University, 5, The Parade, Cardiff, CF24 3AA, Wales; [3]Department of Physics and Astronomy, The Open University, Walton Hall, Milton Keynes, MK7 6AA, England; [4]Division of Earth, Space and Environment, University of Glamorgan, Pontypridd, CF37 1DL, Wales; [5]Department of Astronomy, University of Michigan, 500 Church Street, Ann Arbor, MI 48109, USA; [6]Technische Universität München, Excellence Cluster Universe, Boltzmannstrasse 2, D-85748 Garching, Germany; [7]SRON, Netherlands Institute for Space Research, Sorbonnelaan 2, 3584 CA, Utrecht, the Netherlands; [8]Harvard-Smithsonian Center for Astrophysics, 60 Garden Street, Cambridge, MA 02138, USA; [9]Department of Astrophysics, IMAPP, Radboud University Nijmegen, Toernooiveld 1, 6525 ED, Nijmegen, the Netherlands; [10]MIT Kavli Institute for Astrophysics and Space Research, 70 Vassar Street, Cambridge, MA 02139,USA; [11]Netherlands Institute for Radio Astronomy, Oude Hoogeveensedijk 4, 7991 PD Dwingeloo, the Netherlands; [12]Astronomical Institute of the Romanian Academy, Cutitul de Argint 5, RO-040557 Bucharest, Romania; [13]Research Center for Atomic Physics and Astrophysics, Atomistilor 405, RO-077125 Bucharest, Romania

Abstract. Most accretion-powered relativistic jet sources in our Galaxy are transient X-ray binaries (XBs). Efforts to coordinate multiwavelength observations of these objects have improved dramatically over the last decade. Now the challenge is to interpret broadband spectral energy distributions (SEDs) of XBs that are well sampled in both wavelength and time. Here we focus on the evolution of the jet in their broadband spectra. Some of the most densely sampled broadband SEDs of a neutron star transient (IGR J00291+5934) are used to constrain the optically thick–thin break in the jet spectrum. For the black hole transient XTE J1550-564, infrared – X-ray correlations, evolution of broadband spectra and timing signatures indicate that synchrotron emission from the jet likely dominates the X-ray power law at low luminosities ($\sim (2 \times 10^{-4} - 2 \times 10^{-3})\, L_{\rm Edd}$) during the hard state outburst decline.

Keywords. accretion, accretion discs, X-rays: binaries, ISM: jets and outflows

---

1. Introduction

In X-ray binaries, both black hole (BH) and neutron star (NS) systems are capable of launching powerful jets to relativistic velocities through the process of accretion (e.g. Fender *et al.* 2004). When the X-ray spectrum is hard (the hard state; see Belloni 2010) a continuous jet is thought to be produced, which can radiate synchrotron emission from radio frequencies to at least the optical/near-infrared (OIR) regime (e.g. Hjellming *et al.* 1990, Russell *et al.* 2007, Coriat *et al.* 2009). Multiwavelength studies are required to attempt to isolate the synchrotron jet emission from other components such as thermal emission from the accretion disc and companion star or Comptonized radiation from the corona. Here we focus on two recent results where the jet emission has been revealed and the spectrum constrained at higher energies than the radio; one from a BH system and one from a NS.

2. Evidence for a compact jet dominating the broadband spectrum of the black hole accretor XTE J1550-564

XTE J1550–564 was monitored continuously throughout its outburst in 2000 (Tomsick *et al.* 2001, Jain *et al.* 2001, Corbel *et al.* 2001). We are able to separate the OIR emission from the disc (exponential decay) and jet (excess in hard state, absent in soft state). We find that on the hard state decline of the outburst, the OIR spectral index of the jet component is $\alpha \sim -0.7$, where $F_\nu \propto \nu^\alpha$, which is the same as the measured X-ray and OIR–to–X-ray power law index. Moreover, the OIR jet and X-ray fluxes are linearly correlated. The OIR and X-ray data are consistent with a single power law with a common origin: the synchrotron jet. When the synchrotron jet appears to dominate the X-ray flux, (I) there is an excess in the X-ray light curve over its previous exponential decay, (II) there is possible evidence for a shift in the high energy cut-off to a lower energy, (III) the X-ray timing properties do not change significantly except the possible disappearance of a quasi-periodic oscillation (E. Kalemci, these proceedings). This may be the strongest evidence to date of synchrotron emission from the compact, steady jet dominating the X-ray flux of an XB. For XTE J1550-564, this is likely to occur at $\sim (2 \times 10^{-4} - 2 \times 10^{-3})$ $L_{\rm Edd}$ in the hard state. However, the synchrotron jet can only provide a small fraction ($\sim$ a few per cent) of the X-ray flux at other times in the hard state. Both Comptonization and the synchrotron jet can therefore produce the hard X-ray power law in accreting black holes. These results are published in Russell *et al.* (2010).

3. The double-peaked 2008 outburst of the accreting milli-second X-ray pulsar, IGR J00291+5934

After its 2004 outburst, the NS XB IGR J00291+5934 again became active in 2008. However, instead of returning to quiescence, the system performed a second, more prolonged outburst peak. The double-peaked outburst was monitored extensively, with data collected at X-ray, UV, OIR and radio frequencies. A near-IR excess is commonly seen in outbursts of accreting milli-second X-ray pulsars, which has been attributed to synchrotron emission from the jet. We are able to fit the broadband SEDs with a blackbody (likely the irradiated accretion disc) plus X-ray power law and a simple jet model. If the jet produces the near-IR excess, the models suggest the optically thick–thin break in the jet spectrum resides around the H-band (in the near-IR). This is a higher frequency than previously reported for NS jets (Migliari *et al.* 2010) implying that their jet powers may vary between sources or luminosities. This work is published in Lewis *et al.* (2010).

References

Belloni T. M. 2010, in 'The Jet Paradigm - From Microquasars to Quasars', ed. T. Belloni, *Lect. Notes Phys.*, 794 (arXiv:0909.2474)

Corbel S., Kaaret P., Jain R. K., Bailyn C. D., *et al.* 2001, *ApJ*, 554, 43

Coriat M., Corbel S., Buxton M. M., Bailyn C. D., *et al.* 2009, *MNRAS*, 400, 123

Fender, R., Wu, K., Johnston, H., Tzioumis, T., *et al.* 2004, *Nature*, 427, 222

Hjellming R. M., Stewart R. T., White G. L., Strom R., *et al.* 1990, *ApJ*, 365, 681

Jain R. K., Bailyn C. D., Orosz J. A., McClintock J. E., *et al.* 2001, *ApJ*, 554, L181

Lewis F., Russell D. M., Jonker P. G., Linares M., *et al.* 2010, *A&A*, 517, A72

Migliari, S., Tomsick J. A., Miller-Jones J. C. A., Heinz S., *et al.* 2010, *ApJ*, 710, 117

Russell D. M., Fender R. P., & Jonker P. G. 2007, *MNRAS*, 379, 1108

Russell D. M., Maitra D., Dunn R. J. H. & Markoff S. 2010, *MNRAS*, 405, 1759

Tomsick J. A., Corbel S., & Kaaret P. 2001, *ApJ*, 563, 229

Jets at all Scales
Proceedings IAU Symposium No. 275, 2011
G. E. Romero, R. A. Sunyaev & T. Belloni, eds.

© International Astronomical Union 2011
doi:10.1017/S1743921310016297

Is there a mildly relativistic jet in SN2007gr?

Z. Paragi[1], A. J. van der Horst[2], M. Tanaka[3], G. B. Taylor[4], C. Kouveliotou[5], J. Granot[6], E. Ramirez-Ruiz[7], Y. Pidopryhora[1], S. Bourke[1], R. M. Campbell[1], M. A. Garrett[8] and H. J. van Langevelde[1]

[1] Joint Institute for VLBI in Europe (JIVE), Dwingeloo, the Netherlands

[2] NASA Postdoctoral Program Fellow, NASA/MSFC/ORAU, Huntsville, AL, USA

[3] Institute for the Physics and Mathemathics of the Universe, Univ. Tokyo, Kashiwa, Japan

[4] University of New Mexico, Albuquerque, USA

[5] NASA/MSFC, Huntsville, AL, USA

[6] University of Hertfordshire, UK

[7] University of California, Santa Cruz, USA

[8] Netherlands Institute for Radio Astronomy (Astron), Dwingeloo, the Netherlands

Abstract. SN2007gr was an ordinary type Ic supernova, with a hint of asymmetric explosion seen in the optical polarization spectrum. This type of SNe is occasionally associated with long duration gamma-ray bursts which generate ultra-relativistic jets; no relativistic outflows have yet been found by direct imaging in SNe Ib/c explosions. High resolution very long baseline interferometry (VLBI) data and simultaneous total radio flux density measurements indicated that SN2007gr has expanded mildly relativistically. We performed late time Westerbork Synthesis Radio Telescope (WSRT) observations to measure the level of the underlying extended emission. Comparison of the VLBI and the background-subtracted WSRT and independent VLA data indicate an at least partially resolved source with an average expansion velocity of $\geqslant 0.4c$, although the VLBI data could be consistent with a fainter source with an expansion velocity of $\sim 0.2c$ as well.

Keywords. supernovae: general, supernovae: individual (SN 2007gr), ISM: jets and outflows, radio continuum: general, techniques: interferometric

---

SN 2007gr was discovered on 15 August 2007 in the bright spiral galaxy NGC 1058 at a distance of 10.6 ± 1.3 Mpc. It was one of the closest type Ic supernova detected in the radio band. We carried out real-time e-VLBI observations with the European VLBI Network (EVN) on 6-7 September 2007 at 5 GHz, and found an unresolved source with a peak brightness of 422 μJy/beam at 5.6 times the off-source noise level of 75 μJy/beam, fully consistent with the simultaneous WSRT total flux density measurements. The upper limit of 7 milliarcseconds (mas) for its angular diameter size corresponds to a linear size of $<1.1\times10^{18}$ cm at 10.6 Mpc about 25 days after the explosion, which sets an upper limit of $<8.6c$ to the average isotropic apparent expansion speed of the ejecta (Paragi *et al.* 2010).

To further investigate the evolution of the ejecta on mas scales, we observed again on 5-6 November 2007 (age of about 85 days) with the EVN and the Green Bank Telescope (GBT). We followed exactly the same observing strategy as before, but the data in this case were recorded on disc at a data rate of 1024 Mbps (512 Mbps at the GBT) – these were not only the highest resolution but also the most sensitive radio observations of the source. The measured peak brightness of 60 ± 13 μJy/beam this time was significantly

Figure 1. Left: VLBI images of SN 2007gr on 6 September (colours) and 5 November 2007 (contours). Right: Geometry of the SN explosion from 3D radiative transfer simulations. See the text for more details.

below the total flux density of 260 μJy measured by the WSRT (Paragi *et al.* 2010). To properly remove the background emission from the galaxy, we performed new WSRT observations of NGC 1058 on 10 June 2010. After subtracting the resulting map from the second epoch WSRT image we obtained 150±50 μJy, in agreement with two independent VLA measurements within two weeks of the VLBI observations (see also Soderberg *et al.* 2010). This sets an average constraint of 140±18 μJy to the total flux density.

A point source model fitted to the VLBI *uv*-data is inconsistent with the measured total flux density, and also note that at a distance of 10.6 Mpc the source is in fact expected to be partially resolved at an age of 85 days on mas scales. A $\sim$150 μJy Gaussian (or other type of extended model) with size of $\geqslant$ 1.0 mas gives a reasonable fit to the data. This would correspond to an average expansion speed of $\sim$ 0.4c, somewhat lower than derived from the beamsize as lower limit in Paragi *et al.* 2010. We note that a $\sim$100 μJy source would be marginally consistent with the measured total flux density, therefore our derived average expansion velocity value should be treated with caution.

The possibility that SN 2007gr expanded spherically at mildly relativistic velocities is intriguing because it would require a severe departure from equipartition (relativistic electrons and/or magnetic fields carrying very small fractions of the internal energy). This may be resolved if the mildly relativistic ejecta were collimated into a narrow bipolar jet (Paragi *et al.* 2010). There was also a hint of non-spherical explosion from optical spectropolarimetry data (Tanaka *et al.* 2008). Polarization of $\sim$3% was detected at the absorption feature of the Ca II triplet, which can be understood in the context of a model in which bipolar explosion with an oblate photosphere is being viewed from a slightly off axis direction (although spherical photosphere and clumpy Ca II distribution may be an alternative explanation). In the former case, 3D radiative transfer simulations show that the major axis of the SN photosphere had a projected position angle on the sky of about 60 degrees.

References

Paragi, Z., Taylor, G. B., Kouveliotou, C. *et al.* 2010, *Nature*, 463, 516
Soderberg, A. M., Brunthaler, A., Nakar, E. *et al.* 2010, *arXiv*, 1005.1932
Tanaka, M., Kawabata, K. S., Maeda, K., Hattori, T., Nomoto, K. 2008, *ApJ*, 689, 1191

Jets at all Scales
Proceedings IAU Symposium No. 275, 2011
G. E. Romero, R. A. Sunyaev & T. Belloni, eds.

© International Astronomical Union 2011
doi:10.1017/S1743921310016303

Spectral evolution of the galactic microquasars XTE J1550-564 and GRO J1655-40 during outbursts

Alexander A. Lutovinov[1], Sergei S. Tsygankov[2,1], Vadim A. Arefiev[1] and Mikhail G. Revnivtsev[1]

[1]Space Research Institute, Profsoyuznaya str. 84/32, Moscow 117997, Russia
[2]Max-Plank Institute für Astrophysik, Karl-Schwarzschild str. 1, Garching 85741, Germany
email: aal@iki.rssi.ru

Abstract. We present results of a broadband spectroscopy of the galactic microquasars and black hole candidates XTE J1550-564 and GRO J1655-40, performed with the INTEGRAL and RXTE observatories during strong outbursts in 2003 and 2005, respectively. The spectral parameters evolution was traced during brightening and fading phases of each outburst to search a possible hysteresis and transitions from state to state. We estimated a size and optical depth of different regions around XTE J1550-564, like a hot plasma zone and optically thick accretion disk. Upper limits to the annihilation 511 keV line emission were obtained for both sources using data of the SPI spectrometer onboard the INTEGRAL observatory.

Keywords. X-ray transients, black holes, microquasars, jets

Light curves. Both sources were regularly monitored with the INTEGRAL observatory during 7 years of its operation at the orbit. Two relatively bright outbursts with the flux maximum of $\sim$ 300 mCrab were detected from GRO J1655-40 in 2005, but only one outburst was detected from XTE J1550-564 in 2003 (Fig.1). The total duration of the last in the standard X-ray energy range was about 50 days, and the shape of its light curve proved to be asymmetric with a $\sim$ 10-days rise in intensity to the peak value followed by a $\sim$ 35 − 40-days smooth decay (see, e.g., Arefiev *et al.*, 2004).

Spectra. During both outbursts GRO J1655-40 demonstrated very hard X-ray emission, practically without any cutoff till > 200 keV. The source spectra for 4 extended epochs were presented by Caballero-Garcia *et al.* (2007). Here we made a detailed analysis of its hard X-ray emission (> 20 keV) and found, that the source spectrum could be well

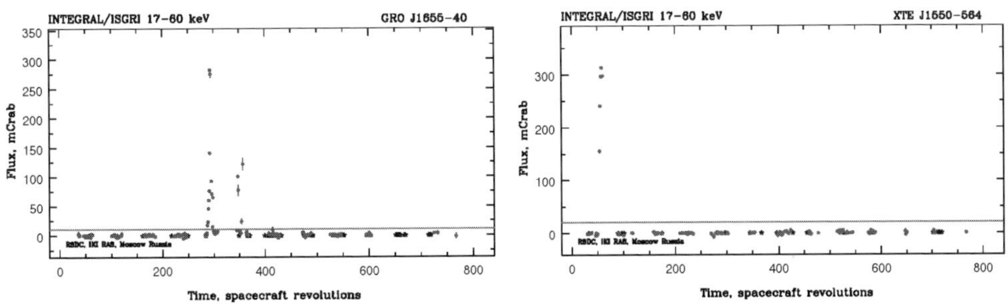

Figure 1. Light curves of GRO J1655-40 (left) and XTE J1550-564 (right) as they were measured by the INTEGRAL observatory in 2003-2009. Strong outbursts are clearly seen. The horizontal axis is given in the INTEGRAL revolutions, starting from Oct 17, 2002. See http://hea.iki.rssi.ru/integral/survey/catalog.php for details.

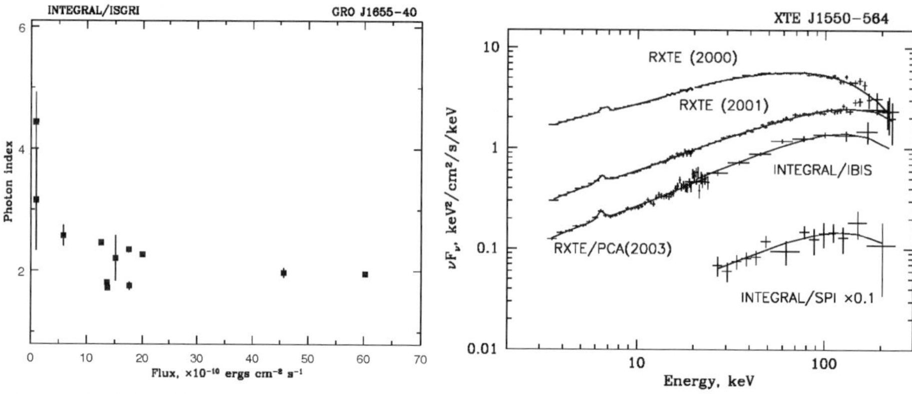

Figure 2. (*left*) The 20-200 keV photon index dependence on the flux measured during outbursts from GRO J1655-40. (*right*) INTEGRAL and RXTE broadband spectrum of XTE J1550-564 obtained during 2003 outburst. Source spectra collected during hard state of 2000 and 2001 outbursts are shown for comparison.

described by a simple powerlaw with a relatively stable photon index ~ 2 during the high state. A some steepening of the spectrum was revealed in the low state (Fig.2, left).

Because XTE J1550-564 demonstrated only subtle spectral variability during the outburst, we averaged its spectrum over all our observations. Figure 2(right) shows the XTE J1550-564 broadband spectra obtained during outburst of 2000, 2001 and 2003. It is seen that both 2001 and 2003 outbursts had much harder spectra than that the 2000 outburst. To describe spectrum, we used the Comptonization model by Poutanen & Svensson (1996), which includes the reflection of a hot radiation from a cold neutral medium. Best-fit parameters are summarized below. The large optical depth may reflect the fact that the source has a rather large inclination, and we look at the central source through the entire thickness of the central cloud (see Arefiev *et al.*, 2004 for details).

kT, keV	τ	R, $\Omega/2\pi$	EW_{line}, eV	Flux, erg/s/cm^2	$\chi2$/d.o.f.
50 ± 10	5 ± 1	0.25 ± 0.13	120 ± 30	3.8×10^{-10}	1.20

The analysis of SPI/INTEGRAL data shown, that the maximum significance of the flux variability in the 511-keV line was observed near the Galactic plane at a Galactic latitude of $\sim 30°$. This is the most interesting feature, since the microquasar GRO J1655-40 can be a potential source of this outburst. The observed outburst in the $508 - 514$ keV energy band coincides in time with an intense outburst in the X-ray energy band (see Fig. 1). But the most conservative upper limit for the 511-keV emission does not exceed $\sim 2 \times 10^3$ phot cm^{-2} s^{-1} (see Tsygankov & Churazov, 2010, for details).

This work was supported by the program of Russian Academy of Sciences "Origin, Structure and Evolution of Objects in the Universe" and grant NSh-5069.2010.2.

References

Arefiev, V., Revnivtsev, M., Lutovinov, A., & Sunyaev, R. 2004, *Astron. Letters*, 30, 669
Caballero-Garcia M., Miller J., Kuulkers E., *et al.*, 2007 *Astrophys. J.*, 669, 534
Poutanen, Yu. & Svensson, R. 1996, *Astrophys. J.*, 470, 249
Tsygankov, S. & Churazov, E. 2010, *Astron. Letters*, 36, 237

Jets at all Scales
Proceedings IAU Symposium No. 275, 2011
G. E. Romero, R. A. Sunyaev & T. Belloni, eds.

© International Astronomical Union 2011
doi:10.1017/S1743921310016315

Evidence of an irradiated accretion disc in XTE J1818−245

J. A. Zurita Heras[1], S. Chaty[1], M. Cadolle Bel[2] and L. Prat[1]

[1] AIM Paris Saclay, CEA-DSM/CNRS-INSU/Université Paris 7 Denis Diderot,
IRFU/Service d'Astrophysique, Bât. 709 L'Orme des Merisiers, 91191 Gif-sur-Yvette, France
email: `juan-antonio.zurita-heras@cea.fr`, `sylvain.chaty@cea.fr`, `lionel.prat@cea.fr`

[2] ESAC, ISOC, Villanueva de la Cañada, Madrid, Spain
email: `Marion.Cadolle@sciops.esa.int`

Abstract. The X-ray transient source XTE J1818−245 went through an outburst in 2005 that was observed during a multi-wavelength campaign from radio to soft γ-rays. We performed new optical observations with the ESO/NTT telescope at La Silla. The broad-band spectral energy distribution revealed that the outer parts of the accretion disc had to be irradiated by its inner parts to explain the optical emission.

Keywords. X-rays: binaries, X-rays: individual (XTE J1818−245).

1. Introduction

XTE J1818−245 was discovered at high-energies with *RXTE* during a bright outburst that started on August 12, 2005 (Levine, Swank, Lin, *et al.* 2005). Cadolle Bel, Prat, Rodriguez, *et al.* (2009) performed several observations (from radio to soft γ-rays) from August to September and concluded that the new X-ray transient source is probably a low-mass X-ray binary and a black hole candidate, located closer to us than the Galactic centre. We carried out new optical observations with the ESO 3.6 m New Technology Telescope at La Silla Observatory on August 24, 2005, when the X-ray flux was still decaying, and when the source was probably in a soft-intermediate state.

2. Spectral energy distribution

Each subset of data (radio/optical/X-rays/hard X-rays) reported in the SED displays different segments, each approximately power law in shape with different spectral indices. We focus our interest to the optical emission where different components of the system may contribute: the outer part of the accretion disc, the companion star, and the jet.

Model 1: Viscously heated accretion disc. The high-energy spectrum of XTE J1818−245 is represented by a combination of an accretion disc modelled with a multi-colour disc, a broad iron emission line at 6–7 keV, and a hard component modelled with a Comptonisation model. The fit of this *model 1* (`gaussian+ezdiskbb+compTT` in `Xspec`), reproduces correctly the high-energy data with a reduced χ^2 of $\chi^2_\nu = 1.21$ with 29 degrees of freedom (dof). However, *model 1* fails to reproduce both visible+HE data, particularly the model being a factor 10 below the visible points ($\chi^2_\nu = 1.45$ with 34 dof, see Fig. 1 *left*). We also added other components to the model to take into account the presence of the secondary star (temperature of $5\,000$ K, radius of 1 $R_\odot$ and distance of 3.5 kpc) and a possible contribution in the visible from a jet (flat power law normalised to the 8.4 GHz flux). This stellar model gives an emission compatible with the upper limits derived from 2MASS, USNO and REM. The sum of all these different components fails to reproduce the optical data that show a steeper shape with a spectral index $\alpha = 1.15$ (in $F_\nu \sim \nu^\alpha$).

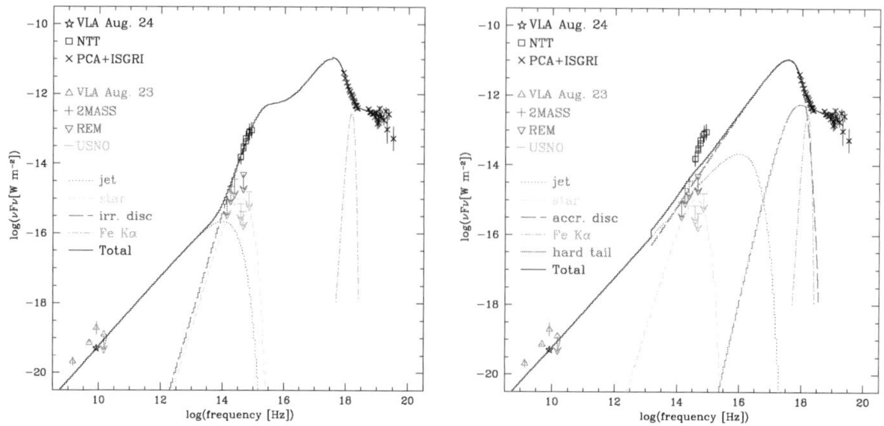

Figure 1. Broad-band SED of XTE J1818−245 using simultaneous VLA(Aug. 24)/NTT/ PCA/ISGRI data. We also show non-simultaneous data from VLA(Aug. 23)/2MASS/REM/ USNO. *left:* Model 1; *right:* Model 2.

<u>*Model 2: Irradiated accretion disc.*</u> Irradiation of the outer parts of the accretion disc by soft X-rays has been proposed to explain the UV-optical emission in soft X-ray transients (Hynes, Haswell, Chaty, *et al.* 2002). We considered an irradiated disc model that takes into account the emission of the disc, the emission of the Compton component, and both the irradiation of the inner part of the disc by the Compton tail and that of the outer part by the inner part (`diskir`, Gierliński, Done, & Page 2008,2009, and references therein for a discussion about the irradiation). We fitted the optical+HE data with *model 2* (`gaussian+diskir`), reproducing well the data with $\chi_\nu^2 = 1.11$ (see Fig. 2 *right*). The addition of the stellar component and the contribution of the jet in the visible does not improve the fit. The emission of the jet may contribute to the redder part of the optical emission, however we cannot assess this statement without simultaneous radio-infrared observations. The Comptonisation-to-disc ratio of XTE J1818−245 is low with $L_c/L_d = 0.07^{+0.03}_{-0.01} << 1$ as expected when the soft component in X-rays is important. Meanwhile, its irradiation fraction is $f_{out} = 2.4^{+1.2}_{-0.8} \times 10^{-3}$ which is $2 \times f_{out}$ of XTE J1817−330 in the intermediate state. The absolute determination of this fraction is uncertain by a factor 2, but a change by a factor 6 between the soft and hard states has been observed in other X-ray transients (Gierliński, Done, & Page 2009). Thus, the irradiation fraction of XTE J1818−245 is compatible with that of XTE J1817−330 when they are in a similar state. We also derive an inner radius of $R_{in} = (31 \pm 9)$ km ($\sim 6 \times R_g$ for a 3.5 $M_\odot$ black hole with $R_g \equiv GM/c^2$), thus an outer radius of $R_{out} \sim 4 \times 10^{10}$ cm which is in agreement with Cadolle Bel, Prat, Rodriguez, *et al.* (2009).

References

Cadolle Bel, M., Prat, L., Rodriguez, J., Ribó, M., Barragán, L., D'Avanzo, P., Hannikainen, D. C., Kuulkers, E., Campana, S., Moldón, J., Chaty, S., Zurita Heras, J., Goldwurm, A., & Goldoni, P. 2009, *A&A*, 501, 1

Gierliński, M., Done, C., & Page, K. 2008 *MNRAS*, 388, 753

Gierliński, M., Done, C., & Page, K. 2009 *MNRAS*, 392, 1106

Hynes, R. I., Haswell, C. A., Chaty, S., Shrader, C. R., & Cui, W. 2002 *MNRAS*, 331, 169

Levine, A. M., Swank, J. H., Lin, D., & Remillard, R. A. 2005, *ATel*, 578

Jets at all Scales
Proceedings IAU Symposium No. 275, 2011
G. E. Romero, R. A. Sunyaev & T. Belloni, eds.

© International Astronomical Union 2011
doi:10.1017/S1743921310016327

Large-scale radio nebula around the Ultra-Luminous X-ray Source IC 342 X-1

Dávid Cseh[1], Cornelia Lang[2], Stéphane Corbel[1], Philip Kaaret[2] and Fabien Grisé[2]

[1]Laboratoire Astrophysique des Interactions Multi-echelles (UMR 7158),

CEA/DSM-CNRS-Universite Paris Diderot, CEA Saclay, F-91191 Gif sur Yvette, France
email: david.cseh@cea.fr

[2]Department of Physics and Astronomy, University of Iowa, Iowa City, IA 52242, US

Abstract. We present discovery of a radio nebula associated with the ultraluminous X-ray source (ULX) IC 342 X-1 using the Very Large Array (VLA). Taking the surrounding nebula as a calorimeter, one can constrain the intrinsic power of the ULX source. We compare the obtained power that is needed to supply the radio nebula with the W50 nebula powered by the microquasar SS433 and with other ULXs. We find that the power required is at least two orders of magnitude greater than that needed to power radio emission from the W50 nebula associated with the microquasar SS433. In addition, we report the detection of a compact radio core at the location of the X-ray source.

Keywords. accretion, accretion disks, black hole physics, radiation mechanisms: general, ISM: kinematics and dynamics

---

ULXs are variable off-nuclear X-ray sources in external galaxies with luminosities greatly exceeding the Eddington luminosity of a stellar-mass compact object, assuming isotropic emission (Colbert & Mushotzky 1999). The irregular variability, observed on time scales from seconds to years, suggests that ULXs are binary systems containing a compact object that is either a stellar-mass black hole with beamed (King *et al.* 2001, Körding *et al.* 2002) or super-Eddington emission (Begelman 2002) or an intermediate mass black hole.

We present the discovery of an associated radio nebula around IC 342 X-1 using the VLA. We detected the source at a 10-σ level with peak intensity of $\sim$120 μJy (Fig.1., left) and the estimated flux density is $\sim$ 2 mJy. We estimated the total energy budget of the radio nebula assuming radiation via synchrotron emission, equipartition between particles and fields, and equal energy in electrons and baryons. A radio spectral index of -0.8 of NGC 5408 X-1 was assumed and we use a lower frequency cutoff of 1.3 GHz and an upper frequency cutoff of 6.2 GHz. For a source diameter of $\sim$220 pc and a filling factor of unity, we find that the total energy required to power the radio nebula is 9×10^{50} erg, (with an equipartition magnetic field of 7 μG).

Only a handful of radio detections of ULXs have been made so far (Kaaret *et al.* 2003, Miller *et al.* 2005, Soria *et al.* 2006, Lang *et al.* 2007). These sources all show large nebulae (>35 pc) that are likely powered by continuous energy input from the ULX, in the same manner as the W50 nebula is powered by the Galactic binary SS 433 (Dubner *et al.* 1998). However, the ULX radio nebulae require very high total energy content, 10^{49} erg as compared with 10^{46} erg for W50. *The energy of the nebula around IC342 X-1 is similar to ULX NGC 5408 X-1 and Holmberg II X-1 but at least 2 orders larger than SS433.*

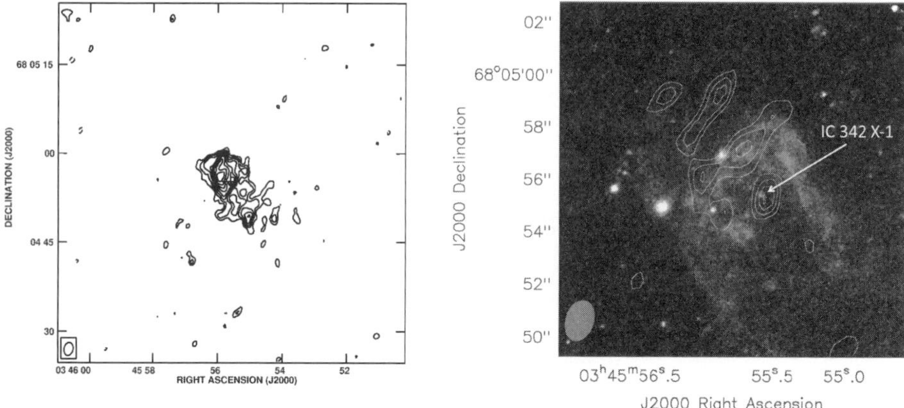

Figure 1. Left: The 5-GHz VLA B- and C-array combined image of IC 342 X-1. The first contours are drawn at 3 σ, at $\pm$ 33 μJy/beam. The positive contour levels increase by a factor of 1 σ. The peak brightness is 122.4 μJy/beam. The Gaussian restoring beam is 2.3" $\times$ 1.6" at PA=$-13°$. The cross marks the Chandra X-ray position of the ULX source. **Right:** The weighted (robust=-2) radio image overlaid on HST Hα image. The first contours are drawn at 3 σ, at $\pm$ 45 μJy/beam. The positive contour levels increase by a factor of 1 σ. The peak brightness is 91.7 μJy/beam. The Gaussian restoring beam is 1.6" $\times$ 1.1" at PA=$-19°$.

We detected a radio point source at the location of IC 342 X-1 (Fig. 1, right) with a peak intensity of $\sim$ 90 μJy at a 7-σ level, which can be consistent with a compact jet (though we have no spectral information). The radio image was weighted (robust=-2) in order to resolve the diffuse nebular emission. Using the fundamental plane of black holes (eg. Körding *et al.* 2006) - which is a relationship between X-ray luminosity, radio luminosity and black hole mass - we can estimate the mass of the ULX. The application of the fundamental plane requires radiatively inefficiently accreting sources, ie. hard state objects. IC 342 X-1 is one of relatively few ULXs usually found in the X-ray hard state (Feng & Kaaret 2009). Substituting the flux of the point source (and $L_X = 6 \times 10^{39}$ erg/s), we obtain for the mass of the black hole $M_{\mathrm{BH}} \simeq 10^4$ M$_\odot$, which should be taken as an order of magnitude under the hypotheses mentioned before (Cseh *et al.* in prep).

Acknowledgements

The research leading to these results has received funding from the European Community's Seventh Framework Programme (FP7/2007-2013) under grant agreement number ITN 215212 "Black Hole Universe".

References

Begelman M. C. 2002, *ApJ*, 568, L97

Colbert E. J. M. & Mushotzky, R. F. 1999, *ApJ*, 519, 89

Dubner, G. M., Holdaway, M., Goss, W. M., & Mirabel, I. F. 1998, *AJ*, 116, 1842

Feng, H. & Kaaret, P. 2009, *ApJ*, 696, 1712

Kaaret, P, Corbel, S, Prestwich, A. H., & Zezas, A. 2003, *Science*, 299, 365

King, A. R., Davies, M. B., Ward, M. J., Fabbiano, G., & Elvis, M. 2001, *ApJ*, 552, L109

Körding, E., Falcke, H., & Markoff, S. 2002, *A&A*, 382, L13

Körding, E., Falcke H., & Corbel S. 2006, *A&A*, 456, 439

Lang, C. C., Kaaret, P., Corbel, S., & Mercer A. 2007,*ApJ*, 666, 79

Miller, N. A., Mushotzky, R. F., & Neff, S. G. 2005, *ApJ*, 623, 109

Soria, R., Fender, R. P., Hannikainen, D. C., Read, A. M., & Stevens, I. R. 2006, *MNRAS*, 368, 1527

Jets at all Scales
Proceedings IAU Symposium No. 275, 2011
G. E. Romero, R. A. Sunyaev & T. Belloni, eds.

© International Astronomical Union 2011
doi:10.1017/S1743921310016339

The determination of the GX 339-4's mass based on its 2010 outburst

Tao Chen

CEA\Saclay, Bât. 709, Orme des Merisiers
91191 Gif sur Yvette Cedex, France
email: tao.chen@cea.fr

Abstract. In this paper we investigate the quasi periodic oscillation (QPO) behavior of the black hole candidate GX 339-4 during its 2010 outburst using RXTE/PCA data. We perform spectral and timing analysis of the observations, where the QPOs are observed. We analyze the relationship between the centroid frequency of QPO and the spectral parameters. The correlation of spectral and timing properties can be used to estimate the mass of black hole with the scaling method. Using this method we estimate a mass of $7.5 \pm 0.8 M_\odot$ of GX 339-4.

Keywords. GX 339, QPO, Black Hole

1. Introduction

The determination of the mass of Galactic black holes (BHs) is one of the most important tasks in modern astronomy. Since it can constrain the maximum mass of a neutron star and the minimum mass of a BH. The knowledge of the BH mass distribution of the Galaxy can also provide important clues on stellar evolution. In particular, the mass of BH can be estimated by the correlation between low-frequency quasi-periodic oscillations (LFQPOs, which observed during low-hard and intermediate states and $\leqslant 10Hz$) and photon index of the power-law spectral component (Shaposhnikov & Titarchuk 2006).

2. Observations & Data Analysis

We analyzed 28 the RXTE observations of GX 339-4 based on its 2010 outburst. For timing analysis, the light curves were extracted from event data using 2–50 KeV energy bands. Power spectra were produced for 256 seconds interval with 0.015625 seconds time resolution. For spectral analysis, the PCA Standard 2 mode (STD2) in energy range 2.5–45.0 keV was used for extraction of the energy spectra. In order to account for residual uncertainties in the instrument calibration a systematic error of 0.8 percent was added to spectra.

For spectral analysis, we fit the energy spectrum where the clear LFQPOs were observed. During the low-hard state, the spectra can be well fitted by the model of exponentially cut-off power law reflected from ionized matter (pexriv) and a Simple Gaussian line profile (gaussian). For the intermediate state we use a simple photon power law (powerlaw), a disk blackbody (diskbb) and a Gaussian (gaussian). For both case the photonelectric absorption was frozen to $0.51 \times 10^{22} cm^{-2}$ (Kong *et al.* 2000). The state transition of this source was reported by Motta *et al.* (2010). The photon indices are very important for determination the mass. The change of photon indexes is shown in Fig. 1.

For the timing analysis, the frequencies of QPOs are obtained by fitting the clear peak on power spectra with a Lorentzian. For each peak we use only one Lorentzian. The change of QPO frequencies is shown in Fig. 2. Comparing this figure with Fig. 1, there are some similar tendency on these two figures. This can be clearly seen in Fig. 3.

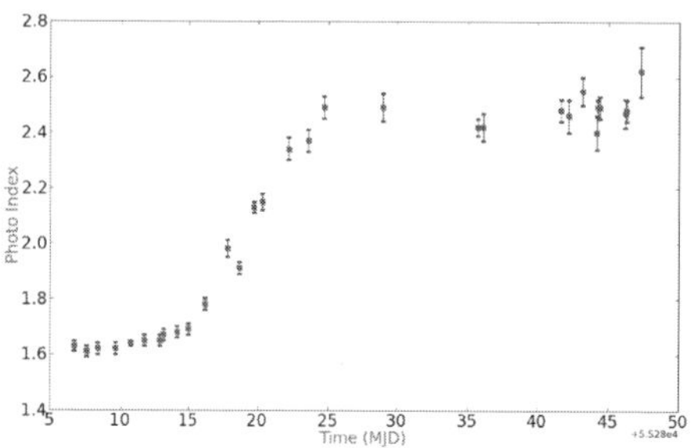

Figure 1. The change of photon indexes with time.

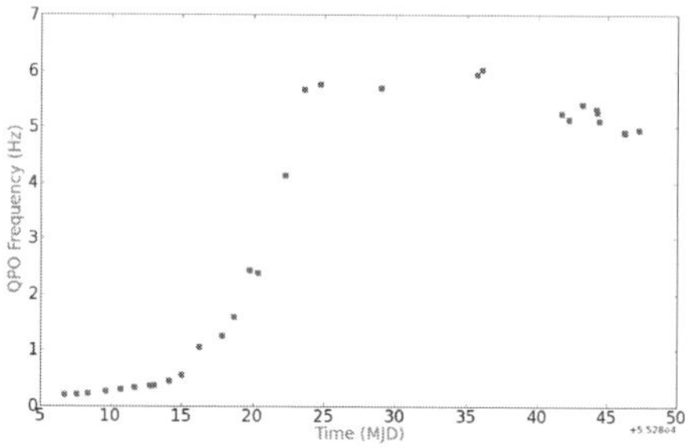

Figure 2. The change of QPO frequencies with time.

3. Scaling method & Conclusion

Shaposhnikov & Titarchuk (2006) have shown that the correlation between photon index and QPO frequency can be used to estimate BH mass: First of all, we fit this correlation with an analytical function (see red line on Fig. 3): $f(\nu) = A - DBln[\exp(\frac{\nu_{tr}-\nu}{D})+1]$.

For determination of the BHs mass M, the parameter B should be knew. It can be obtained by fitting of correlation between photon index and QPO frequency with the above function. Next step is to compare this value with a standard source, which has a well-known mass. GRO J1655-40 (which commonly used for this method as its mass is known by optical method) was used for this purpose. Finally, by analysis spectral and timing properties of GX 339-4, based on its 2010 outburst, we found that the photon index and QPO frequency

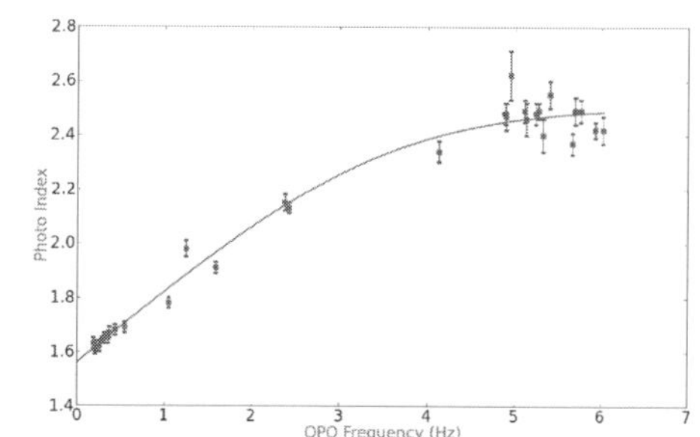

Figure 3. The correlation between photon index and QPO frequency.

showed a strong correlation. Using scaling method based on this correlation we estimated the mass of GX 339-4: $M_{GX339} = B_{GX339}\frac{M_{J1655}}{B_{J1655}} = 7.5 \pm 0.8 M_{\odot}$. This result consistent with previous prediction $M_{GX339} \geqslant 6 M_{\odot}$ (Hynes *et al.* 2003).

4. Acknowledgments

This work is funded by the Europe Communitys Seventh Framework Programme (FP7/2007-2013) under grant agreement number ITN 215212 Black Hole Universe.

References

Hynes R. I. *et al.* 2003, *ApJ*, 583, L95
Kong A. K. H. *et al.* 2000, *MNRAS*, 312, L49
Motta S. *et al.* 2010, *ATel* #2545
Shaposhnikov N. & Titarchuk L. 2006, *ApJ*, 699, 453

Jets at all Scales
Proceedings IAU Symposium No. 275, 2011
G. E. Romero, R. A. Sunyaev & T. Belloni, eds.

© International Astronomical Union 2011
doi:10.1017/S1743921310016340

On the apparent lack of Be X-ray binaries with black holes in the galaxy and in the Magellanic Clouds

Janusz Ziółkowski[1] and Krzysztof Belczyński[2]

[1] Copernicus Astronomical Center, ul. Bartycka 18, 00-716 Warsaw, Poland
email: jz@camk.edu.pl

[2] Los Alamos National Lab, P.O. Box 1663, MS 466, Los Alamos, NM 87545
email: kbelczyn@nmsu.edu

Abstract. In the Galaxy there are 67 Be X-ray binaries known to-date. Out of those, 45 host a neutron star, and for the reminder the nature of a companion is not known. None, so far, is known to host a black hole. This disparity is referred to as a missing Be – black hole X-ray binary problem. The stellar population synthesis calculations following the formation of Be X-ray binaries (Belczyński & Ziółkowski 2009) predict that the ratio of the binaries with neutron stars to the ones with black holes is rather high $F_{\rm NS/BH} \sim 30-50$. A comparison of this ratio with the number of confirmed Be – neutron star X-ray binaries (45) indicates that the expected number of Be – black hole X-ray binaries is of the order of only $\sim 0-2$. This is entirely consistent with the observed Galactic sample. Therefore, there is no problem of the missing Be+BH X-Ray Binaries for the Galaxy

In the Magellanic Clouds there are 94 Be X-ray binaries known to-date. Out of those, 60 host a neutron star. Again, none hosts a black hole. The stellar population synthesis calculations carried out specifically for the Magellanic Clouds (Ziółkowski & Belczyński 2010) predict that the ratio of the Be X-ray binaries with neutron stars to the ones with black holes is only $F_{\rm NS/BH} \sim 10$. This value is rather too low, as it implies the expected number of Be+BH X-ray binaries of the order of ~ 6, while none is observed. We found, that to remove the discrepancy, one has to take into account a different history of the star formation rate in the Magellanic Clouds, with the respect to the Galaxy. New stellar population synthesis calculations are currently being carried out.

Keywords. black holes: binaries, X-rays: binaries, (stars:) binaries: close, stars: evolution, stars: neutron, emission-line, Be

1. Introduction

Be X-ray binaries (Be XRBs) are the most numerous subclass of high mass X-ray binaries. The description of the properties of these systems is given, e.g. in Negueruela *et al.* 2001, Ziółkowski 2002, Belczyński & Ziółkowski 2009 and references therein.

In this work we study the origins of the apparent disparity of number of known Be XRBs with neutron stars (NSs) (105) as compared to no known Be XRBs with black holes (BHs) in Galaxy and in the Magellanic Clouds. This disparity is referred to as a missing Be – black hole X-ray binary problem.

2. Stellar Population Synthesis (SPS) Calculations

We evolve a Galactic population of massive binaries using `StarTrack` stellar population synthesis code (Belczyński *et al.* 2008). We adopt solar metallicity ($Z = 0.02$) for the

Galactic population and low metallicity ($Z = 0.008$) for the Magellanic Clouds. During our SPS calculations, we consider any system a Be X-ray binary if: *(i)* it hosts either a NS or a BH accretor; *(ii)* donor is a main sequence star (burning H in its core); *(iii)* donor mass is higher than 3 $M_\odot$ (O/B star); *(iv)* orbital period is in the range $10 \leqslant P_{\rm orb} \leqslant 300$ day; and *(v)* only a fraction $F_{\rm Be} = 0.25$ of the above systems are designated as hosting a Be star and not a regular O/B star.

3. Galaxy

In the Galaxy there are 67 Be X-ray binaries known to-date. Out of those, 45 host a neutron star, and for the reminder the nature of a companion is not known. None, so far, is known to host a black hole. The stellar population synthesis calculations following the formation of Be X-ray binaries (Belczyński & Ziółkowski 2009) predict that the ratio of the binaries with neutron stars to the ones with black holes is rather high $F_{\rm NS/BH} \sim 30 - 50$. A comparison of this ratio with the number of confirmed Be – neutron star X-ray binaries (45) indicates that the expected number of Be – black hole X-ray binaries is of the order of only $\sim 0 - 2$. This is entirely consistent with the observed Galactic sample. Therefore, there is no problem of the missing Be+BH X-Ray Binaries for the Galaxy

4. Magellanic Clouds

In the Magellanic Clouds there are 94 Be X-ray binaries known to-date. Out of those, 60 host a neutron star. Again, none hosts a black hole. The stellar population synthesis calculations carried out specifically for the Magellanic Clouds (Ziółkowski &Belczyński 2010) predict that the ratio of the Be X-ray binaries with neutron stars to the ones with black holes is only $F_{\rm NS/BH} \sim 10$. This value is rather too low, as it implies the expected number of Be+BH X-ray binaries of the order of ~ 6, while none is observed. We found, that to remove the discrepancy, one has to take into account a different history of the star formation rate in the Magellanic Clouds, with the respect to the Galaxy. Magellanic Clouds are in many ways very different from the Galaxy when comparing the population of XRBs. Let us compare the numbers for three classes of XRBs:

(1) Be XRBs: 67 in the Galaxy vs 94 in the Magellanic Clouds

(2) other High Mass XRBs: 42 vs 10

(3) Low Mass XRBs: 197 vs 2

It is obvious that formation of stars in Magellanic Clouds had to proceed in a completely different way from that in the Galaxy. We are currently carrying out the new stellar population synthesis calculations trying to take into account this fact.

References

Belczyński, K., Kalogera, V., Rasio, F. A., Taam, R. E., Zezas, A., Bulik, T., Maccarone, T. J., & Ivanova, N. 2008, *ApJ Suppl* 174, 223

Belczyński, K. & Ziolkowski, J. 2009, *ApJ* 707, 870

Negueruela, I., Okazaki, A. T., Fabregat, J., Coe, M. J., Munari, U. & Tomov, T. 2001, *A&A* 369, 117

Ziółkowski, J. 2002, *Mem. Soc. Astron. Ital.* 73, 1038

Ziolkowski, J. & Belczyński, K. 2010, *in preparation*

Jets at all Scales
Proceedings IAU Symposium No. 275, 2011
G. E. Romero, R. A. Sunyaev & T. Belloni, eds.

© International Astronomical Union 2011
doi:10.1017/S1743921310016352

The X-ray burster 4U 1608-522 as seen by JEM-X onboard INTEGRAL

Celia Sánchez-Fernández and Erik Kuulkers

INTEGRAL Science Operations Center (ISOC), ESAC, Madrid, SPAIN
email: `Celia.Sanchez@sciops.esa.int`, `Erik.Kuulkers@sciops.esa.int`

Abstract. We present here and overview of the results of a systematic search of Type-I X-ray bursts in the light curves of the transient, atoll system 4U 1608–522 as seen by JEMX onboard *INTEGRAL*.

Keywords. stars: neutron, accretion, nucleosynthesis

1. Introduction

Type I X-ray bursts are thermonuclear explosions on the surface of weakly magnetized accreting neutron stars (NS) in Low-Mass X-ray Binary (LMXB) systems (see reviews in Lewin *et al.* 1993; Strohmayer & Bildsten 2003) that lead to a sudden, rapid, increase of the system X-ray flux. The light curves of X-ray bursts can show a large variety in profiles, but generally they exhibit a fast rise (from less than a second to 10 s) and a longer, usually exponential decay which lasts seconds to minutes. Type-I bursts radiate X-ray spectra with black body shapes and temperatures up to ~ 3 keV, that cool during burst decay.,Most of the X-ray bursts are detected from the "atoll" sources, persistent or transient, with luminosities of the order of 10^{36-37} erg s^{-1}.

The recurrent transient 4U 1608–522 is an atoll system which experiences outbursts that in some cases can reach peak intensities of the order of the Crab (Gottwald *et al.* 1987). The source is also a X-ray burster (Galloway *et al.* 2008 and references therein) from which a superburst has also been detected (Keek *et al.* 2008). In several occasions, burst oscillations at 619 Hz were measured, making this source a member of the group of rapidly spinning NS LMXB.

2. Data analysis

The Joint European Monitor for X-rays (JEMX) on-board *INTEGRAL* provides X-ray spectra and imaging with arc-min angular resolution in the 3 to 35 keV band (Brandt *et al.* 2003). We have analyzed all JEM-X data available in the *INTEGRAL* archive and have conducted a systematic search of Type-I X-ray bursts in the light curves of 4U 1608–522.

3. Results

The light curve of 4U 1608–522 as observed by JEMX over the *INTEGRAL* lifetime and compiled by us is displayed Figure 1. The plot shows the activity from the system over ~ 8 years. In this period, several outbursts have been detected.

On top of the overall variations in the persistent emission, we have detected 58 type-I X-ray bursts from 4U 1608–522 at various levels of accretion rate, both in outburst and during quasi-quiescence (see Fig. 1). Short bursts were detected, with durations of the

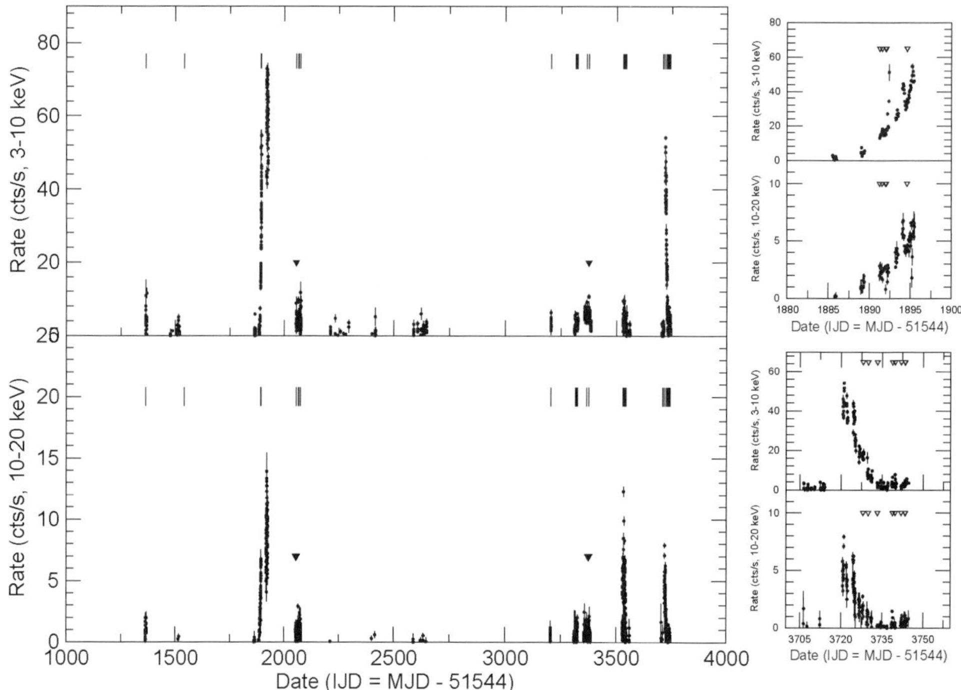

Figure 1. Left panel: The light curve of 4U 1608–522 in two energy bands as seen by JEMX onboard *INTEGRAL* over almost 8 years. Several outbursts were detected from the recurrent transient in this time period. The times of occurence of double and triple bursts are marked by black vertical arrows in this plot. Right panels: closer view of outburst rise and decline. Superimposed to the plot, are marked the times of occurence of type-I X-ray bursts.

order of 10–30 sec, rise times of the order of 1-6 sec, 88–1200 cts/sec peak count rates, and e-folding decay times between 3–14 sec.

A detailed analysis of these data is prsented in Sánchez-Fernández *et al.* (2010).

References

Brandt, S. *et al.* 2003, *A&A*, 411, L243

Lewin, W. H. G., van Paradijs, J., & Taam, R. E. 1993, *SpScRev.*, 62, 223

Galloway, D. K. *et al.* 2008, *ApJS* 179, 360

Gotthelf, E. & Kulkarni, S. 1997, *ApJ*, 490, L161

Gottwald, M., Stella, L., White, N. E., & Barr, P. 1987 *MNRAS*, 229, 359

Sanchez-Fernandez *et al.* 2010, *in preparation*

Strohmayer, T. & Bildsten, L. 2006, *in Compact Stellar X-Ray Sources, eds. W.H.G. Lewin and M. van der Klis, Cambridge University Press, page 113*

Part 4. Gamma-ray bursts

Jets at all Scales
Proceedings IAU Symposium No. 275, 2011
G. E. Romero, R. A. Sunyaev & T. Belloni, eds.

© International Astronomical Union 2011
doi:10.1017/S1743921310016364

Gamma Ray Bursts: basic facts and ideas

Gabriele Ghisellini

Osservatorio Astronomico di Brera, Via Bianchi 46 Merate I–23807 Italy
email: `gabriele.ghisellini@brera.inaf.it`

Abstract. The recent years witnessed a dramatic improvement in our knowledge of the phenomenology and physics of Gamma Ray Bursts (GRBs). However, our "pillars of knowledge" remain a few, while many aspects remain obscure and not understood. There is no general agreement on the radiation mechanism of the prompt emission, nor on the process able to convert the bulk motion of the fireball into random energy of the emitting leptons. The afterglow phase can now be studied at very early phases, showing an unforeseen phenomenology, still to be understood. In this context, the detection of $\sim$GeV emission from $\sim$10% of GRBs, made possible by the *Fermi* satellite, can hopefully shed light on some controversial issues.

Keywords. gamma rays: bursts, radiation mechanisms: nonthermal, supernovae: general

1. Pillars of knowledge

What are the fundamental and not controversial facts characterizing Gamma Ray Bursts? I propose a list of seven "pillars" of knowledge, selected in an admittedly completely subjective way, following this criterion: If we did not know this particular fact, would we lose a basic piece of knowledge?

1.1. *GRBs are cosmological*

One of the major achievements of the *Beppo*SAX satellite was to localize a GRB with enough accuracy to make the pointing of an optical telescope possible, allowing to find the redshift. At the same time, the afterglow was discovered (Costa *et al.* 1997, for GRB 970228). As we know, the first measured redshift was $z = 0.835$ for GRB 970508 (Metzeger *et al.* 1997; the redshift for GRB 970228 was measured later, due to the faintness of its host galaxy).

This ended a long and animated discussion about the origin of GRBs (i.e. "local", i.e. associated to neutron stars in the Galactic halo, or cosmological, as predicted by Paczynski 1986), and finally set the power of these objects: they are indeed the most explosive events of the Universe after the Big Bang. Soft γ–ray repeaters, instead, were found to be "nearby" magnetars undergoing flares, and associated to supernova remnants.

One of the early successes of the *Swift* satellite was to localize short GRBs, and therefore allow the optical follow up leading to establish that they, also, are cosmological events (Gehrels *et al.* 2005).

The top panel of Fig. 1 reports the energetics of the GRBs with measured redshifts, and the bottom panel shows the redshift distribution for long and short GRBs. Most of them have been detected by *Swift*. With the caveat that the shown isotropic energetics $E_{\rm iso}$ are not bolometric ones, nor have been K–corrected, we can see that the largest $E_{\rm iso}$ correspond to more than a solar mass entirely converted into energy. Short GRBs with measured z are still very few, but they seem to lie closer and to be less energetic than long ones.

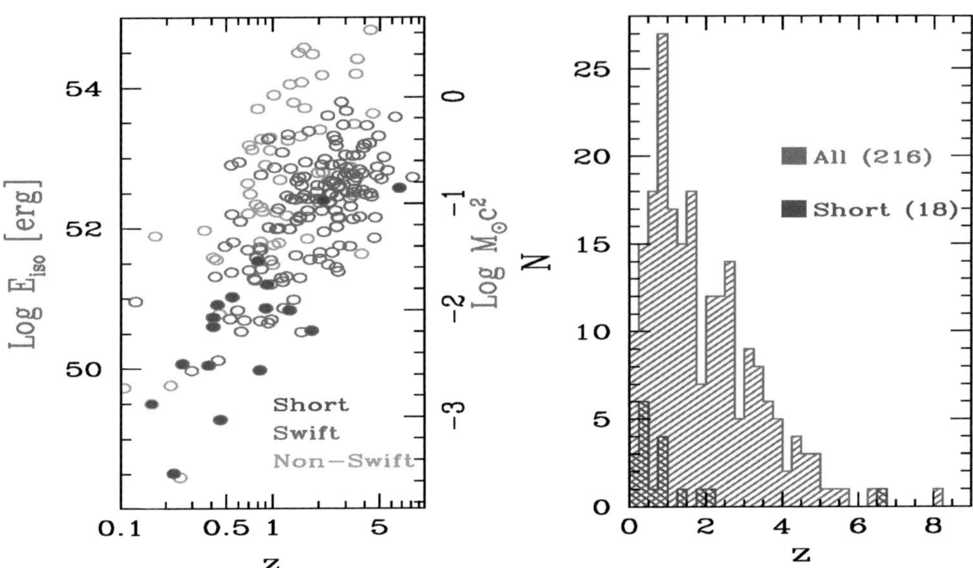

Figure 1. Left: Energetics of the prompt emission of GRBs with measured redshift. Be aware that the plotted energetics are simply the observed fluences multiplied by $4\pi d_{\rm L}^2/(1+z)$, so they are only lower limits to the bolometric $E_{\rm iso}$. Right: Redshift distribution.

1.2. *GRBs have large bulk Lorentz factors*

GRBs are the fastest extended objects of Nature, with bulk Lorentz factors Γ that can exceed 1000. The first evidence came from theory: injecting a colossal amount of energy in a small volume (of the order of a few Schwarzschild radii of size) leads inevitably to the formation of electron–positron pairs that makes the so–called "fireball" opaque to the huge internal pressure. The fireball is then obliged to expand, becoming relativistic with $\Gamma \propto R$ until the internal energy is converted into bulk motion (if the fireball remains opaque).

A nice observational evidence of relativistic speeds came, in the late '90s, from the behavior of the radio afterglow of a few GRBs, whose light curve varied wildly for $\sim$3 weeks, "calming down" after this time. This behavior was immediately interpreted as due to radio scintillation quenched by the increasing size of the radio source. Therefore it was possible to establish the expansion velocity, that turned out to be superluminal, requiring $\Gamma > 4$ after 3 weeks from the trigger (Frail *et al.* 1997).

A very recent evidence came instead from the detection of GRBs in the GeV energy range by the LAT instrument onboard the *Fermi* satellite. The GeV flux partly overlaps with the emission at lower energies detected by the other *Fermi* instrument (the GBM, sensitive in the 8 keV–30 MeV range). If the two emissions are cospatial, then the variability shown in the GBM dictates that the source must have a minimum Γ–factor, to avoid γ–$\gamma \to e^{\pm}$ suppression of high energy photons. For these LAT–detected GRBs (10% of the total), the minimum Γ–values are around 1000. (Abdo *et al.* 2009; Ackermann *et al.* 2010).

If, instead, the GeV emission is not cospatial with the GBM one, then it is very likely that it belongs to the afterglow phase. The short delay between the GBM and LAT emission can be due to the required time for the onset of the afterglow (i.e. the deceleration time of the fireball moving in the circumburst medium). The shorter this time, the higher the Γ–factor. Again, for all the LAT detected bursts, values around 1000 are derived (Ghirlanda, Ghisellini & Nava 2010; Ghisellini *et al.* 2010).

1.3. *Prompt plus afterglow emission phases*

The GRB emission has two phases, the erratic, γ–ray (or hard X–ray) *prompt phase*, and a smoother *afterglow* phase. This means that not all the energy of the fireball is radiated away during the prompt, but some remains. This was predicted before the first observations of the afterglow, but probably at the same time of the detection of long duration GeV emission by the EGRET instrument onboard the *Compton Gamma Ray Observatory* satellite (Meszaros, Rees & Papathanassiuou 1994) and then elaborated by Meszaros & Rees (1997); Vietri (1997); Sari & Piran (1997).

The fact that there are two emission phases suggests that there must be two mechanisms at work, one for the prompt and one for the afterglow.

1.4. *Long and short*

The duration of the prompt emission of GRBs is bimodal, with a minimum around 2 seconds. In astrophysics bimodal distributions are always looked at with suspicion, since malicious selection effects can be at work. The convincing arguments of a real bimodality comes from the spectrum since short GRBs are harder than long ones. This was first evident from the hardness ratio (i.e. the ratio of the flux in two energy bands; Kouveliotu *et al.* 1993) and then substantiated by direct spectral analysis (Ghirlanda, Ghisellini & Celotti 2004). The bimodality suggests that GRBs come in two flavors, in turn suggesting two different operating mechanisms, and possibly two kinds of progenitors. The prevalent idea is that long GRBs originate immediately after the collapse of a massive, Wolf–Rayet star, while short GRBs originate from the merging of two compact objects.

1.5. *Spikes have same durations*

This "pillar" is not very popular, but it was nevertheless crucial for the development of the current leading scenario of "internal shocks" (see below) explaining the prompt emission. The evidence is that the light curves of GRBs (both long and short) often shows spikes of emission, whose duration $\Delta t_{\mathrm{spike}}$ is on average the same (Ramirez–Ruiz & Fenimore 2000). In other words, there is no lengthening of $\Delta t_{\mathrm{spike}}$ with t, the time since the trigger. Emission episodes, on average, should then involve regions of similar sizes, and then probably at the same distance from the central engine.

1.6. *Supernova connection*

We believed that long GRBs are associated to Supernovae Ib,c, but not all SN Ib,c are associated to GRBs (Soderberg *et al.* 2006 estimated a fraction less than 1%). The evidence comes from spectroscopy (for nearby events) and re–brightening of the optical light curve (up to $z \sim 1$). The association strongly indicates that the progenitor of long GRBs is a massive stars, that has lost its hydrogen and helium envelopes. But there are at least two nearby bursts (GRB 060614, Gal–Yam *et al.* 2006, and GRB 060505, Ofek *et al.* 2007) where the SN was not found. If present, it would be at least two orders of magnitude less luminous than SN1998bw (associated to GRB 980425).

1.7. *Common behaviors and trends*

"When you see a GRB, you see just one GRB" was a popular motto in the past, meaning that all GRBs were different, with no common behaviors. Now this is not true any longer, and there are indeed common trends and similarities. Just two examples: the spectral energy relation, linking E_{peak} to the prompt energetics or peak luminosity, and the typical behavior of the early (i.e. less than a day or so) X–ray afterglow, with its characteristic "steep–flat–steep" light curve (Tagliaferri *et al.* 2005), and superimposed on that, $\sim 1/3$ of GRBs show X–ray flares (Burrows *et al.* 2007). These similarities are the starting point for any serious and general modelling: ideas are in fact abundant, but with no clear prevalence of one over the others.

2. Ideas and enigmas

2.1. *Central engine*

The prevalent idea is that long GRBs are caused by the collapse of a Wolf–Rayet star leading to the formation of a black hole of a few solar masses rapidly spinning. This black hole accretes 0.1–1 $M_\odot$ from a dense surrounding torus for a time more or less equal to the duration of the prompt emission. There are several energy reservoirs: neutrinos, the gravitational energy of the infalling matter, and the rotational energy of the newly formed black hole. The latter is the greatest, since it amounts to $\sim 0.29 M_{\mathrm{BH}} c^2 \sim 5.3 \times 10^{53}(M_{\mathrm{BH}}/M_\odot)$. The problem is how to extract it efficiently. The leading idea it to use the Blandford & Znajek (1977) process, for which a super–critical magnetic field of $B \sim 10^{15}$ G is required. For short GRBs, the merging scenario assumes two compact objects (e.g. two neutron stars) forming a $\sim 2M_\odot$ black hole surrounded again by a dense accreting torus. The central engine can then be the same for long and short bursts.

Although prevalent, this is not the only idea. Instead of a black hole, one could have, at least initially, a neutron star, (that collapses into a black hole only later, as a re–edition of the Vietri & Stella (1998) Supranova model). This has been proposed both to explain precursors (Wang & Meszaros 2007, see Burlon *et al.* 2008 for the characterization of precursors). A magnetar has been proposed to explain the flat (plateaux) phase of the early X–ray afterglow (e.g. Lyons *et al.* 2010). An even more radical idea was put forward by Paczynski & Haensel (2005), who suggested a quark star as the central engine. These authors pointed out that the surface of such a star acts as a one–way membrane, since baryons can only enter, but not escape. Leptons and magnetic fields, instead, can escape. This would help to explain the paucity of baryons in the fireball (i.e. the baryon "loading" problem).

2.2. *Magnetic or matter dominated?*

In the most popular scenario a huge amount of energy is injected into a small volume. Due to the colossal internal energy (and the inevitable creation of $e^\pm$ pairs, making the fireball opaque to radiation) the fireball is bound to accelerate with $\Gamma \propto R$. At the same time the comoving temperature ($T' \propto 1/R$) decreases, and when it goes below ~ 20 keV almost all the pairs annihilate without being re–created. Still, a small amount of protons and their accompanying electrons ensures that the fireball continues to be opaque until the internal energy is entirely converted into bulk motion. Thus we need another mechanism to re–convert the bulk energy into radiation. This is provided by collisions of different parts of the outflowing relativistic wind moving with different Γ–factors. This are the so–called "internal shocks", occurring at $R \sim 10^{12}$–10^{14} cm, where the fireball has turned transparent (for Thomson scatterings). Then we have a disorder $\to$ order $\to$ disorder process. Lyutikov & Blandford (2003; see the review by Lyutikov 2006 and references therein) advocated instead a simpler order $\to$ disorder process: the acceleration is due to a dominating magnetic field, allowing for almost matter free fireballs. One clear test to distinguish is the so–called "optical flash", occurring when the fireball starts to be decelerated by the interstellar medium: if it is matter dominated, then a reverse shock develops, that originates an important, and fastly decreasing, emission component (predicted in the optical or in the IR), that would be absent in magnetically dominated fireballs. Indeed optical flashes have been seen, but in a very small fraction of bursts, so the issue is unsettled. Another diagnostic would be polarization of the prompt emission (see the review by Lazzati 2006), that awaits for hard X–rays polarimeters and observations.

2.3. *Internal shocks?*

Collisions between different parts of the relativistic wind (Rees & Meszaros 1994) leading to "internal" dissipation is the leading idea for the dissipation mechanism for the prompt emission. What can be dissipated is only the *relative kinetic energy* of the two colliding parts or shells. The process has then a "built in" low efficiency (e.g. Lazzati *et al.* 1999). Consider also that, as a result of the dissipation, we distribute the available energy to protons, magnetic fields and leptons. Only the energy given to the latter can be efficiently transformed into radiation. After the different parts of the fireball have collided and "merged", the fireball runs into the circumburst medium, originating a forward "external" shock. This collision is with not–moving material, and should be much more efficient, since the entire kinetic energy of the fireball can be used. The foreseen densities and energies ensure that leptons initially radiatively cool rapidly (fast cooling regime). Thus, at least initially, the resulting afterglow is an efficient radiator. The energy radiated by the afterglow should then be greater than the energy radiated during the prompt phase. We observe just the opposite (e.g. Willingale *et al.* 2007): $E_{\rm prompt}/E_{\rm afterglow} \sim 10$.

2.4. *Radiation process of the prompt*

The popular choice is that it is synchrotron emission. Shocks should indeed accelerate electrons to relativistic energies, and amplify magnetic fields, making the synchrotron option a natural one. On the other hand, we also require any emission process to be efficient, making the electron to cool completely in a timescale much shorter than any conceivable dynamical or integration time. The spectrum produced by a cooling electron population cannot be harder than $F(\nu) \propto \nu^{-1/2}$. The spectra of practically all bursts are harder than that, and a minority are even flatter than $\nu^{1/3}$, the low frequency tail of the spectrum produced by a non cooling electron population (Preece *et al.* 1998; Ghisellini *et al.* 2000). Continuous re–acceleration or heating of the electrons is not compatible with the idea of internal shocks (in which each electron is energized only once), and other possible "way–outs" (adiabatic expansion, steep gradients of the magnetic field, contribution from synchrotron self–Compton) face severe problems as well. Alternatives have been proposed [such as jitter radiation (Medvedev 2000); quasi thermal Comptonization (Ghisellini & Celotti 1999; Giannios 2008); bulk Compton (Lazzati *et al.* 2000); multicolor blackbody (Peer & Ryde 2010); effects of Compton cooling in the Klein Nishina limit (Daigne *et al.* 2010)], but there is no prevalent idea yet.

2.5. *Spectral energy correlations*

The time integrated spectrum of the prompt, in νF_ν, has a well defined peak at $E_{\rm peak}$, that correlates with the isotropic energy of the prompt $E_{\rm iso}$: $E_{\rm peak} \propto E_{\rm iso}^{1/2}$ (Amati *et al.* 2002). When it is possible to measure the jet opening angle of the jet, it is possible to estimate the collimated energy E_γ, that correlates more tightly with $E_{\rm peak}$ ($E_{\rm peak} \propto E_\gamma^b$, with $b = 0.7$ for an homogeneous circumburst density and $b = 1$ for a wind-like profile; Ghirlanda *et al.* 2004; Nava *et al.* 2006). This correlation is tight enough to allow to "standardize" GBRs for their use as standard candles to constrain the cosmological parameters. The physical reality of these correlations has been hotly disputed (see Ghirlanda, these proceedings, and reference therein), because selection effects could play a role. On the other hand, the observations of the same correlations *within single GRBs* (Firmani *et al.* 2009; Ghirlanda *et al.* 2010) proves that a physical robust mechanism, still to be understood, is responsible for these spectral–flux (or spectral–energy) correlations.

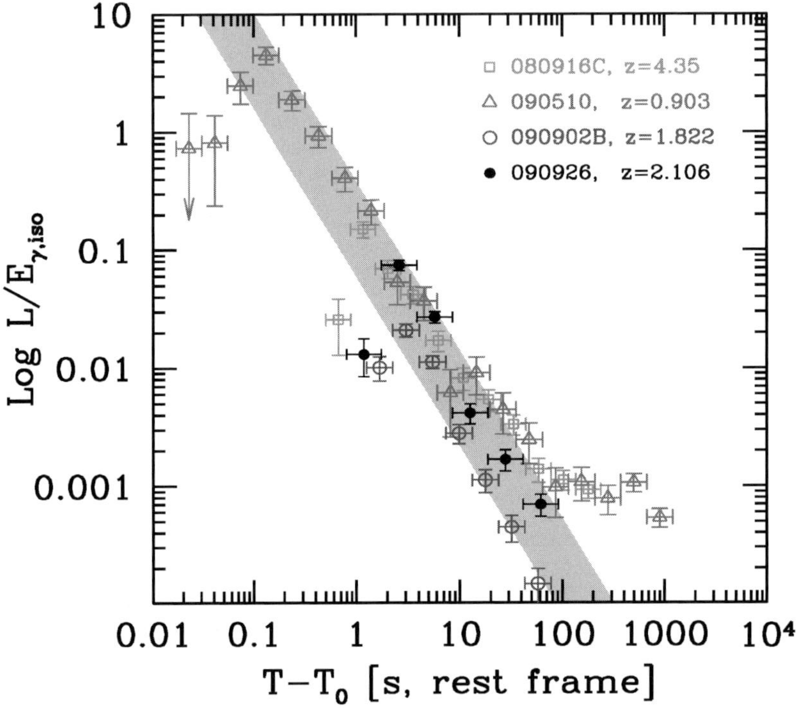

Figure 2. The light curve of the four brightest bursts detected by the *Fermi*/LAT at GeV energies. The luminosity has been divided by the energetics of the emission detected by the *Fermi*/GBM instrument ($\sim$MeV). The time is in the rest frame of the sources. The grey stripe indicates a $t^{-10/7}$ slope. The flattening at long times for GRB 090510 and GRB 080916C indicates the level of the background. Note the similarity of the different light curves.

3. High energy emission

EGRET, in the '90s, detected a handful of GRBs above 100 MeV, and since then we have been left with the question: does this emission belong to the prompt phase or is it afterglow emission produced by the fireball colliding with the circum–burst medium? Or has it still another origin? A puzzling feature of the EGRET high energy emission was that it was long lasting, yet it started during the prompt phase as seen by BATSE. *Fermi*/LAT is $\sim$ 20 times more sensitive, and indeed it detected a dozen GRBs just in its first year of life.

3.1. *Common behaviors*

From the analysis of the first 12 GRBs detected by the LAT we have found these properties (Ghisellini *et al.* 2010):

Time delay – Usually, the LAT emission lags the emission detected by the GBM (from fractions of seconds, especially for short bursts, to a few seconds).

Long lasting – As already shown by the first EGRET detections, the emission seen by the LAT lasts for a longer time than the emission in the GBM.

No spectral evolution – The average, time integrated LAT photon index is close to 2, with no evidence of strong spectral evolution.

LAT and GBM spectral slopes are often different – The GBM data can be fitted with a Band function, composed of two smoothly joining power laws. All but two bursts (GRB 080916C and GRB 090926) have LAT slopes intermediate between the two slopes of the GBM fit.

LAT fluences are smaller than GBM ones – The majority of bursts have LAT fluences smaller than the GBM ones. The two short bursts GRB 081024B and GRB 090510 and GRB 090902B have comparable LAT and GBM fluences.

Common decay – Fig. 2 shows the light curves of the 4 brightest GRBs with redshift, once the 0.1–100 GeV luminosity is divided by the energetics $E_{\gamma,\mathrm{iso}}$ of the flux detected by the GBM. The shaded stripe with slope $t^{-10/7}$ is shown for comparison. These four GRBs are all consistent, within the errors, with the same decay, both in slope and in normalisation. Note that GRB 090510, a short burst, behaves similarly to the other 3 bursts, that belong to the long class, but its light–curve begins much earlier.

3.2. *A radiative fireball?*

The above properties are just what expected by the afterglow emission due to an external shock. The short LAT–GBM delay can be caused by a large Γ (close to 1000), making the fireball to decelerate at early observed times. *The earlier the onset of the afterglow, the brighter the afterglow at early times:* this explains why only 10% of GRBs have been detected by the LAT: they correspond to bursts having the largest Γ–factors.

The relatively steep decay of the LAT light curves is very close to what expected if the fireball is radiative (i.e. most of the dissipated energy is radiated). In this case the bolometric flux in fast cooling decays as $F(t) \propto t^{-10/7} \sim t^{-1.43}$ (Sari *et al.* 1998; Ghisellini *et al.* 2010). The slopes of the LAT spectra, being close to unity (in energy), are indeed a good proxy for the bolometric fluxes.

The large peak energy of the GBM flux suggests that electron–positron pairs might play a crucial role for the setting of the radiative regime: a tiny fraction of the prompt photons, scattered (i.e. "decollimated") by the circumburst electrons, are immediately converted into pairs by the high energy photons of the prompt, largely increasing the lepton to proton ratio (Beloborodov 2002). When the shock comes, the dissipated energy can then be given mostly to leptons rather than to protons, and this makes the fireball radiative. Note a key ingredient of this scenario: to produce pairs efficiently, the prompt emission should have a sufficient number of photons above threshold, i.e. above 511 keV.

4. Conclusions

The increase of knowledge can be represented as the volume of an expanding sphere, so it goes like R^3, the cube of the radius. The surface of the sphere is the at the frontier with the unknown, the unexplained, and usually goes like R^2. But for GRBs it seems that that surface is a fractal, whose dimensionality is greater than 2... Is it good or bad? It may be perceived as depressing at first, after all these years of hectic studies, but at a second sight it should be taken as a good opportunity, especially for the younger scientists: there is still something really fundamental to be found out.

References

Abdo A. A., Ackermann M., Arimoto, M. *et al.*, 2009 *Science*, 323, 1688
Ackermann M., Asano K., Atwood W. B. *et al.*, 2010, *ApJ*, 716, 1178

Amati L., Frontera F., Tavani M. *et al.*, 2002, *A&A*, 390, 81

Beloborodov A. M., 2002, *ApJ*, 565, 808

Blandford R. D. & Znajek R. L., 1977, *MNRAS*, 179, 433

Burlon D., Ghirlanda G., Ghisellini G., Lazzati D., Nava L., Nardini M. & Celotti A., 2008, *ApJ*, 685, L19

Burrows D. N., Falcone A., Chincarini G. *et al.*, 2007, *Phys. Trans. A*, 365, 1213

Costa E., Frontera F., Heise J. *et al.*, 1997, *Nature*, 387, 783

Daigne F., Bosnjak Z. & Dubus G., 2010, subm to *A&A* (astro–ph/1009.2636)

Firmani C., Cabrera J. I., Avila–Reese V., Ghisellini G., Ghirlanda G., Nava L., Bosnjak Z., 2009, *MNRAS*, 393, 1209

Frail D. A., Kulkarni S. R., Nicastro L., Feroci M. & Taylor G. B., 1997, *Nature*, 389, 261

Gal–Yam A., Fox D. B., Price P. A. *et al.*, 2006, *Nature*, 444, 1053

Gehrels N., Sarazin C. L., O'Brien P. T. *et al.*, 2005, *Nature*, 437, 851

Ghirlanda G., Ghisellini G. & Celotti A., 2004, *A&A*, 422, L55

Ghirlanda G., Ghisellini G. & Lazzati D., 2004, *ApJ*, 616, 331

Ghirlanda G., Nava L., Ghisellini G., Firmani C. & Cabrera J. I. 2008, *MNRAS*, 387, 319

Ghirlanda G., Nava L. & Ghisellini G., 2010 *A&A*, 511. 43

Ghirlanda G., Ghisellini G. & Nava L., 2010, *A&A*, 510, L7

Ghisellini G. & Celotti A., 1999, *ApJ*, 511, L93

Ghisellini G., Lazzati D. & Celotti A., 2000, *MNRAS*, 313, L1

Ghisellini G., Ghirlanda G., Nava L. & Celotti A., 2010, *MNRAS*, 403, 926

Giannios D., 2008, *A&A*, 480, 305

Kouveliotou K., Meegan C. A., Fishman G. J., *et al.* 1993, ApJ, 413, L101

Lazzati D., Ghisellini G. & Celotti A., 1999, *MNRAS*, 309, L13

Lazzati D., Ghisellini G., Celotti A. & Rees M. J., 2000, *ApJ*, 529, L17

Lazzati D., 2006, *New J. Phys.*, 8, 131

Lyons N., O'Brien P. T., Zhang B., Willingale R., Troja E. & Starling R. L. C., 2010, *MNRAS*, 402, 705

Lyutikov M. & Blandford R., 2003, Preprint astro–ph/0312347

Lyutikov M., 2006, *New J. Phys.*, 8, 119

Medvedev M. V., 2000, *ApJ*, 540, 704

Meszaros P., Rees M. J. & Papathanassiou H., 1994, *ApJ*, 432, 181

Meszaros P. & Rees M. J., 1997, *ApJ*, 476, 232

Metzger M. R., Cohen J. G., Chaffee F. H. & Blandford R. D., 1997, *IAU Circ*, 6676

Nava L., Ghisellini G., Ghirlanda G., Tavecchio F. & Firmani C., 2006, *A&A*, 450, 471

Ofek E. O., Cenko S. B., Gal–Yam A. *et al.*, 2007, *ApJ*, 662, 1129

Paczynski B., 1986, *ApJ*, 308, L43

Paczynski B. & Haensel P., 2005, *MNRAS* 362, L4

Peer A. & Ryde F., 2010, subm to *ApJ* (astro–ph/1008.4590)

Preece R. D., Briggs M. S., Mallozzi R. S., Pendleton G. N., Paciesas W. S. & Band D. L., 1998, *ApJ*, 506, L23

Ramirez–Ruiz E. & Fenimore E. E., 2000, *ApJ*, 539, 712

Rees M. J. & Meszaros P., 1994, ApJ, 430, L93

Sari R. & Piran T., 1997, *MNRAS*, 287, 110

Sari R., Piran T. & Narayan R., 1998, *ApJ*, 497, L17

Soderberg A. M., Nakar E., Berger E. & Kulkarni S. R., 2006, *ApJ*, 638, 930

Tagliaferri G., Goad M., Chincarini G. *et al.*, 2005, *Nature*, 436, 985

Vietri M., 1997, *ApJ*, 478, L9

Vietri M. & Stella L., 1998, *ApJ*, 507, L45

Wang X.–Y. & Meszaros P., 2007, *ApJ*, 670, 1247

Willingale R., O'Brien P. T., Osborne J. P. *et al.* 2007, *ApJ*, 662, 1093

Discussion

MIRABEL: What is the fraction of long GRBs with no Supernovae?

GHISELLINI: Difficult to say, since there is a clear observational bias (only the nearby ones can be found). I can answer the symmetric question: only ∼1% of SN Ibc are associated to a GRB.

MIRABEL: What is the range of masses for quark stars?

GHISELLINI: One is tempted to associate quark stars to the 2–5 solar mass range, because the observed masses of neutron stars are below 2 solar masses, and the estimates for Galactic black hole masses are larger than 5 solar masses.

DE GOUVEIA: A comment and a question. The comment: you mention the work of Paczynski (2005) proposing a quark star model to explain the engine of GRBs. In 2002 Lugones, Ghezzi, myself and Horvath published an ApJ Letter suggesting the same. I'm glad to see that the GRB community is starting considering this alternative possibility.

The question: you have mentioned also the magnetar model for the engine but we know that there are major theoretical constraints on the production of magnetars. Could you comment on that?

GHISELLINI: The Soft Gamma Ray Repeaters are rather convincingly associated to magnetars. These systems can undergo major flares (once per century?) as the one observed on December 27 2004 from SGR 1806–20. If we put SGR 1806–20 at a few tens of Mpc, then we would classify its giant flare as a short GRB. There has been discussion about what fraction of short GRBs are giant flares from magnetars, but both spectral studies (the spectrum should be a blackbody) and correlations with nearby galaxies suggested that this fraction must be small.

Jets at all Scales
Proceedings IAU Symposium No. 275, 2011
G. E. Romero, R. A. Sunyaev & T. Belloni, eds.

© International Astronomical Union 2011
doi:10.1017/S1743921310016376

Gamma Ray Bursts Spectral–Energy correlations: recent results

Giancarlo Ghirlanda

Osservatorio Astronomico di Brera, Via Bianchi 46 Merate I–23807 Italy
email: `giancarlo.ghirlanda@brera.inaf.it`

Abstract. The correlations between the rest frame peak of the νF_ν spectrum of GRBs (E_{peak}) and their isotropic energy (E_{iso}) or luminosity (L_{iso}) could have several implications for the understanding of the GRB prompt emission. These correlations are presently founded on the *time–averaged* spectral properties of a sample of 95 bursts, with measured redshifts, collected by different instruments in the last 13 years (pre–*Fermi*). One still open issue is wether these correlations have a physical origin or are due to instrumental selection effects. By studying 10 long and 14 short GRBs detected by *Fermi* we find that a strong *time–resolved* correlation between E_{peak} and the luminosity L_{iso} is present within individual GRBs and that it is consistent with the *time–integrated* correlation. This result is a direct proof of the existence in both short and long GRBs of a similar physical link between the hardness and the luminosity which is not due to instrumental selection effects. The origin of the $E_{\mathrm{peak}} - L_{\mathrm{iso}}$ correlation should be searched in the radiation mechanism of the prompt emission.

Keywords. gamma rays: bursts, radiation mechanisms: nonthermal, supernovae: general

1. Spectral–Energy correlations

One of the key properties of the prompt emission of gamma ray bursts (GRBs) that is still poorly understood concerns the spectral–energy correlations found when considering the time–integrated spectra of bursts of known redshift. The peak energy of the spectrum, E_{peak} in the νF_ν representation, is strongly correlated with the isotropic luminosity L_{iso} (Yonetoku *et al.* 2004) or with the isotropic energy E_{iso} (Amati *et al.* 2002), and more tightly with the collimation–corrected energy E_γ (Ghirlanda, Ghisellini & Lazzati 2004).

There are two strong motivations for studying these correlations: understand their physics and use them to standardize the GRB energetics, making them cosmological tools.

However, the $E_{\mathrm{peak}} - E_{\mathrm{iso}}$ and $E_{\mathrm{peak}} - L_{\mathrm{iso}}$ correlations have been derived considering long GRBs (Fig. 1 filled grey and open red circles) due to the lack of measured redshifts of short GRBs. It has been shown recently (Ghirlanda *et al.* 2009) that the few short GRBs with measured z (filled blue squares in Fig. 1) do not follow the $E_{\mathrm{peak}} - E_{\mathrm{iso}}$ correlation (Fig. 1, left panel) but they are consistent with the $E_{\mathrm{peak}} - L_{\mathrm{iso}}$ correlation (Fig. 1, right panel).

A still debated issue is wether these correlations have a physical origin or they are the result of instrumental selection effects (Nakar & Piran 2005; Band & Preece 2005; Butler *et al.* 2007, Butler, Kocevski & Bloom 2009; Shahmoradi & Nemiroff 2009 but see Ghirlanda *et al.* 2005, Bosnjak *et al.* 2008, Ghirlanda *et al.* 2008; Nava *et al.*, 2008; Krimm *et al.* 2009; Amati, Frontera & Guidorzi 2009).

A completely orthogonal possibility (with respect to the debate opened in the literature - e.g. Butler, Kocevski & Bloom 2009 but see e.g. Ghirlanda *et al.* 2008) to answer this question is to study *individual* bursts to see whether the luminosity and peak energy

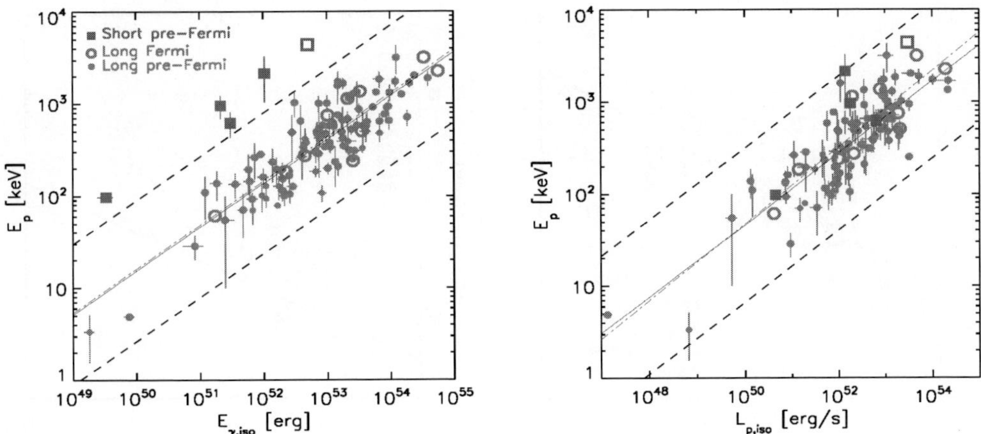

Figure 1. Rest frame peak energy versus isotropic energy (left) and luminosity (right) of 95 long GRBs detected before *Fermi*, i.e. pre–*Fermi* sample (filled grey circles). The 10 *Fermi* long GRBs with measured redshift and $E_{\rm peak}^{\rm obs}$ (up to July 2009) are shown by the open (red) circles, respectively. The only short *Fermi* GRB with measured redshift and measured $E_{\rm peak}^{\rm obs}$ (090510) is shown by the open (blue) square. The filled (blue) squares are the 4 short GRBs with measured redshift and $E_{\rm peak}^{\rm obs}$ detected by other instruments than *Fermi*. The solid line is the best fit to the pre-*Fermi* sample, while the dashed lines represent its 3σ scatter. The dot–dashed line is the best fit including the 10 *Fermi* long GRBs. (Figure adapted from Ghirlanda *et al.* 2009, 2010, 2010a)

at different times during the prompt phase correlate. If they do, and furthermore if the slope of this *time–resolved* correlation (indicated $E_{\rm peak}^{t} - L_{\rm iso}^{t}$ hereafter) is similar to the *time–integrated* $E_{\rm peak} - L_{\rm iso}$ correlation found among different bursts, then we should conclude that the spectral energy correlations are surely a manifestation of the physics of GRBs and not the result of instrumental selection effects.

The two questions we want to answer are: (1) is there a time resolved spectral energy correlation within individual GRBs? and (2) is this correlation present in both short and long GRBs?

To answer these questions the *time–resolved* spectral analysis of GRB is required. Moreover, in order to follow the evolution of the spectrum and its peak energy $E_{\rm peak}$ in time all over the burst duration, a large spectral energy window is desirable like that of the Gamma Burst Monitor (GBM, 8keV-40MeV) onboard the *Fermi* satellite. Here we present the main results of the study of the spectral evolution of 10 long GRBs and 14 short GRBs detected by *Fermi* (these results have been published in Ghirlanda *et al.* 2010, 2010a).

2. Time resolved spectral–energy correlations

We have analyzed the time resolved spectra of 10 long GRBs with measured redshifts detected by *Fermi* up to July 2009. Their spectral evolution shows that the peak energy tracks the flux and a strong time–resolved $E_{\rm peak}^{t} - L_{\rm iso}^{t}$ correlation is present within individual GRBs (Fig. 2 left panel). The time resolved $E_{\rm peak}^{t} - L_{\rm iso}^{t}$ correlation is similar to the time–integrated one (solid line in Fig. 2 left panel). The same analysis was applied to short GRBs detected by *Fermi*. Except for GRB 090510 at $z = 0.9$ (open blue square in Fig. 1), the redshifts of the other *Fermi* short bursts is not measured. However, we can still study the correlation between the observer frame peak energy and the bolometric flux (Fig. 2 right panel). Also in short bursts there is strong hardness–intensity correlation.

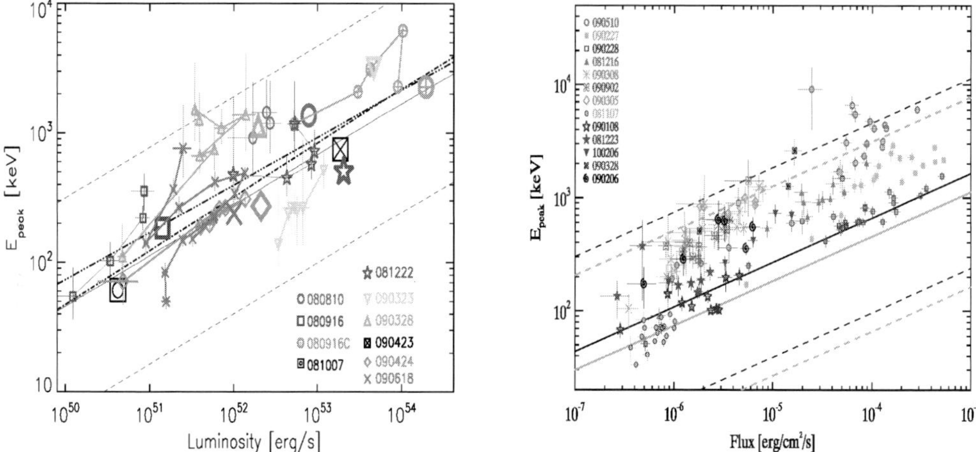

Figure 2. Left: time–resolved $E_{\mathrm{peak}} - L_{\mathrm{iso}}$ correlation of 10 *Fermi* **long** GRBs with known redshift (shown with open red circles in Fig. 1). Small symbols show the evolution of E_{peak} vs L_{iso}. Large symbols are the location in the $E_{\mathrm{peak}} - L_{\mathrm{iso}}$ plane of the corresponding bursts when the time–integrated spectra are considered (for GRB 081007 and GRB 090423, only the time–integrated spectrum is available). The solid and dotted lines represent the $E_{\mathrm{peak}} - L_{\mathrm{iso}}$ correlation and its 3σ scatter, respectively, as obtained with the pre–Fermi GRBs (filled grey circles in Fig. 1. The dot–dashed line represents the t to the 10 Fermi GRBs (time–integrated) and the triple–dot–dashed line is the t to the 51 time–resolved spectra (Ghirlanda *et al.* 2010). Right: time–resolved correlation between the peak energy E_{peak} and the flux of the 163 time resolved spectra of the 13 **short** *Fermi* GRBs analyzed in Ghirlanda *et al.* 2010a. Different symbols/colors correspond to different bursts (as shown in the legend). For all but one short GRBs (090510 - open blue square in Fig. 1) the redshift is not known, this is why their spectral evolution is represented in the observer frame. The solid line (dashed lines) is the $E_{\mathrm{peak}} - L_{\mathrm{iso}}$ correlation (and its 3σ scatter) of long GRBs transformed in the observer frame assuming $z = 1$ ($z = 0.5$ for the grey solid and dotted lines).

It is consistent with the $E_{\mathrm{peak}} - L_{\mathrm{iso}}$ transformed in the observer frame by assuming a typical redshift $z = 0.5$ (grey solid line in Fig. 2) or $z = 1$ (black solid line in Fig. 2).

In order to verify that the time resolved hardness–intensity correlation found is not induced by the spectral model used to fit the time resolved spectra, we show in Fig. 3 that there is no correlation between the low energy spectral index α and the peak energy E_{peak} in both short (filled blue squares) and long GRBs (open red circles). Fig. 3 also shows that short GRBs have harder low energy spectral index than long events as a confirmation of what already found from the comparison of their time integrated spectra (Ghirlanda *et al.* 2004, 2009 - where it was instead shown that the peak energies are similar in short and long events).

Recently, Guirec *et al.* (2010), by studying the spectral evolution of the three short brightest *Fermi* GRBs, find that they have larger E_{peak} than long ones, claiming that this (apparently) contradicts what found in Ghirlanda *et al.* (2004, 2009). Not surprisingly, this is simply due to the existence of a correlation between E_{peak} and the flux in both short and long GRBs. Indeed, the three brightest short GRBs analyzed by Giuriec *et al.* 2010, which are also present in our sample of 14 events, reach values of E_{peak} significantly harder than long events but they are also significantly brighter. For the same fluxes short and long GRBs have similar E_{peak}.

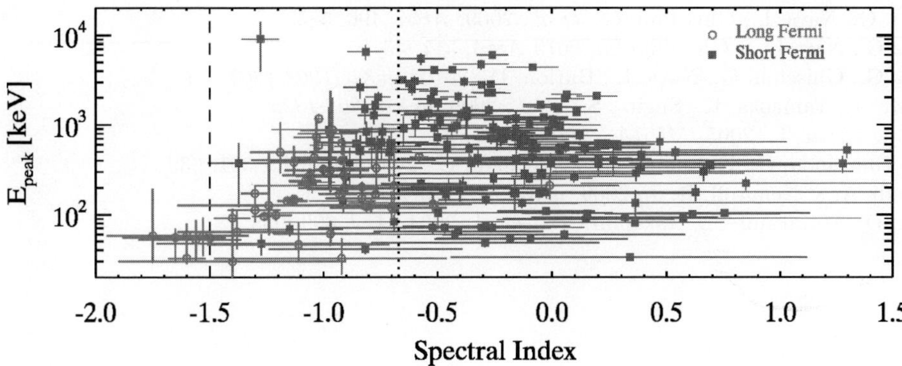

Figure 3. Observer frame peak energy versus low energy spectral index of the time resolved spectra of the 10 long GRBs (open red circles) and 14 short GRBs detected by *Fermi* and analyzed in Ghirlanda *et al.* 2010 and Ghirlanda *et al.* 2010a, respectively. The dotted (dashed) vertical line is the synchrotron limit (with cooling).

3. Conclusions

All the spectral–energy correlations in GRBs have been derived considering the time–integrated GRB spectral properties. The spectral analysis of 10 *Fermi* GRBs with measured z and 14 short *Fermi* GRBs reveals the existence of a time–resolved $E^t_{\mathrm{peak}} - L^t_{\mathrm{iso}}$ within individual bursts. The evolutionary tracks lie in the upper part of the $E_{\mathrm{peak}} - L_{\mathrm{iso}}$ correlation (Fig. 2). This could be caused by a systematic underestimate (overestimate) of the luminosity (peak energy) in time–resolved spectra with respect to time–integrated spectra because the former are more frequently fitted with a CPL model that lacks the high–energy power law component (Band function) typically fitted to time–integrated spectra. The existence of a $E^t_{\mathrm{peak}} - L^t_{\mathrm{iso}}$ correlation within individual GRBs, consistent with the $E_{\mathrm{peak}} - L_{\mathrm{iso}}$ correlation defined by time–integrated spectra, is the strongest argument in favour of a physical origin of this correlation (or the other way round, the strongest argument against selection effects). If the origin of the spectral energy correlations is to be found in the emission mechanism, then it should be the same for short and long GRBs.

Acknowledgements

I am grateful to G.Ghisellini, L. Nava, D. Burlon, M. Nardini, A. Celotti for useful discussions. ASI is thanked for grant I/088/06/0. A PRIN-INAF grant is acknowledged for funding. This research made use of the *Fermi*–Gamma Burst Monitor data publicly available via the NASA-HEASARC data center.

References

Amati L., Frontera F., Tavani M. *et al.*, 2002, *A&A*, 390, 81
Amati, L., Frontera, F., & Guidorzi, C., 2009, *A&A*, 508, 173
Band, D. L. & Preece, R., 2005, *ApJ*, 627, 319
Bosnjak, Z., Celotti, A., Longo, F., *et al.*, 2008, *MNRAS*, 384, 599
Butler N. R., Kocevski D., Bloom J. S., & Curtis J. L., 2007, *ApJ*, 671, 656
Butler N. R., Kocevski D., & Bloom J. S., 2009, *ApJ*, 694, 76
Ghirlanda G., Ghisellini G., & Celotti A., 2004, *A&A*, 422, L55
Ghirlanda G., Ghisellini G., & Lazzati D., 2004, *ApJ*, 616, 331
Ghirlanda, G., Ghisellini, G., Firmani, C., Celotti, A., & Bosnjak, Z., 2005, *MNRAS*, 360, 45
Ghirlanda G., Nava L., Ghisellini G., Firmani C., & Cabrera J. I. 2008, *MNRAS*, 387, 319

Ghirlanda, G., Nava, L., Ghisellini, G., *et al.*, 2009, *A&A*, 496, 585
Ghirlanda G., Nava L. & Ghisellini G., 2010 *A&A*, 511, 43
Ghirlanda G., Ghisellini G., Nava, L., Burlon, D., 2010a *arXiv:1008.4767*
Krimm, H., A., Yamaoka, K., Sugita, S., *et al.*, 2009, *arXiv0908.1335*
Nakar, E. & Piran, T., 2005, *MNRAS*, 360, L73
Nava, L., Ghirlanda, G., Ghisellini, G., & Firmani, C., 2008, *MNRAS*, 391, 639
Shahmoradi, A. & Nemiroff, R. J., 2009, *arXiv0904.1464*
Yonetoku, D., Murakami, D., Nakamura, T., *et al.*, 2004, *ApJ*, 609, 935

Jets at all Scales
Proceedings IAU Symposium No. 275, 2011
G. E. Romero, R. A. Sunyaev & T. Belloni, eds.

© International Astronomical Union 2011
doi:10.1017/S1743921310016388

Instabilities in the Gamma Ray Burst central engine. What makes the jet variable?

Agnieszka Janiuk[1], Ye-Fei Yuan[2], Rosalba Perna[3] and Tiziana Di Matteo[4]

[1] Center for Theoretical Physics, Polish Academy of Sciences, Al. Lotnikow 32/46, 02-668 Warsaw, Poland

[2] Department of Astronomy, University of Science and Technology of China, Chinese Academy of Sciences, Hefei, Anhui 230026, P.R. China

[3] JILA and Department of Astrophysical and Planetary Sciences, University of Colorado, Boulder, CO 80309 USA

[4] Physics Department, Carnegie Mellon University, 5000 Forbes Avenue, Pittsburgh, PA 15232

Abstract. Both types of long and short gamma ray bursts involve a stage of a hyper-Eddington accretion of hot and dense plasma torus onto a newly born black hole. The prompt gamma ray emission originates in jets at some distance from this 'central engine' and in most events is rapidly variable, having a form of sipkes and subpulses. This indicates at the variable nature of the engine itself, for which a plausible mechanism is an internal instability in the accreting flow. We solve numerically the structure and evolution of the neutrino-cooled torus. We take into account the detailed treatment of the microphysics in the nuclear equation of state that includes the neutrino trapping effect. The models are calculated for both Schwarzschild and Kerr black holes. We find that for sufficiently large accretion rates ($> \sim 10 M_\odot$ s^{-1} for non-rotating black hole, and $> \sim 1 M_\odot$ s^{-1} for rotating black hole, depending on its spin), the inner regions of the disk become opaque, while the helium nuclei are being photodissociated. The sudden change of pressure in this region leads to the development of a viscous and thermal instability, and the neutrino pressure acts similarly to the radiation pressure in sub-Eddington disks. In the case of rapidly rotating black holes, the instability is enhanced and appears for much lower accretion rates. We also find the important and possibly further destabilizing role of the energy transfer from the rotating black hole to the torus via the magnetic coupling.

Keywords. physical data and processes: accretion, black hole physics, instabilities, neutrinos, nuclear reactions, gamma-rays:bursts

---

1. Introduction

Roughly 80 per cent of the observed gamma ray bursts exhibit a substructure in their time profiles and the timescales of the sub-pulses are about 10^3-10^4 times shorter than the total duration of the event, estimated by T_{90} (see Piran (2005) for a review). This effect is most probably connected with the rapidly variable conditions within the jet plasma, such as the Lorentz factor. Therefore, since the energy input into the jet is required to vary, the activity of the central engine that produces this energy should not be stationary but rather change in much shorter timescale than the engine lifetime.

The central engine of the gamma ray burst is presumably a newly born black hole, surrounded by an accretion disk built from either a remnant of the disrupted companion star in the compact binary system, or the fallback material from the hypernova envelope. Such disk is extremely hot and dense due to the large accretion rate which can be on the order of 0.1 up to a few solar masses per second. Such physical conditios make the plasma totally opaque to photons and partially opaque to neutrinos (Di Matteo *et al.* 2002), which are produced in the nuclear reactions and provide a source of cooling to the disk (Popham *et al.* 1999).

The gravitational instability of the outer parts of the hyperaccreting disk in the GRB central engine was suggested first by Perna *et al.* (2006) and studied in the numerical model of Chen & Beloborodov (2007). Such instability may be a source for the late time activity and longer duration flares from the central engine. The prompt emission however is more plausibly explained with the instability discussed in Janiuk *et al.* (2007), and more recently in Janiuk & Yuan (2010), which arises in the regions of the disk closest to the central black hole.

2. Instability mechanism

The time dependent model of the hyperaccreting disk in the GRB central engine, presented in Janiuk *et al.* (2004), was based on a simplified equation of state, where the total pressure was given by a sum of the ideal gas, radiation and degenerate electron pressure. The more elaborate EOS, used in Janiuk *et al.* (2007), invokes self-consistently a chemical equilibrium of species such as protons, neutrons, electron-positron pairs and alpha particles. The reaction rates are computed under the assumption of beta equilibrium and charge neutrality, and the neutrino opacities are due to their absorption and scattering. To the total pressure contribute both partially trapped neutrinos and alpha particles, which are continuously photodissociated and recombined in the plasma.

As results from our first calculations, at very high accretion rates, which for a non-rotating black hole should be at least 10 $M_\odot$ s^{-1}, a thermal instability arises in the innermost disk region close to the inner edge. This instability is associated with helium photodisintegration, which was found to lead to the local accretion rate fluctuations already by e.g. MacFadyen & Woosley (1999). What can be seen from our Fig. 1, bottom panel, the helium photodissociacion in a narrow strip of the accreting disk leads to a phase

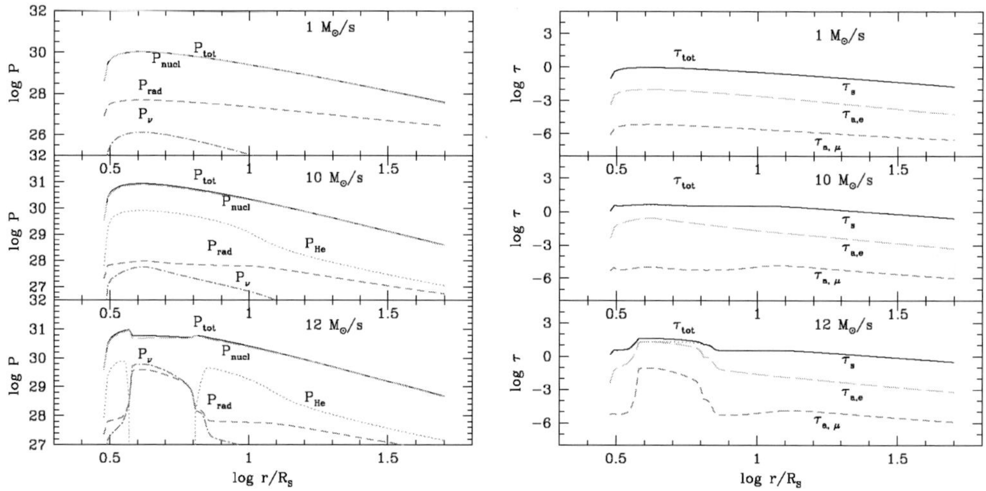

Figure 1. Radial profiles of the total pressure and its components: nuclear, helium, neutrino and radiation pressure (left), and neutrino total opacity and its components due to absorption, for muon and electron neutrinos, and scattering (right). The models were computed for a non-rotating black hole, with three values of accretion rate, as indicated in the panels. The black hole mass is 3 $M_\odot$ and viscosity $\alpha = 0.1$

transition in the nuclear material. Due to this transition, the vanishing helium pressure is compensated by the neutrino pressure. The plasma becomes opaque to neutrinos and the neutrino absorption optical depths increase locally for both muon and electron flavor species. However, the net total pressure drops somewhat in the unsstable region and the resulting density profile in the disk is inversed. The unstable strip is much hotter and less dense than its surroundings. This leads to a faster replenishing of the strip than would be implied by the constant accretion rate in a stable situation. The accretion rate varies in both time and radius, which in turn leads to the neutrino luminosity variations and variable energy output from the disk.

The instability mechanism is much similar to the well known radiation pressure instability, operating in the accretion disks of microquasars and AGN (e.g., Janiuk *et al.* (2002)). In case of gamma ray burst disks, the role of radiation is taken by the neutrino pressure, which has a similar dependency on temperature. However, the mechanism stabilizing the disk in its hot phase, which in the radiation pressure dominated disk is due to advection, is not sufficient in the neutrino cooled disks.

3. Effect of black hole rotation and its energy extraction via magnetic fields

The accretion rates needed for the thermal instability to appear do not have to be extremely large if the disk surrounds a rotating black hole. Such a black hole would be a natural outcom e.g. of a collapsing Wolf Rayet star with a rapid rotation and is much more plausible for powering the GRB jets than a Schwarzschild black hole. As estimated in Janiuk *et al.* (2008), the longest duration GRBs require for their lifetimes large black hole spins, $a > 0.9$. This is because otherwise the realistic distributions of angular momentum within the collapsing envelope do not provide a sufficient condition for a long duration central engine lifetime.

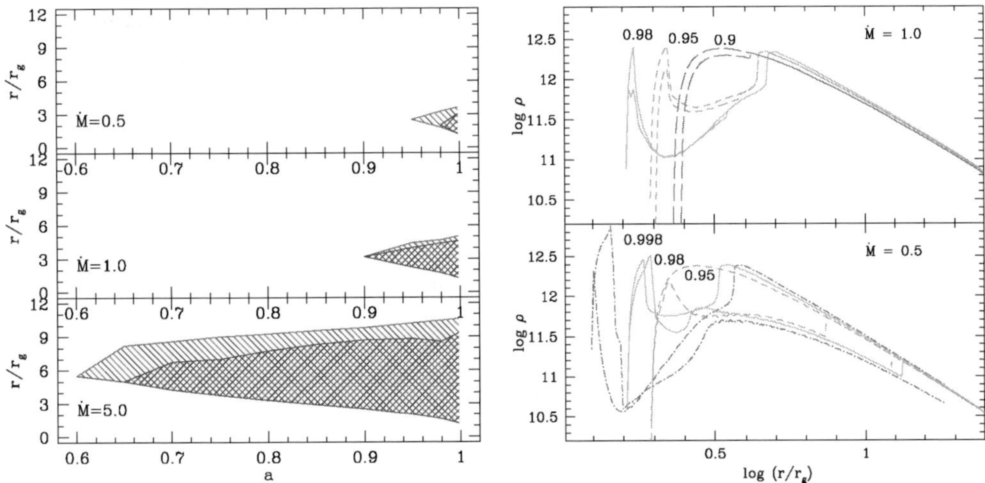

Figure 2. Extension of the unstable zones, depending on the black hole spin (left), and radial density profiles (right). The values of the accretion rate are marked in the panels and the black hole mass is $4M_\odot$. The visocsity in the left figure is $\alpha = 0.1$ (line shades regions) or $\alpha = 0.3$ (cross-shaded). The right figure is for $\alpha = 0.1$, and the spin parameters are indicated at the top of each line. The thicker lines show the results for the disk magnetically coupled with a rotating black hole and the thinner lines are for neglected magnetic coupling.

In Figure 2 we show that the thermal instability occurs in the Kerr black hole disks at quite low accretion rates, e.g. at 0.5 $M_\odot$ s^{-1}, as is plausible for the collapsar scenario, or a few $M_\odot$ s^{-1}, as more plausible in case of merging neutron stars. This result slightly depends on the adopted disk viscosity (α parameter) and is very sensitive to the black hole spin.

The additional effect which needs to be taken into account in case of the Kerr black hole, is the transfer of its rotational energy to the disk and vice-versa. In other words, the black hole can be spun up by accretion, or spun down if the angular velocity in the inner radii of the differentially rotating disk exceeds that of the black hole. The energy transfer proceeds via the closed magnetic field lines and leads to the additional torque due to the black hole coupling (Wang *et al.* 2002). At the same time, the open magnetic field lines can transfer the black hole rotation energy to the remote load and power the jet via the Blandford - Znajek process. We found that such a coupling does not stabilize the disk against the thermal instability discussed above. The unstable region can be shrunken but also shifted outwards, while the density drop in this region can even be deeper. Another effect is, that in case of the stable disk with a moderate neutrino pressure and some helium nuclei, the additional heating due to the rotating black hole coupling may lead to another type of unstable behaviour (Lei *et al.* 2009).

4. Conclusions

According to our results, the longest duration GRBs are powered by the fast rotation of the black hole. They sould be rapidly variable at the beginning of their prompt emission, due to the instabilities in the accretion process. Later on, the accretion rate drops and the black hole spins down, so that the instability is no longer operating and the burst profile will smoothen. On the other hand, also a moderately rotating black hole can be produced after the hypernova collapse. If the accretion rate is large, the black hole spin up may later allow for sufficient conditions for the instability. The GRB observed properties may therefore serve to test the role of black hole spin in the jet production in accreting black hole systems, other than X-ray binaries and AGN (Fender *et al.* 2010).

Acknowledgments This work was supported in part by grant NN 203 512638 from the Polish Ministry of Science.

References

Chen W. X. & Beloborodov A. 2007, *ApJ*, 657, 383
Di Matteo, T., Perna, R., Narayan, R. 2002, *ApJ*, 579, 706
Fender, R. P., Gallo, E., Russell, D. 2010, *MNRAS*, 406, 1425
Janiuk, A., Czerny, B., Siemiginowska, A. 2002, *ApJ*, 576, 908
Janiuk, A., Perna, R., Di Matteo, T., Czerny, B. 2004, *MNRAS*, 355, 950
Janiuk, A., Yuan, Y.-F., Perna, R., Di Matteo, T. 2007, *ApJ*, 664, 1011
Janiuk, A., Moderski, R., Proga, D. 2008, *ApJ*, 687, 433
Janiuk, A., Yuan, Y.-F. 2010, *A&A*, 509, 55
Lei, W. H., Wang, D. X., Zhang, L., *et al.* 2009, *ApJ*, 700, 1970
MacFadyen, A., Woosley S. E. 1999, *ApJ*, 524, 262
Perna, R., Armitage, P., Zhang, B. 2006, *ApJ* 636, L29
Piran, T. 2005, *Rev. Mod. Phys.* 76, 1143
Popham, R., Woosley, S. E., Fryer, C. 1999, *ApJ*, 518, 356
Wang, D. X., Xiao, K., Lei, W.-H. 2002, *MNRAS*, 335, 655

Discussion

KOIDE: You noted the energy transport from the rotating black hole to the disk through the closed magnetic field lines. As far as we assume ideal MHD or a force-free condition, it is difficult to store the energy in the disk. I think the energy extracted from the black hole is stored in magnetic field outside the disk, where the magnetic configuration changes. Could you comment on this issue?

JANIUK: The disk plasma is fully ionized so we assume the disk to be perfectly conducting and the magnetic filed lines frozen in. When the accretion proceeds, the field lines are advected and its configuration indeed may be changed. Our assumption is in fact a steady state approximation.

Jets at all Scales
Proceedings IAU Symposium No. 275, 2011
G. E. Romero, R. A. Sunyaev & T. Belloni, eds.

© International Astronomical Union 2011
doi:10.1017/S174392131001639X

Simulation of relativistic shocks and associated radiation from turbulent magnetic fields

K.-I. Nishikawa[1], J. Niemiec[2], M. Medvedev[3], B. Zhang[4], P. Hardee[5], Y. Mizuno[1], A. Nordlund[6], J. Frederiksen[6], H. Sol[7], M. Pohl[8], D. H. Hartmann[9] and G. J. Fishman[10]

[1] Center for Space Plasma and Aeronomic Research, University of Alabama in Huntsville, NSSTC, 320 Sparkman Drive, Huntsville, AL 35805, USA
email: `ken-ichi.nishikawa-1@nasa.gov`

[2] Institute of Nuclear Physics PAN, ul. Radzikowskiego 152, 31-342 Krakow, Poland

[3] Department of Physics and Astronomy, University of Kansas, KS 66045, USA

[4] Department of Physics and Astronomy, University of Nevada, Las Vegas, NV 89154, USA

[5] Department of Physics and Astronomy, The University of Alabama, Tuscaloosa, AL 35487, USA

[6] Niels Bohr Institute, Juliane Maries Vej 30, 2100 Kbenhavn , Denmark

[7] LUTH, Observatore de Paris-Meudon, 5 place Jules Jansen, 92195 Meudon Cedex, France

[8] Institut fuer Physik und Astronomie, Universitaet Potsdam, 14476 Potsdam-Golm, Germany

[9] Department of Physics and Astronomy, Clemson University, Clemson, SC 29634, USA

[10] NASA/MSFC, 320 Sparkman Drive, Huntsville, AL 35805, USA

Abstract. Recent PIC simulations of relativistic electron-positron (electron-ion) jets injected into a stationary medium show that particle acceleration occurs in the shocked regions. Simulations show that the Weibel instability is responsible for generating and amplifying highly nonuniform, small-scale magnetic fields and for particle acceleration. These magnetic fields contribute to the electron's transverse deflection behind the shock. The "jitter" radiation from deflected electrons in turbulent magnetic fields has different properties from synchrotron radiation calculated in a uniform magnetic field. This jitter radiation may be important for understanding the complex time evolution and/or spectral structure of gamma-ray bursts, relativistic jets in general, and supernova remnants. In order to calculate radiation from first principles and go beyond the standard synchrotron model, we have used PIC simulations. We will present detailed spectra for conditions relevant to various astrophysical sites of collisionless shock formation. In particular we will discuss application to GRBs and SNRs.

Keywords. instabilities, magnetic fields, shock waves, radiation mechanisms: general

1. Relativistic PIC Simulations

Particle-in-cell (PIC) simulations can shed light on the physical mechanism of particle acceleration that occurs in the complicated dynamics within relativistic shocks. Recent PIC simulations of relativistic electron-ion and electron-positron jets injected into an ambient plasma show that acceleration occurs within the downstream jet (e.g., Nishikawa *et al.* 2009b). In general, these simulations have confirmed that the Weibel instability, which generates current filaments and associated magnetic fields, mediates the relativistic collisionless shock (Medvedev 1999), and accelerates electrons (Hededal 2005; Nishikawa *et al.* 2003, 2005, 2006; Ramirez-Ruiz, Nishikawa, & Hededal 2007; Chang, Spitkovsky, & Arons 2008; Spitkovsky 2008a,b; Sironi & Spitkovsky 2009a; Nishikawa *et al.* 2009a). Therefore, the investigation of radiation resulting from accelerated particles (mainly electrons and positrons) in turbulent magnetic fields is essential to understand the radiation and its observable spectral properties. In this report we present a numerical method for obtaining spectra from particles self-consistently traced in our PIC simulations that goes beyond the standard synchrotron model (Waxman 2006).

2. Calculating Emission from Electrons Moving in Self-consistently Generated Magnetic Fields

We calculate the radiation spectra directly from our simulations by integrating the expression for the retarded power, derived from Liénard-Wiechert potentials for a large number of representative particles in the PIC representation of the plasma (Jackson 1999; Hededal 2005; Nishikawa *et al.* 2009a,c; Sironi & Spitkovsky 2009b; Frederiksen *et al.* 2010). In order to obtain the spectrum of the synchrotron/jitter emission, we consider an ensemble of electrons selected in the region where the Weibel instability has fully grown and where the electrons are accelerated in the self-consistently generated magnetic fields.

We have validated our numerical method by performing simulations using a small system with $(L_\mathrm{x}, L_\mathrm{y}, L_\mathrm{z}) = (645\Delta, 131\Delta, 131\Delta)$ ($\Delta = 1$: grid size) and a total of ~ 0.5 billion particles (12 particles /cell/species for the ambient plasma) in the active grid zones (Nishikawa *et al.* 2006). We first performed simulations without calculating radiation up to $t = 450\omega_\mathrm{pe}^{-1}$ when the jet front is located at about $x = 480\Delta$. We randomly selected 16,200 jet electrons near the jet front and calculated the emission during the sampling time $t_\mathrm{s} = t_2 - t_1 = 75\omega_\mathrm{pe}^{-1}$ with Nyquist frequency $\omega_\mathrm{N} = 1/2\Delta t = 200\omega_\mathrm{pe}$ where $\Delta t = 0.005\omega_\mathrm{pe}^{-1}$ is the simulation time step and the frequency resolution $\Delta\omega = 1/t_\mathrm{s} = 0.0133\omega_\mathrm{pe}$.

The spectra shown in Figure 1 are obtained for head-on ($\theta = 0°$) emission from the jet electrons. The radiation from shows a Bremsstrahlung-like spectrum for the three cases shown (Hededal 2005). Since the magnetic fields generated by the Weibel instability are rather weak and jet electron acceleration is modest, the electron trajectories bend only slightly. For the case with $\gamma = 15$ and $B_\mathrm{x} = 3.7$ (simulation units), the spectrum is similar to that of a case without magnetic field. However, for the cases with $\gamma = 7.08$ the spectrum in the case with $B_x = 3.7$ shows an enhanced amplitude at higher frequencies. At this smaller Lorentz factor the Weibel instability grows to larger amplitude (the growth time is proportional to $(\gamma)^{1/2}$) than for the $\gamma = 15$ and $B_\mathrm{x} = 3.7$ case. In this case the transverse magnetic fields generated by the Weibel instability combined with

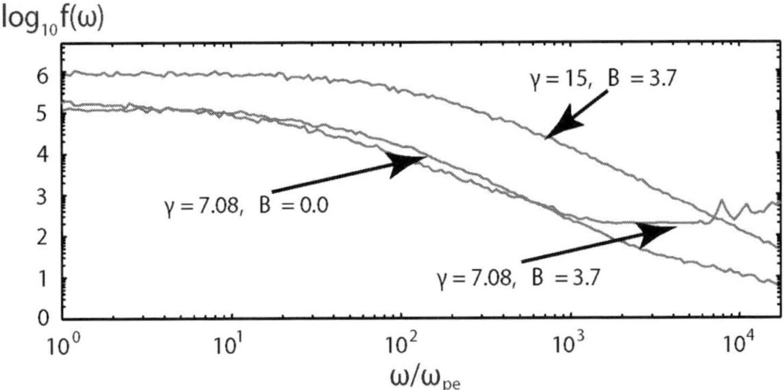

Figure 1. Spectra obtained from hot jet electrons ($v_\mathrm{th}^\mathrm{jet} = 0.33c$) for three different cases ($\gamma = 15, B_x = 3.7$, $\gamma = 7.08, B_x = 3.7$, and $\gamma = 7.08, B_x = 0.0$) for viewing angle $0°$.

the (longitudinal) ambient magnetic field bend the jet electron trajectories significantly to generate the higher frequencies.

Like a Bremsstrahlung spectrum, the lower frequencies have flat spectra and the higher frequencies decrease monotonically (Nishikawa *et al.* 2009C). However, the higher frequency slopes in Figure 1 are less steep than that in the Bremsstrahlung spectrum. This is due to the fact that the spread of Lorentz factors of jet electrons is larger and the average Lorentz factor is larger as well. Furthermore, even though the magnetic field strength is not so large, the spectra seem to be extended to higher frequency. This is explained that as shown in Fig. 7.16 (left) in Hededal?s Ph. D. thesis (Hededal, 2005) the turbulent magnetic field shifts the frequency higher with shorter wave length (smaller μ). We obtained several different parameters with jet electrons and ambient magnetic field using a small system as in this report. However, the strength of the magnetic fields generated by the Weibel instability is small, therefore the spectra for these cases are very similar to the Bremsstrahlung spectrum.

Figure 2 shows how our synthetic spectrum matches with spectra obtained from Fermi observations. Figure 2a shows the model spectra for five time intervals (Abdo *et al.* 2009). The red line shows the slope ($a = 1$) in Fig. 2a and except the spectrum for the interval "a" the slopes for all other time intervals are approximately 1. This is similar to a Bremsstrahlung-like spectrum at least for the low frequency side. As shown in Fig. 2b the slope in the low frequency is very similar to the observed spectra. The peaks and slopes at high frequencies change over time.

3. Discussion

Emission obtained using the method described above is obtained self-consistently, and automatically accounts for magnetic field structures on the small scales responsible for jitter emission. By performing such calculations for simulations using different parameters, we can investigate and compare the different regimes of jitter- and synchrotron-type emission (Medvedev 2000, 2006). Thus, we should be able to address the low frequency GRB spectral index violation of the synchrotron spectrum line of death (Medvedev 2006).

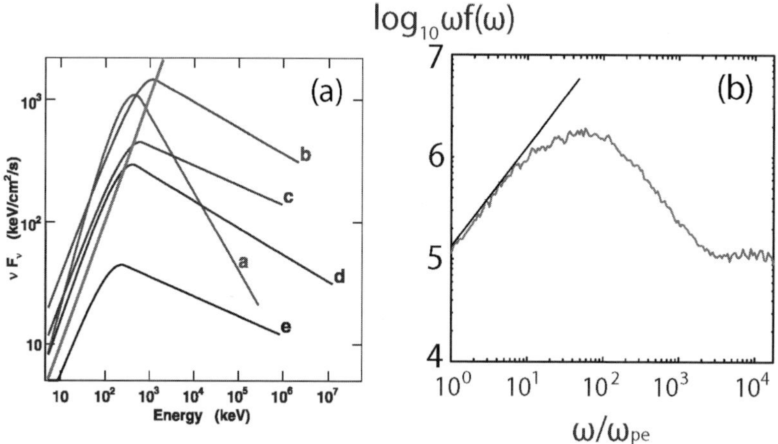

Figure 2. Brief comparison of a synthetic spectrum with spectra obtained from Fermi observations. Figure 2a shows the modeled Fermi spectra in νF_ν units for five time intervals. A flat spectrum would indicate equal energy per decade in photon energy. The changing shapes show the evolution of the spectrum over time. Figure 2b shows the spectrum for the case of $\gamma = 7.08$ with $B_x = 3.7$. The low frequency slope is approximately 1 and very similar to those spectra except interval "a".

Recently, synthetic radiation has been calculated from simulations of laser-wakefield electron acceleration (Martins *et al.* 2009), counter-streaming jet (Frederiksen *et al.* 2010), and in shocks reflected at the wall (Sironi & Spitkovsky 2009). In this study we inject relativistic jets into ambient plasma (from the left side of the computational box; see also Nishikawa *et al.* 2009b). A Weibel instability-mediated shock is formed and electrons become accelerated. We calculate the radiation from jet electrons in the observer frame, therefore calculated spectra can be directly compared with observations.

Behind the trailing shock the electrons are accelerated and strong magnetic fields are generated (Nishikawa *et al.* 2009b). Therefore, this region seems to produce the emission that is observed by satellites. We will examine the observed spectrum changes over time using different plasma conditions such as jet Lorentz factors, jet thermal temperatures, plasma compositions and other parameters.

Achknowledgments

This work is supported by NSF-AST-0506719, AST-0506666, AST-0908040, AST-0908010, NASA-NNG05GK73G, NNX07AJ88G, NNX08AG83G, NNX08AL39G, and NNX09AD 16G. JN was supported by MNiSW research project N N203 393034, and The Foundation for Polish Science through the HOMING program, which is supported through the EEA Financial Mechanism. Simulations were performed at the Columbia facility at the NASA Advanced Supercomputing (NAS). and IBM p690 (Copper) at the National Center for Supercomputing Applications (NCSA) which is supported by the NSF. Part of this work was done while K.-I. N. was visiting the Niels Bohr Institute. Support from the Danish Natural Science Research Council is gratefully acknowledged. This report was finalized during the program "Particle Acceleration in Astrophysical Plasmas" at the Kavli Institute for Theoretical Physics which is supported by the National Science Foundation under Grant No. PHY05-51164.

References

Abdo, A. A., *et al.* 2009, *Science*, 323, 1688

Chang, P., Spitkovsky, A., & Arons, J. 2008, *ApJ*, 674, 378

Frederiksen, J. T., Haugboelle, T., Medvedev, M. V., & Nordlund, Å. 2010, *ApJ*, 722, L114

Hededal, C. B., Ph.D. thesis, 2005, (arXiv:astro-ph/0506559)

Hededal, C. B. & Nishikawa, K.-I. 2005, *ApJ*, 623, L89

Jackson, J. D., *Classical Electrodynamics*, Interscience, 1999.

Martins, J. L., Martins, S. F., Fonseca, R. A. Silva, L. O. 2009, *Proc. of SPIE*, 7359, 73590V-1

Medvedev, M. V. & Loeb, A. 1999, *ApJ*, 526, 697

Medvedev, M. V. 2000. *ApJ*, 540, 704

Medvedev, M. V. 2006, *ApJ*, 637, 869

Nishikawa, K.-I., *et al.* 2003, *ApJ*, 595, 555

Nishikawa, K.-I., *et al.* 2005, *ApJ*, 623, 927

Nishikawa, K.-I., Hardee, P., Hededal, C. B., & Fishman, G. J. 2006, *ApJ*, 642, 1267

Nishikawa, K. -I., *et al.* 2009a, AIPCP, 1085, 589

Nishikawa, K.-I., *et al.* 2009b, *ApJ*, 689, L10

Nishikawa, K.-I., *et al.* 2009c, *AdvSR*, submitted, (arXiv:0906.5018)

Ramirez-Ruiz, E., Nishikawa, K.-I., & Hededal, C. B. 2007, *ApJ*, 671, 1877

Sironi, L. & Spitkovsky, A. 2009a, *ApJ*, 698, 1523

Sironi, L. & Spitkovsky, A. 2009b, *ApJ*, 707, L92

Spitkovsky, A. 2008a, *ApJ*, 673, L39

Spitkovsky, A. 2008b, *ApJ*, 682, L5

Waxman, E. 2006, *Plasma Phys. Control. Fusion*, 48, B137 (arXiv:astro-ph/0607353)

Jets at all Scales
Proceedings IAU Symposium No. 275, 2011
G. E. Romero, R. A. Sunyaev & T. Belloni, eds.

© International Astronomical Union 2011
doi:10.1017/S1743921310016406

Afterglow light curves
from magnetized GRB flows

Petar Mimica[1], Dimitrios Giannios[2] and Miguel Ángel Aloy[1]

[1]Departamento de Astronomía y Astrofísica, Universidad de Valencia, 46100, Burjassot, Spain
email: `Petar.Mimica@uv.es`
[2]Department of Astrophysical Sciences, Peyton Hall, Princeton University, Princeton, USA

Abstract. Using the RMHD code *MRGENESIS* and the radiative transfer code *SPEV* we compute multiwavelength afterglow light curves of magnetized ejecta of gamma-ray bursts interacting with a uniform circumburst medium. We are interested in the emission from the reverse shock when ejecta magnetization varies from $\sigma_0 = 0$ to $\sigma_0 = 1$. For typical parameters of the ejecta, the emission from the reverse shock peaks for magnetization $\sigma_0 \sim 0.01 - 0.1$, and is suppressed for higher σ_0. We fit the early afterglow light curves of GRB 990123 and 090102 and discuss the possible magnetization of the outflows of these bursts. Finally we discuss the amount energy left in the magnetic field which is available for dissipation at later afterglow stages.

Keywords. hydrodynamics (magnetohydrodynamics:) MHD, radiation mechanisms: nonthermal, radiative transfer, shock waves, methods: numerical, gamma rays: bursts

1. Introduction

Two alternatives for creating gamma ray burst (GRB) outflows are usually considered today: a thermal energy dominated fireball (Paczynski 1986, Goodman 1986) or a Poynting-flux dominated flow (PDF; Usov 1992, Thompson 1994, Meszaros & Rees 1997). The flow magnetization parameter $\sigma_0 \ll 1$ in the fireball, and the radiation pressure accelerates the flow up to bulk Lorentz factors $\Gamma_0 \simeq h_{in}$, latter being the flow initial specific enthalpy (see e.g., Aloy, Janka & Müller 2005). PDF, on the other hand, results from a jet launched with $\sigma_{in} \gg 1$. During the process of magnetized flow acceleration a GRB emission may result from the magnetic dissipation (Thompson 1994, Spruit, Daigne & Drenkhahn 2001, Giannios 2008) or the internal shocks (Fan, Wei & Zhang 2004, Mimica & Aloy 2010). These processes are not expected to be ultra-efficient consumers of magnetic energy and we expect that the flow can still be considerably magnetized at the end of the acceleration phase ($\sigma_0 \simeq 1$) while its bulk Lorentz factor is expected to be $\Gamma_0 \simeq \sigma_{in}$ (Komissarov *et al.* 2009, Tchekhovskoy *et al.* 2009, Lyubarski 2010). However, there is currently no consensus on the exact value of σ_0 at the onset of the afterglow phase (there are scenarios where $\sigma_0 \gg 1$, see e.g., Lyutikov & Blandford 2003). Regardless of this, the mere fact that the flow is magnetized at the beginning of the afterglow should leave an imprint on the early optical light curve: if $\sigma_0 \ll 1$ a reverse shock (RS) is expected to form in the ejecta shell and a characteristic feature (a peak or a flattening of the slope) is expected in the light cuve (Zhang *et al.* 2003); if $\sigma_0 \simeq 1$ then the RS might not form (Giannios *et al.* 2008) and the optical light curve is expected to lack the corresponding signature (Mimica *et al.* 2009a).

In this work we use numerical simulations to study the sensitivity of the afterglow emission on the value of σ_0. In Sec. 2 we outline our model, and discuss the generic results of a parametric study. Application to two GRBs is detailed in Sec. 3.

2. Dependence of afterglow dynamics and emission on flow magnetization

It is generally thought that the afterglow emission of a GRB originates after the flow has been collimated and accelerated, and after the GRB prompt emission has finished. We idealize the GRB flow assuming it is composed of cold ejecta moving in a radial direction. If the ejecta is magnetized we assume that magnetic field is dominated by a toroidal component. The initial jet magnetization is parameterized via the expression $\sigma_0 \equiv B'^2/4\pi\rho c^2$, where B' and ρ are the comoving magnetic field strength and fluid density; c is the speed of light. We furthermore simplify the model by assuming that the ejecta geometry is a spherical shell of initial thickness Δ_0 at the initial radius r_0, where we set $\Delta_0 \ll r_0$. Finally, we assume that the shell has a bulk Lorentz factor $\Gamma_0 \gg 1/\theta$, where θ is the opening angle of the jet. This means that, at least in the early phases we are interested in, the ejecta can be modeled as spherically symmetric, and two-dimensional effects such as lateral spreading can be ignored.

The difference in dynamics between the non-magnetized and magnetized ejecta evolution has been discussed in detail in Mimica *et al.* (2009a). Two crucial parameters are σ_0 and $\xi \equiv \sqrt{l/\Delta_0}\Gamma_0^{-4/3}$, where l is the Sedov length, defined as $l = (3E/4\pi n_{\text{ext}} m_p c^2)^{1/3}$, with n_{ext} being the numer density of the external medium, m_p the proton mass and E the total initial ejecta energy. It has been shown analytically (Giannios *et al.* 2008) and confirmed numerically (Mimica *et al.* 2009a) that there exist regions in the $\xi - \sigma_0$ parameter space where the RS is very weak or does not form at all, and thus its observational signature is absent. Recently Mimica *et al.* (2010) we have performed a number of simulations of afterglow ejecta with σ_0 taking values 0, 10^{-4}, 10^{-3}, 10^{-2}, 10^{-1} and 1. We have computed evolution of both thin ($\xi = 1.1$) and thick ($\xi = 0.5$) shells, as in Mimica *et al.* (2009a).

These simulations, combined with the newly available *SPEV* code for computing non-thermal emission (Mimica *et al.* 2009b) has enabled us to study dependence of both dynamics and emission on the value of σ_0. We refer to Mimica *et al.* (2010) for the technical details of the hydrodynamical simulations and the calculation of emission. In the following two subsections we discuss the observational signature of σ_0 and the long-term evolution of the magnetic energy.

2.1. *Observational signatures of varying magnetization*

We compute the emission from the forward shock (FS) and the RS (if present) resulting from the ejecta-medium interaction. We assume that the random magnetic field energy density is a fraction $\epsilon_B = 5 \times 10^{-3}$ of the thermal energy density and that electrons are accelerated at both shocks with a power-law energy distribution. Left panel of Fig. 1 illustrates the importance of the RS emission at early times in the optical bands (the peak seen at $\simeq 10$ seconds).

On the right panel of Fig. 1 we see total light curves for four different initial values of σ_0. Although numerical simulations show that the RS is progressively weaker as σ_0 is increased, this effect is compensated by the brighter emission due to the increase in the magnetic field. The result is that the brightest RS flash appears for moderately magnetized ejecta (typical with $0.1 \lesssim \sigma \lesssim 0.1$). For very magnetized ejecta ($\sigma_0 = 1$) the RS optical emission is several times weaker than in the non-magnetized case.

2.2. *Magnetic energy evolution*

Now we look into the issue of the residual magnetic energy after the ejecta has finished interacting with the external medium. If present and dissipated at later stage, this energy

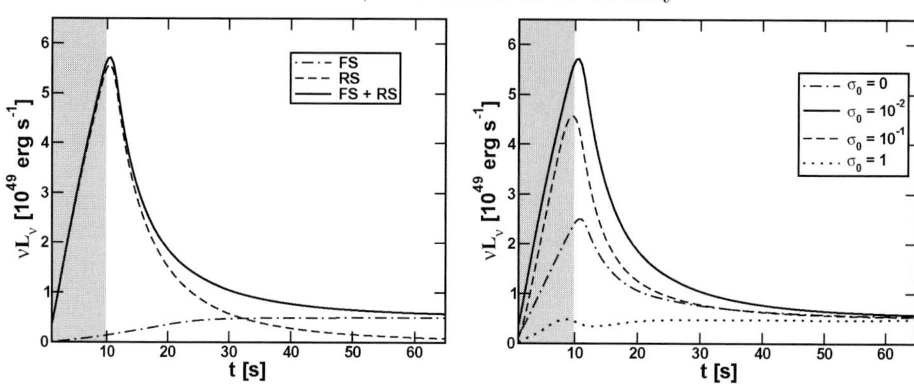

Figure 1. Left panel: R-band light curve in the GRB rest frame for the thick shell ejecta with initial magnetization $\sigma_0 = 0.01$ and Lorentz factor $\Gamma_0 = 750$. Dot-dashed and dashed lines show the FS and RS contributions, while the full line shows the total emission. The shaded region shows the observational time interval which is simultaneous with the prompt emission. Right panel: total light curves for the thick shell models $\sigma_0 = 0$, 0.01, 0.1 and 1 (dot-dashed, full, dashed and dotted lines, respectively).

may have important observational consequences for the late time emission. As is argued by Giannios (2006), ejecta deceleration might lead to the revival of current driven instabilities. Our simulations show that more than $\simeq 10$ per cent of the total ejecta energy is still in the magnetic form at the time when the RS crosses the ejecta, and is still $\simeq 2$ per cent at about 10 times the burst duration. This means that potentially there is sufficient magnetic energy at late times to be dissipated in localized reconnection regions. These might explain the afterglow X-ray flaring (Burrows *et al.* 2005) with no need for late time central engine activity.

3. Applications

In this section we focus on the optical emission from GRB 990123 (Akerlof *et al.* 1999, Briggs *et al.* 1999) and GRB 090102 (Gendre *et al.* 2009, Steele *et al.* 2009). The burst 990123 is famous because of a ninth magnitude optical flare detected few seconds after the end of the prompt emission and is one of the brightest ever detected. A recent 090102 does not show a flare, but has a $\simeq 10$ per cent polarization (an indication of large-scale magnetic fields). The duration of both bursts in the rest frame is approximately 10 seconds and, assuming the efficiency of the prompt emission of ≈ 20 per cent we estimate the isotropic equivalent ejecta energy to be 1.5×10^{55} erg for 990123 and 3×10^{54} for 990102. We use the thick shell model ($\xi < 1$) to model these bursts because the RS peak is observed close to the end of the prompt emission.

Using results of our simulations and the rescaling relations of Mimica *et al.* (2009a), we have looked for combinations of Γ_0 and σ_0 which best fit the observations. We find that for 990123 a model with $\Gamma_0 = 640$ and $\sigma_0 = 0.01$ best fits the observations, while for 090102 we get $\Gamma_0 = 940$ and $\sigma_0 = 0.1$. For both bursts we had to assume $\epsilon_B = 4 \times 10^{-7}$. The results are illustrated in Fig. 2.

The optical polarization of GRB 090102 is hard to understand as coming from the small-scale magnetic field synchrotron emission. Furthermore, it is smaller than expected from a coherent field. The polarization measured by Steele *et al.* (2009) is at a time interval $\simeq 60 - 90$ seconds in the GRB frame. As can be seen on Fig. 2, that is where the FS emission starts to dominate over the RS emission, which might indicate that 10 per

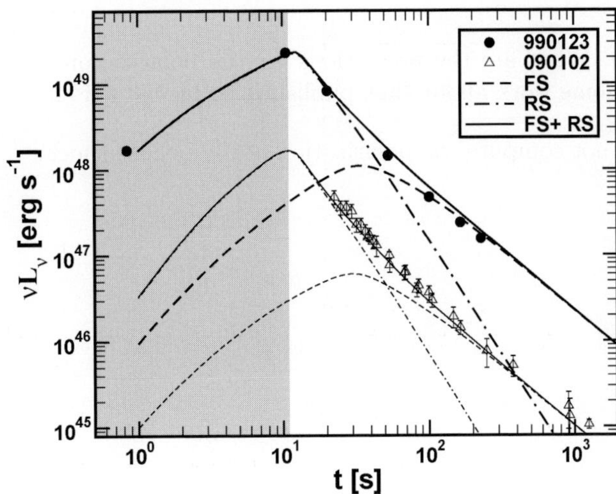

Figure 2. GRB frame R-band light curves for GRB 990123 (filled circles) and GRB 090102 (triangles), and the best fit models (lines). FS, RS and the total emission is showed by dashed, dot-dashed and full lines, respectively.

cent polarization comes from the combination of a highly polarized RS and the weakly polarized FS emission.

References

Akerlof C., *et al.*, 1999, *Nature*, 398, 400

Aloy M. A., Janka, H. T., Müller, E., 2005, *A&A*, 436, 273

Briggs M. S., *et al.*, 1999, *ApJ*, 524, 82

Burrows D. N., *et al.*2005, Science, 309, 1833

Fan Y. Z., *et al.*, 2004b, *MNRAS*, 354, 1031

Gendre B., *et al.*, 2009, *MNRAS*, 405, 2372

Giannios D., 2006, *A&A*, 455, L5

Giannios D., 2008, *A&A*, 480, 305

Giannios D., *et al.*, 2008, *A&A*, 478, 747

Goodman J., 1986, *ApJ*, 308, L47

Komissarov S. S., *et al.*, 2009, *MNRAS*, 394, 1182

Lyubarsky Y. E., 2010,*MNRAS*, 402, 353

Lyutikov M. & Blandford R., 2003, ArXiv:astro-ph/0312347

Meszaros P. & Rees M. J., 1997, *ApJ*, 482, L29

Mimica P. & Aloy M. A., 2010, *MNRAS*, 401, 525

Mimica P., *et al.*, 2009a, *A&A*, 494, 879

Mimica P., *et al.*, 2009b *ApJ*, 696, 1142

Mimica P., *et al.*, 2010, *MNRAS*, 407, 2501

Paczynski B., 1986, *ApJ*, 308, L43

Spruit H. C., *et al.*, 2001, *A&A*, 369, 694

Steele I. A., *et al.*, 2009, *Nature*, 462, 767

Tchekhovskoy A., *et al.*, 2009, *ApJ*, 699, 1789

Thompson C., 1994, *MNRAS*, 270, 480

Usov V. V., 1992, *Nature*, 357, 472

Zhang B., *et al.*, 2003, *ApJ*, 595, 950

Discussion

CASTRO-TIRADO: What are the predictions for the polarization of the reverse shock? There is already some work about this, published by D. Lazatti *et al.*

MIMICA: We did not compute the polarization of the reverse shock.

Jets at all Scales
Proceedings IAU Symposium No. 275, 2011
G. E. Romero, R. A. Sunyaev & T. Belloni, eds.

© International Astronomical Union 2011
doi:10.1017/S1743921310016418

Cosmology and the subclasses of the gamma-ray bursts

Attila Mészáros[1], István Horváth[2], Jakub Řípa[1], Zsolt Bagoly[3], Lajos G. Balázs[4], and Péter Veres[2,3]

[1] Charles University, Faculty of Mathematics and Physics,
Astronomical Institute, Prague, Czech Republic,
email: `meszaros@cesnet.cz`; `ripa@sirrah.troja.mff.cuni.cz`

[2] Dept. of Physics, Bolyai Military University, Budapest, POB 15, H-1581, Hungary,
email: `horvath.istvan@zmne.hu`

[3] Dept. of Physics of Complex Systems, Eötvös University, H-1117 Budapest, Pázmány P. s.
1/A, Hungary, email: `zsolt.bagoly@elte.hu`; `veresp@elte.hu`

[4] Konkoly Observatory, Budapest, POB 67, H-1525, Hungary, email: `balazs@konkoly.hu`

Abstract. Several statistical studies - done also by the authors of this contribution - show that there are three subclasses of gamma-ray bursts. They can be called as short, intermediate and long ones, because they can be separated with respect to their durations. The short and intermediate bursts are distributed anisotropically on the sky. This behavior is highly remarkable, and can have a cardinal impact on the cosmology. The subject of this contribution is a survey of this topic.

Keywords. gamma rays: bursts, stars: Wolf-Rayet, supernovae: general, cosmology: early universe

1. Subgroups of the gamma-ray bursts: Impact on the cosmology

The authors have shown that there are three subgroups of gamma-ray bursts (GRBs). They were confirmed in the datasets of different satellites: Compton (BATSE) - Horváth (1998), Mukherjee *et al.* (1998) Horváth (2002), Hakkila *et al.* (2003), Horváth *et al.* (2006); Swift - Horváth *et al.* (2008), Huja *et al.* (2009); Horváth *et al.* (2010), Veres *et al.* (2010); RHESSI - Řípa *et al.* (2009); BeppoSAX - Horváth (2009). GRBs can be separated with respect to their durations (short, intermediate and long subgroups). In this contribution we briefly discuss the cosmological implications of this separation. The short and long GRBs are physically different phenomena (Balázs *et al.* 2003).

The physics of the intermediate GRBs is unclear. The angular sky distribution of the subgroups are also different for the GRBs in the BATSE database: The short and intermediate GRBs are distributed anisotropically, but no such behavior was confirmed for the long ones (Balázs *et al.* 1998, Balázs *et al.* 1999, Mészáros *et al.* 2000, Vavrek *et al.* 2008). The long GRBs can be at extremely high redshifts (up to $z \simeq 20$) (Mészáros & Mészáros 1996, Horváth *et al.* 1996, Mészáros *et al.* 2006); the short and intermediate GRBs should be at moderate redshifts (see Figure 1) - i.e. up to $z \simeq 1 - 3$.

But even at these redshifts, together with the fact that they are distributed anisotropically, the short and intermediate GRBs can have a cardinal impact on the cosmology, because the cosmological principle requires any class of objects at redshift $z > 0.1$ to be distributed homogeneously and isotropically (Mészáros *et al.* 2009a, Mészáros *et al.* 2009b).

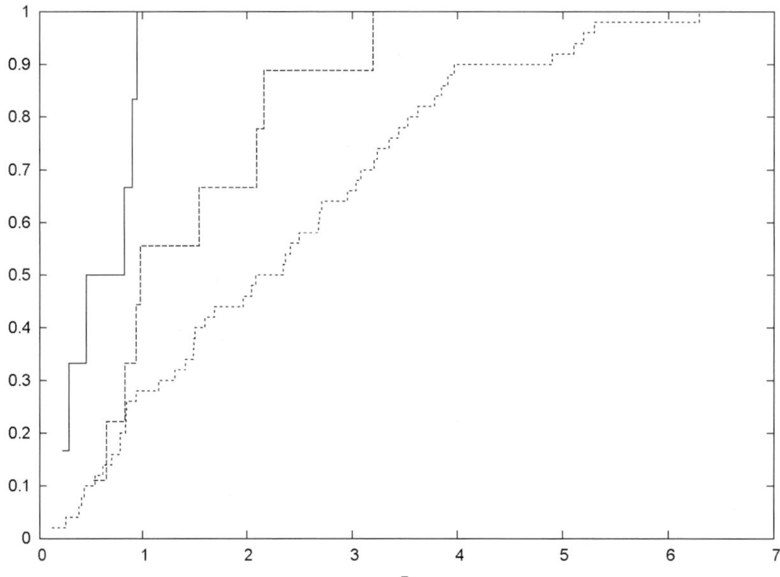

Figure 1. The redshift distribution of the three subgroups of GRBs with known redshifts and detected by the Swift satellite; z means redshift, and on the vertical axis any step gives a GRB similarly to the KS statistical test; for more details see Horváth *et al.* (2010).

This study was supported by the OTKA grant K77795, by OTKA/NKTH A08-77719 and A08-77815 grants (Z.B.), by a Bolyai Scholarship (I.H.), by the Grant Agency of the Czech Republic grants No. 205/08/H005 and P209/10/0734, by the project SVV 261301 of the Charles University in Prague, and by the Research Program MSM0021620860 of the Ministry of Education of the Czech Republic (A.M. and J. Ř.).

References

Balázs, L. G., Mészáros, A. & Horváth, I. 1998, *A&A*, 339, 1

Balázs, L. G. *et al.* 1999, *A&A. Suppl.Ser.*, 138, 417

Balázs, L. G. *et al.* 2003, *A&A*, 401, 129

Hakkila, J. *et al.* 2003, *ApJ*, 582, 320

Horváth, I. 1998, *ApJ*, 508, 757

Horváth, I. 2002, *A&A*, 392, 791

Horváth, I. 2009, *Ap&SS*, 323, 83

Horváth, I., Mészáros, P. & Mészáros, A. 1996, *ApJ*, 470, 56

Horváth, I. *et al.* 2006, *A&A*, 447, 23

Horváth, I. *et al.* 2008, *A&A*, 489, L1

Horváth, I. *et al.* 2010, *ApJ*, 713, 552

Huja, D., Mészáros, A. & Řípa, J. 2009, *A&A*, 504, 67

Mészáros, A. & Mészáros, P. 1996, *ApJ*, 466, 29

Mészáros, A. *et al.* 2000, *ApJ*, 539, 98

Mészáros, A. *et al.* 2006, *A&A*, 455, 785

Mészáros, A. *et al.* 2009a, *GRB: Sixth Huntsville Symp., AIP Conf. Proc.*, 1133, 483

Mészáros, A. *et al.* 2009b, *Baltic Astronomy*, 18, 293

Mukherjee, S. et al 1998, *ApJ*, 508, 314

Řípa, J. *et al.* 2009, *A&A*, 498, 399

Vavrek, R. *et al.* 2008, *MNRAS*, 391, 1741

Veres, P. *et al.* 2010, *ApJ*, forthcoming, astro-ph/1010.2087

Part 5. Young stellar objects

Jets at all Scales
Proceedings IAU Symposium No. 275, 2011
G. E. Romero, R. A. Sunyaev & T. Belloni, eds.

© International Astronomical Union 2011
doi:10.1017/S174392131001642X

Radio observations of jets from massive young stars

Luis F. Rodríguez

Centro de Radiostronomía y Astrofísica, UNAM
A. P. 3-72, (Xangari), 58089 Morelia, Michoacán, México
email: l.rodriguez@crya.unam.mx

Abstract. The formation of low mass stars takes place with the assistance of an accretion disk that transports gas and dust from the envelope of the system to the star, and a jet that removes angular momentum and allows accretion to proceed. In the radio, these ionized jets can be studied very close to the star via the thermal (free-free) emission they produce and at larger scales by the molecular outflows that result from their interaction with the surrounding medium. Is the same disk-jet process responsible for the formation of massive stars? I will review recent evidence for the presence of collimated jets and accretion disks in association with forming massive stars. The jets in massive protostars have large velocities that could produce a synchrotron component and I discuss the evidence for the presence of this non-thermal process in the jet associated with the HH 80-81 system.

Keywords. radiation mechanisms: nonthermal, stars: formation, ISM: jets and outflows

---

1. Introduction

A successful model, based on accretion via a circumstellar disk and a collimated outflow in the form of jets (Shu, Adams, & Lizano 1987), has been developed and found to be consistent with the observations for the case of low-mass star formation. An important question related to star formation is whether or not this model can simply be scaled up for the case of high-mass protostars or if other physical processes (i.e., merging; Bonnell, Bate, & Zinnecker 1998; Stahler, Palla, & Ho 2000; Bally & Zinnecker 2005) are present.

Since processes such as merging are expected to destroy or severely alter a disk-jet system, an observational approach to this issue has been to search, has has been found in low-mass stars (e. g. Rodríguez *et al.* 1998), for disk-jet systems also in massive protostars. In this contribution, I will concentrate on recent radio (centimeter, millimeter, and sub-millimeter) results obtained with interferometers. The forming massive stars are usually very heavily obscured and only radio and far-infrared observations can penetrate the surrounding dust and allow studies to be made. The interferometry is requiered because we are interested in the smallest possible scales of the phenomenon. Of course, star formation is now a field with observational and theoretical results coming from many angles and I recommend reading the recent reviews by McKee & Ostriker (2007), and Zinnecker & Yorke (2007).

2. Searching for Jets Associated with Massive Protostars

The study of massive forming stars is difficult because, in comparison to low mass stars, few massive stars form per unit time. This implies that one has to go to larger distances to find massive protostars. In addition, they form in clusters and stellar multiplicity is a problem in that it difficults identifying which of the stars is, for example, responsible for a large scale molecular outflow.

One can search for jets in the radio continuum. However, this in not easy because the compact, thermal (i.e., free-free) radio sources found at centimeter wavelengths in regions of star formation can have different natures. Some of them are ultracompact or hypercompact H ii regions, photoionized by an embedded hot luminous star. Other sources are clumps of gas or even circumstellar disks that are being externally ionized by a nearby star, such as the Orion proplyds (e.g., O'Dell & Wong 1996; Zapata *et al.* 2004) and the bright-rimmed clouds (e.g., Carrasco-González *et al.* 2006). An additional class of sources is formed by the thermal jets, collimated outflows that emanate from young stars and whose ionization is most probably maintained by the interaction of the moving gas with the surrounding medium (Eisloffel *et al.* 2000). Finally, there are also some examples of "radio Herbig-Haro" objects, obscured knots of gas that are being collisionally ionized by the shock produced by a collimated outflow (e.g., Curiel *et al.* 1993). There are also compact non-thermal sources, of which the most common are the young low-mass stars with active magnetospheres that produce gyrosynchrotron emission (e.g., Feigelson & Montmerle 1999). There are also emission regions where fast shocks may produce synchrotron emission (e.g., Henriksen *et al.* 1991; Garay *et al.* 1996).

To advance in the understanding of the nature of these compact radio sources it is necessary to have good quality data that allows the observer to establish the angular size, the morphology, the spectral index, the time variability, the polarization, and the presence of proper motions in them. Only a handful of the known star-forming regions have been studied carefully enough to clearly establish the nature of their compact radio sources. An example of a recent detection of a jet associated with a massive protostar is the case of the W75N region (Carrasco-González *et al.* 2010). These authors found that the radio continuum source Bc, previously believed to be tracing an independent star in the region, exhibits important changes in total flux density, morphology, and position. These results suggest that source Bc is actually a radio Herbig-Haro object, one of the brightest known, powered by the VLA 3 jet source (see Fig. 1). This result consolidates VLA 3 as a jet source, that given its morphology and spectral index was long suspected to be such a source.

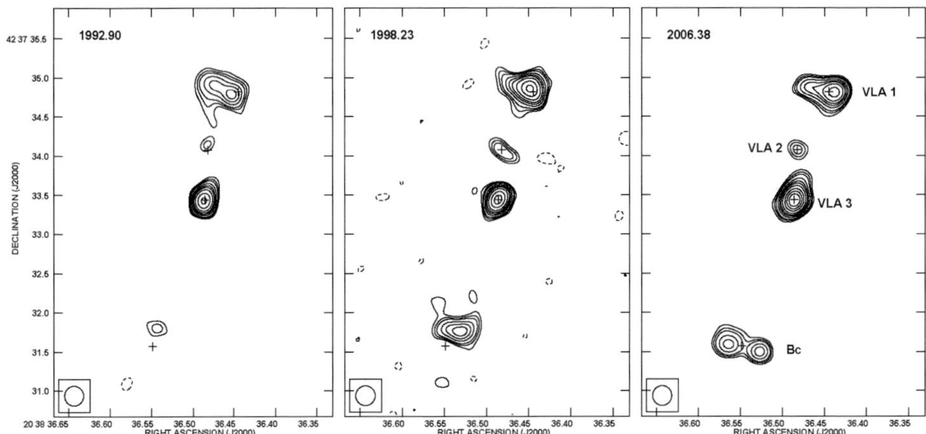

Figure 1. VLA contour images of the 3.6 cm emission from W75N for the three epochs (1992.90, 1998.23, and 2006.38) analyzed by Carrasco-González *et al.* (2010). Contours are –4, 4, 5, 6, 8, 10, 12, 15, 20, 25, and 30 times 80 μJy beam–1. The images have been restored with a circular beam of 0.''25, shown in the bottom left corner of the images. The crosses mark the centroids of the radio sources for epoch 2006.38. Note the morphology changes and proper motions of VLA Bc, that suggest it is a radio HH object produced by the jet source VLA 3.

3. IRAS 16547-4247: The Most Massive Protostar with a Jet and a Disk

The source IRAS 16547-4247 is the best example of a highly-collimated outflow associated with an O-type protostar studied so far, and its study may reveal important information about the way high-mass stars form. The systemic LSR velocity of the ambient molecular cloud where IRAS 16547-4247 is embedded is -30.6 km s^{-1} (Garay *et al.* 2007). Adopting the galactic rotation model of Brand & Blitz (1993) and assuming that the one-dimensional root-mean square (rms) velocity dispersion among molecular clouds is 7.8 km s^{-1} (Stark & Brand 1989), Rodríguez *et al.* (2008) estimate a distance of 2.9±0.6 kpc for the source. IRAS 16547-4247 has a bolometric luminosity of 6.2×10^4 $L_\odot$, equivalent to that of a single O8 zero-age main-sequence star, although it is probably a cluster for which the most massive star would have a somewhat lower luminosity. Garay *et al.*

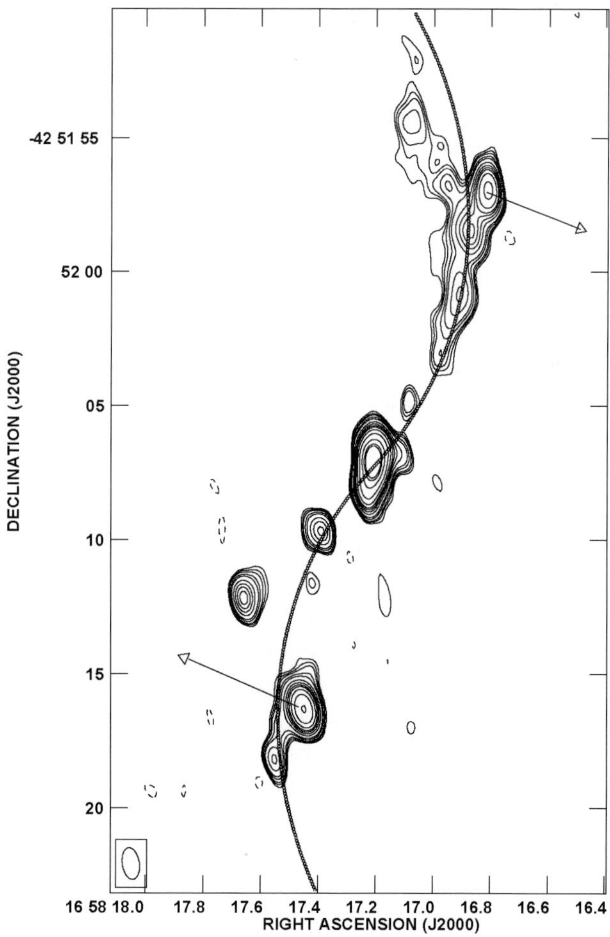

Figure 2. VLA contour image at 8.46 GHz toward IRAS 16547-4247 for epoch 2003.74. Contours are $-25, -20, -15, -10, -8, -6, -5, -4, 4, 5, 6, 8, 10, 15, 20, 25, 30, 40, 60, 100, 140, 160,$ and 200 times 27 μJy beam^{-1}. The half power contour of the synthesized beam ($1\rlap{.}''20 \times 0\rlap{.}''65$; P.A. = $9°$) is shown in the bottom-left corner. The solid line indicates the position of the spiral model discussed by Rodríguez *et al.* (2008). The arrows indicate the proper motions of components N-1 and S-1 for a period of 300 years.

(2003) detected an embedded triple radio continuum source associated with the IRAS 16547-4247. The three radio components are aligned in a northwest-southeast direction, with the outer lobes symmetrically separated from the central source by an angular distance of $\sim 10''$, equivalent to a physical separation in the plane of the sky of about 0.14 pc. The positive spectral index of the central source is consistent with that expected for a radio thermal (free-free) jet (e.g., Anglada 1996; Rodríguez 1997), while the spectral index of the lobes suggests a mix of thermal and nonthermal emission.

Rodríguez *et al.* (2008) do not detect proper motions along the axis of the outflow in the outer lobes of this source at a 4-σ upper limit of ~ 160 km s^{-1}. This suggests that these lobes are probably working surfaces where the jet is interacting with a denser medium. However, the brightest components of the lobes show evidence of precession, at a rate of $0\overset{\circ}{.}08$ yr^{-1} clockwise in the plane of the sky (see Figure 2). It may be possible to understand the distribution of almost all the identified sources in this region as the result of ejecta from a precessing jet.

This source also has good evidence favoring the existence of an associated disk. Franco-Hernández *et al.* (2009) found a rotating structure associated with IRAS 16547-4247, traced at small scales (~ 50 AU) by water masers. At large scales (~ 1000 AU), they find a velocity gradient in the SO_2 molecular emission with a barely resolved structure that can be modeled as a rotating ring or two separate objects. The velocity gradients of the masers and of the molecular emission have the same sense and may trace the same structure at different size scales. The position angles of the structures associated with the velocity gradients are roughly perpendicular to the outflow axis observed in radio continuum (see Fig. 2) and several molecular tracers. Franco-Hernández *et al.* (2009) estimate the mass of the most massive central source to be around 30 solar masses from the velocity gradient in the water maser and SO_2 emissions. They conclude that their results suggest that the formation of this source, one of the most luminous protostars known, is taking place with the presence of ionized jets and disk-like structures.

Evidence for disks in association with other massive forming stars has been provided recently by Beuther & Walsh (2008; IRAS 18089–1732), Zapata *et al.* (2009; W51 North), and Galván-Madrid *et al.* (2010).

4. A Magnetized Jet in IRAS 18162–2048

The highly collimated HH 80-81 radio jet is driven by the massive protostar IRAS 18162–2048, with a luminosity of ~ 17000 $L_\odot$ (Aspin & Geballe 1992). The jet consists of a chain of radio sources aligned in a SW-NE direction and terminates at both ends in optical/radio Herbig-Haro objects. With a total extension of 5.3 pc (for an assumed distance of 1.7 kpc; Rodríguez *et al.* 1980), this is the largest and most collimated protostellar radio jet known so far. Early radio observations showed that the spectrum of the emission from the central source is characterized by a positive spectral index, suggesting that it is dominated by thermal free-free emission (Martí *et al.* 1993). In contrast, the spectral indices of the emission from HH 80-81, HH 80 N, as well as from some of the knots in the jet lobes, suggest that an additional non-thermal component could be present in these sources (Martí *et al.* 1993).

Theoretical models for the formation of jets from young stellar objects require magnetic fields, on one hand close to the star to accelerate centrifugally gas off a rotating accretion disk and on the other at larger distances to collimate the outflow. Evidence of synchrotron emission has been found in a handful of objects, as suggested by negative (non-thermal) spectral indices. In contrast with free-free emission that is unpolarized, synchrotron emission brings information on the strength and direcion of the magnetic field. Carrasco-González *et al.* (2010, in preparation) have found polarized synchrotron emission arising from the lobes of the jet associated with the massive protostar IRAS 18162−2048 (see Fig. 3). The direction of the apparent magnetic field and the increase of the polarization at the projected edges of the jet are consistent with what is predicted for a confining helical magnetic field configuration. This result could open a new way of investigating the nature and relevance of magnetic fields in star formation.

Furthermore, recent theoretical works (e.g. Bosch-Ramón *et al.* 2010), suggest that protostellar jets could be a source of gamma rays. These models postulate, as a working hypothesis, the presence of relativistic electrons in such jets. Our detection of synchrotron emission in the HH 80-81 jet demonstrates the presence of relativistic electrons in the jets from massive protostars, providing an observational ground to the theoretical conjecture, making these objects a potential target for future high-energy studies.

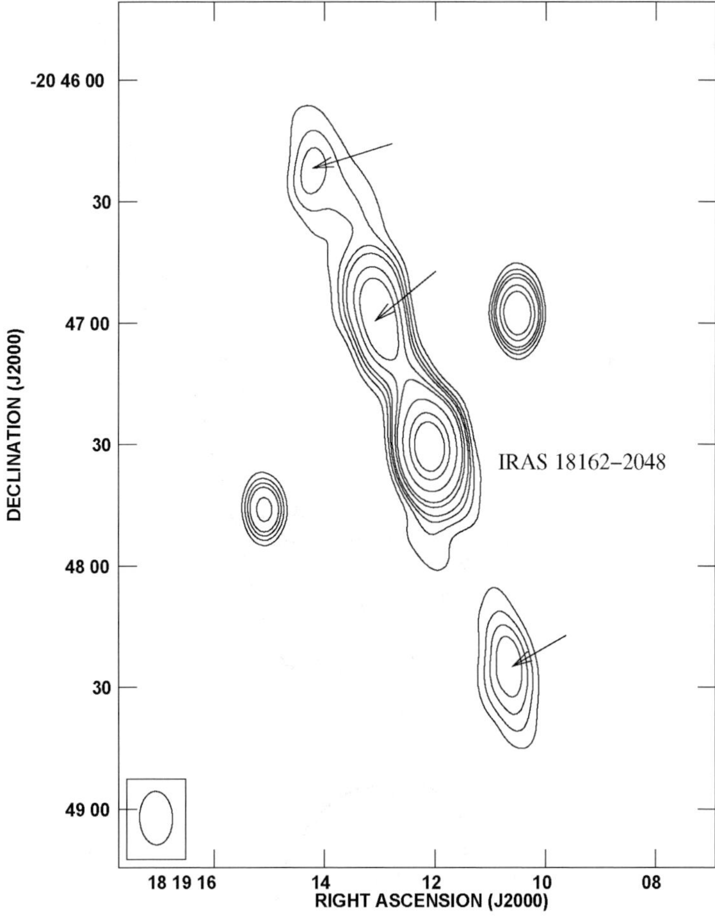

Figure 3. VLA contour image at 4.9 GHz toward IRAS 18162–2048. Contours are −4, 4, 6, 8, 10, 15, 20, 40, 80, and 120 times 26 μJy beam^{-1}, the rms noise of the image. The half power contour of the synthesized beam (13$.\!\!''$2 × 8$.\!\!''$4; P.A. = 2°) is shown in the bottom-left corner. The components where Carrasco-González *et al.* (2010, in preparation) detected linear polarization are indicated with arrows. The central source is indicated as IRAS 18162–2048 and has a thermal spectrum and no detectable polarization. The other two sources in the field probably trace stars in the cluster associated with IRAS 18162–2048.

References

Anglada, G. 1996, *Radio Emission from the Stars and the Sun (ASP Conf. Ser. 93)*, ed. A. R. Taylor & J. M. Paredes (San Francisco, CA: ASP), 3

Aspin, C. & Geballe, T. R. 1992, *Astronomy and Astrophysics*, 266, 219

Bally, J. & Zinnecker, H. 2005, *The Astronomical Journal*, 129, 2281

Beuther, H. & Walsh, A. J. 2008, *The Astrophysical Journal*, 673, L55

Bonnell, I. A., Bate, M. R., & Zinnecker, H. 1998, *Monthly Notices of the Royal Astronomical Society*, 298, 93

Bosch-Ramon, V., Romero, G. E., Araudo, A. T., & Paredes, J. M. 2010, *Astronomy and Astrophysics*, 511, A8

Brand, J., & Blitz, L. 1993, *Astronomy and Astrophysics*, 275, 67

Carrasco-González, C., López, R., Gyulbudaghian, A., Anglada, G., & Lee, C. W. 2006, *Astronomy and Astrophysics*, 445, L43

Carrasco-González, C., Rodríguez, L. F., Torrelles, J. M., Anglada, G., & González-Martín, O. 2010, *The Astronomical Journal*, 139, 2433

Curiel, S., Rodriguez, L. F., Moran, J. M., & Cantó, J. 1993, *The Astrophysical Journal*, 415, 191

Eisloffel, J., Mundt, R., Ray, T. P., & Rodriguez, L. F. 2000, *Protostars and Planets IV*, 815

Feigelson, E. D., & Montmerle, T. 1999, *Annual Review of Astronomy & Astrophysics*, 37, 363

Franco-Hernández, R., Moran, J. M., Rodríguez, L. F., & Garay, G. 2009, *The Astrophysical Journal*, 701, 974

Galván-Madrid, R., Zhang, Q., Keto, E., Ho, P. T. P., Zapata, L. A., Rodríguez, L. F., Pineda, J. E., & Vázquez-Semadeni, E. *2010, arXiv:1004.2466*

Garay, G., Ramirez, S., Rodriguez, L. F., Curiel, S., & Torrelles, J. M. 1996, *The Astrophysical Journal*, 459, 193

Garay, G., Brooks, K. J., Mardones, D., & Norris, R. P. 2003, *The Astrophysical Journal*, 587, 739

Garay, G., *et al.* 2007, *Astronomy and Astrophysics*, 463, 217

Henriksen, R. N., Mirabel, I. F., & Ptuskin, V. S. 1991, *Astronomy and Astrophysics*, 248, 221

Marti, J., Rodriguez, L. F., & Reipurth, B. 1993, *The Astrophysical Journal*, 416, 208

McKee, C. F., & Ostriker, E. C. 2007, *Annual Review of Astronomy & Astrophysics*, 45, 565

O'dell, C. R. & Wong, K. 1996, *The Astronomical Journal*, 111, 846

Rodriguez, L. F., Moran, J. M., Ho, P. T. P., & Gottlieb, E. W. 1980, *The Astrophysical Journal*, 235, 845

Rodríguez, L. F. 1997, *Herbig-Haro Flows and the Birth of low-mass Stars (Proc. of IAU Symp. 182)*, ed. B. Reipurth & C. Bertout (Dordrecht: Kluwer), 83

Rodríguez, L. F., *et al.* 1998, *Nature*, 395, 355

Rodríguez, L. F., Moran, J. M., Franco-Hernández, R., Garay, G., Brooks, K. J., & Mardones, D. 2008, *The Astronomical Journal*, 135, 2370

Shu, F. H., Adams, F. C., & Lizano, S. 1987, *Annual Review of Astronomy & Astrophysics*, 25, 23

Stahler, S. W., Palla, F., & Ho, P. T. P. 2000, *Protostars and Planets IV*, 327

Stark, A. A. & Brand, J. 1989, *The Astrophysical Journal*, 339, 763

Zapata, L. A., Rodríguez, L. F., Kurtz, S. E., & O'Dell, C. R. 2004, *The Astronomical Journal*, 127, 2252

Zapata, L. A., Ho, P. T. P., Schilke, P., Rodríguez, L. F., Menten, K., Palau, A., & Garrod, R. T. 2009, *The Astrophysical Journal*, 698, 1422

Zinnecker, H. & Yorke, H. W. 2007, *Annual Review of Astronomy & Astrophysics*, 45, 481

Discussion

DE GOUVEIA DAL PINO: What is the typical extinction of the jets coming out of massive stars and at what distance have you observed the non-thermal components?

RODRIGUEZ: They can extend over several parsecs. However, the two lobes where polarization was detected are relatively close to the central source, at about 0.5 pc.

BENAGLIA: Which is the degree of polarization of the Carrasco-Gonzalez + 2010 source? And the threshold for detection?

RODRIGUEZ: The degree of polarization in this source goes from 10% to 40%. This type of sources are weak and we can detect polarization above 10%. Ammonia is very useful to study dense ($\gtrsim 300\,\mathrm{cm}^{-3}$) molecular gas, so it will usually be detecxtable in massive star-forming regions.

BENAGLIA: Which is the ratio of number of sources with detected polarization to total number studied?

RODRIGUEZ: We studied in detail for polarization six jets rom massive young stars. Only HH 80-81 showed detectable polarization.

BENAGLIA: Which will be a time scale to look for source variability (when nothing but emission with a negative spectral index is what you see)?

RODRIGUEZ: There is clear evidence of variability in time scales of years. Faster variability may be present, but the very high signal-to-noise ratios and with the required short time separations.

Jets at all Scales
Proceedings IAU Symposium No. 275, 2011
G. E. Romero, R. A. Sunyaev & T. Belloni, eds.

© International Astronomical Union 2011
doi:10.1017/S1743921310016431

Molecular and atomic jets
in young low-mass stars: Properties
and origin

Sylvie Cabrit[1] and J. Ferreira[2] and C. Dougados[2]

[1]LERMA, Observatoire de Paris, CNRS
61 Avenue de l'Observatoire, F-75014 Paris
email: sylvie.cabrit@obspm.fr

[2] LAOG, Observatoire de Grenoble, CNRS
email: ferreira@obs.ujf-grenoble.fr, dougados@obs.ujf-grenoble.fr

Abstract. The key characteristics of molecular and atomic ejection from young stars are summarized, with emphasis on similarities across evolutionary stages, and the need for efficient magnetic collimation and ejection extracting a large fraction of the accretion power. The jet kinematics, and its dust and molecular content, are confronted to steady MHD jet models, and the probable contribution of non-steady processes is pointed out.

Keywords. stars: pre–main-sequence; stars: winds, outflows; ISM: jets and outflows; ISM: molecules

1. Introduction

Bipolar ejection of matter in molecular and/or atomic form is a ubiquitous phenomenon in accreting young stars of all evolutionary phases (see Section 2). However, the ejection mechanism(s), and the region(s) that dominate the outflow mass-flux and power (stellar surface / magnetosphere / inner disk edge / extended disk surface) are still heavily debated and the subject of intense research (cf. Tsinganos *et al.* 2009 for recent views). One important clue is that the ejection/accretion efficiency and the jet collimation remain similar over decades in source age, requiring a highly efficient *magnetic* ejection and collimation process (see Sections 3,4). Additional constraints on the ejection mechanism are provided by the recent discovery of transverse velocity decrease, tentative rotation signatures, metal depletion, and molecules in jets (see Section 5). Among proposed *steady* MHD ejection models, disk winds seem more promising to explain these properties. However, there is mounting evidence that non-steady interaction between the disk and the stellar magnetosphere also contributes to the observed outflow features (see Section 6).

2. Ubiquity of jets in accreting young stars

Jets are observed in *all* classes of young stellar objects where accretion is occuring, although their strength and molecular fraction declines with age (see Cabrit 2002 for a detailed comparative description):

- In the youngest embedded protostars ("Class 0") of ages $\leqslant 10^4$ yr, where the infall rate is highest: outflow activity is most apparent in molecular tracers and takes two aspects, illustrated in Fig. 1: (1) slow, moderately collimated bipolar CO outflow cavities of swept-up ambient gas, moving at $V \leqslant 10$ km s^{-1}, and (2) fast, highly collimated molecular "jets" seen in SiO, CO, and H$_2$ at $V > 10 - 100$km s^{-1}, appearing as molecular "bullets" on larger scales > 0.1 pc (Bachiller & Tafalla 1999). The atomic counterpart

LE FLOT MOLECULAIRE DE HH 211
Resolution angulaire : 1.5″

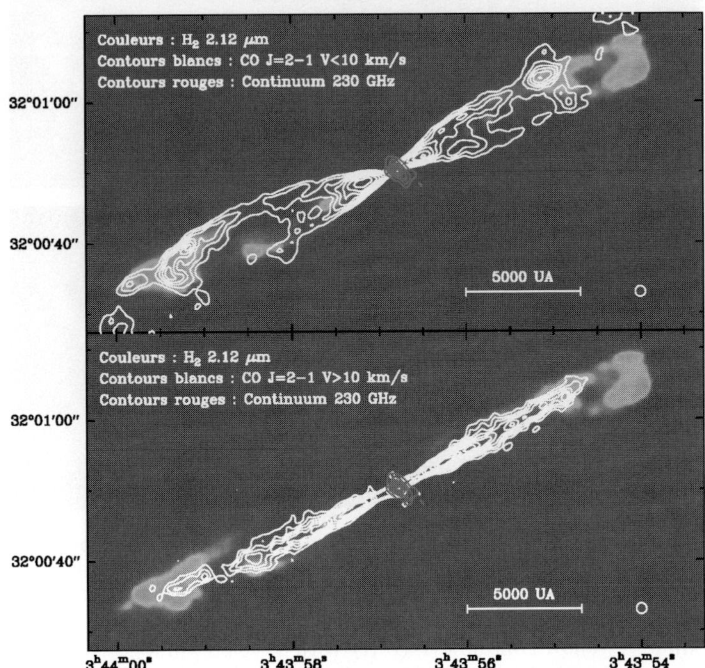

Figure 1. Bipolar ejection tracers in the young Class 0 protostar HH211; White contours of CO(2-1) emission (Gueth & Guilloteau 1999) delineate a swept-up bipolar cavity at low speed (top panel), and a narrow axial jet at high $V \geqslant 10$ km s^{-1} (bottom panel). The background image shows shock-excited H$_2$ in the leading jet bowshocks (Mc Caughrean *et al.* 1994).

appears to carry a lower mass-flux than the molecular jet (Dionatos *et al.* 2009, 2010). Only a handful of such jets have been mapped in detail so far, but the collimation and kinematics of outflow cavities appear consistent with jet-driven bowshocks in young low-luminosity Class 0 sources, while a wider-angle component is needed in massive / older flows (Cabrit, Raga, & Gueth 1997; Downes & Cabrit 2003; Arce *et al.* 2007 and refs. therein).

• In evolved infrared protostars ("Class I"), of age $\simeq 10^5$ yr, where residual infall is occuring at much slower rate: jets become brighter in optical ionic lines out to several pc (see top panel of Fig. 2, and Reipurth & Bally 2001 for a review) and H$_2$ carries a lower mass-flux than the atomic gas (Nisini *et al.* 2005, Podio *et al.* 2006). Swept-up CO outflows are weaker than in the Class 0 stage (Bontemps *et al.* 1996),

• In "Class II" pre-main sequence stars of age $\simeq 10^6$ yr, which have no more envelope but retain an active accretion disk: High-velocity atomic jets are much fainter, with the jet beam traced out to $\leqslant 500$ AU (Hartigan *et al.* 1995, Hirth *et al.* 1997). H$_2$ counterparts tend to be slower and less collimated than the jet (see bottom panel of Fig 2; Beck *et al.* 2008). An important finding is that jets are absent in young stars with no accretion disks (Hartigan *et al.* 1995). Hence, *accretion is clearly fundamental for the jet process.* In contrast, jets are seen over a wide range of stellar masses, from brown dwarfs up to $M_\star > 2M_\odot$ (see Whelan, this volume, Ray *et al.* 2007, Bacciotti 2009), indicating a robust universal mechanism.

A universal property of stellar jets at all stages is their "knotty" appearance, with very similar knot spacings in Class 0 and Class I (Cabrit 2002). These features trace internal

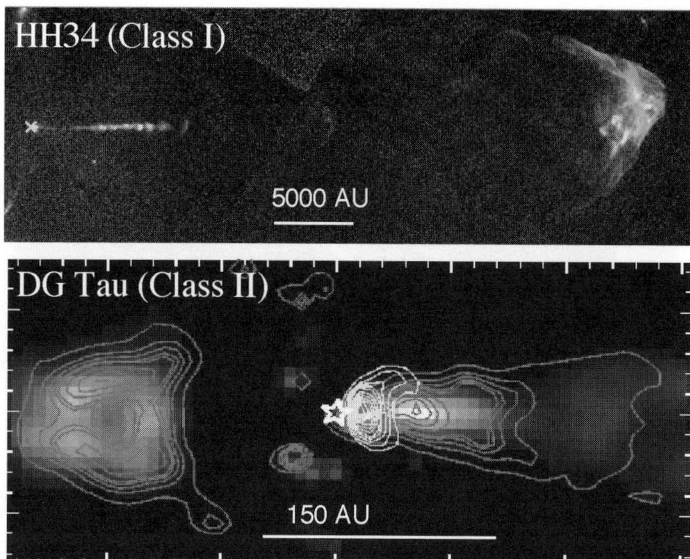

Figure 2. Bipolar ejection in late phases of accreting young stars: (i) top panel: evolved infrared Class I protostar (HH34), with jet image in [S II] and Hα (from Reipurth *et al.* 1997); the knotty structure arises from time variability; (ii) pre-main sequence Class II star (DG Tau): The colour image traces fast [Fe II] with $V > 150$ km s^{-1}, while red contours trace slower [Fe II] with $V \simeq 50 - 150$ km s^{-1}; yellow contours trace H$_2$ with $V < 20$ km s^{-1}. Note the wider opening angle at lower velocity, and the "bubble" features near the tip of both lobes (adapted from Agra-Amboage *et al.* 2010).

shocks driven by velocity variability on several timescales $\simeq$ 30–60 yrs, 300 yrs, and 1000 yrs, with longer modes having larger amplitudes (Raga *et al.* 2002). An even faster mode of period $\simeq$ 3–5 yrs is identified in Class II jets (Hartigan *et al.* 2007; Agra-Amboage *et al.* 2010). The origin of such variations is yet unclear, although jet wiggling suggests orbital motions of period 50 yrs in some sources (Anglada *et al.* 2007; Lee *et al.* 2010).

3. Jet power compared to the accretion power

While both accretion and ejection strongly decrease over time, they remain tightly correlated with one another:

• Estimates of the ejection power in the early Class 0 phase rely heavily on observations of the larger, slower swept-up CO cavities, for which a rich body of data is available. The momentum injection rate in swept-up Class 0 molecular outflows shows a clear correlation with the source bolometric luminosity over 5-6 orders of magnitude, probably tracing an underlying ejection-accretion correlation (see Fig. 3). The momentum efficiency is very high, $F_{CO}c/L_{bol} \simeq 1000(L_{bol}/10L_{\odot})^{-0.3}$, and clearly rules out radiative or thermal ejection mechanisms (Lada 1985; Cabrit & Bertout 1992). Assuming that the outflows are momentum-driven by an underlying wind/jet of speed $V_w \simeq 200$ km s^{-1}, the inferred wind mechanical luminosity, $L_w = 0.5\ F_w V_w$, is 50%–100% of L_{bol} in low-luminosity sources, indicating that *the ejection mechanism extracts a considerable fraction of the accretion power* (Cabrit & Bertout 1992; Cabrit 2002). Direct measurements of the jet mass-flux in a few Class 0 sources confirm this result (Lee *et al.* 2010). The lower momentum efficiency $F_{CO}c/L_{bol} \simeq 10 - 100$ for $L_{bol} \geqslant 10^4 L_{\odot}$ could be due to the increasing contribution of the photospheric luminosity $L_\star$ in massive protostars. Indeed, reasonable assumptions about

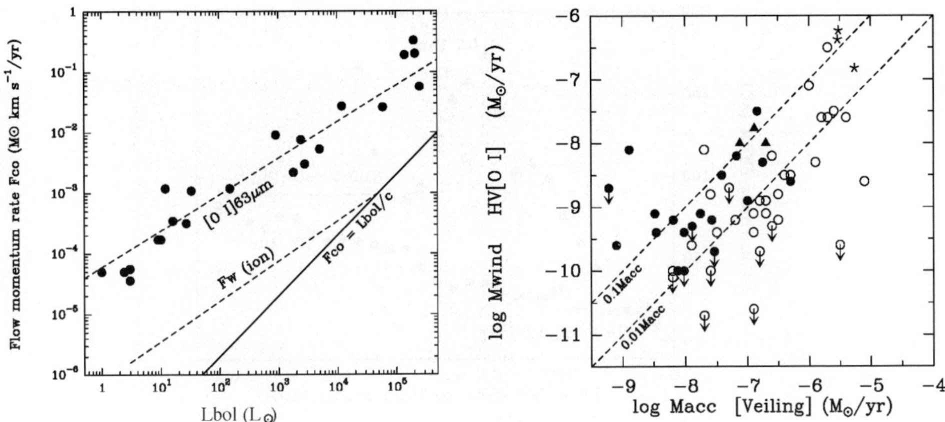

Figure 3. *Left:* Correlation between the momentum flux F_{CO} in the swept-up CO outflow and the source bolometric luminosity L_{bol} in Class 0 sources. The two dashed lines show the typical momentum flux in the ionized part of the jet (Fw(ion)) and in dissociative shocks ([OI]63μm). Adapted from Richer *et al.* (2000). *Right:* mass-loss rate in high-velocity atomic gas versus disk accretion rate in Class II stars. Open circles are from Hartigan *et al.* (1995), while filled circles use revised $\dot{M}_{acc}$ from Muzerolle *et al.* (1998) and filled triangles use revised $\dot{M}_j$ from spatially resolved jet data. Star symbols denote Class I sources. From Cabrit (2007b).

the stellar mass or age suggest a roughly similar ratio $(\dot{M}_{w}/\dot{M}_{acc})(V_{w}/V_{K,*}) \simeq$ 0.1-0.3 in Class 0 sources of all luminosities (Richer *et al.* 2000, Beuther *et al.* 2002).

• In Class II sources, statistical studies clearly indicate that ejection is correlated with accretion, not with L_* (Hartigan *et al.* 1995). Although uncertainties in jet mass-flux $\dot{M}_j$ and $\dot{M}_{acc}$ for individual sources can reach a factor 3–10, the most reliable estimates to date yield average mass, momentum, and energy efficiencies (two-sided): $(2\dot{M}_j)/\dot{M}_{acc}\simeq$ 0.1-0.2, $(2F_j)c/L_{acc}\simeq$ 100–300, $(2L_j)/L_{acc}\simeq$ 0.1 (see Fig. 3; Cabrit 2007b; Bacciotti 2009). Similar values are obtained in Class I sources, from observations of CO swept-up cavities and atomic jets (Bontemps *et al.* 1996; Hartigan, Morse, & Raymond 1994; Antoniucci *et al.* 2008).

4. Jet collimation, and collimating agent

Measures of jet collimation have greatly progressed over the last decade, thanks to the sub-arcsecond resolution afforded by adaptive optics and the *Hubble Space Telescope* in atomic jets, and by radio interferometers in molecular jets.

Studies of nearby atomic Class II stars reveal that the jet axis is closely aligned with the disk axis at the base, even if jet wiggling is present further out (eg. Anglada *et al.* 2006). The apparent half-opening angle of the jet beam drops from $\simeq 10 - 15^o$ initially to 2–3° beyond 50 AU, with a typical jet radius of 10 AU at this distance (see Hartigan *et al.* 2007; Ray *et al.* 2007 and references therein). Hence, the bright fast jet core appears to be collimated *on small, inner disk scales*.

Another striking result is that jet widths in Class 0 protostars, which possess a very dense infalling envelope, are similar to those in Class II stars, with no envelope and a thin disk (see Fig. 4, and Cabrit *et al.* 2007). Thus, jet collimation cannot rely on ambient thermal or ram pressure. Indeed, ambient densities around Class II stars seem too low by several orders of magnitude to provide jet collimation on the observed scale (see Cabrit

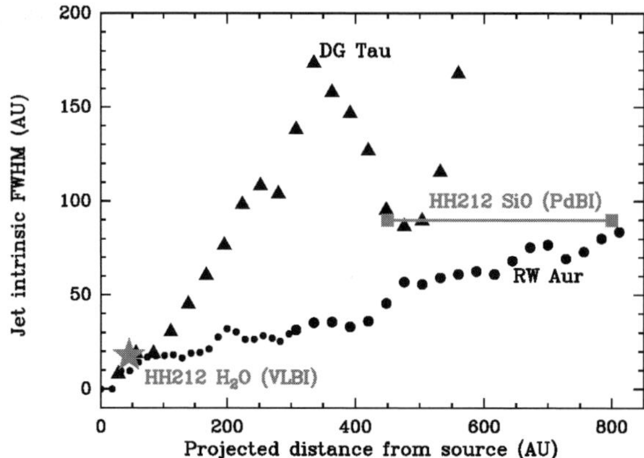

Figure 4. Width of the jet from the young Class 0 protostar HH 212 versus distance from the source, as observed in H_2O masers (red star; from Claussen *et al.* 1998) and in SiO (red bar). The similarity to jet widths in T Tauri stars with no circumstellar envelope (black symbols) favors a magnetic collimation process. From Cabrit *et al.* (2007).

2007a for a detailed discussion of this issue). A *magnetic* collimation process thus appears called for.

5. New constraints: Jet transverse kinematics and composition

5.1. *Transverse velocity decrease*

The flow speed appears to drop away from the jet axis. A wider-angle, slower "sheath" surrounding the fast axial jet is clearly seen in the DG Tau jet (see bottom panel of Fig. **??**; Lavalley-Fouquet *et al.* 2000; Bacciotti *et al.* 2000); A transverse velocity decrease is also detected in other Class II jets such as CW Tau or Th 28 (Coffey *et al.* 2007), and in the Class 0 jet HH 212 (Lee *et al.* 2008). Whether this slower "sheath" around the fast jet beam is intrinsic to the launch mechanism, or develops from interaction with ambient gas or internal shocks or instabilities, is yet unclear (Garcia-Lopez *et al.* 2008; Agra-Amboage *et al.* 2010). In any case, it appears challenging to ejection models that predict a *fast* wide-angle wind around the jet (see eg., Cai *et al.* 2008).

5.2. *Tentative jet rotation signatures*

Transverse velocity shifts of 5–10 km s^{-1} suggestive of jet rotation have been reported in DG Tau and several other Class II and Class I jets (see Bacciotti 2009 for a recent review). These data set important constraints on models where the jet traces a steady centrifugal MHD disk wind extracting most of the angular momentum from the accretion flow. The observed shifts imply an outer MHD launch radius $r_e \simeq 0.15$–3 AU for *atomic streamlines* and a magnetic lever arm parameter $\lambda = r_A^2/r_o^2 \simeq 4$–13 (see Bacciotti *et al.* 2002; Anderson *et al.* 2003; Pesenti *et al.* 2004; Ferreira *et al.* 2006). Such low values of λ predict moderate maximum flow speeds of $\simeq 200 - 500$ km s^{-1} and an ejection efficiency $\simeq 0.05$–0.1, in line with observations (Ferreira *et al.* 2006; Agra-Amboage *et al.* 2010). The r_e and λ values are in fact upper limits for MHD disk winds, since transverse shifts may include other effects than rotation, eg. jet precession, binary orbital motion, or shock asymmetries (eg. Cerqueira *et al.* 2006; Cabrit *et al.* 2006).

Tentative rotation has also been reported in two molecular Class 0 jets in Orion (Lee *et al.* 2007, 2008, 2009). However, the inferred small r_e values are highly uncertain, because the transverse velocity gradient occurs on scales $\simeq$ 2–3 times smaller than the beam size. Under such circumstances, any real rotation signature is expected to be *strongly beam-smeared*, so r_e would be underestimated (Pesenti *et al.* 2004; see Cabrit 2009 for a detailed discussion). The closer-by and better resolved CO jet from CB26 shows a clear pattern suggestive of rotation (Launhardt *et al.* 2009). The inferred launch radius is 30 AU if angular momentum is conserved, and 0.5–5 AU for a steady centrifugal MHD wind (Cabrit 2009). But it is intringuing that the similar CO jet from HH30 does not show any clear rotation signature (Pety *et al.* 2006). The enhanced resolution brought by the ALMA interferometer will be critical to progress on this topic.

5.3. *Metal depletion in jets*

The gas phase depletion of refractory species such as Fe, Si, C, Ni, Ca, has been measured in several Class 0 and Class I jets by comparing their emission line fluxes with other species that do not significantly deplete into grains, such as hydrogen, sulfur, phosphorus, and oxygen (Nisini *et al.* 2002, 2005; Podio *et al.* 2006, 2009; Dionatos *et al.* 2009, 2010). Typical gas-phase depletions of 20%-90% with respect to solar abundances are inferred, with a trend for stronger depletion closer to the source. Similar iron depletion levels are suggested in the inner 150 AU of the Class II jet from DG Tau, with higher depletion at lower speed (Agra-Amboage *et al.* 2010). These data suggest that the bulk of the emitting jet material does not originate from the star, but from a more external region where 20–90% of the dust grains can survive, with slower gas ejected from more dusty regions.

5.4. *Molecules in jets*

The presence of molecules in jets at speeds $\geqslant$ 100km s^{-1} well exceeding the typical dissociative shock limit in the ISM of $\simeq$ 25-35 km s^{-1}, has been a challenge since its discovery. Entrainment of ambient molecules by an underlying atomic jet does not appear efficient enough to produce the observed H_2 and CO columns, as any turbulent entrainment layer should be very thin (Taylor & Raga 1995), and there is little ambient material left to sweep-up behind the leading bowshock. Synthesis of molecules in a dust-free stellar wind can be efficient for CO and SiO, but excessive high mass-fluxes $> 10^{-5}$ $M_\odot$yr^{-1} appear to be required to reach substantial H_2 abundances (Ruden *et al.* 1990). H_2 reformation behind dust-free shocks is efficient only at low shock speeds < 15 km s^{-1} (Raga *et al.* 2005).

Alternatives avoiding these caveats would be that H_2 reforms in a dusty jet (Garcia-Lopez *et al.* 2008) or that molecules are gently magnetically accelerated from the disk surface. We have recently performed detailed calculations of the chemical and thermal structure in an MHD disk wind with $\lambda \simeq 13$ and launch radii $\simeq$ 0.2-10 AU, including heating by ambipolar diffusion, and ionization and dissociation by coronal X-rays and UV radiation from the accretion shock (Panoglou *et al.* 2010). As shown in Fig. 5, we find that (1) adiabatic and H_2 cooling can efficiently limit the gas heating and collisional dissociation, yielding asymptotic temperatures $\simeq$ 700 K in Class 0 to 2500 K in Class I, II in agreement with observations (Dionatos *et al.* 2010; Beck *et al.* 2008); (2) H_2 is more efficiently screened against photo-dissociating UV at high mass accretion rate. This would explain the trend for a higher molecular fraction in Class 0 jets compared to Class I, II. The chemical effect of internal shocks propagating in such disk winds, in particular the release of SiO from grains, should now be investigated to compare against observed jet chemistry (Cabrit *et al.* 2007; Tafalla *et al.* 2010).

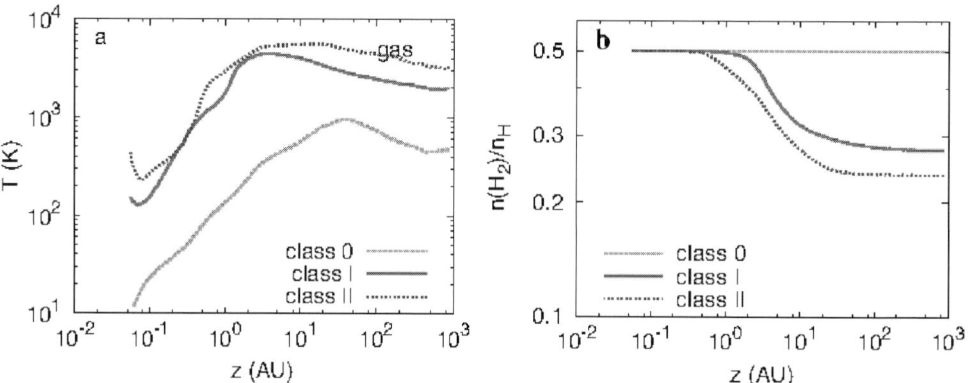

Figure 5. *Left*: Predicted gas temperature along an MHD disk wind streamline launched from 1 AU, for an accretion rate of 5×10^{-6} $M_\odot \mathrm{yr}^{-1}$ (Class 0), 10^{-6} $M_\odot \mathrm{yr}^{-1}$ (Class I), and 10^{-7} $M_\odot \mathrm{yr}^{-1}$ (Class II). Ambipolar diffusion heating is eventually balanced by H_2 cooling, with a lower asymptotic temperature at higher accretion rate. *Right:* H_2 abundance relative to H nuclei on MHD disk wind streamlines launched from 1 AU. Abundance increases with accretion rate, as the denser flow is better self-screened against stellar UV photons. From Panoglou *et al.* (2010).

6. Conclusions

The high ratio of jet to accretion power in young low-mass stars (10% in disk-dominated sources, and $\geqslant 50\%$ in protostars) suggests that jets are extracting a large fraction of the accreting angular momentum down to a few stellar radii. Magnetic fields clearly play an important role in this process, as indicated by the tight jet collimation operating within 50 AU. The similarity in $\dot{M}_\mathrm{w}/\dot{M}_\mathrm{acc}$ among all evolutionary stages and masses suggests that a universal ejection mechanism might be at work.

However, the growing set of observational constraints are very challenging to most currently proposed *steady* MHD ejection models (see Cabrit 2009 for a detailed discussion): Wave-driven *stellar winds* may have difficulties in explaining the high ejection/accretion ratio $\simeq 0.1$ (Cranmer 2009, Ferreira *et al.* 2006), and the depletion of refractory elements. The magneto-centrifugally driven *"X-wind" model* from the inner disk edge (Cai *et al.* 1998) predicts a fast wide-angle wind that is not seen. Magneto-centrifugal disk winds from an extended disk region may explain the drop in velocity away from the jet axis and the rotation signatures (Pesenti *et al.* 2004); apparent jet widths (Ray *et al.* 2007; Stute *et al.* 2010), presence of molecules in jets (Panoglou *et al.* 2010), and ejection/accretion ratio (Ferreira *et al.* 2006). However, the small magnetic lever arms $\lambda \leqslant 4-13$ compatible with constraints on jet rotation require strong heating within the disk (Casse & Ferreira 2000), of yet unclear origin.

Another challenge to steady disk winds is the recent evidence for velocity variability on a timescale of $\simeq 3$–10 yrs in Class II jets (Hartigan *et al.* 2007; Agra-Amboage *et al.* 2010). This is 500 times the orbital period at the inner disk edge but close to the orbital period at 1–3 AU, hence any extended disk wind would not be steady in its outer regions. Also intriguing is the observation in XZ and DG Tau of episodic inflating "bubbles" (Krist *et al.* 2008; Bacciotti *et al.* 2000; Agra-Amboage *et al.* 2010; see Fig. 2,bottom), reminiscent of laboratory experiments of episodic "magnetic towers" driven by a strong B_ϕ pressure gradient (Ciardi *et al.* 2009). Such variable features are probably related to the non-steady interaction between the disk and the twisted stellar magnetosphere, a process which is necessarily present, and crucial to spin down the accreting star (Ferreira *et al.* 2006, Zanni 2009, Fendt 2009; Romanova 2009; Fendt *et al.*, this volume). The relative contribution of unsteady/steady ejections to the total jet power and mass-flux remains to be established.

References

Agra-Amboage, V., Dougados, C., Cabrit, S., Reunanen, J. 2010, submitted

Anderson, J. M., Li, Z.-Y., Krasnopolsky, R., Blandford, R. 2003, *ApJ* 590, L107

André, P., Ward-Thompson, D., Barsony, M. 2000, in Protostars and Planets IV, eds. V. Mannings, A. P. Boss & S. S. Russell (U. of Arizona Press, Tucson), p. 59

Anglada, G. *et al.* 2007, *AJ*, 133, 2799

Antoniucci, S., Nisini, B., Giannini, T., Lorenzetti, D. 2008, *A&A* 479, 503

Arce, H. G. *et al.* 2007, in *Protostars & Planets V*, ed. by B. Reipurth, D. Jewitt, K. Keil, (University of Arizona Press, Tucson), pp. 245-260

Bacciotti, F., Mundt, R., Ray, T. P., Eisloeffel, J., Solf, J., Camenzind, M. 2000, ApJ 537, L49

Bacciotti, F., Ray, T. P., Mundt, R., Eisloeffel, J., Solf, J. 2002, ApJ 576, 222

Bacciotti. F., 2009, in Protostellar Jets in Context, eds. K. Tsinganos, T. Ray, M. Stute. *ApSS Proceedings Series* (Berlin: Springer), pp. 231-240

Bachiller, R., Tafalla, M. 1999, in *The Origin of Stars and Planetary Systems*, eds. C. J. Lada and N. D. Kylafis (Kluwer Academic Publishers), p. 227

Beck, T. L., McGregor, P. J., Takami, M., Pyo, T-S. 2008, ApJ 676, 472

Beuther, H., *et al.* 2002, *A&A* 383, 892

Cabrit, S., 2002, in Star Formation and the Physics of Young stars, eds. J. Bouvier and J. P. Zahn, *EAS Pub. Series*, Vol. 3, pp. 147–182

Cabrit, S. 2007a, in *MHD jets and winds from young stars*, Lecture Notes in Physics (Spinger-Verlag Berlin Heidelberg), vol. 723, p. 21

Cabrit, S. 2007b, in *IAU Symp.* vol. 243 (Cambridge University Press), pp. 203–214

Cabrit, S., 2009, in Protostellar Jets in Context, eds. K. Tsinganos, T. Ray, M. Stute. *ApSS Proceedings Series* (Berlin: Springer), pp.247-257

Cabrit, S., Bertout, C. 1992, A&A, 261, 274

Cabrit, S., Raga, A., Gueth, F. 1997, in *IAU Symp.* vol. 182 (Kluwer: Dordrecht) p.163

Cabrit, S., Pety, J., Pesenti, N., Dougados, C. 2006, *A&A*, 452, 897

Cabrit, S., *et al.* 2007, A&A 468, L29

Cai, M. J., Shang, H. Lin, H. H., Shu, F. H. 2008, ApJ 672, 489

Casse, F. & Ferreira, J. 2000, *A&A* , 361, 1178

Cerqueira, A. H., Velázquez, P. F., Raga, A. C., Vasconcelos, M. J., de Colle, F. 2006, *A&A* 448, 231

Ciardi, A., *et al.* 2009, *Ap.J. Lett.* , 691, L147

Claussen, M. J., Marvel, K. B., Wootten, A., Wilking, B. A. 1998, *ApJ* 507, L79

Coffey, D., Bacciotti, F., Ray, T. P., Eisloeffel, J., Woitas, J. 2007, ApJ 663, 350

Cranmer, S. R. 2009, *Ap.J.* , 706, 824

Dionatos, O., Nisini, B., Cabrit, S., Kristensen, L., & Pineau Des Forêts, G. 2010, *A&A* , 521, A7

Dionatos, O., *et al.* 2009, *Ap.J.* , 692, 1

Downes, T. P., Cabrit, S., 2003, *A&A* 403, 135

Fendt, C. 2009, *Ap.J.* , 692, 346

Ferreira, J., Dougados, C., Cabrit, S. 2006, *A&A* 453, 785

Garcia Lopez, R., *et al.* 2008, *A&A* , 487, 1019

Gueth, F., Guilloteau, S. 1999, *A&A* 343, 571

Hartigan, P., Morse, J., Raymond,. J. 1994, *ApJ* 436, 125

Hartigan, P., Edwards, S., Gandhour, L. 1995, *Ap.J.* 452, 736 (HEG95)

Hirth, G., Mundt, R., Solf, J. 1997, *A&A Supp. Ser.*, 126, 437

Krist, J. E., *et al.* 2008, *A.J.* , 136, 1980

Lada, C. J. 1985, *ARAA*, 23, 267

Launhardt, R. *et al.* 2009, *A&A* , 494, 147

Lavalley-Fouquet, C., Cabrit, S., Dougados, C. 2000, *A&A* 356, L41

Lee, C.-F., *et al.* 2007, *ApJ* 670, 1197

Lee, C.-F., *et al.* 2008, *ApJ* 685, 1026

Lee, C.-F., *et al.* 2009, *Ap.J.* 699, 1584

Lee, C.-F., *et al.* 2010, *Ap.J.* 713, 731

McCaughrean, M. J., Rayner, J. T., & Zinnecker, H. 1994, *Ap.J. Lett.* , 436, L189

Muzerolle, J., Hartmann, L., Calvet, N. 1998, AJ 116, 2965

Nisini, B., *et al.* 2005, *A&A* , 441, 159

Panoglou, D., Cabrit, S., Pineau des Forêts, G., Garcia, P. *et al.* 2010, submitted

Pety, J., Gueth, F., Guilloteau, S., & Dutrey, A. 2006, *A&A* , 458, 841

Pesenti, N., *et al.* 2004, A&A 416, L9

Podio, L., *et al.* 2006, *A&A* , 456, 189

Podio, L., Medves, S., Bacciotti, F., Eislöffel, J., & Ray, T. 2009, *A&A* , 506, 779

Raga, A., Velasquez, P. F., Canto, J., Masciadri, E. 2002, *A&A* 395, 647

Raga, A., Williams, D. A., Lim, A., J. 2005, RMxAA 41, 137

Ray, T. P. *et al.* 2007, In *Protostars & Planets V*, B. Reipurth, D. Jewitt, and K. Keil (eds.), (University of Arizona Press, Tucson), pp. 231–244

Reipurth, B., Bally, J. 2001, *ARA&A* 37, 403

Reipurth, B., Hartigan, P., Heathcote, S., Morse, J. A., Bally, J. 1997, AJ 114, 757

Richer, J. S., Shepherd, D. S., Cabrit, S., Bachiller, R., Churchwell, E. 2000, in *Protostars and Planets IV* eds Mannings, V., Boss, A. P., Russell, S. S. (University of Arizona Press, Tucson) pp 867–894

Romanova, M. M., Ustyugova, G. V., Koldoba, A. V., & Lovelace, R. V. E. 2009, *MNRAS*, 399, 1802

Stute, M., Gracia, J., Tsinganos, K., & Vlahakis, N. 2010, *A&A* , 516, A6

Tafalla, M., Santiago-Garcia, J., Hacar, A., & Bachiller, R. 2010, arXiv:1007.4549

Taylor, S. D., Raga, A. C. 1995, *A&A* 296, 823

Tsinganos, K., Ray, T., & Stute, M. 2009, Protostellar Jets in Context, Astrophysics and Space Science Proceedings Series (Berlin: Springer).

Zanni, C. 2009, *Rev. Mex. Astron. Astroph. Conf. Series*, 36, 284

Discussion

DE GOUVEIA DAL PINO: The steady-state models with implicit chemistry you showed to explain molecular jets are kind of ideal because actually if we include intermittency then we have shocks and knots and then the shocks may destroy the molecules: so, I don't think a steady-state model may solve this question on molecular jets survival. Could you comment on this ?

CABRIT: Indeed, you are completely right that internal shocks might modify the chemical abundances, possibly dissociating molecules and/or reforming them in dense and cool postshock layers. Hence, shocks should be included in any model aiming at reproducing the jet composition. However, the effect of shocks on chemistry depends critically on whether the shocks are discontinuous (J-type) or continuous (C-type). This in turn depends on the (unknown) preshock H_2 content, ionization fraction, and transverse magnetic field in the jet. Our steady-state chemistry calculations are thus a required first step in that they provide these initial pre-shock conditions in the case of MHD disk winds. Using these conditions, we find that internal shocks in class 0 disk winds should develop a magnetic precursor (C-type) and be much less dissociative than under ISM conditions. The predicted H_2 spectrum is in excellent agreement with *Spitzer* observations of class 0 jets (V. Taquet, in prep.). Other molecules, eg. SiO, could in principle be formed in such shocks.

Jets at all Scales
Proceedings IAU Symposium No. 275, 2011
G. E. Romero, R. A. Sunyaev & T. Belloni, eds.

© International Astronomical Union 2011
doi:10.1017/S1743921310016443

MHD simulations of jet formation - protostellar jets & applications to AGN jets

Christian Fendt, Bhargav Vaidya,
Oliver Porth and Somayeh Sheikh Nezami†

Max Planck Institute for Astronomy, Königstuhl 17, D-69117 Heidelberg, Germany
email: fendt@mpia.de

Abstract. Jet formation MHD simulations are presented considering a variety of model setups. The first approach investigates the interrelation between the disk magnetisation profile and jet collimation. Our results suggest (and quantify) that outflows launched from a very concentrated region at the inner disk tend to be weakly collimated. In the second approach, jet formation is investigated from a magnetic field configuration consisting of a stellar dipole superposed by a strong disk field. We find that the central dipole considerably de-collimates the disk wind. In addition, reconnection flares are launched in the interaction region of disk and stellar magnetic field, subsequently changing the outflow mass flux by factors of two. The time interval between flare ejection is about 1000 Keplerian periods - surprisingly similar to the observed time lag between jet knots. The third approach considers radiative pressure effects on jet collimation - an environment which is interesting mainly for outflows from massive young stars (but also for relativistic jets). Finally we present relativistic MHD simulations of jet formation from accretion disks extenting the previous non-relativistic approaches.

Keywords. accretion, accretion disks, stars: magnetic fields, stars: mass loss, stars: winds, outflows, ISM: Herbig-Haro objects, galaxies: jets

1. Introduction

Highly collimated outflows are one of the most striking signatures of young stars. Jets are observed, however, also in other sources – among them micro-quasars or X-ray binaries (MQs, XRBs), or active galactic nuclei (AGN). In stellar sources, a central stellar magnetic field is surrounded by a disk carrying its own magnetic flux. Such a geometrical setup can be found young stars, cataclysmic variables, high-mass and low-mass X-ray binaries, and other micro-quasar systems. The current understanding of jet formation is that outflows are launched by *magnetohydrodynamic* (MHD) processes in the close vicinity of the central object – an accretion disk surrounding a protostar or a compact object Blandford & Payne (1982), Pudritz *et. al.* (2007), Shang *et al.* (2007). The details of the physical processes involved are, however, not completely understood.

Jet formation simulations can be distinguished in those taking into account the evolution of the disk structure and others considering the disk surface as a fixed-in-time boundary condition for the jet (see below). The case of superposed stellar/disk magnetic fields is rarely treated in simulations, although the first models were discussed by Uchida & Low (1981). Simulations of a dipole with aligned vertical disk field were presented first by Hayashi *et al.* (1996), Hirose *et al.* (1997), or Miller & Stone (1997). The stellar magnetosphere impacts jet formation by enhancing the magnetic flux available, imposing a central pressure, or providing excess angular momentum in the launching area.

† Home institute: Ferdowsi University of Mashhad, Iran

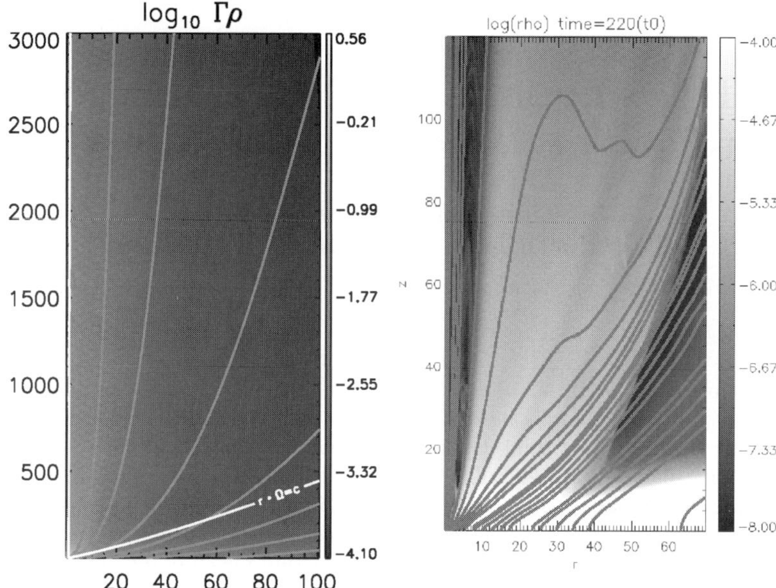

Figure 1. Jet formation simulations considering the disk as fixed-in-time boundary condition (left) and evolving the disk structure together with the outflow (right). The left figure shows a relativistic MHD simulation of jet formation by Porth & Fendt (2010) reaching Lorentz factors of about $\Gamma = 10$ (grid size 100x3000 inner disk radii). The outflow in the right figure originates in a $(h/r) = 0.1$ disk with density contrast 10^4 to the initial corona and disk resistivity $\eta = 0.01$ (grid size 812x1500 cells, resp. 80x120 inner disk radii).

Jet formation in massive young stars is a rather new topic. The conditions for jet launching as well as important jet parameters such as mass fluxes, velocities, or magnetic field strengths are hardly known. However, we know that massive stars have a strong radiation field which is expected to affect accretion as well as outflow processes.

2. Jet formation - the standard MHD model

The principal processes involved in jet formation can be summarized as follows.

(i) Magnetic flux is provided by the star-disk (or black-hole - disk) system - possibly by a disk or stellar dynamo, or by advection of the interstellar field. The star-disk system also drives an electric current.

(ii) Accreting material is diverted and launched as a plasma wind (from the stellar or disk surface), couples to the magnetic field, and is flung out magneto-centrifugally.

(iii) Inertial forces wind up the poloidal field inducing a toroidal component.

(iv) The plasma becomes accelerated magnetically (conversion of Poynting flux).

(v) The toroidal field tension collimates the outflow into a high speed jet beam.

(vi) The plasma velocities subsequently exceed the speed of the magnetosonic waves. The super-fast magnetosonic regime is causally decoupled from the surrounding medium.

(vii) Where the outflow meets the ISM, a shock develops, thermalizing the jet energy.

(viii) The underlying hypothesis is that jets can only be formed in a system with a high degree of axi-symmetry. This might be the reason why only few jets are found in cataclysmic variables or pulsars (e.g. see Fendt & Zinnecker (1998)).

The model of a dipole-plus-disk magnetic flux configuration were introduced for protostellar jets by Uchida & Low (1981). MHD simulations of jets from dipolar magnetospheres were performed by Uchida & Shibata (1984), Uchida & Shibata (1985). However,

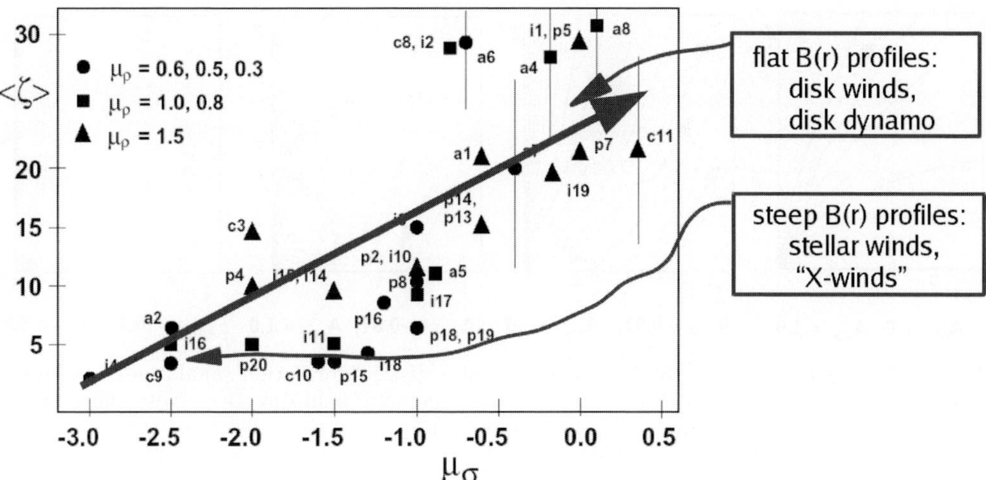

Figure 2. Collimation degree as function of the disk wind magnetic field, resp. magnetization profile. Error bars indicate flows which have not reached a steady state.

probably due to the success of MHD disk-jet models (Blandford & Payne (1982), Pudritz & Norman (1983)) - in particular for extragalactic jets - and the limitations of the early numerical simulations, this concept was somewhat repressed until it became evident that the young star itself does carry a substantial large-scale magnetic field.

For the numerical simulations of jet formation in general two approaches were made (see Fig.1). One is to prescribe the accretion disk properties as a boundary condition for the jet. Studying the acceleration and collimation of a disk/stellar wind requires essentially to follow the jet dynamical evolution for i) very long time ii) on a sufficiently large grid with iii) appropriate resolution. For such a goal, this second approach is better suited (Ustyugova *et al.* (1995), Ouyed & Pudritz (1997), Krasnopolsky *et al.* (1999), Fendt & Čemeljić (2002), Fendt (2006), Fendt (2009), Porth & Fendt (2010)). Naturally, the mass flux from disk to jet cannot be determined by such an approach. The other approach includes the disk structure in the simulation, in particular investigating the launching mechanism which lifts matter from the disk into the outflow, determining the mass flux from disk to jet (Uchida & Shibata (1984), Miller & Stone (1997), Goodson *et al.* (1997), Casse & Keppens (2002), Romanova *et al.* (2002), von Rekowski & Brandenburg (2004), Meliani *et al.* (2007), Zanni *et al.* (2007)). This approach is computationally expensive and still somewhat limited by spatial and time resolution. Thus, in many simulations published so far, the disk model underlying the jet launching is rather simple.

3. MHD simulations: disk jets with of different magnetic flux profiles

Here we present jet formation simulations where jets are formed from pure disk winds (see Fendt (2006)). The physical grid size corresponds to $(r \times z) = (150 \times 300) \, r_{\rm in}$.

We start from a force-free initial field distribution in a hydrostatic equilibrium gas. The simulation evolves under the boundary condition of a (spatially and temporarely) fixed mass flux from the disk surface into the outflow. We run models covering a wide range of disk magnetic flux profiles and disk wind mass flux profiles, typically parameterized by power laws, $B_{\rm p,wind}(r) \sim r^{-\mu}, \rho_{\rm wind}(r) \sim r^{-\mu_\rho}$. Both quantities can be combined in the disk wind magnetization parameter (Michel (1969)), $\sigma_{\rm wind} \sim B_{\rm p}^2 r^4 \Omega_F^2 / \dot{M}_{\rm wind} \sim r^{\mu_\sigma}$.

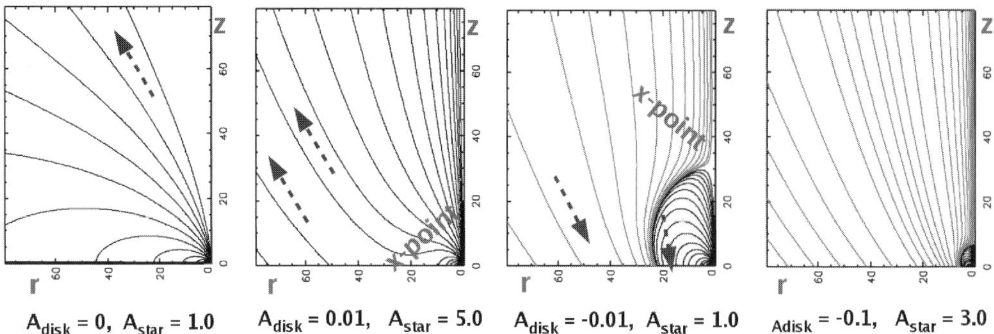

Figure 3. Initial magnetic field distribution for star-disk jet formation simulations, shown are poloidal magnetic field lines. Arrows indicate the magnetic field direction. Note the different location of the X-points. Different strength and orientation of the superposed stellar and disk magnetic field component, $A_{disk} = 0.0, 0.01, -0.01, -0.1$, resp. $A_{star} = 1.0, 5.0, 1.0, 3.0$ (from left to right). Note that here we show the left hemisphere (rotation axis directs upwards).

We quantify the collimation degree by comparing the axial and lateral mass fluxes (see Fendt & Čemeljić (2002), Fendt (2006)). Figure 2) shows the collimation degree $\zeta \simeq (\dot{M}_z/\dot{M}_r)$ plotted against the power law exponent of the magnetization μ_σ. The main result is that steep magnetization profiles (e.g. the X-wind), resp. disk magnetic field profiles, are unlikely to generate highly collimated outflows. Flat profiles - generally leading to a higher collimation - tend to be unstable, i.e. do not establish a steady state.

4. MHD simulations: outflows from disk-star magnetospheres

We now discuss simulations considering the co-evolution of a stellar magnetosphere with a disk magnetic field. Both components are fed by a mass flux injected from the underlying physical boundary - the stellar surface and the accretion disk. We have used the ZEUS-3D codes extended for magnetic diffusivity (Stone & Norman (1992), Fendt & Čemeljić (2002)). The field direction of both components can be aligned or anti-aligned. Similar configurations were considered early by Uchida & Low (1981).

Applying cylindrical coordinates (r, ϕ, z), we divide the equatorial plane in three parts - the stellar surface $r < r_\star = 0.5r_{\rm in}$, the disk $r > r_{\rm in} = 1.0$, and the gap between star and disk. The stellar magnetospheric co-rotation radius is at the disk inner radius. The grid size is $(r \times z) = (80 \times 80)$ inner disk radii which refers to different physical scales when applied to e.g. protostars or XRBs. The initial (force-free) magnetic field is composed of a stellar (dipolar) field and a disk field (see Fig. 3),

$$\Psi_{\rm total}(r, z) = A_{\rm disk}f_{\rm disk}(r, z) + A_{\rm star}f_{\rm star}(r, z), \qquad (4.1)$$

where $\Psi_{0,\rm disk}$ and $\Psi_{0,\rm star}$ measure the strength of both components and the functions $f(r, z)$ describe the initial field distribution (see Fendt (2009)).

Figure 4 shows how the outflow evolves for the example with $\Psi_{0,\rm disk} = -0.1$ and $\Psi_{0,\rm star} = 3.0$. In this case, disk and stellar magnetic field direction (along the equatorial plane) are aligned. We evolve this simulation for 2800 rotations at the inner disk radius. At intermediate times (700 rotations) a quasi-stationary state emerges. One clearly sees the de-collimating effect of the central stellar wind component. Such quasi-stationary states may appear again on much longer time scales. We observe a cyclic behavior of the outflow opening angle with a periodicity of about 500 rotations.

Independent of the alignment, the central dipole does not survive on the large scale. A two-component outflow emerges: a stellar wind surrounded by a disk wind. For a

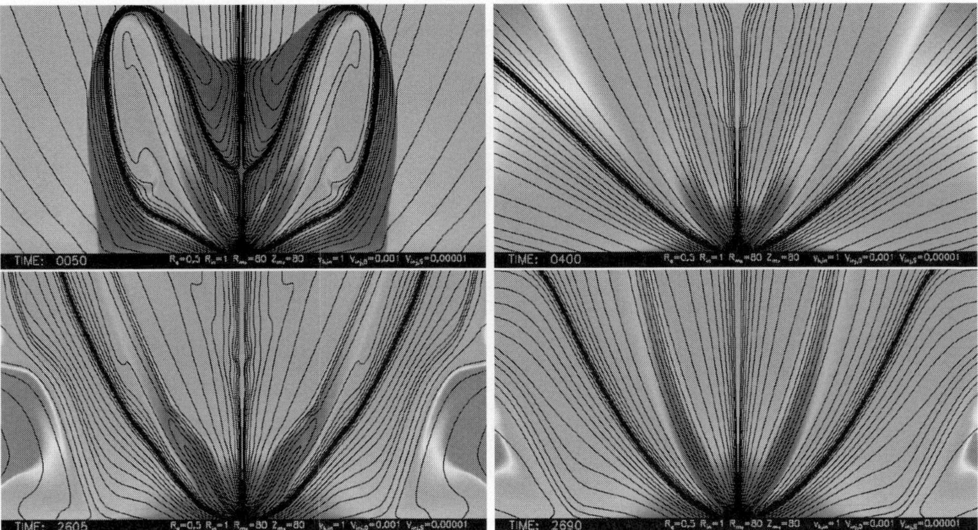

Figure 4. Time evolution of a star-disk magnetosphere from initial state of Fig. 3, middle. Time step is 50, 400, 2606, 2700 rotations of the inner disk (from top left to bottom right). Colors show logarithmic density contours, black lines are poloidal field lines (magnetic flux contours). Note that here we show both upper hemispheres (rotation axis directs upwards).

reasonably strong disk magnetic flux a collimated outflow emerges. If the overall flow is dominated by the stellar outflow, the disk wind remains un-collimated. Thus, the favorable setup to launch collimated jets from a star-disk magnetosphere is that of a heavy disk wind with strong magnetic flux.

We also observe reconnection events emerging from the X-point between the remaining inner dipole and the disk magnetic field, leading to sudden large-scale flares (see also Goodson *et al.* (1999)). The flare expansion is rapid and within few rotational periods. The flare events are accompanied by a temporal change in the outflow mass flux. Figure 5 shows the mass loss rate in axial direction integrated across the jet. We see two flares with a 20%-increase in the mass flux followed by a sudden decrease of mass flux by a factor of two. This behavior is also seen in the poloidal velocity profile.

Considering the ejection of large-scale flares and the follow-up re-configuration of outflow dynamics, we hypothesize that the origin of jet knots is triggered by such flaring events. Our time-scale for flare generation is of 1000 rotational periods and longer than the typical dynamical time at the jet base, but similar to the observed knots in protostellar jets. The flare itself for about 30-40 inner disk rotation times (see Fendt (2009) for a comparison to the Sweet-Parker reconnection time scale which turns out to be of the same order for the simulation parameters applied).

5. Jets from massive young stars

For massive young stars, the outflow dynamics is affected by the strong radiation field. Both the central object and the surrounding inner hot accretion disk may contribute to the radiative forces. Therefore, outflows from massive stars need to be modeled considering radiative forces and MHD. Vaidya *et al.* (2009) have shown that the inner disk can be internally heated up to 10^5 K while being sufficiently stable to launch an outflow.

Using the PLUTO code (Mignone *et al.*(2007)) we have run MHD simulations of jet formation under the influence of radiative forces from stellar and disk luminosity. The

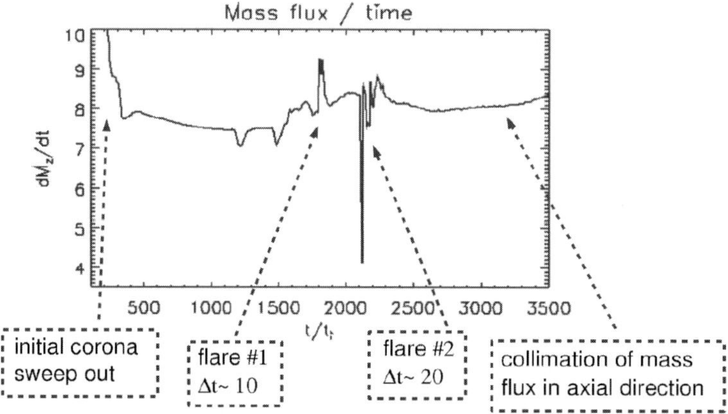

Figure 5. Time evolution of the axial mass flux close to the upper boundary. The mass flux changes during the initial evolution (sweep-out of the initial corona), but also during the flaring events.

radiative force $\mathbf{F}^{\mathrm{rad}} = \mathbf{f}_{\mathrm{cont},*}+\mathbf{f}_{\mathrm{cont,disk}}+\mathbf{f}_{\mathrm{line},*}+\mathbf{f}_{\mathrm{line,disk}}$ considered in our study comprises of accelerations due to continuum radiation from star ($\mathbf{f}_{\mathrm{cont},*}$) and disk ($\mathbf{f}_{\mathrm{cont,disk}}$) and those due to lines forces from disk ($\mathbf{f}_{\mathrm{line,disk}}$) and star ($\mathbf{f}_{\mathrm{line},*}$), For the line forces we apply the well known CAK theory by Castor *et al.*(1975). This approach has been also explored for AGN disk winds by Proga (2003). The line force can be expressed as a product of force due to continuum radiation and a force multiplier $M(\mathrm{t})$, summarized over all lines, where t is the optical depth parameter. The parameter t is related to gradient of velocity along the l.o.s., the wind density, and the ion thermal speed v_{th}, thus $t = (\rho\sigma_e v_{th})/|\hat{\mathbf{n}}\cdot\nabla(\hat{\mathbf{n}}\cdot\mathbf{v})|$.

An empirical form for $M(t)$ as a sum over all lines for model atmospheres for massive OB stars has been defined as $M(t) \sim kt^{-\alpha}$, where k and α are line force parameters Abbott(1982). Depending on the selection of lines, typical values obtained for k range from 0.4-0.6 and for α between 0.3-0.7. A force multiplier parameterization independent of arbitrary v_{th}, was introduced by Gayley(1995). The parameter $\bar{Q}$ is related to parameter k initially introduced by Castor *et al.*(1975) as

$$M(t) = \left[\frac{\bar{Q}^{1-\alpha}}{1-\alpha}\left(\frac{|\hat{\mathbf{n}}.\nabla(\hat{\mathbf{n}}.\mathbf{v})|}{\sigma_e c\rho}\right)^{\alpha}\right] \tag{5.1}$$

Figure 6 show preliminary results (Vaidya & Fendt 2010, to be submitted). We start off with a pure MHD simulation similar to Ouyed & Pudritz (1997) or Fendt (2006). When a steady outflow is established (200 inner orbital periods), we switch-on the radiative forces (here only stellar line forces are considered). The radiative forces disturb the pure MHD jet structure, and a new quasi-steady state is reached after another 200 rotations.

The degree of collimation of the MHD jet under stellar radiation is higher. This is surprising, as one would naturally expect a *de-collimation* by the central flux. Our preliminary interpretation is that the stellar flux heats the lower (and inner) disk wind, by that changes the disk inflow boundary condition, resulting in an enhanced jet mass flux (by 20%). The higher mass load (and thus the larger outflow inertia) leads to a stronger toroidal field and a higher degree of collimation.

6. Relativistic MHD jet formation

We extended the previous non-relativistic simulations to the (special) relativistic regime. (Newtonian) gravity was added to enable a realistic Keplerian disk boundary

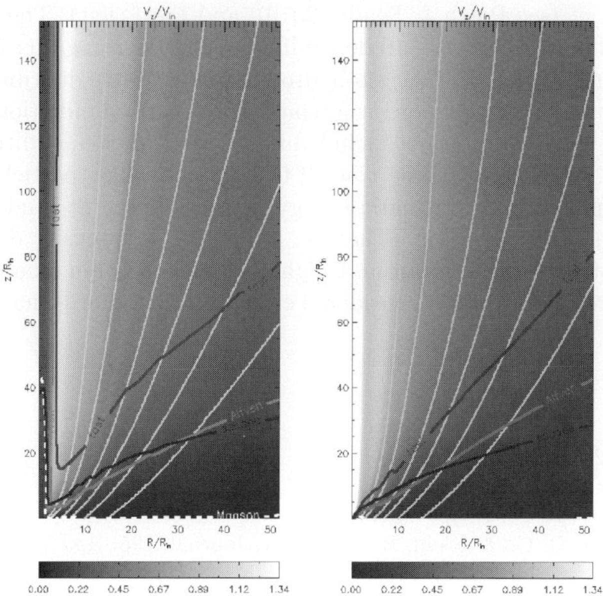

Figure 6. Jet formation simulations for massive young stars considering radiative forces. Pure MHD simulation for 200 disk periods (left), followed by a simulation including the stellar radiative force up to 400 disk periods (right). Mass fluxes are $\dot{M}_{\rm out} = 2.7 \times 10^{-4}\,{\rm M}_\odot/{\rm yr}$ (left), and $\dot{M}_{\rm out} = 3.9 \times 10^{-4}\,{\rm M}_\odot/{\rm yr}$ (right). The collimation degree derived from vertical/lateral mass fluxes is 0.86 (left), and 1.26 (right). The radiative forces is parametrized by the continuum flux force / gravity ratio $\Gamma_e = 0.14$, and the line force multiplier $Q_o = 2000$, $\alpha = 0.39$.

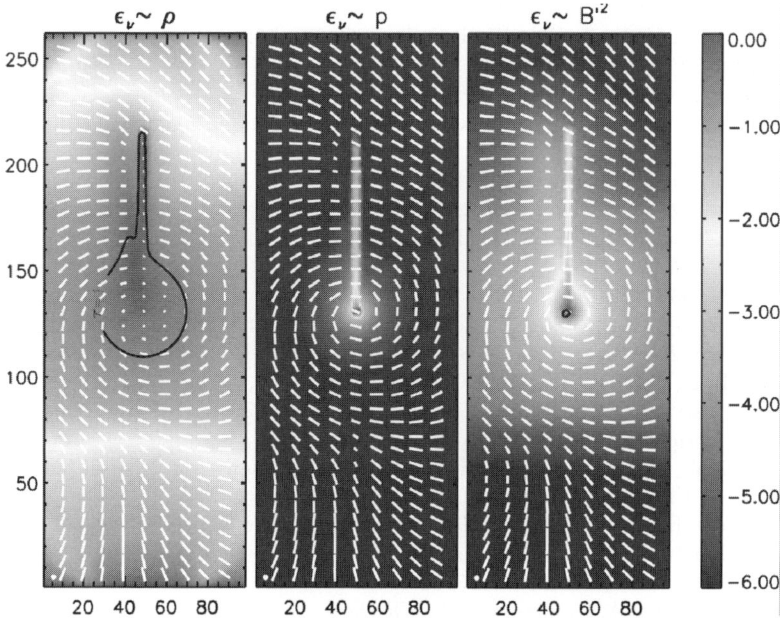

Figure 7. Ideal resolution synchrotron maps considering relativistic MHD simulations of jet formation and polarized radiation transfer. Results for three tracers for the particle acceleration are shown: density (left), thermal pressure (middle), magnetic energy (right). Shown are polarization vectors and the (normalized) 43 GHz intensity distribution for a viewing angle of 30 deg. The radial box size corresponds 400 inner disk radii.

condition for the jet (see Porth & Fendt (2010), and Fig.1, left). The resulting Lorentz factors of the jets reaches $\Gamma = 8 - 10$ depending on the Poynting flux injected.

The numerical results for the magnetohydrodynamic variable distribution were used to derive radio synchrotron maps following a relativistic polarized radiation transfer (Fig. 7). Since the particle acceleration is not included in the MHD model, additional assumptions have to be made. We applied three tracers for the power-law particle acceleration, ie. density, thermal pressure, or magnetic energy density (Porth & Fendt, submitted). All tracers give a similar polarization structure, although the intensity distribution differs. The spiky axial structure results from a high temperature, thin axial outflow from the compact object area and could be considered either as a Blandford-Znajek jet or a coronal wind.

References

Abbott, D. C. 1982, *ApJ* 259, 282

Blandford, R. D. & Payne, D. G. 1982, *MNRAS* 199, 883

Casse, F. & Keppens, R. 2002, *ApJ* 581, 988

Castor, J. I., Abbott, D. C., & Klein, R. I. 1975, ApJournal, 195, *ApJ* 195, 157

Fendt, Ch. & Zinnecker, H. 1998, *A&A* 334, 750

Fendt, Ch. & Čemeljić, M. 2002, *A&A* 395, 1045

Fendt, Ch. 2006, *ApJ* 651, 272

Fendt, Ch. 2009, *ApJ* 692, 346

Gayley, K. G. 1995, *ApJ* 454, 410

Goodson, A. P., Winglee, R. M., & Böhm, K-.H. 1997, *ApJ* 489, 199

Goodson, A. P., Böhm, K.-H., & Winglee, R. M. 1999, *ApJ* 524, 142

Hayashi, M. R., Shibata, K., & Matsumoto, R. 1996, *ApJ* 468, L37

Hirose, S., Uchida, Y., Shibata, K., & Matsumoto, R. 1997, *PASJ* 49, 193

Krasnopolsky, R., Li, Z.-Y., & Blandford, R. D. 1999, *ApJ* 526, 631

Meliani, Z., Casse, F., & Sauty, C. 2007, *A&A* 460, 1

Michel, F. C. 1969, *ApJ* 158, 727

Mignone, A., Bodo, G., Massaglia, S., Matsakos, T., Tesileanu, O., Zanni, C., & Ferrari, A. 2007, *ApJS* 170, 228

Miller, K. A. & Stone, J. M. 1997, *ApJ* 489, 890

Ouyed, R. & Pudritz, R. E. 1997, *ApJ* 482, 712

Porth, O. & Fendt, C. 2010, *ApJ*, 709, 1100

Proga, D. 2003, *ApJ*, 585, 406

Pudritz, R. E. & Norman, C. A. 1983, *ApJ* 274, 677

Pudritz, R. E., Ouyed, R., Fendt, Ch., & Brandenburg, A. 2007, in: B. Reipurth, D. Jewitt, & K. Keil (eds.), *Protostars & Planets V*, University of Arizona Press, Tucson, 2007, p.277

von Rekowski, B. & Brandenburg, A. 2004, *A&A* 420, 17

Romanova, M., Ustyugova, G., Koldoba, A., & Lovelace, R. 2002, *ApJ* 578, 420

Stone, J. M. & Norman, M. L. 1992, *ApJS* 80, 753

Shang, H., Li, Z.-Y., & Hirano, N. 2007, in: B. Reipurth, D. Jewitt, & K. Keil (eds.), *Protostars & Planets V*, University of Arizona Press, Tucson, 2007, p.261

Uchida, Y. & Low, B. C. 1981, *Journal of Astroph. and Astron.* 2, 405

Uchida, Y. & Shibata, K. 1984, *PASJ* 36, 105

Uchida, Y. & Shibata, K. 1985, *PASJ* 37, 515

Ustyugova, G., Koldoba, A., Romanova, M., Chechetkin, V., & Lovelace, R. 1995, *ApJ* 439, 39

Vaidya, B., Fendt, C., & Beuther, H. 2009, *ApJ* 702, 567

Zanni, C., Ferrari, A., Rosner, R., Bodo, G., & Massaglia, S. 2007, *A&A* 469, 811

Discussion

FALCKE: Does the radiation pressure (from the disk) potentially also play a role in AGN jets?

FENDT: I would guess so. We intend to apply the model also to AGN jets. Disk radiation forces will, however, mainly accelerate the material. But MHD will dominate. Second order effects could be an enhanced mass flux compard to pure MHD jets.

Jets at all Scales
Proceedings IAU Symposium No. 275, 2011
G. E. Romero, R. A. Sunyaev & T. Belloni, eds.

© International Astronomical Union 2011
doi:10.1017/S1743921310016455

On the time variability of the HH jet ejection process

Fabio De Colle

Astronomy & Astrophysics Department, University of California,
Santa Cruz, CA 95064, USA. email: `fabio@ucolick.org`

Abstract. Two-dimensional emission line images of the HH30 jet were recently used (De Colle *et al.* 2010) to recover the three-dimensional structure of the jet by applying standard tomographic technique ("Tikhonov regularization techniques"). In this paper I show that it is possible to determine the ejection history of the HH30 jet by directly comparing the outcome of numerical simulations with the results of the tomographic inversion. In particular, it is shown that the HH30 jet electron density map is best reproduced by assuming a velocity variation at the base of the jet with a large scale periodicity (with a period of $\sim$ 3 yrs) added to small scales velocity variation (with periods $\lesssim$ months).

Keywords. hydrodynamics, methods: data analysis, methods: numerical, stars: winds, outflows, ISM: Herbig-Haro objects, ISM: jets and outflows.

---

1. Introduction

Herbig-Haro (HH) jets play a crucial role in the star formation process (e.g. Reipurth & Bally 2001). For instance, they are important in the angular momentum evolution of the star-disk system (e.g. Chrysostomou *et al.* 2008), and they produce an important feedback on the star formation process itself (e.g. Wang *et al.* 2010).

Collimated jets are present at different scales and around very different classes of objects, ranging from young stars to compact objects (neutron stars and black holes). Assuming that the jets at all scales share a common origin, the study of the HH jets is particular useful, as the amount of observational information available for HH jets is in several cases much larger than in any other jet-driving systems.

HH jets present a characteristic knotty structure with a typical periodicity of $\sim$ a few to several years, emitting a rich radiation spectrum ranging from radio wavelength to, in a limited number of cases, x-ray. While there are evidences indicating that the knots are due to velocity variations in the star-disk system, the origin of these velocity variations is unkown (see De Colle *et al.* 2008 for a discussion). From the ratios of forbidden emission lines, the electron density, temperature and hydrogen ionization fraction can be easily determined down to a distance $\gtrsim$ 50 AU from the star.

Recently, observations with information on emission profiles across the jet have been presented by several authors (e.g. Beck *et al.* 2007, Hartigan & Morse 2007). One example is the largely studied HH30 jet. The HH30 jet moves nearly on the plane of the sky, has a clear side-to-side symmetry in the region close to the central star, and the cooling region is resolved spatially with Hubble Space Telescope (HST) observations (e.g. Burrows *et al.* 1996, Ray *et al.* 1996, Bacciotti *et al.* 1999, Hartigan & Morse 2007). Assuming that the HH30 jet is axisymmetric, we showed in a previous work (De Colle *et al.* 2010) that standard tomographic techniques may be employed to recover the three-dimensional structure of the jets.

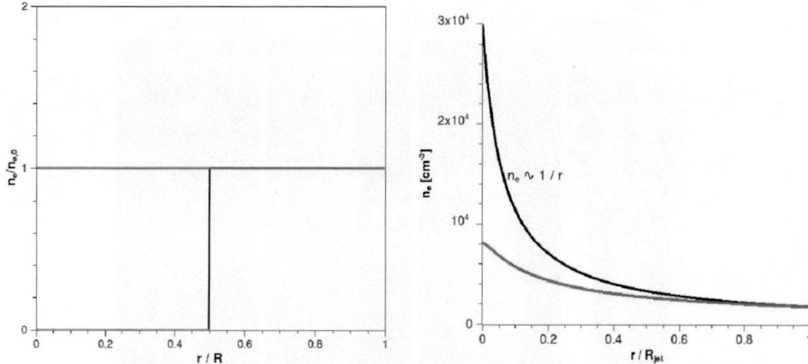

Figure 1. Effect of the hypothesis of homogeneity on the electron density determination. Assuming a certain n_e profile (black curves), [SII]$\lambda 6716, 6731$ emission coefficients were determined solving the 5-levels atom problem (assuming a constant temperature $T_e = 10^4$ K). These are later integrated along the line of sight, and from the ratio of the intensities the electron density (equivalent to the one obtained directly assuming a jet homogeneous along the line of sight) is recovered (red curves). *Left panel*: the density profile is taken as $n_e = n_{e,0}$ for $r > 0.5$ and 0 otherwise. The resulting n_e strongly overestimate the real n_e. *Left panel*: the density profile is taken as $n_e \propto 1/r$. The resulting n_e in this case underestimate the real n_e.

These results, together with a brief discussion of the error introduced in the determination of the physical parameters when assuming a jet as homogeneous along the line of sight, are reviewed in Section 2, while Section 3 present a comparison between the results of the inversion technique and the prediction obtained by running detailed hydrodynamics simulations, showing that the data are best fitted by a model assuming a velocity variation at the base of the jet with a large scale periodicity (with a period of ~ 3 yrs) added to small scales velocity variation (with periods $\lesssim$ months).

2. Observations

Usually, observations can be used to extract two-dimensional electron density, temperature and ionization fraction images. The resulting maps of the physical parameters are accurate (assuming that the effect of the instrumental response is negligible) is the jet is homogeneous along the line of sight, and the the presence of dishomogeneity can introduce large errors in the determination of the physical parameters (see De Colle *et al.* 2008). Furthermore, the examples of Figure 1 illustrate that the density profile determined from the homogeneity assumption may over- or under-estimation the correct profile. In a previous paper (De Colle *et al.* 2010) we showed that, actually, replacing the homogeneity hypothesis with that of axisymmetric medium, standard tomographic techniques may be employed to recover the three-dimensional structure of the jets. The main result was that the reconstructed density, temperature, and ionization fraction present much steeper profiles than those inferred using the assumption of homogeneity. In particular, the HH30 jet shows a much more fragmented and irregular jet structure with several small scale knots present along the main jet axis, together with large scale knots.

As a detailed description of the tomographic techniques and its application to the HH30 jet has been already presented previously (De Colle *et al.* 2010), we focus here on the consequences of those results on the ejection models. In the next Section, in particular, we show by using detailed numerical simulation that the best fit to the observations is obtained when assuming a chaotic variation in the velocity ejection from the star-disk system.

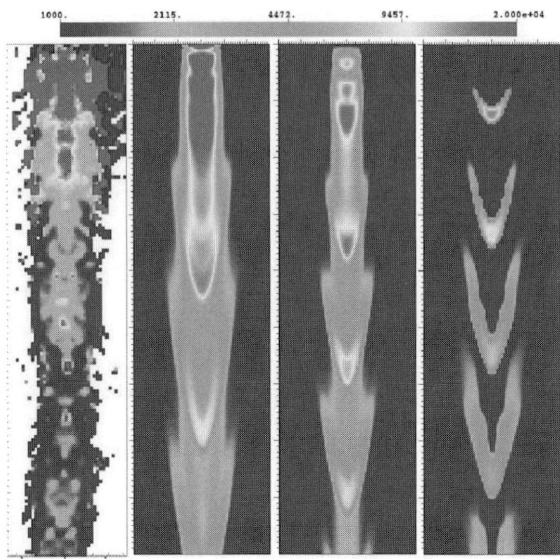

Figure 2. Electron density maps determined from the tomographic inversion of the HH30 observations (left panel) and from hydrodynamics simulations using four, two and a single variability (right three panels) in the injection velocity of the jet (see eq. 3.1) from the boundary of the computational grid (located on the upper $x-$axis; the jet is moving from up to down). The variabilities are on scales of years (for the three models) added to variabilities on scales of months (models shown in the II and III panels from the left) and days (model shown in the II panel from the left).

3. Numerical Simulations

We use the hydrodynamics version of the Mezcal code (De Colle 2010, in preparation). The code includes rate equations for the calculation of the ionization and recombination of a set of 17 atomic/ionic species: [HI], [HII], [HeI], [HeII], [HeIII], [NI], [NII], [NIII], [OI], [OII], [OIII], [OIV], [SII], [SIII], [CII], [CIII], [CIV]. The cooling term in the energy equation is calculated explicitly, as the sum of the contributions coming from the 17 species. The details of the implementation of the cooling term and the rate equations are given in Raga *et al.* (2007).

It is commonly accepted that the HH objects (the knots along the jet flow) are produced by supersonic velocity variations at the base of the jet. For instance, numerical simulations, using a velocity variations of about 10–20% of the average velocity, have been able to reproduce the morphology and the emission property of the knots in HH objects. Following the same recipe, the velocity at the boundary of the computational grid is changed periodically as

$$
v_{jet} = v_0 + \sum_{i=1}^{N} A_i \sin \frac{2\pi t}{\tau_i} .
\tag{3.1}
$$

The jet density is 10^4 cm^{-3}, the jet temperature 1000 K, and the average velocity $v_0 = 200$ km/s (with a radial "top hat" profile). To reproduce the observed change in the FWHM (see De Colle *et al.* 2010), the jet is assumed to be conical, with an opening angle of $2°$. The ambient medium values have been choose as $\rho_{amb} = 10^3$ cm^{-3}, and $T = 1000$ K. The jet is therefore slightly overpressure. This overpressure is likely to be matched by a strong toroidal component of the magnetic field (that collimates the jet). As in this study we are not interested in reproducing the observed HH30 jet in detail, but only to explore qualitatively the theoretical implications of the results obtained from the application of the tomographic reconstruction technique to the HH30 jet, we neglect the effect of the magnetic field.

We run a series of simulations changing N, τ_i, and A_i. In the first model, we use $N = 1$, $\tau_1 = 3$ yrs, and $A_1 = 0.5$. In the second model we add a second perturbation with $\tau_2 = 6$ months and $A_1 = 0.25$, $A_2 = 0.5$. In the third simulation we add other two sinusoidal component, with $A_{3,4} = 0.43, 0.59$ and $\tau_3 = 2$ months and $\tau_4 \approx 10$ days. As τ_4 is lower than the timestep of the simulation itself, the $i = 4$ periodic component represents substantially a chaotic contribution to the velocity variation.

The volumetric electron density map $n_e = n_e(r)$ determined from the tomographic inversion techniques and corresponding to the models with $N = 4, 2, 1$ is shown in Figure 2 (left to right panels respectively). In the model with $N = 1$ the electron density presents strong post-shock peaks, with large electron density drops between the shock traveling downstream. This is in clear contradiction with observations, where elongated structures with large electron density (and emission line intensities) are present also in the regions between the main knots. The observed substructures and the relatively high degree of ionization (of order of ~ 0.1) present in the observation are best matched by the model with $N = 4$.

These results imply that the mechanism responsible for the generation of a velocity variation at the base of the jet is producing a chaotic or at least a multi-component variation, with a typical large timescale of 3 yrs (necessary to reproduce correctly the position of the main knots) added to a low scale (less than 6 months) velocity variation.

References

Bacciotti, F., Eislöffel, J., & Ray, T. P 1999, A&A, 350, 917

Beck, T. L., Riera, A., Raga, A. C., & Reipurth, B. 2007 *AJ*, 133, 1221

Burrows, C. J., Stapelfeldt, K. R., Watson, A. M., *et al.* 1996, *ApJ*, 473, 437

Chrysostomou, A., Bacciotti, F., Nisini, B., *et al.* 2008 *A&A*, 482, 575

De Colle, F., del Burgo, C., & Raga, A. C. 2008, *A&A*, 485, 765

De Colle, F., del Burgo, C., & Raga, A. C. 2008, *A&A*, 721, 929

De Colle, F., Gracia, J., & Murphy, G. 2008, *ApJ*, 688, 1137

Hartigan, P. & Morse, J. 2007 *ApJ*, 660, 426

Raga, A. C., De Colle, F., Kajdic, P., Esquivel, A., & Cantó, J. 2007, *A&A*, 465, 879

Ray, T. P., Mundt, R., Dyson, J. E., *et al.* 1996, ApJ, 468, 103

Reipurth, B. & bally, J. 2001, *ARA&A*, 39, 403

Wang, P. and Li, Z.-Y. and Abel, T., & Nakamura, F., *ApJ*, 709, 27

Discussion

L.F. RODRIGUEZ: Is it understood why on larger scales uou see longer periodicities? Is it real or some observational bias?

DE COLLE: This can be explained by assuming a multi-periodic ejection from the source. The small scale knots present close to the source collide and leave, at larger distances, the knots with the larger periodicity.

DE GOUVEIA DAL PINO: Can you use the same tomographic technique to reconstruct the 3D structure of light relativistic jets?

DE COLLE: Well, the necessary condition to apply the tomographic technique is to have high resolution images of a jet that is also located in the plane of the sky and is axisymmetric. Do we have a light relativistic jet with these characteristics?

Jets at all Scales
Proceedings IAU Symposium No. 275, 2011
G. E. Romero, R. A. Sunyaev & T. Belloni, eds.

© International Astronomical Union 2011
doi:10.1017/S1743921310016467

Brown dwarf jets: Investigating the universality of jet launching mechanisms at the lowest masses

Emma Teresa Whelan[1] and Francesca Bacciotti[2], and Tom Ray[3], and Catherine Dougados[1]

[1]Laboratoire d'Astrophysique de Grenoble, UMR 5571, BP 53, 38041 Grenoble Cedex 09, France
email: `whelane@obs.ujf-grenoble.fr`

[2]INAF-Osservatorio Astrofisico di Arcetri, Largo E. Fermi 5, 50125 Firenze, Italy, Box 515, SE-75120 Uppsala, Sweden
email: `fran@arcetri.astro.it`

[3]Dublin Institute for Advanced Studies, Ireland
email: `tr@cp.dias.ie`

Abstract. Recently it has become apparent that proto-stellar-like outflow activity extends to the brown dwarf (BD) mass regime. While the presence of accretion appears to be the common ingredient in all objects known to drive jets fundamental questions remain unanswered. The more prominent being the exact mechanism by which jets are launched, and whether this mechanism remains universal among such a diversity of sources and scales. To address these questions we have been investigating outflow activity in a sample of protostellar objects that differ considerably in mass and mass accretion rate. *Central to this is our study of brown dwarf jets.* To date Classical T Tauri stars (CTTS) have offered us the best touchstone for decoding the launching mechanism. Here we shall summarise what is understood so far of BD jets and the important constraints observations can place on models. We will focus on the comparison between jets driven by objects with central mass $< 0.1 M_\odot$ and those driven by CTTSs. In particular we wish to understand how the the ratio of the mass outflow to accretion rate compares to what has been measured for CTTSs.

Keywords. stars: low-mass, brown dwarfs, mass loss, ISM: jets and outflows

1. Brown Dwarf Outflows

Jets from CTTSs are traditionally probed at forbidden emission line (FEL) wavelengths and long-slit and integral field spectroscopic techniques have been hugely important in their study (Dougados *et al.* 2000). That CTT-like outflows could be launched by actively accreting BDs was first suggested when high quality spectra revealed the presence of FEL regions in the optical spectra of BDs known to be accretors (Fernández & Comerón 2001). While the FEL regions were easily identified they were considerably fainter than those detected in the spectra of CTTSs and any extension in the form of an outflow was not apparent (Fernández & Comerón 2001). Hence their origin in an outflow could not be established. As the most intense forbidden emission coincides with the critical density region it is currently challenging to directly resolve a BD outflow in FELs and this goal has not yet been achieved. Our approach to this problem has been to obtain high quality spectra using the UV-Visual Echelle Spectrometer (UVES) on the European Southern Observatory's (ESO) Very Large Telescope (VLT) and to recover the spatial offset in the region of forbidden emission using spectro-astrometry (Whelan & Garcia 2008).

To date we have observed 5 optical outflows driven by BDs (Table 1). While overall these jets are T Tauri-like in nature each object has its own unique properties. ISO-Oph 102 was the first BD confirmed to be driving a jet. Interesting follow-up observations with the SMA conducted by Phan-Bao *et al.* (2008) also detected a molecular outflow driven by this BD. The properties of the outflow agreed with what was observed in the optical by Whelan *et al.* (2005). This is only one of a small number of objects (including protostellar objects) known to be driving both an optical jet and a molecular outflow. 2MASS1207-3932 at only $24M_{JUP}$ is the lowest mass galactic object known to drive a jet. The detection of a jet was first reported by Whelan *et al.* (2007) and follow-up observations have constrained the position angle (PA) of the jet at $\sim 220°$. 2MASS1207-3932 is an intriguing object as it has a planetary mass companion which was imaged with the VLT (Chauvin *et al.* 2005). Our recent observations show the jet PA to be approximately perpendicular to the PA of the companion suggesting the presence of circumbinary structure (Whelan *et al.* 2010, in prep). This observation is relevant to the formation of the companion. In the case of LS-RCr A1 spectro-astrometry confirmed that both the line-wings of the Hα line and the forbidden emission originated in a jet. What is interesting about this object is that while only blue-shifted forbidden emission is observed both lobes of the outflow are detected in the Hα line. This is taken as evidence of a dust hole in the disk of LS-RCr A1 which suggests the onset of planet-forming processes (Whelan *et al.* 2009). Finally we will mention the jet driven by ISO-ChaI 217. We detected a bipolar outflow from this object however what was notable was the strong asymmetry between the two lobes of the outflow. This asymmetry is revealed in the relative brightness of the two lobes (red-shifted lobe is brighter), the factor of two difference in radial velocity (the red-shifted lobe is faster) and the difference in the electron density (again higher in the red lobe). Such asymmetries are common in jets from low mass protostars and the observation of a marked asymmetry at such a low mass ($< 0.1M_\odot$) supports the idea that BD outflow activity is scaled down from low mass protostellar activity. Also note that although asymmetries are unexceptional, it is uncommon for the red-shifted lobe to be the brightest as some obscuration by the accretion disk is assumed.

2. Constraining Models

Studies of BD outflows can offer constraints to both models of jet launching and BD formation. Current models describing the launching and collimation of protostellar jets predict how various jet parameters, including the ratio of the mass outflow to accretion rate, scale with mass and mass accretion rate ($\dot{M}_{out}/\dot{M}_{acc}$). Thus we have identified $\dot{M}_{out}/\dot{M}_{acc}$ as an important observational constraint and have thus begun the work of measuring this for sub-stellar outflows. To date we have estimated $\dot{M}_{out}/\dot{M}_{acc}$ for three of the known BD outflows. This was derived using two methods. Method A is based on the equation

$$\dot{M}_{out} = \mu m_H n_H \pi r_J^2 v_J \qquad (2.1)$$

where n_e, x_e and thus n_H are derived using the Bacciotti & Eisloffel (BE) technique (Bacciotti *et al.* 1999). Reasonable estimates of the jet velocity v_J and jet radius r_J come from measured radial velocities and a previously derived relation between jet width and distance. Method B is based on the observed luminosity *L(line)* of an optically thin line such as [SII] or [OI]. The mass of the flow M is derived using the equations of Hartigan *et al.* 1995. $\dot{M}_{out}$ is estimated as follows, $\dot{M}_{out} = MV_{tan}/l_{tan}$ where n_c, V_{tan} and l_{tan} are the critical density, outflow tangential velocity and the size of the aperture in the plane of the sky. Table 2 compares the values derived for $\dot{M}_{out}$ using both methods with previously known estimates of $\dot{M}_{acc}$. What is clear from these results is that can be stated at present is that the mass outflow and mass accretion rates are comparable. To further constrain their ratio the number of substellar outflows investigated must be greatly increased. It is also important to simultaneously derive the two rates and to better understand the sources of error in the various methods for estimating the mass accretion rate. See Whelan *et al.* (2009b) for further discussion of the methods used and the results. We also plan to investigate the occurrence of episodic jets in the BD regime and the frequency of molecular outflows. For low mass stars episodic jets are taken as evidence of variable accretion. If BDs jets are also found to episodic this will help to drive models of BD accretion activity. Projects to increase the sample of BD jets, further investigate $\dot{M}_{out}/\dot{M}_{acc}$ and search for BD molecular outflows are currently underway.

References

Bacciotti, F., Eislöffel, J., & Ray, T. P. 1999, *A&A*, 350, 917

Barrado y Navascués, D., Mohanty, S., & Jayawardhana, R. 2004, *ApJ*, 604, 284

Camenzind, M. 2005, Memorie della Societa Astronomica Italiana, 76, 98

Chauvin, G., Lagrange, A.-M., Dumas, C., Zuckerman, B., Mouillet, D., Song, I., Beuzit, J.-L., & Lowrance, P. 2005, *A&A*, 438, L25

Comerón, F., Fernández, M., Baraffe, I., Neuhäuser, R., & Kaas, A. A. 2003, *A&A*, 406, 1001

Dougados, C., Cabrit, S., Lavalley, C., & Ménard, F. 2000, *A&A*, 357, L61

Fernández, M. & Comerón, F. 2001, *A&A*, 380, 264

Gatti, T., Testi, L., Natta, A., Randich, S., & Muzerolle, J. 2006, *A&A*, 460, 547

Hartigan, P., Edwards, S., & Ghandour, L. 1995, *ApJ*, 452, 736

Herczeg, G. J., Cruz, K. L., & Hillenbrand, L. A. 2009, *ApJ*, 696, 1589

Lada, C. J. 1987, Star Forming Regions, 115, 1

Mohanty, S., Jayawardhana, R., & Basri, G. 2004, *ApJ*, 609, 885

Mohanty, S., Jayawardhana, R. & Basri, G. 2005, *ApJ*, 626, 498

Mohanty, S., Jayawardhana, R., Huélamo, N., & Mamajek, E. 2007, *ApJ*, 657, 1064

Muzerolle, J., Luhman, K. L., Briceño, C., Hartmann, L., & Calvet, N. 2005, *ApJ*, 625, 906

Natta, A., Testi, L., Comerón, F., Oliva, E., D'Antona, F., Baffa, C., Comoretto, G., & Gennari, S. 2002, *A&A*, 393, 597

Natta, A., Testi, L., Muzerolle, J., Randich, S., Comerón, F., & Persi, P. 2004, *A&A*, 424, 603

Natta, A., Testi, L., & Randich, S. 2006, *A&A*, 452, 245

Pascucci, I., Apai, D., Luhman, K., Henning, T., Bouwman, J., Meyer, M. R., Lahuis, F., & Natta, A. 2009, *ApJ*, 696, 143

Phan-Bao, N., *et al.* 2008, *ApJL*, 689, L141

Ray, T., Dougados, C., Bacciotti, F., Eislöffel, J., & Chrysostomou, A. 2007, Protostars and Planets V, 231

Reipurth, B. & Bally, J. 2001, *A&AR*, 39, 403

Scholz, A., & Jayawardhana, R. 2006, *ApJ*, 638, 1056

Stahler, S. W. 1983, *ApJ*, 274, 822

Whelan, E. T., Ray, T. P., Bacciotti, F., Natta, A., Testi, L., & Randich, S. 2005, *Nature*, 435, 652

Whelan, E. T., Ray, T. P., Bacciotti, F., & Jayawardhana, R. 2006, New Astronomy Review, 49, 582

Whelan, E. T., Ray, T. P., Randich, S., Bacciotti, F., Jayawardhana, R., Testi, L., Natta, A., & Mohanty, S. 2007, *ApJL*, 659, L45

Whelan, E. & Garcia, P. 2008, Lecture Notes in Physics, Berlin Springer Verlag, 742, 123

Whelan, E. T., Ray, T. P., & Bacciotti, F. 2009, *ApJL*, 691, L106

Whelan, E. T., Ray, T. P., Podio, L., Bacciotti, F., & Randich, S. 2009, *ApJ*, 706, 1054

Source	RA (J2000)	Dec (J2000)	Spectral Type	Mass (M_{JUP})
ISO-ChaI 217	11 09 52.0	-76 39 12.0	M6.2	80^1
2MASS1207-3932	12 07 33.4	-39 32 54.0	M8	24^2
DENIS-P J160603.9-205644	16 06 03.90	-20 56 44.6	M7.5	40^3
ISO-Oph 32	16 26 22.05	-24 44 37.5	M8	40^4
ISO-Oph 102	16 27 06.58	-24 41 47.9	M6	60^4
LS-RCr A1	19 01 33.7	-37 00 30.0	M6.5	$35\text{-}72^5$

Table 1. The spectral type and predicted mass of the BD candidates investigated by us to date. All sources except DENIS-P J160603.9-205644 are found to drive outflows.

Object	$\dot{M}_{out}$ ($M_\odot yr^{-1}$)	Method
LS-RCr A1		
	2.4×10^{-9}	A
	6.1×10^{-10}	B [OI]λ6300
	2.0×10^{-10}	B [SII]λ6731
ISO-Oph 102		
	$1.7\text{-}11.8 \times 10^{-10}$	B [SII]λ6731
	1.4×10^{-9}	CO molecular outflow^6
ISO-ChaI 217 Red flow		
	3.1×10^{-10}	B [SII]λ6731
ISO-ChaI 217 Blue flow		
	1.8×10^{-10}	B [SII]λ6731

Object	$\dot{M}_{acc}$ ($M_\odot yr^{-1}$)	Method
LS-RCr A1		
	2.8×10^{-10}	CaII(λ8542)^7
	$10^{-10}\text{-}10^{-9}$	Optical Veiling^5
	10^{-9}	CaII(λ8662)^8
	10^{-10}	Hα 10% width^9
ISO-Oph 102		
	10^{-9}	Hα 10% width^{10}
	4.3×10^{-10}	J and K band spectra^{11}
ISO-ChaI 217		
	1.0×10^{-10}	Hα emission^1

Table 2. Measurements of $\dot{M}_{out}$ and $\dot{M}_{acc}$ for the objects studied to date. For the BDs in our sample $\dot{M}_{out}$ is comparable to $\dot{M}_{acc}$.

Table References: 1=Muzerolle *et al.* 2005, 2=Mohanty *et al.* 2007, 3=Mohanty *et al.* 2004, 4=Natta *et al.* 2002 and 5=Barrado y Navascués *et al.* 2004, 6=Phan-Bao *et al.* (2008), 7=Comerón *et al.* 2003, 3=Barrado y Navascués *et al.* 2004, 8=Mohanty *et al.* 2005, 9=Scholz & Jayawardhana 2006, 10=Natta *et al.* 2004, 11=Natta *et al.* 2006, 8=Muzerolle *et al.* 2005

Jets at all Scales
Proceedings IAU Symposium No. 275, 2011
G. E. Romero, R. A. Sunyaev & T. Belloni, eds.

© International Astronomical Union 2011
doi:10.1017/S1743921310016479

Side-entrainment in a jet embedded in a sidewind

D. López-Cámara and A. C. Raga

Instituto de Ciencias Nucleares, Universidad Nacional Autónoma de México, Ap. 70-543,
04510 D.F., México
email: diego.lopez@nucleares.unam.mx

Abstract. In this study, we present the results from 3D simulations in which a side-streaming motion pushes the post-bow shock into direct contact with the jet beam. This is a possible mechanism for modeling well collimated molecular jets as an atomic/ionic flow which entrains molecules initially present only in the surrounding environment.

Keywords. hydrodynamics – stars: formation – ISM: jets and outflows

1. Introduction

The main problem when trying to incorporate molecular, environmental material (EM), into a collimated jet is that the leading head of the jet pushes away the EM. The possibility that we study here is that a side-streaming environment pushes the bow shock wing against the jet beam.

2. Results

The simulations were carried out with the "Yguazú-a" code (Raga *et al.* 2000), solving the 3D gasdynamic equations together with a continuity/rate equation for neutral H. We integrate an equation for a normalized passive scalar with which we distinguish between the ambient and jet medium. The jet was injected at $x = 0$, with a velocity parallel to the x-axis. An initially neutral, top-hat jet of velocity v_j, density n_j, radius $r_j = 10^{15}$ cm and temperature $T_j = 10^3$ K moves into an initially uniform, neutral environment with a density $n_a = 200$ cm^{-3}, temperature $T_a = 10$ K and sidestreaming vertical velocity v_a (set from $y = 0$). We ran various simulations with different values of v_j (150 km s^{-1} for models a1;b1;c1;a3;b3;c3; and 300 km s^{-1} all else), n_j (1000 cm^{-3} for a1;b1;c1;a2;b2;c2, and 5000 cm^{-3} all else), and v_a (2 km s^{-1} for a1;a2;a3, 5 km s^{-1} for b1; b2; b3, and 10 km s^{-1} for the rest).

As an example of the flows resulting from our simulations, in Figure 1 we show xy-cuts showing the time-evolution of the mid-plane density stratification obtained from model a1 at various times, and for two other models at the same time. It is clear that there is a side-to-side asymmetry. This asymmetry is seen as a distortion of the leading bow shock due to the presence of the sidewind, and as a penetration of environmental material to regions close to the jet beam.

Since the total entrained material is a composition of material which was entrained directly through the bow shock, and also due to the sidewind. So, in order to recognize the material which was entrained due to the sidewind we computed the of the entrained material with $> v_j/2$. The spatial distribution of the such entrained material, as well as its correspondent mass rate ($\dot{M}_{AM}$), are shown in Figure 2. The side-entrainment

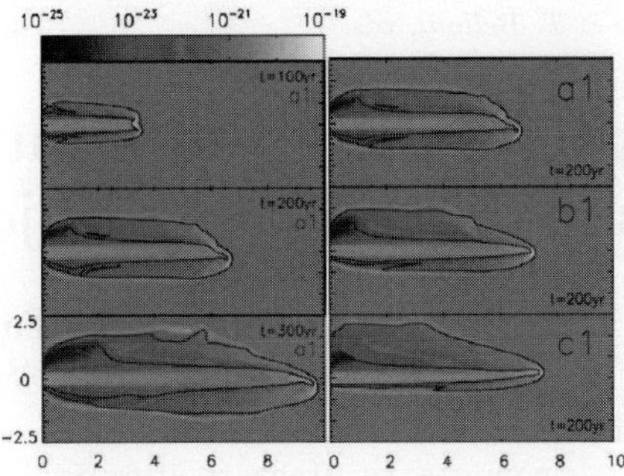

Figure 1. Density stratifications (g cm^{-3}). The axes are labeled in units of 10^{16} cm.

results in $\dot{M}_{AM} \sim 5 \times 10^{14}$ g s^{-1}, corresponding to ~ 0.5 % of the mass loss rate of the jet. If the molecular, environmental gas is not dissociated during the process of side-entrainment, this would result in a molecular fraction of ~ 0.5 % within the jet beam, which would result in molecular column densities high enough to produce observable molecular emission.

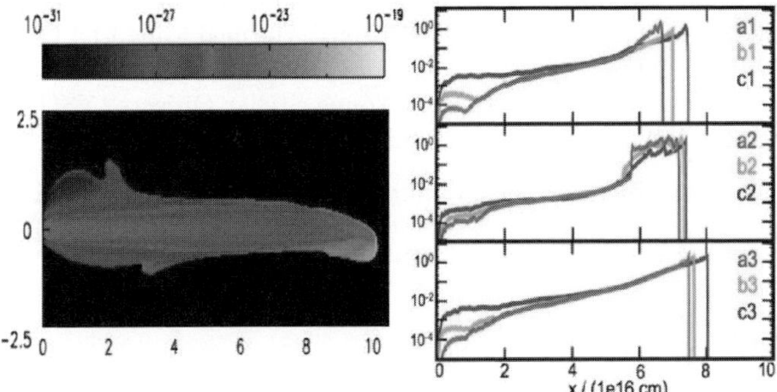

Figure 2. Left: Density stratifications (g cm^{-3}) of the environmental mass fraction for model a1. Right: $\dot{M}_j$ and $\dot{M}_{AM}$ for all the models.

3. Conclusions

We present 3D numerical simulations of a jet in a sidewind. We find that for many parameter combinations the sidewind pushes the post-bow shock shell into direct contact with the jet beam. In this region of contact, side-entrainment of environmental material into the jet beam does take place. Our simulations do not include the chemistry of the entrained material, so that we are not able to see whether or not the molecules in the side-entrained material actually survive the entrainment process. However, the fact that the side-entrained material has been shocked by the slow-moving far bow shock wings, and that the region of contact between the shocked environment and the jet remains cool ($\sim 10^3$ K) indicates that molecules indeed might be entrained into the jet beam without being dissociated. We show that a side-streaming environment will push the post-bow shock shell into direct contact with the jet beam.

References

Raga, A. C., Navarro-González, R., & Villagrán-Muniz, M. 2000, RMxAA, 36, 67

Jets at all scales
Proceedings IAU Symposium No. 275, 2011
G. E. Romero, R. A. Sunyaev & T. Belloni, eds.

© International Astronomical Union 2011
doi:10.1017/S1743921310016480

Astrophysical outflows simulated by laser-driven plasma jets

C. Michaut[1], C. D. Gregory[2], B. Loupias[3], E. Falize[1,3], A. Ravasio[2], A. Dizière[2], T. Vinci[2], M. Koenig[2] and S. Bouquet[1,3]

[1]LUTH, Observatoire de Paris, CNRS, Université Paris-Diderot, 92190 Meudon, France
email: `claire.michaut@obspm.fr`

[2]LULI, CNRS, CEA, Université Paris VI, Ecole Polytechnique, 91128 Palaiseau, France

[3]CEA-DAM-DIF, F-91297 Arpajon, France

Abstract. Within the framework of laboratory astrophysics, we form a qualified multidisciplinary group in radiative hydrodynamics. Since 10 years, we have developed laboratory experiments as radiative shocks and plasma jets in connection to astrophysics. Such laboratory experiments provide a unique opportunity to validate models and numerical schemes introduced in radiative hydrodynamics codes. Here we summarize our experimental researches about plasma jets. Laboratory astrophysical experiments have been performed using LULI2000 (France), VULCAN (UK) and GEKKO XII (Japan) intense lasers. The goal of these experiments is to investigate some of the complex features of jets from Young Stellar Objects (YSO), and in particular its interaction with the interstellar medium (ISM).

Keywords. jet, laboratory astrophysics, radiative hydrodynamics, laser-plasma interaction

1. Introduction

For studying YSO jet propagation, experimental research related to laboratory astrophysics using an intense laser were performed. The relevance of these experiments to astrophysics is measured through scaling laws (Falize *et al.* 2008, Bouquet *et al.* 2010). They ensure the similarity between astrophysics and laboratory provided that the dimensionless numbers are equivalent. These non-exhaustive numbers are Mach M, Reynolds Re, Peclet Pe and Mihalas R numbers, the cooling coefficient χ, the aspect ratio and the jet-to-ambient density ratio η. Some of these numbers are estimated at the time of the experiment design while others are measured during the experiment. The range of values for these parameters represent upper and lower limits given by the accuracy of the experimental measurements which are thus crucial. An important number of temporally-resolved optical diagnostics are implemented and in some experiments, X-ray radiography and proton shadowgraphy are added. X-rays are able to diagnose the dense plasma part, but the more tenuous regions are below the diagnostic sensitivity. In contrast, the optical probe is unable to penetrate high-density jet cores, but is well suited to imaging relatively low plasma densities. Proton radiography is able to image small-scale density structures in the dense parts with high resolution. The capability of fielding these diagnostics on a single shot is beneficial, and allows a more complete understanding of the jet dynamics. We designed different targets: metallic V-foils, foam cones and conical metallic shells. A jet-like structure is observed and its time evolution is studied by varying either the foam density or the metal nature. Nominally identical jets were propagated into vacuum. The interaction with ambient medium is realized using a gas outlet nozzle ensuring a ratio η close to astrophysical conditions ($2 \leqslant \eta \leqslant 50$).

2. Summarized Experimental Results

At the beginning, a serie of experiments was driven by using V-foil targets irradiated by two laser beams. The thin aluminum foils explode and collide, driving a narrow plasma flow. The major results are presented in Gregory *et al.* (2008), (2009). The jet propagating in gas has a reduced width when compared to the vacuum case, and shock structures appears at the leading edge of the flow. This is consistent with numerical simulations indicating that an ambient medium should increase the collimation of laser produced jets. A velocity was recorded around 300 km/s and a temperature of 10 eV was measured for these flows. Dimensionless numbers are satisfied, but as these targets have no cylindrical geometry, the jet is not cylindrical. These preliminary experiments allowed to determine the optimal irradiated angle to obtain a good collimation and to design cone targets.

Numerical simulations (Vinci *et al.* 2008) allowing to understand the collimated plasma generation have permit to design an original target using a cone filled with foam. In this case the larger cone side is irradiated and a collimated plasma is ejected by the smaller side. The full experimental set-up and results are readable in Loupias *et al.* (2007a), (2007b), (2009a), Gregory *et al.* (2010a). The jet velocity ($\sim [90-190]$ km/s) into vacuum accords with analytical predictions (Falize *et al.* 2008b), and slows down of 10% in ambient medium due to the bow shock creation. The measured envelope temperature is 2 eV since the estimated central one is 15 eV. The cocoon is clearly shown comparing visible and X-ray diagnostics, but it has not a large aspect ratio instead of the high-collimated dense core. In addition, the proton radiography shows a perturbed shape of the interface jet/ambient medium which is not yet explained. Physical reasons lead to the fact this interface is Rayleigh-Taylor unstable. As the foam is a hydrogen/carbon mixture, it gives a jet more scalable to astrophysical outflow with dimensionless numbers displaying good similarities except for χ showing the radiation escaping is too low.

An third target type is used to investigate the radiative cooling effects on the jet morphology. Jets are created through laser irradiation of thin conical shells of either gold, copper or aluminum. In choosing different atomic numbers, the degree of radiative losses is altered and plays a role on collimation. The data in Gregory *et al.* (2010b) show the importance of the conical geometry creating high-collimated outflow. The results suggest the jet collimation is increased for higher-atomic number targets and are consistent with computer simulations showing a radial collapse. Conical shells give the better collimated jet but their temperature is not very well known, therefore the radiative cooling can not be estimated.

Roughly these laboratory experiments are in a regime scalable to YSO jets.

References

Bouquet, S., Falize, E., Michaut, C., Gregory, C. D., and 3 coauthors, 2010, *HEDP* doi:10.1016/j.hedp.2010.03.001

Falize, E., Bouquet, S., Michaut, C., 2008a, *JPCS*, 112, 042015

Falize, E., Bouquet, S., Michaut, C., 2009, *Ap&SS*, 322, 107

Gregory, C. D., Loupias, B., Waugh, J., Barroso, P., and 20 coauthors, 2008, *PPCF*, 50, 124039

Gregory, C. D., Howe, J., Loupias, B., Myers, S., and 9 coauthors, 2009, *Ap&SS*, 322, 37

Gregory, C. D., Loupias, B., Waugh, J., Dono, S., and 9 coauthors, 2010a, *PoP*, 17, 052708

Gregory, C. D., Dizière, A., Aoki, H., Besio, M., and 13 coauthors, 2010b, submitted in *Ap&SS*

Loupias, B., Koenig, M., Falize, E., Bouquet, and 10 coauthors, 2007a, *PRL* 99, 265001

Loupias, B., Falize, E., Koenig, M., Bouquet, S., and 9 coauthors, 2007b, *Ap&SS*, 307, 103

Loupias B., Falize E., Gregory C. D., Vinci, T., and 14 coauthors, 2009a, *PPCF*, 51, 124027

Loupias, B., Gregory, C. D., Falize, E., Waugh and 13 coauthors, 2009b, *Ap&SS*, 322, 25

Vinci, T., Loupias B., Koenig, M., Benuzzi-Mounaix, and 7 coauthors, 2008, *JPCF*, 112, 042012

Jets at all Scales
Proceedings IAU Symposium No. 275, 2011
G. E. Romero, R. A. Sunyaev & T. Belloni, eds.

© International Astronomical Union 2011
doi:10.1017/S1743921310016492

Non thermal emission from T Tauri stars

María V. del Valle and Gustavo E. Romero

Instituto Argentino de Radiastronomía (IAR),
CCT La Plata (CONICET), C.C.5, 1894 Villa Elisa, Buenos Aires, Argentina
Facultad de Ciencias Astronmicas y Geofsicas,
Paseo del Bosque s/n, 1900 La Plata, Buenos Aires, Argentina
email: maria@iar-conicet.gov.ar

Abstract. T Tauri stars are low mass, pre-main sequence stars. These objects are surrounded by an accretion disk and present strong magnetic activity. T Tauri stars are copious emitters of X-ray emission which belong to powerful magnetic reconnection events. Strong magnetospheric shocks are likely outcome of massive reconnection. Such shocks can accelerate particles up to relativistic energies through Fermi mechanism. We present a model for the high-energy radiation produced in the environment of T Tauri stars. We aim at determining whether this emission is detectable. If so, the T Tauri stars should be very nearby.

Keywords. Radiation mechanisms: nonthermal, stars: pre–main-sequence, gamma rays: theory

1. Introduction

T Tauri stars are low-mass stars in their early stages of evolution. They are surrounded by an accretion disk and are actively accreting material from the disk (e.g. Feigelson & Montmerle 1999).

Variable thermal keV X-ray emission is detected from T Tauri stars. This emission comes from a high density plasma at a typical temperature of $\sim 10^7$. X-rays flares are considered as upscaled versions of solar flares, related to magnetic reconnection. Several works have been done on particle acceleration in magnetic reconnection (e.g. Zenitani & Hoshino 2001, de Gouveia Dal Pino *et al.* 2010).

Non-thermal radio emission from T Tauri stars have been detected (e.g. Ray *et al.* 1997, Loinard *et al.* 2008). Then, an acceleration mechanism for non-thermal particles must operate in these systems.

2. The model

We consider that a power-law population of relativistic particles (electrons and protons) is injected in the magnetosphere. These particles interact with the magnetic field, with the various radiation fields, and with the magnetosphere plasma. The values adopted for the different parameters in our model are: R_m (magnetosphere radius) 0.1 AU, a (hadron-to-lepton energy ratio) 100, q_rel (content of relativistic particles) 10^{-4}, α (particle injection index) 2, v_w (wind velocity) 2×10^8 cm s^{-1}, B (magnetic field) 5×10^2 G, n (maximum magnetospheric density) 10^{11} cm^{-3} and L_x (X-ray luminosity) 10^{30} erg s^{-1}. The available power is estimated as $L = B^2/8\pi Ac$ where A is the magnetosphere area.

The relativistic electrons lose energy mainly through synchrotron emission, inverse Compton (IC) scattering with the star radiation field and the X-ray radiation field, and through relativistic bremsstrahlung with the magnetosphere plasma. The relativistic protons lose energy mainly through synchrotron emission and $p - p$ inelastic collisions with ambient matter. The particles can escape from the acceleration region due to wind convection. The maximum energy for both types of particles is obtained equating the cooling rates with the acceleration rate $t_\mathrm{acc}^{-1} = \eta ecB/E$.

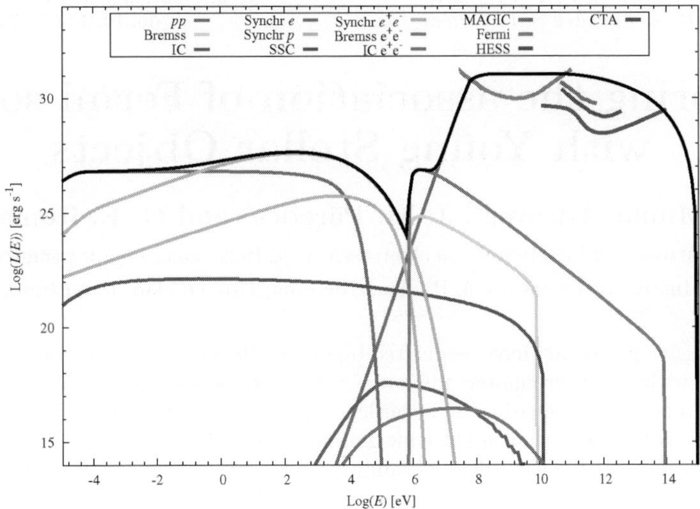

Figure 1. Computed SED for a source at d $\sim$150 pc and the sensitivity curves for CTA, Fermi, MAGIC, and HESS.

The particles steady state distribution is calculated using the standard transport equation. We also consider the population of secondary $e^{\pm}$ pairs injected by charge pion decay (e.g. Orellana *et al.* 2007).

We calculate the processes of interaction of the relativistic particles with the fields in the magnetosphere (e.g. Vila & Aharonian 2009). The particles will collide with the accretion plasma columns which occupy a not well-established volume of the magnetosphere. We consider a small filling factor $f \sim 10^{-4}$. We also calculate the opacity from internal and external photon-photon absorption (e.g. Gould & Schréder 1967).

3. Results and conclusions

Figure 1 shows the computed spectral energy distribution (SED) and the sensitivity curves from gamma-ray detectors. We consider a source at a distance $d \sim 150$ pc, similar to that of the nearest T Tauri stars.

Some of the existing gamma-ray instruments should be able of detecting T Tauri stars with the characteristic adopted in this work. If so, these sources might be the closest gamma-ray sources ever observed.

References

de Gouveia Dal Pino, E. M., Piovezan, P. P., & Kadowaki, L. H. S. 2010, *A&A*, 518, id. A5
Feigelson, E. D. & Montmerle, T. 1999, *Annu. Rev. A&A*, 37, 363
Gould, R. J. & Schréreder, G. P. 1967, *Pys. Rev.*, 155, 1404
Loinard, L. *et al.* 2008, *ApJ*, 675, L29
Ray, T. P. *et al.* 1997, *Nature*, 385, 415
Orellana, M. *et al.* 2007, *A&A*, 476, 9
Vila, G. S. & Aharonian, F. A. 2009, in *Compact Objects and their Emission*, Romero, G. E. & Benaglia, P. (eds.), Paideia, La Plata, P. 1
Zenitani, S. & Hoshino, M. 2001, *ApJ*, 562, L63

IAU Symposium 275: Jets at all Scales
Proceedings IAU Symposium No. 275, 2011
G. E. Romero, R. A. Sunyaev & T. Belloni, eds.

© International Astronomical Union 2011
doi:10.1017/S1743921310016509

Exploring the association of Fermi sources with Young Stellar Objects

P. Munar-Adrover[1], **J. M. Paredes**[1] **and G. E. Romero**[2]

[1] Dept. d'Astronomia i Meteorologia, Universitat de Barcelona. email: `pmunar@am.ub.es`

[2] Instituto Argentino de Radioastronomía, Universidad de La Plata

Abstract. Massive protostars have associated bipolar outflows which can produce strong shocks when interact with the surrounding medium. Some theoretical models predict that particle acceleration at relativistic velocities can occur leading to gamma ray emission. In order to identify young stellar objects (YSO) that might emit gamma rays, we have crossed the Fermi First Year Catalog with catalogs of known YSOs, obtaining a set of candidates by spatial correlation. We have conducted Montecarlo simulations to find the probability of chance coincidence. Our results indicate that $\sim$70% of the candidates should be gamma-ray sources with a confidence of $\sim$5σ.

Keywords. stars: early-type, gamma rays: observations, ISM: jets and outflows

1. Introduction

Massive YSOs show collimated outflows and thermal radiation has been detected up to distances of $10^{16} - 10^{18}$ cm from the central star. These are strongly supersonic jets and in some cases, non-thermal radio lobes have been detected at distances of $Z_j \sim 1$pc (Garay *et al.* 2003). These radio lobes are probably generated by strong terminal shocks of the jets, which also ionize the shocked material. The possibility of YSOs to be γ-ray emitters has already been discussed in Araudo *et al.* (2007) and Bosch-Ramon *et al.* (2010). The action of the jet head on the external medium leads two shocks, the bow shock and the reverse shock. The observed non-thermal radio emission would be generated at the shocks where the particles are accelerated.

2. Numerical simulations

With the aim of finding the positional coincidences between Fermi sources and Young Stellar Objects we have crossed the First Fermi-LAT Catalog (1FGL) (Abdo *et al.* 2010) and a catalog of MYSOs from the RMS Survey (Urquhart *et al.* 2009) by using a computer code that determines the angular distance between two points in the sky, taking into account the positional uncertainties in each of them. We ran the code with the 1392 sources of the 1FGL that have not been firmly identified, and the 556 sources identified as YSOs in the RMS survey. In order to estimate the statistical significance of these coincidences, we have simulated a large number of sets of Fermi sources. Specifically, we have simulated 1500 populations of 1392 Fermi sources, through rotations on the celestial sphere, displacing a source with original galactic coordinates (l,b) to a new position (l_0,b_0) as done in Romero *et al.* (1999). The separation between the Fermi source and the YSO is calculated in each case using the statistical parameter R (Allington-Smith *et al.* 1982)

$$R = \sqrt{\frac{(\Delta\alpha\cos\delta)^2}{\sigma_{i_\alpha}^2 + \sigma_{j_\alpha}^2} + \frac{\Delta\delta^2}{\sigma_{i_\delta}^2 + \sigma_{j_\delta}^2}} \qquad (2.1)$$

where $\sigma_{\alpha_i}, \sigma_{\delta_i}$ is the uncertainty in the position of the source, and i and j are Fermi and RMS sources, respectively.

3. Results

We have found 13 Fermi sources being positionally coincident with 24 YSOs (see Table 1), and 8 of these Fermi sources have not any proposed counterpart (like SNR, PWN, pulsar, etc.) to the gamma emission. In Table 2 we present the statistical results obtained from simulations with a random distribution in galactic longitude. As can be seen, the estimated probability of a pure chance association is as low as 2.0×10^{-8} for the 2°-binning simulations (2.2×10^{-6} for the 1°-binning), and there is a correlation at $\sim 5\sigma$. When we considered the restrictions in both l and b (see Table 3), the chance probability raised, but still shows a quite negligible values ($\sim 10^{-6}$).

Fermi Name	95% Semi Major Axis	Spectral Index Γ $F \propto E^{-\Gamma}$	Energy Flux erg cm^{-2}s^{-1}	MSX Name	Freq. GHz	Int.Flux mJy	Angular dist. °	R
1FGL J0541.1+3542	0.1397	2.41±0.13	$1.61 \times 10^{-11} \pm 4.9 \times 10^{-12}$	G173.6328+02.8064	5	< 0.7	0.12	0.846
				G173.6339+02.8218	5	< 0.7	0.13	0.903
				G173.6882+02.7222	5	< 0.8	0.05	0.364
				G173.7215+02.6924	5	< 0.8	0.05	0.386
1FGL J0647.3+0031	0.2150	2.41±0.11	$1.89 \times 10^{-11} \pm 5.4 \times 10^{-12}$	G212.0641−00.7395	5	< 0.9	0.10	0.467
1FGL J1256.9−6337	0.1955	2.26±0.12	$4.97 \times 10^{-11} \pm 1.1 \times 10^{-11}$	G303.5990−00.6524	4.8	< 0.6	0.12	0.589
1FGL J1315.0−6235	0.1860	2.31±0.12	$6.86 \times 10^{-11} \pm 0.0$	G305.4840+00.2248	8.6	< 2.1	0.18	0.958
				G305.5610+00.0124	8.6	7.2	0.16	0.869
1FGL J1651.5−4602	0.2258	2.21±0.07	$1.39 \times 10^{-10} \pm 3.4 \times 10^{-11}$	G339.8838−01.2588	8.6	2.6	0.14	0.638
1FGL J1702.4−4147	0.0800	2.39±0.07	$8.7 \times 10^{-11} \pm 2.0 \times 10^{-11}$	G344.4257+00.0451B	8.6	< 13.7	0.05	0.631
				G344.4257+00.0451C	8.6	< 13.7	0.05	0.638
1FGL J1846.8−0233	0.1262	2.21±0.06	$9.3 \times 10^{-11} \pm 2.3 \times 10^{-11}$	G030.1981−00.1691	5	< 0.8	0.08	0.646
1FGL J1848.1−0145	0.0859	2.23±0.04	$9.5 \times 10^{-11} \pm 3.2 \times 10^{-11}$	G030.9726−00.1410	5	< 0.7	0.07	0.763
				G030.9959−00.0771	5	< 1.2	0.00	0.036
1FGL J1853.1+0032	0.5207	2.18±0.07	$5.7 \times 10^{-11} \pm 1.7 \times 10^{-11}$	G032.8205−00.3300	5	< 0.7	0.34	0.658
				G033.3891+00.1989	5	< 1.0	0.40	0.768
				G033.3933+00.0100	5	< 1.1	0.26	0.496
				G034.0126−00.2832	5	< 0.7	0.43	0.830
				G034.0500−00.2977	5	< 0.8	0.47	0.908
1FGL J1925.0+1720	0.1443	2.28±0.12	$2.37 \times 10^{-11} \pm 1.02 \times 10^{-11}$	G052.2078+00.6890	5	< 0.7	0.08	0.533
				G052.2025+00.7217A	5	< 0.8	0.07	0.453
1FGL J1943.4+2340	0.1118	2.23±0.11	$2.62 \times 10^{-11} \pm 6.8 \times 10^{-12}$	G059.7831+00.0648	8.6	1.0	0.08	0.743
1FGL J2032.8+3928	0.2507	2.59±0.07	$5.1 \times 10^{-11} \pm 1.4 \times 10^{-11}$	G078.4705−00.1830	5	< 1.2	0.23	0.908
1FGL J2040.0+4157	0.1970	2.66±0.06	$7.9 \times 10^{-11} \pm 1.2 \times 10^{-11}$	G081.5168+00.1926	5	< 1.1	0.04	0.195

Table 1. Positional coincidence between Fermi sources and MYSOs.

Actual coincidence	Simulated 1°-bin	Probability 1°	Simulated 2°-bin	Probability 2°
1 3	3.9± 1.9	2.2×10^{-6}	3.3±1.7	2.0×10^{-8}

Table 2. Statistical results obtained from simulations with a random distribution in galactic longitude.

Actual coincidence	Simulated 20°-bin	Probability 20°	Simulated 40°-bin	Probability 40°
13	5.17± 2.07	1.5×10^{-4}	3.8± 1.9	1.6×10^{-6}

Table 3. Statistical results obtained from simulations constrained both in galactic latitude and galactic longitude.

Acknowledgements. J.M.P. and G.E.R. acknowledge support by DGI of the Spanish Ministerio de Educación y Ciencia (MEC) under grant AYA2007-68034-C03-01 and FEDER funds. In addition, this work has been supported by the Consejería de Innovación, Ciencia y Empresa (CICE) of Junta de Andalucía as research group FQM-322 and excellence grant FQM-5418.

References

Abdo A. A. *et al.*, 2010, *ApJ Supplement Series*, 188, 405-436
Allington-Smith J. R., Perryman M. A. C., Longair M. S. *et al.*, 1982, *MNRAS*, 201, 331-344
Araudo A. T., Romero G. E., Bosch−Ramon V., and Paredes J. M., 2007, *A&A*, 476, 1289-1295
Bosch-Ramon V., Romero G. E., Araudo A. T. and Paredes J. M., 2010, *A&A*, 511, A8
Garay G., Brooks K. J., Mardones D., and Norris R. P., 2003, *ApJ*, 587, 739-747
Romero G. E., Benaglia P. and Torres D. F., 1999, *A&A*, 348, 868-876
Urquhart J. S., Horae M. G., Purcell C. R. *et al.*, 2009, *A&A*, 501, 539-551

Jets at all Scales
Proceedings IAU Symposium No. 275, 2011
G. E. Romero, R. A. Sunyaev & T. Belloni, eds.

© International Astronomical Union 2011
doi:10.1017/S1743921310016510

Modelling of T Tauri jets with low mass accretion rate

N. Globus[1], C. Sauty[1], V. Cayatte[1], Z. Meliani[1,2], J. J. G. Lima[3,4], K. Tsinganos[5] and C. Michaut[1]

[1]LUTh, Observatoire de Paris, F-92190 Meudon, France
email: noemie.globus@obspm.fr
[2]Centrum voor Plasma Astrofysica, Leuven, Belgium
[3]Centro de Astrofisica da Universidade do Porto, Portugal
[4]Departamento de Fisica e Astronomia da Faculdade de Ciencias, Porto, Portugal
[5]IASA and Section of Astrophysics, Astronomy & Mechanics, Department of Physics,
University of Athens, Greece

Abstract. We show that low mass accreting T Tauri stars may have a strong stellar jet component which can effectively brake the star to the observed rotation speed. By means of meridional self similarity, we construct semi analytical solutions describing the complete dynamics and topology of the stellar component of the jet emerging from the corona of the star. We show two typical solutions with the same mass loss rate but different magnetic lever arms and jet radius, corresponding to differente phases of T Tauri star activity.

Keywords. MHD, stars: winds, outflows, stars: mass loss, stars: rotation

1. Meridionally self-similar model

The basic equations governing plasma outflows in the framework of ideal MHD for steady, axisymmetric flows are the momentum, mass and magnetic flux conservation, the frozen-in law for infinite conductivity and the first law of thermodynamics. By assuming self-similarity in the meridional (-θ) direction, we can obtain solutions of the MHD equations. The complete description of the model can be found in Sauty & Tsinganos (1994). This model presents a double component jet structure for cylindrically collimated solutions. One part of the jet comes from the star itself while the other comes from the inner boundary of the disk which is connected with the stellar magnetosphere.

Disk winds can remove most of the angular momentum of the accreting plasma. However, they cannot describe the inner stellar jet and explain the slowing down of the star itself. Moreover, more evolved T Tauri stars like RY Tau seem to have weaker jets with accretion rates of the order of only a few $10^{-8}\,M_\odot/\mathrm{yr}$. Our aim is to examine the contribution of the stellar component to the overall jet and how efficiently it brakes the star.

2. Constraining the parameters from the observations

We have constructed two different solutions with typical values close to the ones observed for RY Tau and other low mass accreting objects for which we have assumed the same initial mass loss rate and stellar parameters. In both cases, the jet radius is defined as the last flux tube connected to the star. The first solution (*Solution 1*) has a large lever arm, corresponding to our initial guess, and a large opening angle for the jet. This should be close to the observed values for CTTS having a jet, as RY Tau in 2003. This is a solution where the asymptotic structure is collimated only by the magnetic hoop stress. This type of solution gives the largest possible angle for the jet. The star braking

408

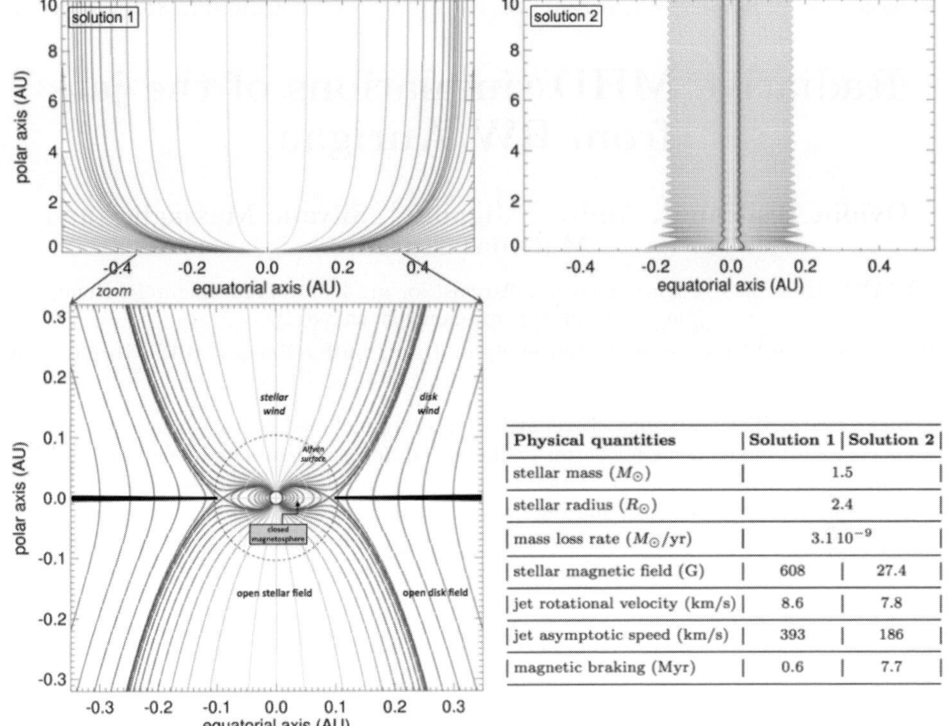

Physical quantities	Solution 1	Solution 2
stellar mass ($M_\odot$)	1.5	
stellar radius ($R_\odot$)	2.4	
mass loss rate ($M_\odot$/yr)	$3.1\,10^{-9}$	
stellar magnetic field (G)	608	27.4
jet rotational velocity (km/s)	8.6	7.8
jet asymptotic speed (km/s)	393	186
magnetic braking (Myr)	0.6	7.7

Figure 1. Topologies of the non-oscillating collimated solution (left panel) and the recollimating solution (right panel). The pink part represents the streamlines connected to the star, i.e. the stellar component of the jet solution. The first solution with a large lever arm and jet radius effectively brakes the star and can be applied to CTTS jets such as the one of RY TAU. The second solution with a smaller lever arm and a very narrow jet may explain why some CTTS do not have visible jets. RY Tau for instance seems to have different phases depending probably on the activity of the star.

time in this case is around $0.6\,10^6$ yr, within the typical lifetime for a CTTS ($\approx$1 Myr). The initial challenge with modelling RY Tau jet is the presence of a UV shock. We know from previous studies (Pelletier & Pudritz, 1992) that shocks may occur in our analytical solutions at the recollimating point where the jet starts oscillating. Thus, we have looked for a solution that starts recollimating at the UV shock observed by Gomez & Verdugo (2001). By slightly changing the parameters we were able to get a completely different solution (*Solution 2*). The lever arm is reduced as well as the opening angle. The solution exhibits strong recollimation and oscillations. It may either reproduce CTTS such as RY Tau in 2001 when the jet was not visible, but a UV shock was detected close to the star possibly due to the recollimation. In this case the jet is not sufficient to brake the star in the lifetime of a CTTS. It may also correspond to a WTTS (Weak-line T Tauri Star) where jets are not detected. The braking time in this case is around $7.7\,10^6$ yr which is roughly the typical lifetime for a WTTS ($\approx$10 Myr).

References

Sauty C. & Tsinganos K. 1994, *A&A*, 287, 893
Pelletier G. & Pudritz R. E. 1992, *ApJ*, 394, 117
Gomez de Castro A. I. & Verdugo E. 2001, *ApJ*, 548, 976

Jets at all Scales
Proceedings IAU Symposium No. 275, 2011
G. E. Romero, R. A. Sunyaev & T. Belloni, eds.

© International Astronomical Union 2011
doi:10.1017/S1743921310016522

Radiative MHD simulations of the jets from RW Aurigae

Ovidiu Teşileanu[1], Andrea Mignone[2], Silvano Massaglia[2] and
Matthias Stute[2]

[1]RCAPA, University of Bucharest, Str. Atomistilor nr. 405, 077125 Magurele, Romania
email: ovidiu.tesileanu@ph.unito.it

[2]Dipartimento di Fisica Generale, University of Turin, via P. Giuria 1, 10125 Torino, Italy

Abstract. The MHD simulations of stellar jets recently included complex models of radiative
emission computation, allowing for better predictions in terms of emission line ratios. Employing
also Adaptive Mesh Refinement, the large-scale propagation of jets could be followed. The simulation of multiple shockwaves originating in perturbations close to the jet origin and travelling
along the jet beam allows for the construction of synthetic emission maps at various wavelengths,
to be directly compared to observations. We apply this procedure for the jets originating from
RW Aurigae.

Keywords. atomic processes, MHD, plasmas, radiation mechanisms: thermal, shock waves,
stars: winds, outflows

1. Framework

During star formation there is a high probability for the development of jets. Simulating
the radiative processes in YSO jets will provide a valuable tool for model discrimination.
Employing methods that balance the required accuracy with efficiency for the MHD
simulations of jet propagation and emission mechanisms, we were able to construct synthetic emission maps for these objects, to be compared with observations (mainly from
the Hubble Space Telescope).

The MHD simulation code we use - PLUTO, is developed and maintained at the
Turin University by A. Mignone (http://plutocode.to.astro.it, see Mignone *et al.* 2007).
The newly developed cooling module (MINEq) permits a detailed treatment of non-
equilibrium cooling losses and is much more accurate than the previously employed
models (See Teşileanu *et al.* (2008) for details).

1D simulations were used to explore the parameter space that defines the jet characteristics – mean flow velocity, density, velocity and density perturbation producing traveling
shocks in the jet. Emission line ratios are similar in the 1D and 2D simulations, for the
same set of parameters (see Table 1).

Table 1. Comparison of emission line ratios for the same parameters, with MINEq cooling.

Simulation	Out	[OI]/[NII]	[SII]/[OI]	Log(OI/NII)	Log(SII/OI)
MINEq 1D	20	6.099	0.954	0.785	-0.021
MINEq 1D	40	11.870	0.899	1.074	-0.046
MINEq 2D	20	12.275	1.055	1.089	0.023
MINEq 2D	40	16.158	1.032	1.208	0.014

2. Numerical Setup

The very high resolution required by the necessity to resolve the post-shock zone where the gradients are steep, as much as the distance scales of the propagation, leaded to the idea of simulating the propagating shock in a reference frame co-moving with the average velocity of the jet material. This was done previously in 1D as a first approximation to simulate highly collimated jets. For the case without magnetic fields the setup was designed and implemented, the tests showing consistency with simulations done in the reference frame of rest. Adaptive Mesh Refinement was also used.

3. 2D Radiative MHD Simulations

The observational data for RW Aurigae jets is from Woitas *et al.* (2002), reconstructed from STIS spectra. Estimating from observations the time employed by the knots to arrive from the base of the jet to their current positions, the relevant output maps from the simulation can be selected. One can apply this procedure for the several knots in the jet, considering their estimated age and various models from the numerical simulations. We have thus begun to generate an array of such dynamical shock models by varying the basic parameters - the jet velocity and density, and the perturbation amplitude.

For the case of best agreement presented in Fig.1, the density was $n_0 = 5 \times 10^4 cm^{-3}$, the jet base velocity $110 km \cdot s^{-1}$ and the amplitude of the initial perturbation in velocity was $40 km \cdot s^{-1}$.

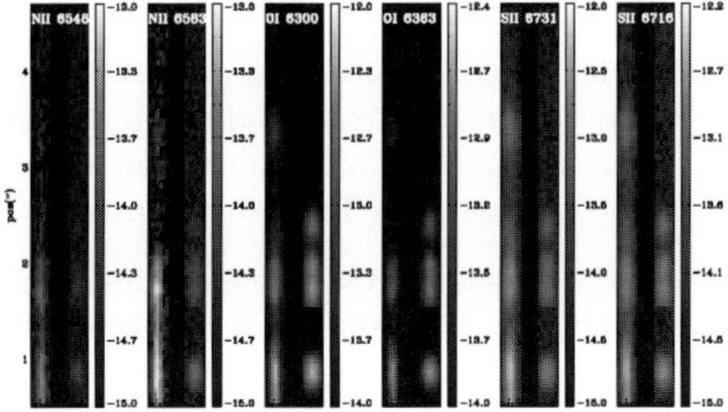

Figure 1. Emission maps in the emission lines of interest. In each frame, on the left is the observed and on the right the synthetic jets. Log scale, in $ergs \cdot s^{-1} \cdot arcsec^{-2} cm^{-2} \overset{\circ}{A}^{-1}$.

4. Conclusions

The agreement with the observational data was satisfactory, the values in the emission maps being similar in the observed and simulated cases. Several C and IDL routines were developed in order to ease the post-processing of the MHD simulation data and automatically compute average line ratios from the synthetic emission files.

Acknowledgements

OT acknowledges the support of UEFISCSU-CNCSIS, contract RP8/2009. Simulations were performed using an HPC2 Europa grant at CINECA Bologna, and at CASPUR.

References

Mignone, A., Massaglia, S., Bodo, G., *et al.*. 2007, *ApJS*, 170, 228
Teşileanu, O., Mignone, A., & Massaglia, S. 2008, *A & A*, 488, 429
Woitas, J., Ray, T. P., Bacciotti, F., Davis, C. J., & Eislöffel, J. 2002, *ApJ*, 580, 336

Jets at all Scales
Proceedings IAU Symposium No. 275, 2011
G. E. Romero, R. A. Sunyaev & T. Belloni, eds.

© International Astronomical Union 2011
doi:10.1017/S1743921310016534

VLT/NACO detection of a proplyd/jet candidate in the core of Trumpler 14

Sílvia Vicente[1], João Alves[2], Isamu Matsuyama[3],
Hervé Bouy[4], Loredana Spezzi[1], Joana Ascenso[5], Filipe D. Santos[6]
and Timo Prusti[1]

[1]ESA (ESTEC), Research and Scientific Support Department,
Keplerlaan 1, P.O. Box 299, 2200 AG Noordwijk, The Netherlands
email: svicente@rssd.esa.int

[2]Institut für Astronomie, Universität Wien,
Dr.-Karl-Lueger-Ring 1, 1010 Wien, Austria

[3]Department of Earth and Planetary Sciences, University of California Berkeley,
307 McCone Hall, Berkeley, CA 94720, USA

[4]CAB (INTA/CSIC), LAEFF,
P.O. Box 78, E-28691 Villanueva de la Cañada, Madrid, Spain

[5]European Southern Observatory,
Karl-Schwarzschild Straße 2, D-85748 Garching bei München, Germany

[6]Departamento de Física da Faculdade de Ciências da Universidade de Lisboa,
Ed. C8, Campo Grande, 1749-016 Lisboa, Portugal

Abstract. This paper reports the discovery and presents the results of a first analysis of the observed morphology of a candidate external irradiated circumstellar disk/jet system found in the deep core of Trumpler 14, a cluster an order of magnitude more massive than the only cluster where bona-fide proplyds have been found, the Trapezium cluster in the Orion Nebula.

Keywords. HII regions, planetary systems: protoplanetary disks, ISM: globules, ISM: jets and outflows, stars: formation, open clusters and associations: individual (Trumpler 14)

The proplyd/jet candidate was discovered during a VLT/NACO $JHKsL'$Brα survey of the core of Trumpler 14 (Vicente *et al.* 2010, Ascenso *et al.* 2007), a young cluster in the Carina Nebula (NGC 3372), and is similar in size and morphology to the numerous proplyds found in the Trapezium cluster (Bally *et al.* 2000, Vicente & Alves 2005). Archival HST/ACS/HRC optical images, together with the adaptive optics near-IR images and existing photoevaporation theories, were used to investigate the morphology of the object and the possible scenarios for its nature, origin and expected lifetime.

The evaporating globule is located very close to the O2If* supergiant star, HD 93 129Aa, at a projected distance of ~ 0.024 pc. Visible as a tailless object in the optical emission lines, [OIII] and Hα, as a faint point source in H and with a faint tail in K_s, it appears in the L' band and 4.05 μm images as an extended object with a bright "head" and a long irregular, clumpy tail pointing nearly radially outwards from HD 93129A (Fig. 1). In the L'-band image, the object's head has a diameter of 560 AU and a head-to-tail length of 3 080 AU, for the adopted distance of 2.8 kpc to Trumpler 14. A spike emerging from the head and a 2-tail morphology, also visible in the deconvolved image, are suggestive of the presence of a jet. Are the two sections of the tail associated with different physical processes? Possible scenarios for the observed morphology include: 1) a jet + proplyd evaporative flow, 2) a precessing jet/disk, and 3) a non-resolved binary driving two jets. With the present data we cannot state beyond any doubt that this object is a true

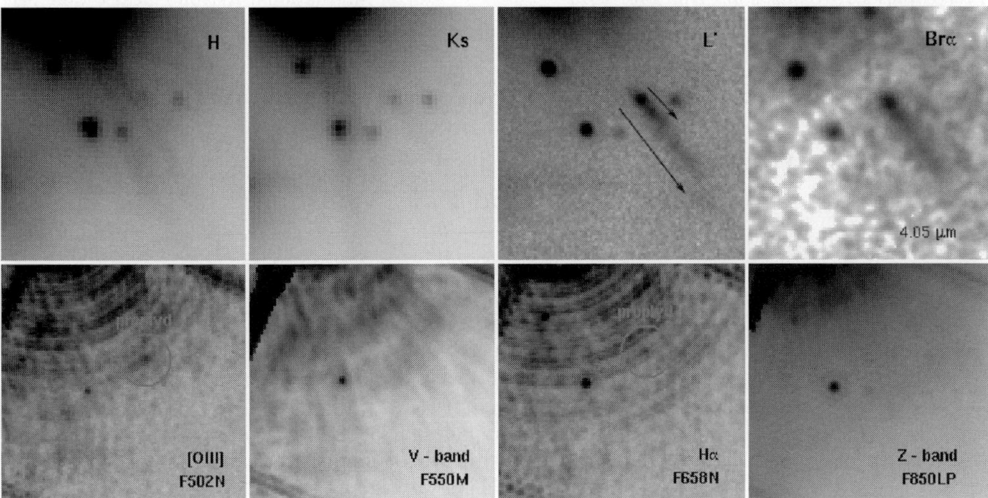

Figure 1. VLT/NACO near-IR images (upper row, 2".5 x 2".5) and HST/ACS/HRC optical images (bottom row, 2".3 x 2".3) of the proplyd/jet candidate in the core of Trumpler 14. North is up and east to the left. A spike emerging from the globule's head and a 2-tail morphology in the L'-band image are suggestive of the presence of an externally illuminated bipolar jet where dust is entrained. Tail1 is shorter and brighter with a head-to-tail extent of 0.5" (1400 AU), and tail2 is longer (1.1" or 3 080 AU), fainter, and nearly parallel to the former. Their length and direction are indicated by the black arrows in the L'-band image.

proplyd/jet as opposed to an evaporating gaseous globule but photoevaporation mass-loss rates, predicted at that location, favor a very compact and dense system, and hence, the proposition that it is indeed a proplyd. A 560 AU spherical globule of molecular hydrogen would be disrupted and completely evaporated in less than 50 yr. Future detailed high-resolution multi-wavelength observations are required to determine accurately the physical parameters of this intriguing object and will lead to a better understanding of its nature.

Other proplyd candidates have been discovered in the core of Trumpler 14 in HST/ACS images (S. Vicente PhD thesis 2009, Smith *et al.* 2010). This discovery is surprising since Trumpler 14 is older (1-2 Myr, Smith *et al.* 2006) than the Trapezium (0.1–1 Myr), a much more extreme environment ($Q_H \sim 22$ times larger) and has a PDR located at 2 pc from the cluster core (Brooks *et al.* 2003). The survival of these candidate protoplanetary disks raises many questions related to the current photoevaporation theories developed for the Trapezium proplyds. Can we apply the same models to the harsh environment of Trumpler 14? What is missing in the whole picture?

References

Ascenso, J., Alves, J., Vicente, S., & Lago, M. T. V. T. 2007, *A&A*, 476, 199
Bally, J., O'Dell, C. R., & McCaughrean, M. J. 2000, *AJ*, 119, 2919
Brooks, K. J., Cox, P., Schneider, N., *et al.* 2003, *A&A*, 412, 751
Smith, N. 2006, *MNRAS*, 367, 763
Smith, N., Bally, J., & Walborn, N. R. 2010, *MNRAS*, 405, 1153
Vicente, S. & Alves, J. 2005, *A&A*, 441, 195
Vicente, S. 2009, *PhD Thesis, Faculty of Sciences of Lisbon University*
Vicente, S., Alves, J., Matsuyama, *et al.* 2010, *submitted to A&A*

Jets at all Scales
Proceedings IAU Symposium No. 275, 2011
G. E. Romero, R. A. Sunyaev & T. Belloni, eds.

© International Astronomical Union 2011
doi:10.1017/S1743921310016546

What is shaping the planetary nebula K3-35?

Y. Gómez[1], D. Tafoya[2], G. Anglada[3], L. F. Miranda[3], L. Uscanga[3], J. M. Torrelles[4] and P. F. Velázquez[5]

[1]Centro de Radioastronomía y Astrofísica, UNAM, A.P. 3-72, 58089 Morelia, Michoacán
email: y.gomez@crya.unam.mx

[2]Graduate School of Science and Engineering, Kagoshima University, 1-21-35 Korimoto, Kagoshima, Kagoshima 890-0065, Japan
email: dtafoya@astro.sci.kagoshima-u.ac.jp

[3]Instituto de Astrofísica de Andalucía (CSIC), Apartado 3004, E-18080Granada, Spain
[4]Instituto de Ciencias del Espacio (CSIC)-UB-IEEC, Facultat de Física, Universitat de Barcelona, Planta 7a, Martí i Franqués 1, E-08028 Barcelona, Spain
[5]Instituto de Ciencias Nucleares, Universidad Nacional Autónoma de México, A.P.70-543, 04510 México.

Abstract. K 3-35 is a very young planetary nebula (PN) with a characteristic S-shaped radio emission morphology. It is the first PN where water vapor maser was detected: the emission is located in a torus-like structure with a radius of 100 AU and also at the surprisingly large distance of 5000 AU from the star, in the tips of the bipolar lobes. Several mechanism have been proposed to explain the bipolar morphology of PNe, and in the case of K 3-35 we believe we may be observing several of them at the same time: i) a disk-like structure traced by the H_2O masers, ii) a precessing bipolar jet probably due to the presence of a binary companion and iii) circular polarization in the OH 1665 MHz masers, which suggests the presence of a magnetic field. Additional observations and modeling are needed to establish what mechanisms are shaping K 3-35.

Keywords. magnetic fields: masers - stars: planetary nebulae: individual (K 3-35), ISM: jets and outflows

1. Summary

It is known that intermediate mass stars ($0.8 < M_\star < 10\ M_\odot$), spend most of their life converting hydrogen into helium. However, when the hydrogen in the core is exhausted, they start a short phase of much more rapid evolution to the planetary nebula (PN) phase. A planetary nebula consists of a hot ($T_{eff} > 20,000$ K) central star, surrounded by an expanding ionized nebula that in most cases is not spherically symmetric. Recently, a considerable effort has been put to understand the precise shaping mechanisms responsible for the different morphologies observed in PNe that go from spherical round to bipolar with collimated jets (Balick & Frank 2002).

K 3-35 was the first, of three confirmed PNe, that exhibit water maser emission (Miranda *et al.* 2001; de Gregorio-Monsalvo *et al.* 2004; Gómez *et al.* 2008). The bipolar morphology, and the presence of large amounts of neutral molecular gas in these three PNe not only suggest that these objects are young, but also that they have followed a similar evolutionary track corresponding to massive progenitors (Gómez 2007; Tafoya *et al.* 2007; Tafoya *et al.* 2009).

Several mechanisms have been proposed to explain bipolar collimated jets in PNe, and in the particular case of K 3-35 we note that three of them are present at the same time. *i) A disk-like structure traced by the H_2O masers at the center of K3-35.* From a kinematic study of the H_2O masers, we suggest the presence of an expanding

414

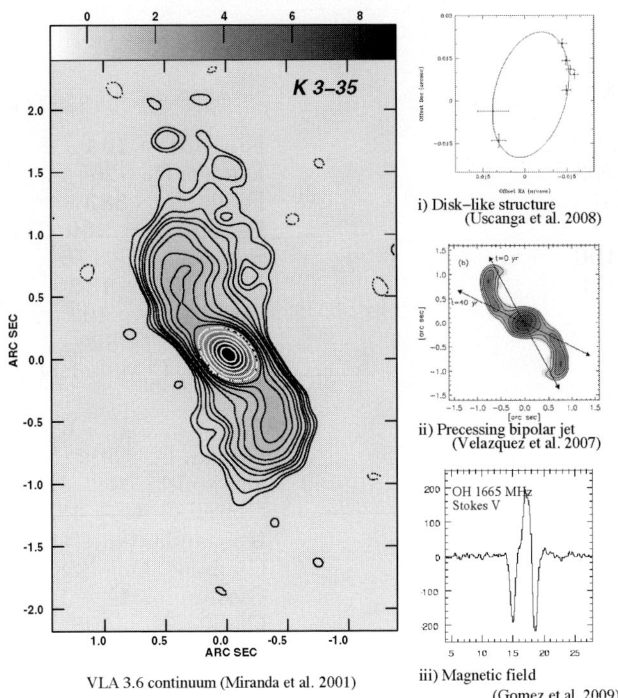

Figure 1. Left: VLA 3.6 cm continuum of K 3-35. Right: the three different mechanism that can be contributing in the jet collimation.

(V$\simeq$1.4 km s^{-1}) and rotating (V$\simeq$3.1 km s^{-1}) disk (Fig. 1), with a $\simeq$100 AU radius (Uscanga *et al.* 2008). *ii) A precessing bipolar jet.* The 'S' morphology shown by K 3-35 in the 8.3 GHz continuum image (see left side of Fig. 1) can be reproduced by hydro-dynamical simulations, if the jet precesses with a period of 100 yrs on a cone with a half-opening angle of 20° (Velázquez *et al.* 2007). *iii) A magnetic field.* Circular polar-ization in the OH 1665 MHz masers traces a Zeeman pair, supporting the presence of a magnetic field of $\sim$0.9 mG, at a radius of 150 AU (Gómez *et al.* 2009).

References

Balick, B. & Frank, A. 2002, *ARA&A.*, 40, 439

de Gregorio-Monsalvo, I., Gómez, Y., Anglada, G., Cesaroni, R., Miranda, L. F., Gómez, J. F., & Torrelles, J. M. 2004, *ApJ* 601, 921

Gómez, J. F., Suárez, O., Gómez, Y., Miranda, L. F., Torrelles, J. M., Anglada, G., & Morata, Ó. 2008, *AJ*, 135, 2074

Gómez, Y. 2007, *IAU Symposium*, 242, 292

Gómez, Y., Tafoya, D., Anglada, G., Miranda, L. F., Torrelles, J. M., Patel, N. A., & Hernández, R. F. 2009, *ApJ*, 695, 930

Miranda, L. F., Gómez, Y., Anglada, G., & Torrelles, J. M. 2001, *Nature* 414, 284

Tafoya, D., *et al.* 2007, *AJ*, 133, 364

Tafoya, D., Gómez, Y., Patel, N. A., Torrelles, J. M., Gómez, J. F., Anglada, G., Miranda, L. F., & de Gregorio-Monsalvo, I. 2009, *ApJ*, 691, 611

Uscanga, L., Gómez, Y., Raga, A. C., Cantó, J., Anglada, G., Gómez, J. F., Torrelles, J. M., & Miranda, L. F. 2008, *MNRAS*, 390, 1127

Velázquez, P. F., Gómez, Y., Esquivel, A., & Raga, A. C. 2007, *MNRAS*, 382, 1965

Author Index

Subject Index

Object Index